Voluntary feed intake in pigs

Voluntary feed intake in pigs

edited by:

David Torrallardona

Eugeni Roura

Wageningen Academic
Publishers

ISBN 978-90-8686-096-8

Cover design: Jordi Armengol

First published, 2009

Table of contents

Preface

Many of the challenges of modern pig production systems are closely related to feed intake. Initiation of feeding at weaning is probably one of the most critical points in pig production. Delayed feeding at this stage may result in digestive disorders and growth check that may condition the pig's whole productive period. Also, for the precision feeding of growing-finishing pigs an accurate prediction of voluntary feed intake is essential to ensure a sufficient, but not excessive, supply of nutrients that optimise performance at each stage of physiological development, and to minimise the environmental impact of pig manure. Finally, an adequate feed intake in highly productive sows during lactation is also crucial for the proper development of the litter, and to maintain the sow's body condition, its reproductive performance and its overall longevity. All this must be achieved in compliance with the latest and continuously evolving environmental and welfare legislation.

This book has been prepared to complement the "Physiological basis of voluntary feed intake in swine. Pre and post-absorptive regulation" workshop, which was organised on 19 May 2009, to precede the XIth Symposium on Digestive Physiology of Pigs, in Montbrió del Camp, Spain on 20-22 May 2009. The book includes a total of fourteen chapters, eight of which have been based on presentations given at the workshop. The book was completed with six additional invited chapters dealing with other highly relevant aspects of feed intake that are not directly related to digestive physiology. While the aim is to give an overview of the main factors affecting feed intake in pigs, special emphasis has been given to cover those that have received little attention in the past such as the role of peripheral sensing.

After many decades of research in farm animal physiology and nutrition, a unified synthesis of peripheral biology has been beyond reach until molecular biologists identified the genes that code for the chemo/mechano-sensing receptors (e.g. taste, olfactory and temperature receptors; TRs/ORs/TRPs). These oronasal sensors recognize nutrients such as sugars, amino acids and lipids as well as plant derived volatile compounds indicative of nutritional value and other food characteristics. The field of peripheral sensing biology has undergone a large increase in research activity over the past decade particularly since the discovery of the smell receptor family by 2004 Nobel prize winners Axel and Buck. More than 100 peer-reviewed scientific papers are published yearly related to taste and smell and chemosensing signal transduction events including neurobiological studies. It has been estimated that genes related to peripheral sensing may account for as much as 2 to 3% of the mammalian genome (i.e. more related genes than the immune system).

The chapters in the book uncover several disputes among the scientific community related to feed intake regulation in swine. The predominance of the long-term central mechanisms orchestrating feed intake is challenged by the novel understanding of short-term based peripheral sensing mechanisms. Geneticists, nutritionists, physiologists, endocrinologists and behaviouralists review how animal physiology copes with feed intake control. Throughout chapters, the main factors determining meal size and/or frequency are reviewed with emphasis on nutritional status and particularly energy balance as the driving force. Nevertheless, amino acid and protein metabolism tries to make its way through. Age related aspects affecting feed intake regulation are also addressed across chapters. Perinatal and post-weaning events that determine feeding behaviour later in life gather significant attention. Plant physiology related to feedstuffs composition is also assessed in the view of the impact on voluntary feed intake and further relates to how feed technology may impact dietary appetence. Finally, an updated discussion on the intricacies between the immune status and food consumption are put in prospective. Overall, the book is a multidisciplinary compendium that confronts short with long term effects and integrates peripheral with central mechanisms that coordinate the cycles of hunger and satiety in swine.

Support from the Spanish Government (INIA) and the generous financial contribution of Lucta S.A. are gratefully acknowledged. Without their support, the workshop and this book would have not been possible. We would like to thank all the contributors to the book, not only for the scientific excellence that brought to the publication and to our community, but also for their diligence in helping us make the book a reality in due time. We also want to thank Jordi Armengol for designing the book cover.

Chapter 1. Voluntary feed intake and pig breeding

P.W. Knap
PIC International Group, Ratsteich 31, 24837 Schleswig, Germany; pieter.knap@pic.com

Abstract

Pig voluntary feed intake (VFI) decreased by 30 g/d annually during the early 1980s. This was mainly due to reduced energy requirements through reduced body fatness and improved efficiency, but caused concern that feed intake *capacity* (FIC) was being reduced which might (1) compromise improvement of lean tissue growth rate (LTGR) and (2) constrain lactation feed intake (LFI) and reproductive performance. However, timetrends show that VFI levels stabilised around 1990, whereas LTGR continued to increase by 4 g/d annually. Attempts to experimentally quantify FIC and relate it to the level required for maximum LTGR show no indication of LTGR being limited by FIC. Breeding goals in modern pig breeding aim for diverse combinations of genetic trend in production traits; with proper selection methods this can be achieved without substantial genetic changes in VFI, either up or down – genetic antagonisms can then be neutralised. The use of electronic VFI recording equipment led to attempts to exploit intake patterns over time, and intake behaviour traits, for strengthening breeding value estimation for growth and efficiency – largely unsuccessfully. But options for quantifying environmental sensitivity based on the within-animal variation of VFI over time are underexploited, and deserve more attention. The same holds for social interaction effects among penmates on each other's feed intake. Increased LFI is related to (1) higher reproductive output in terms of litter weight gain, and (2) reduced maternal tissue catabolism. The latter is beneficial because it reduces reproductive problems (weaning-mating interval, litter size, longevity) in subsequent parities. Possible genetic antagonisms within this system can be neutralised by properly designed selection methods, as above. LFI, maternal tissue loss, and/or litter weight gain will then become formal breeding goal traits in dam line breeding. Sow nutrition and management strategies will have to be adapted as well.

Keywords: feed intake capacity, optimum feed intake, lactation feed intake, environmental sensitivity, genetic antagonisms

Introduction

Most of the other chapters in this book describe the influence of nutritional, management-related and/or health-related factors on voluntary feed intake. By contrast, in this chapter we assume that all such factors are constant and we consider the variation in voluntary feed intake that is caused by animal-intrinsic effects, i.e. by differences between breeds, by variation among animals within a breed, and by variation over time within individual animals.

Twenty years before this book was published, John Webb (1989) wrote an influential paper about the *Genetics of food intake in the pig*. Briefly summarised, he argued as follows: up to that time, pig breeding had concentrated on improving lean content and feed efficiency, with little concern for declining or static feed intake levels; but a genetic *increase* in feed intake is required to avoid constraints to lean growth rate and/or sow productivity; the subsequent genetic challenge would be to create pigs with a large appetite combined with a high lean growth efficiency.

Since then, the following developments have taken place in pig breeding:

- BLUP was widely implemented. This has given breeders much more control of the genetic change of traits in their populations than previously. As a consequence, less has to be left to chance in pig breeding.
- Electronic feeding equipment was widely introduced. This allows for the recording of individual voluntary feed intake in group-housed pigs. It has increased feed intake data volumes from performance testing, and it also made those data more applicable to real-life conditions than the previous records from individual housing.
- Specialised sire and dam line breeding was widely formalised. This leads to very different breeding goals, so it has allowed breeders to focus on different resource allocation patterns in different lines.
- Molecular genetics technology is currently uncovering QTL that explain a sizeable proportion of the variation in grower-finisher feed intake (e.g. Mohrmann *et al.*, 2006; Duthie *et al.*, 2008) – these can be used to further increase the power of BLUP estimated breeding values.

Therefore, pig breeders are now much better equipped to target the voluntary feed intake of their breeding populations than twenty years ago.

In technified pig production, voluntary feed intake plays a role of interest in two complicated dimensions:

- Grower-finisher pigs are required to deposit lean tissue at a fast rate (which calls for a high feed intake) and in a feed-efficient way (which calls for a low feed intake).
- Sows are required to sustain a high reproductive output (which calls for high body energy reserves and/or a high feed intake) in a feed-efficient way (which calls for a low feed intake), and at the same time to supply half the genetic potential of lean, fast-growing and feed-efficient slaughter pigs (as in the previous point).

In terms of pig breeding, this leads to a distinction between feed intake as a sire line breeding goal trait, and feed intake as a dam line breeding goal trait – which provides useful structure to this chapter.

Feed intake in growing pigs: past developments

Sire line breeding goals focus on traits expressed by the grower-finisher pig: growth and carcass traits, meat quality traits, and robustness traits. The growth and carcass traits most commonly recorded in practice are bodyweight growth rate, feed intake, backfat depth, and loin muscle depth. The latter two traits are commonly measured *in vivo* with ultrasound equipment. Feed intake is commonly measured with electronic equipment that recognises individual animals in group housing; this equipment registers individual meal details that are commonly combined into an average daily feed intake statistic throughout performance testing.

A useful example of long-term phenotypic time trends of voluntary feed intake in grower-finisher pigs is given in Figure 1, which shows how this trait has developed in two Swiss breeds since 1980. Following growing awareness of the importance of the trait, the breeding goals of these lines were changed in 1988 with the expectation of slightly negative subsequent genetic trends for feed intake (A. Hofer, personal communication, 2008). Two to three years (i.e. one or two pig generations) later, this expected pattern is actually realised phenotypically and the previous reduction of feed intake by about 27 g/d per year comes to a halt.

The two breeds in Figure 1 show not only very similar timetrends, but also very similar levels at any point in time. This is by no means typical: different pig breeds and their crosses can differ considerably in their voluntary feed intake, which is illustrated in the left graph of Figure 2 (the graph on the right will be dealt with below). These are results of public Commercial Product Evaluation trials – each datapoint in the graph represents the mean feed intake of a group of grower-finisher pigs from a particular commercially available terminal cross. Hence these datapoints are disconnected over time: most of these crosses appear in such a test only once, because the breeding companies that produce them usually enter a different cross next time they participate, or they may participate only once. In spite of this disconnection, and although we have not adjusted these data for any possible differences in diet composition, housing system, end weight, gender or lean content prediction equation, the overall timetrend is remarkably similar to the one in Figure 1: an initial reduction of feed intake by about 35 g/d per year comes to a halt just before 1990. More importantly, Figure 2 shows a considerable variation in feed intake between the various crosses: the range indicated by the 95% confidence interval spans about 0.7 kg/d. These datapoints represent terminal crossbreds that were commercially available at the time, and many of the more recent crosses eat as much as the ones from the 1970s – what has truly changed over time is that there are now also pigs available with a clearly lower feed intake.

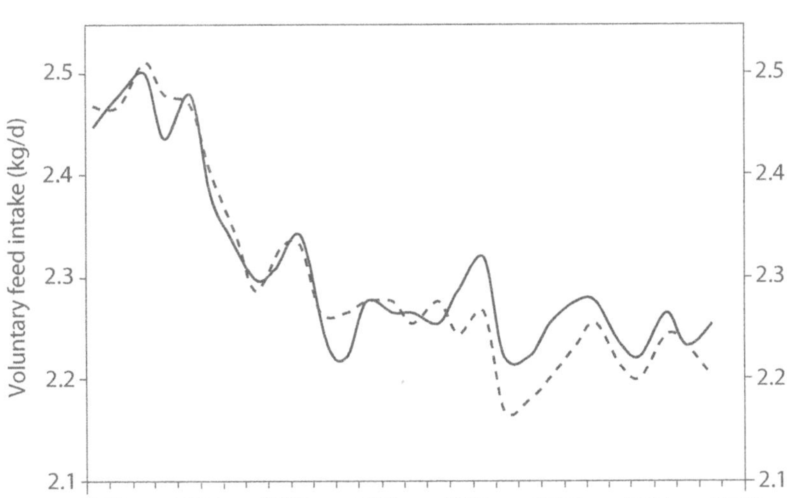

Figure 1. Time trends of voluntary feed intake in growing Swiss Landrace (solid line) and Swiss Large White (broken line) pigs. Annual phenotypic population means, data from SUISAG (2007).

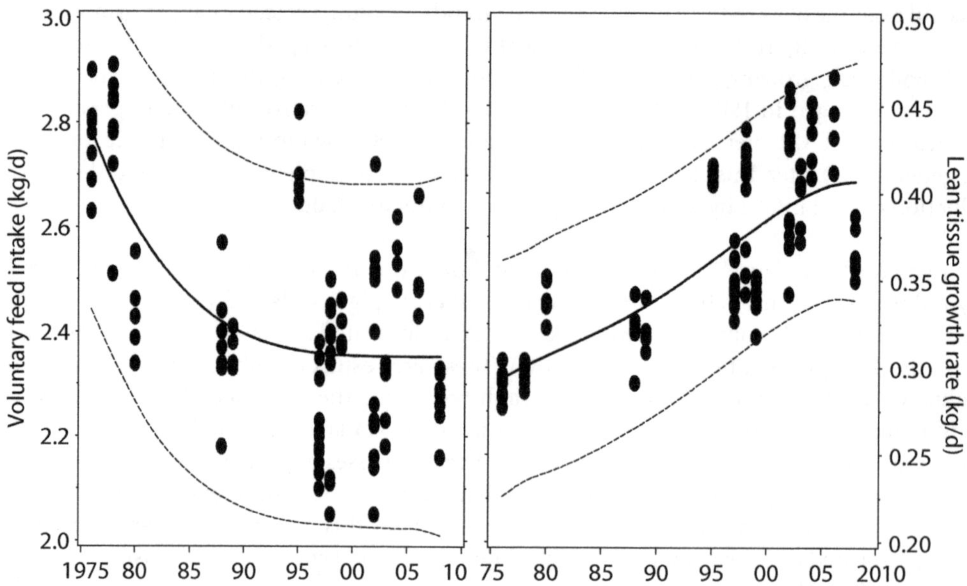

Figure 2. Time trends of voluntary feed intake and lean tissue growth rate (calculated as live body weight growth rate × 0.75 × estimated carcass lean content) in grower-finisher pigs of 103 terminal crosses, as recorded in public Commercial Product Evaluation trials in Denmark, France, Germany, the Netherlands, UK and USA. Unadjusted phenotypic population means, data from 18 literature and Internet sources. The trendlines are spline interpolation plots through the data, with their 95% confidence limits.

The variation in feed intake *between* genotypes (between crosses, in this case) illustrated in Figure 2 is complemented by the genetic variation among individual animals *within* a genotype (a pure line, in this case) as shown in Figure 3. Each datapoint in this Figure represents the estimated breeding value for voluntary feed intake of an individual pig from this line. While the average genetic value for this trait remains unchanged over time in this population (similar to the trends since 1998 in Figures 1 and 2), individual animals within this line show a range of about 0.5 kg/d in genetic value and the span of this range is very stable over time. This magnitude of genetic variation is consistent with a trait heritability of 0.3 to 0.5. Hence the non-genetic variation between animals (which adds to the genetic variation to form phenotypic variation) is at least as large (0.5 to 0.7) as the genetic variation shown here.

Webb (1989, as cited in the Introduction section) noticed a cause-and-effect relationship between voluntary feed intake and lean growth rate: 'selection for efficiency of lean growth appears to lead to a reduction in rate of fat deposition *via a decline in intake*' (our emphasis). In other words, voluntary feed intake ('appetite' as Webb accentuates it elsewhere in his text) is seen as the driving factor from which the deposition of lean and (particularly) fat derive. This is essentially a nutritiocratic concept (presented by a geneticist), which is not surprising against the background of a production system such as the pre-1985 pig sector where feed intake was commonly restricted

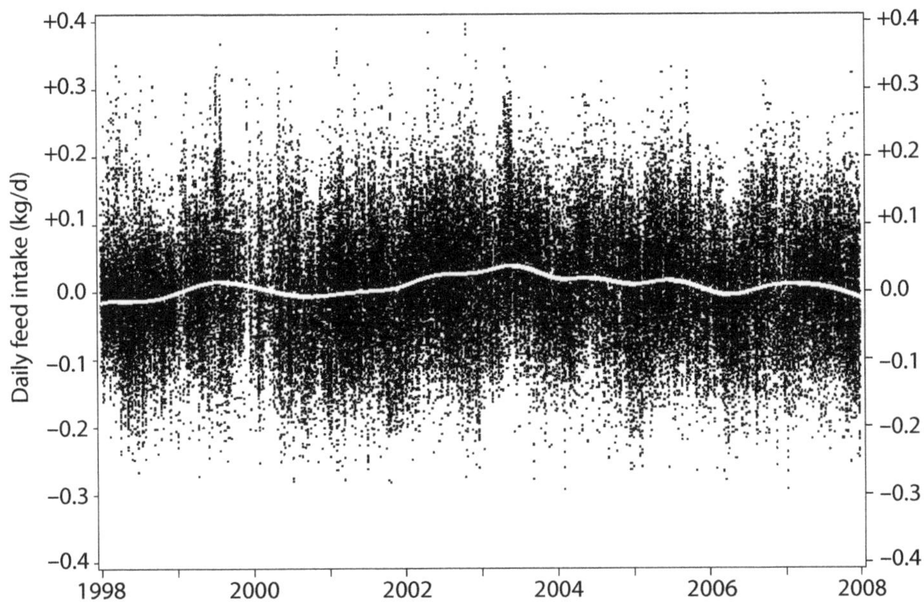

Figure 3. Estimated breeding values for voluntary feed intake (grower-finisher stage) of 84,434 pigs of a PIC line. The white trendline represents the genetic trend for this trait in this population, the same as line F in Figure 5.

to prevent the pig from expressing its full potential for fat deposition. In such a system feed intake is indeed the driving (or rather, the limiting) factor – but *voluntary* feed intake is not.

At that same time, the opposite approach was described by Emmans and Fisher (1986) and Emmans (1988, 1997) whose models predict voluntary feed intake based on the animal's intrinsic maintenance requirements and its (equally intrinsic) 'desire' to deposit body protein and lipid. This essentially geneticratic concept (presented by a nutritionist) implies that the pig has a genetic drive for body maintenance and a genetic drive to grow particular amounts of fat and lean, and that it tries to consume nutrients to the extent required to support these drives. From this point of view, voluntary feed intake in growing pigs went down during the pre-1990 period because pigs had been selected for lower fat deposition rates, so that they required less energy.

The above view of 'loss of appetite' has been consolidated into the traits *feed intake capacity* (Kanis, 1990) and *optimum feed intake*. Kanis and De Vries (1992) made use of Whittemore and Fawcett's (1976) linear-plateau model to parameterise this concept. In that model, body protein deposition in a growing animal increases linearly with increasing energy consumption – up to a certain point beyond which protein deposition plateaus, because the animal's genetic potential for deposition has been reached. De Lange *et al.* (2001) give a useful summary of these patterns in several pig genotypes. The optimum feed intake for an animal is then defined as the level where energy consumption is just adequate to make the animal realise its full potential protein deposition (and not higher, to prevent it from depositing excess lipid). Clearly this is 'optimum' from an economic point of view, not necessarily from a biological one. The animal's feed intake capacity plays an important role here because it may put a constraint on feed intake and prevent the animal from realising its optimum feed intake, so that it cannot achieve its potential protein deposition.

Kanis and De Vries (1992) and, in more detail, Hermesch *et al.* (2003) present a way to derive economic values for these traits, so that they can be included in a formal breeding goal together with Whittemore's model parameters; the novel element here is that voluntary feed intake has a positive economic value (so selection should aim to increase its level) as long as its level is below the optimum.

Eissen (2000) estimated optimum feed intake according to this framework, in grower-finisher pigs from nine breed/cross by gender combinations in three body weight ranges. His results are in Figure 4, where most datapoints are not significantly above the diagonal, which means that most of these pigs had an estimated optimum feed intake that was not significantly higher than their voluntary intake (which serves here as a proxy for their feed intake capacity). Eissen concludes that 'selection procedures as used by both breeding organisations [that produced his pigs] over the last decades have not led to large discrepancy between *ad libitum* and optimum feed intake'. This is in line with the findings of McCracken *et al.* (1994), who subjected 'high lean growth' pigs of that time to forced feeding and failed to measure an increased protein deposition rate in these overfed pigs relative to unrestricted controls.

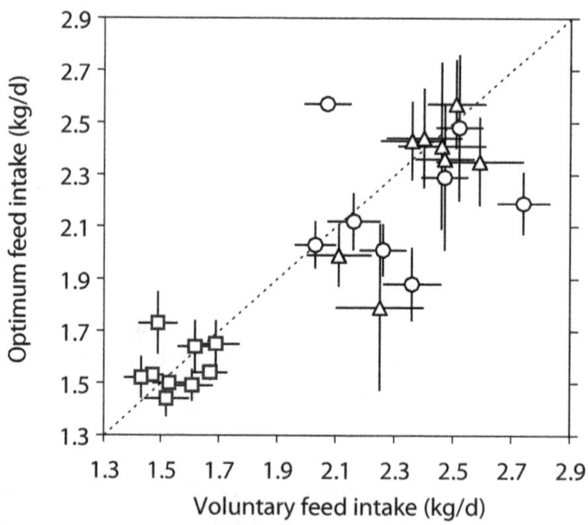

Figure 4. Recorded voluntary feed intake and estimated optimum feed intake in grower-finisher pigs, measured in 25-65 kg (□), 65-95 kg (○) and 95-125 kg (Δ) liveweight ranges. Data from Eissen (2000). Horizontal and vertical bars represent one standard error at each side of the estimate. The dotted line is the diagonal.

Egbert Kanis's framework would require a very elaborate performance testing regime (see Hermesch, 2004, for reflections on that), so we may never see it implemented in routine pig breeding. But it integrates the above approaches of Webb and Emmans, in two steps:

- It recognises that the potential for body protein deposition (i.e. lean tissue growth) is the driving factor against which energy consumption (i.e. feed intake) is to be targeted.
- It adds to this the notion that the animal may not be able to realise that feed intake, because of capacity ('appetite') constraints, which form another factor to be targeted.

This removes the above cause-and-effect conflict, and brings the two approaches to the same level, as usefully summarised by Ellis and Augspurger (2001): 'The genotype of an animal influences both its feed intake and its lean growth potential'. Black (2009) gives more detail. For pig breeders this leads to the (currently obvious, but at that time novel) strategy to bring voluntary feed intake under control and to keep it in balance with the other production traits. Breeding goals were widely adapted this way since the late 1980s, see Andreas Hofer's comment around Figure 1.

The result of this policy, as shown in Figure 2, is that terminal crossbreds that have been commercially available since that time continue to increase their lean tissue growth rate by almost 4 g/d per year (right-hand graph) while their voluntary feed intake (left-hand graph) does not go down anymore.

It does not go up either – this is not self-evident, as is illustrated by the results of the 1990s Edinburgh-Wye selection experiment for lean growth component traits (Cameron and Curran, 1994), which has produced much more useful detail than we can cover here. Of the populations that were divergently selected for increased lean growth rate (in the form of an index of bodyweight growth rate and backfat depth), one showed a non-significant reduction of voluntary feed intake (–19±40 g/d) and the other a significant increase (+69±45 g/day). And the populations that were selected for increased voluntary feed intake both showed an increase in lean tissue growth rate (+20 and +24 g/d, calculated from Cameron's reported results similarly as in Figure 5), arriving there via very different combinations of growth rate and body composition. It follows that the genetic connection of voluntary feed intake with lean tissue growth and its components is very much population-specific and cannot be generalised.

Obviously, the concern that further genetic improvement of lean tissue growth rate would be constrained by voluntary feed intake limitations has not come true. With hindsight, it can now be inferred that the reduction of voluntary feed intake up to the late 1980s was indeed caused by selection against body fatness (i.e. those pigs consumed less feed because they had a lower drive to deposit body lipid) and that this correlation was broken afterwards. Before the widespread implementation of electronic recording equipment and BLUP, genetic control of voluntary feed intake was very limited. Also, the main focus in pig breeding before the mid-1980s was on the reduction of fatness, often combined with a desire to reduce feed conversion ratio. The consequences of this for voluntary feed intake were not seen as important, if only because most pigs were fed restrictedly at that time. Clearly this has changed, and voluntary feed intake in growing pigs has become a serious breeding goal trait. This is illustrated in Figure 5, and it was foreseen by Fowler *et al.* (1976) – at a time when it had not even been established yet if the trait is heritable at all, they wrote: 'a restricted index to maximise change in the overall objective [i.e.

breeding goal], subject to the constraint of zero change in mean feed intake [...] treats intake as an objective [i.e. a breeding goal trait] in its own right and not just part of the selection criterion'.

Figure 5 shows realised ten-year genetic trends for lean tissue growth rate, voluntary feed intake and feed conversion ratio in eight PIC lines (the trends in litter weight gain will be dealt with in

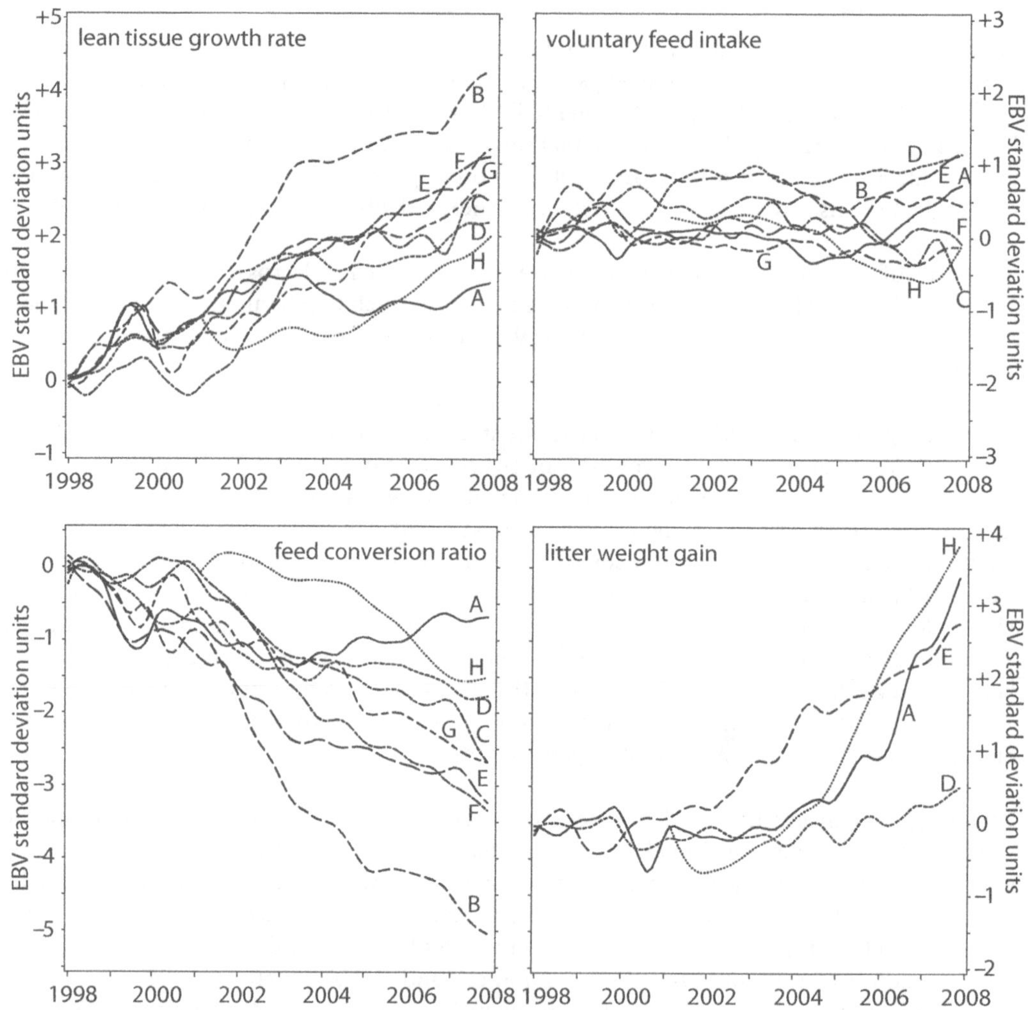

Figure 5. Genetic trends of lean tissue growth rate, voluntary feed intake, feed conversion ratio (all in the grower-finisher stage) and litter weight gain in eight PIC lines. All lines for each trait were forced through the same origin in 1998 to make the trends comparable across lines. All traits were scaled to the standard deviation of their estimated breeding values (EBVs) to make the vertical axes comparable across traits. Lean tissue growth rate was calculated as live body weight growth rate × 0.75 × carcass lean content, with lean content = 59 – 0.8 × backfat depth + 0.2 × loin muscle depth.

the final section of this chapter). Together with other lines not shown here, these pig lines form the basis of crossbred products that are sold into widely different production systems worldwide; to support this, their breeding goals aim for line diversification. Hence the developmental patterns shown here are not due to chance – they are the consequence of deliberate selection strategies: the weighting factors for growth rate, backfat depth, loin muscle depth, voluntary feed intake, and any other traits in the selection index for each line aim at a specific combination of trait selection responses.

The connections among all the patterns in Figure 5 deserve more consideration than we can cover here, but lines A, B and C (indicated by the labels in the graphs) provide a useful example of how differently feed intake and its correlated production traits can co-develop over time:

- The breeding goal of line A does not focus on grower-finisher traits. As a consequence, lean tissue growth rate shows little change over time, and the same holds for feed intake and feed conversion ratio.
- Among these eight lines (and according to its breeding goal), line B shows the largest increase in lean tissue growth rate (due to a strong increase in bodyweight growth rate and intermediate changes of backfat and loin muscle depth) and the largest reduction in feed conversion ratio. But its feed intake has changed just as little as in line A.
- Lean tissue growth rate of line C shows an intermediate change (due to a very small increase in bodyweight growth rate, and the largest changes of backfat and loin muscle depth among these lines) and so does feed conversion ratio; but this line shows just as little change in feed intake as lines A and B, this time into the other direction.

It follows that the genetic levels of production traits in different lines are evolving in very different ways, simply because these lines are selected for different breeding goals. Attempts to explain a change in voluntary feed intake from changes in lean tissue growth rate or its component traits and/or feed efficiency without knowledge of the line's breeding goal are therefore pointless. Most importantly, once voluntary feed intake is under genetic control, there is clearly no need for this trait to move up or down to any important extent because of changes in production trait levels. Perhaps the best illustration of this is the fact that the range of feed intake trends in Figure 5 spans only two EBV standard deviation units, whereas the other traits show ranges of three to 4.5 units.

The conclusion from this section is simple: in spite of possible genetic antagonisms between voluntary feed intake and the other production traits under selection, breeding goals that keep feed intake under control while at the same time increasing lean growth (and its efficiency) are perfectly feasible – genetic antagonisms can be neutralised this way. Figure 2 shows that, starting in the late 1980s, such breeding goals have also been implemented in practice. And for future development in whichever direction, Figure 3 shows that there is a very large amount of genetic variation available to sample from.

Feed intake in growing pigs: new developments

Feed intake patterns and environmental sensitivity

Most modern electronic feeding equipment delivers day-to-day measurements of not only feed intake itself but also its timing and its subdivision among visits to the feeder, and among meals. This has led to several analyses of traits such as the shape of the feed intake curve over time, feeding frequency and timing, and meal size and duration, mostly with the hypothesis that such traits might provide added value to the breeding value estimation of the conventional performance traits. Such studies were done by De Haer and De Vries (1993), Labroue *et al.* (1996), Von Felde *et al.* (1996), Hyun *et al.* (1997), Hall *et al.* (1999), Schnyder *et al.* (2001), Schulze *et al.* (2003), Lorenzo Bermejo *et al.* (2003) and Rauw *et al.* (2006), among others. They typically led to conclusions similar to Labroue's: 'it seems more advisable to increase the economic weight of average daily gain in the global selection index rather than to include a feeding behaviour criterion among selected traits'. Apparently in a system where the performance traits themselves are recorded as such, information on the behaviour patterns that are taking place around them has little to add to their estimation. It might be different if the target would not be those same performance traits but characteristics such as physical activity, group behaviour, net metabolic efficiency, etc. – but such connections have rarely been studied in livestock species up to now. This might work out as follows.

Figure 6 shows daily voluntary feed intake records from two pigs that were performance-tested (with an alternating feed intake recording scheme, hence the missing values) simultaneously. The linear regression lines through the data are practically the same, but the day-to-day variability of

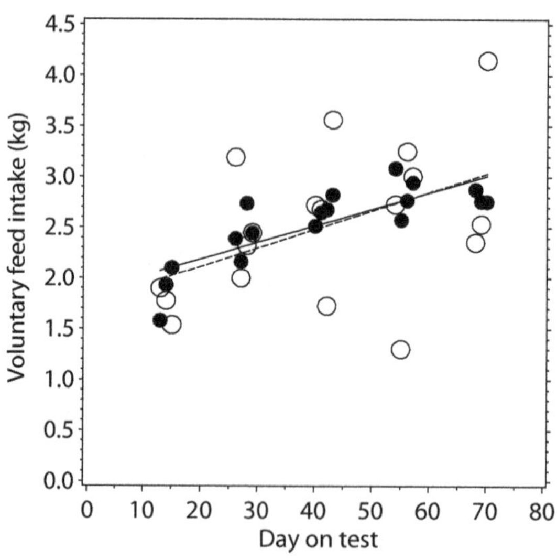

Figure 6. Daily voluntary feed intake records of two grower-finisher pigs. Data from D.S. Casey (personal communication, 2008). The residual standard deviations around the linear regression lines are 0.235 (●) and 0.678 kg (○).

the datapoints is very different for these two pigs. Ellis and Augspurger (2001) present similarly different patterns and suggest that 'the significance of this striking daily variability in feeding behaviour and feed intake level in terms of animal performance and nutritional programs is not obvious but requires further investigation'. Voluntary feed intake is probably the most easily measurable trait to reflect the day-to-day dynamics of the animal's metabolism. Therefore the variation of feed intake of a particular pig over time (i.e. its dispersion around the average course of the regression lines in Figure 6) may be used as a measure of the animal's sensitivity to environmental changes.

To illustrate this, consider Figure 7 which shows the familiar course of mammalian heat production in relation to ambient temperature. This is shown here for 500 simulated pigs of one particular genotype at a bodyweight of 100 kg. The lower critical temperature shows variation between animals (from this simulation, a standard deviation of 0.55 degrees at a mean value of 15 °C), so different animals have different threshold levels for the environmental factor (i.e. ambient temperature, in this example) beyond which they must change their heat production and therefore their feed intake.

During the grower-finisher period, a large number of environmental factors (microclimate, nutrition, health, other pigs, internal factors, etc.) is changing regularly. Therefore every day, several metabolic thresholds may be crossed in some animals with sensitive threshold levels, and not in other animals with more robust levels. Hence corrective metabolic action is (or is not) taken, and as a result feed intake is (or is not) temporarily increased. Metabolic threshold levels differ between animals as in Figure 7, so the changes in corrective metabolic activity can be expected to differ as well. This should create variation between animals in the degree of dispersion around

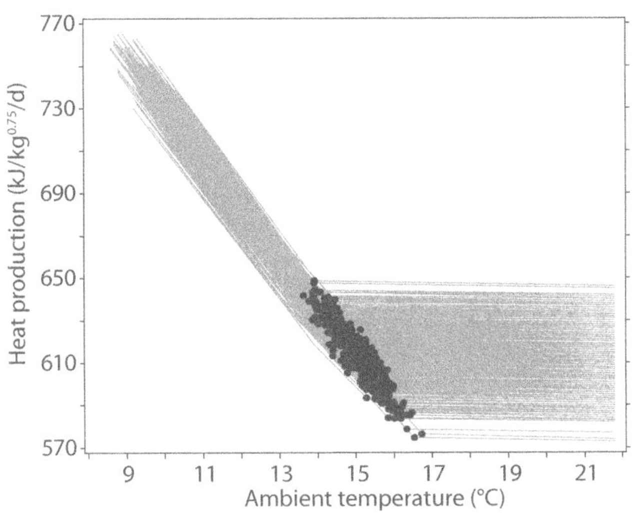

Figure 7. Heat production in relation to ambient temperature in 500 simulated grower-finisher pigs of one genotype. Each animal is represented by its gray broken trendline and by a black dot that shows its lower critical temperature and the associated heat production. Data from Knap (2000).

the average course of feed intake, as in Figure 6. Animals with sensitive metabolic threshold levels have a high *environmental sensitivity*. Such animals react strongly to improvements and deteriorations of the environmental factors that they must live in. And because the within-animal feed intake variation over time will reflect a large part of the corrective metabolic action of such animals, this variation may be used as an indirect measurement of environmental sensitivity.

Eissen (2000) analysed the course of daily feed intake (DFI) of 494 grower-finisher pigs over time, by fitting a linear regression (DFI = a + b × time) for each individual animal – the same approach as in Figure 6. For each animal, this produces a value for the residual standard deviation (RSD) around the fitted regression line. In this analysis, the average value of the RSD was 0.43 kg, and its standard deviation between animals was 0.11 kg (so David Casey's two pigs of Figure 6 would have neatly fitted Eissen's extremes). Its heritability was estimated at 0.46, and its genetic correlations with growth rate, feed intake and body composition traits were estimated at absolute values from 0.54 to 0.75. Notice that this within-animal variation complements the between-animal variation of Figure 3.

This opens possibilities for adding a measurement for environmental sensitivity (i.e. RSD) to the conventional selection criterion of growth rate, feed intake and body composition traits. Most of the genetic parameters that would be required for this were estimated by Eissen, as above – the missing elements are (1) the genetic correlation between RSD and 'environmental sensitivity' in general terms, and (2) an economic value to provide weighting of the trait relative to the performance traits. Both will require some innovative action, see Knap (2005) for a framework to derive economic values for environmental sensitivity. But the issue seems much more promising from the point of view of added value than the above mentioned work on intake and meal patterns. As Schulze *et al.* (2003) concluded, 'this residual standard deviation [...] will be of increasing interest for future breeding programs in order to obtain high performances in a wide range of environments'.

Social interactions among pigs

The introduction of electronic feeding equipment made it clear that voluntary feed intake of group-housed grower-finisher pigs is not the same trait as in individual housing. One of the earliest reports about this is by De Haer and Merks (1992), who showed that when pigs must compete for their feed, the mean intake level is reduced and between-animal variation is strongly increased – they reported means ± phenotypic standard deviations of voluntary feed intake of 2.20±0.20 kg/d in individual housing versus 2.04±0.29 kg/d in groups of eight pigs. Von Felde (1996: 88) confirmed this pattern with corresponding values of 2.87±0.29 kg/d (individual housing) and 2.46±0.48 kg/d (groups of 15). Hence in group housing, many pigs eat much less than they would have done in individual housing, whereas some pigs eat much more.

De Haer *et al.* (1993) estimated the phenotypic correlation between feed intake and growth rate in their group-housed pigs at +0.47, 'lower than in individual housing' (De Haer, 1992). Likewise, an unpublished Norsvin study produced a genetic correlation estimate of +0.31 between feed intake and growth rate in groups of twelve pigs, compared to a value of +0.70 in the previous individual housing system (E. Sehested, personal communication, 1995). If the correlation between feed intake and growth rate goes down when moving from individual to group housing, then the group environment must present an additional factor that directs nutritional resources away from growth metabolism. Social activity is an obvious candidate factor.

By contrast, Von Felde (1996) reported equal estimates for the genetic correlation between feed intake and growth rate in individual housing (+0.56) and in groups (+0.57). But he estimated the genetic correlation between feed intake in individual housing and feed intake in groups at –0.06±0.12, and the corresponding correlation for growth rate at +0.45; these are very different traits indeed.

The above estimates from De Haer, Sehested and Von Felde are pairwise-connected, in the sense that the pigs in individual and group housing in each study belonged to the same genetic population. Labroue (1995) reviewed the literature and lists results from twelve other studies with group sizes from one to twelve pigs, showing no clear effect of group size on the correlation between feed intake and growth rate. But each of those estimates derives from a different pig population, so they are disconnected and therefore difficult to compare – genetic correlations are much more tricky statistics than population means such as in Figure 2.

Labroue *et al.* (1996) analysed grower-finisher feed intake behaviour traits (e.g. meal frequency, duration and size) of Landrace and Large White pigs, housed either in single-breed pens or in mixed pens. She concluded from that study: 'Meal characteristics from single-breed pens show no difference between the breeds. By contrast, in mixed pens, Landrace animals seem to modify their feeding behaviour: they realise fewer meals per day but these meals last longer and are larger than for Landrace animals in single-breed pens. This change of behaviour is probably a reflection of more intense competition phenomena' (Labroue, 1995).

It follows that we must expect voluntary feed intake in group housing to be influenced by the pig's penmates, through social interaction. Chen *et al.* (2007) reviewed the literature on social interaction and competition among pigs, focusing on the implications for selection for body weight growth rate. They distinguished the individual's own genetic potential for the trait (the 'direct genetic effect') versus its genetic potential to influence the expression of the trait in its penmates (the 'competitive effect' or 'social effect'), which goes back to Muir and Schinckel (2002). Based on that model, Chen *et al.* (2008) analysed data on growth rate in grower-finisher pigs and produced results similar to those of Bergsma *et al.* (2008a), who analysed similar data on growth rate and voluntary feed intake, and who concluded that 'social effects contribute the vast majority of heritable variance in growth rate and feed intake in this population'. Two striking features of Bergsma's results are:

- The heritabilities of the direct effects on growth rate and voluntary feed intake were estimated at 0.25 and 0.18, respectively. The associated statistics for the combined direct and social effects were estimated at 0.71 and 0.70, respectively.
- The correlation estimate between the direct genetic effect and the social effect was slightly positive, both for growth rate (+0.20±0.10) and for voluntary feed intake (+0.15±0.14). Hence Bill Muir's earlier term 'competitive effect' is misleading in this case: in Bergsma's data, pigs with a positive genetic potential for feed intake and growth rate do *not* have a negative influence on the expression of those traits by their penmates – which was unexpected.

Canario *et al.* (2008) made use of Bergsma's parameter estimates to characterise another pig population for its social effects on body weight growth rate, and revealed significant correlations with aggressive behaviour traits. Andersen *et al.* (2004) studied aggressive behaviour in relation to group size and concluded that pigs 'alter their behaviour according to how the competitive situation changes with group size. With an increasing number of competitors, the probability of success of an aggressive strategy in gaining priority to important resources decreases'. This raises the notion that the easiest way for a pig to influence its penmates' growth rate (or any other production trait) is likely to be through an influence on their feed intake. Bergsma's and Chen's results show not only that such an influence is present, but also that it has a very strong genetic component. Both studies suffered from a data structure that had not been designed for this type of analysis, particularly with regard to the family relationships among the animals in a pen. Bergsma concludes that 'estimation of social genetic effects seems most powerful when [performance-tested] populations consist of many small groups of unrelated individuals', which would require rather elaborate performance testing regimes – so it may take some time for this approach to be implemented in routine pig breeding. But it seems obvious that voluntary feed intake, and the way pigs influence each other's level of that trait, must play a central role here. See Manteca and Edwards (2009) for more details.

This opens possibilities for exploiting genetic variation that has so far been ignored in pig breeding. Interestingly, the 'social equivalent' of the thermoregulation patterns of Figure 7 was modelled by Wellock *et al.* (2003) in terms of *excitability*, which expresses 'both the intensity of a stressor at which the animal becomes stressed [i.e. the threshold value equivalent to critical temperature in Figure 7] and the extent to which stress reduces performance and increases energy expenditure (activity) at a given stressor intensity'. This suggests that the two issues in this section on 'new developments' could usefully be connected.

Feed intake in sows

Dam line breeding goals focus on two elements:
- The traits that are expressed by the grower-finisher pig, similar to sire line breeding goals.
- The reproductive performance of the sow.

Since the advent of specialised sire and dam lines in pig breeding, the second element tends to receive more and more focus in dam line breeding – much of the improvement of the first element is often left to the sire line part of the breeding program.

Again in the late 1980s, Dourmad (1988) wrote: 'Reviewing the literature on appetite regulation in growing pigs, Henry (1985) found a lower voluntary feed intake in lean pigs than in fat pigs. This genetic effect is also found in the lactating sow, but apparently less markedly so [...] This raises the question if current selection programs, where the main goal is an improvement of carcass lean content, are leading to a reduction of appetite in the sow'. At that same time, Lynch (1989) wrote: 'The lactating sow requires 5 to 7 kg food daily if it is to maintain body weight and condition [...] but even on *ad libitum* intake sows fail to achieve the required food consumption and so lose [weight]. The problem is most acute in primiparous sows, and low lactation food intake is recognised as a major factor in [...] impaired reproductive performance'.

Lynch's concerns were confirmed several years later. Feed intake and body tissue mobilisation of lactating sows were shown to influence reproductive performance both in the current parity and in subsequent ones, see for example Rydhmer *et al.* (1992), Koketsu and Dial (1997), Clowes *et al.* (2003), Grandinson *et al.* (2005), Bunter *et al.* (2008) and Hermesch *et al.* (2008). And in line with Dourmad's concern, this has been extended to possibly unfavourable correlations between lean growth performance and reproductive performance (e.g. Kerr and Cameron, 1996; Ten Napel and Johnson, 1997 – see also Figures 12.3 and 12.4 in Knap and Rauw, 2008), mediated through grower-finisher feed intake (e.g. Karsten *et al.*, 2000; Holm *et al.*, 2004) and its correlation with lactation feed intake.

Successful selection for increased litter size together with ongoing selection for increased lean growth (see Figure 12.1 in Knap and Rauw, 2008) has led to increasingly prolific sows with fewer and fewer body energy reserves. Since the early 1990s, the literature is full of texts that mention the increased resource demands of this type of animal, particularly in hot climatic conditions with their constraining effect on voluntary feed intake. Very similar to grower-finisher feed intake in the 1980s, lactation feed intake has become a subject of concern. But as is more often the case, much of the issue is anecdotal – all these literature sources produce little solid evidence on the matter, particularly when it comes to genetic effects. In fact, any data on voluntary feed intake in lactating sows is very scarce, and most of it comes from experimental settings.

Eissen *et al.* (2003) and Bergsma *et al.* (2008b) analysed lactation feed intake data from some of those experimental settings. Eissen gives estimates of multiple regression coefficients that lead to the patterns shown in Figure 8. This is an input-output system, the output represented by litter weight gain during lactation, and the input represented by sow feed intake plus sow tissue catabolism (measured as sow body weight loss and backfat depth loss). Obviously, a higher feed intake leads to (1) a higher output and (2) a lower required level of the other input factors, i.e. savings on tissue catabolism. In Figure 8, those savings are more pronounced at lower litter

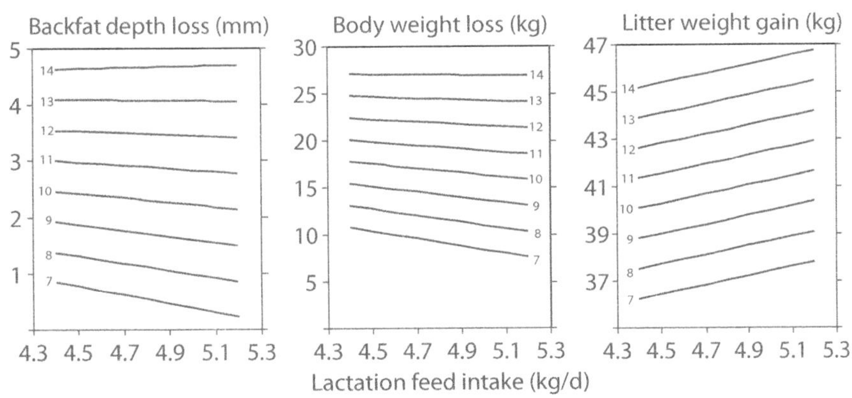

Figure 8. Sow backfat depth loss and body weight loss, and litter weight gain, in relation to sow voluntary feed intake (all from 10 to 28 days in lactation, parity 1 litters) and in relation to standardised litter size (as indicated by the trendline labels). Based on linear regression coefficient estimates by Eissen et al. *(2003). Average day 10 levels were 16 mm backfat depth, 170 kg body weight and 36 kg litter weight.*

size levels, which is counter-intuitive; Eissen explains this in terms of increased maintenance requirements of sow and piglets at larger litter sizes, which adds to the output.

Bergsma takes this approach a step further by integrating the input factors (and the output factors in terms of the various functional components of litter weight gain) into a simulation model, so that the dynamics of resource allocation from nutrients and catabolised maternal tissue to progeny tissue, and to maternal and progeny maintenance, can be quantified. This leads to a more consistent description of input and output, which is then used to calculate 'lactation efficiency', i.e. units of output per unit of input, analogous to feed conversion efficiency in growing animals. Heritabilities were estimated at 0.12 to 0.17 for input, output and efficiency, at 0.30 for lactation feed intake, and at 0.18 for litter weight gain. Genetic correlations were estimated at +0.69 between input and output, –0.67 between input and efficiency, and +0.20±0.15 between output and efficiency. Litter weight gain (from cross-fostering to weaning) came out as genetically the same trait as Bergsma's 'output', with a correlation of +0.99. Genetic correlations with lactation feed intake were estimated at –0.62 to –0.85 for sow body weight loss, fat loss and protein loss, and at +0.48 for litter weight gain.

A similar experiment is being carried out by Bunter *et al.* (2008) and Hermesch *et al.* (2008), who are currently reporting on their parity-1 results. They estimated the heritability of lactation feed intake at around 0.16, with an associated repeatability of 0.32. Litter weight gain was reported here up to 10 days in lactation, 'when lactation output is not at its peak', as Bunter notices. This may explain its low heritability (0.09) and its low genetic correlation estimate with lactation feed intake (+0.10±0.24), which was recorded up to day 35. But phenotypically, Jaco Eissen's patterns of Figure 8 are neatly reflected in these data: 'for every additional kg/day of feed eaten during lactation, there was an extra [...] 0.72 pigs per litter at 21 days, and piglets weighed on average 0.2 kg extra. Litter weight gains [...] by day 21 were of larger magnitude than is suggested by changes in piglet numbers and average weights. Sows were approximately 9 kg heavier and 0.7 mm fatter at weaning per additional 1 kg of daily intake during lactation' (Bunter *et al.*, 2008).

The total reproductive performance of a sow can then be described in terms of litter size at farrowing, pre-weaning survival rates and litter weight gain, plus weaning-mating interval and longevity. Figure 9 presents a subjective summary of trait correlations from the literature sources cited earlier in this section, and from references provided there. Those sources frequently disagree about the value (and sometimes about the sign) of several of these correlations – but generally, lactation feed intake influences reproductive performance in two ways: it provides resources to litter weight gain, and it reduces sow tissue catabolism during lactation, which has favourable effects on subsequent weaning-mating interval, on subsequent litter size, and through these factors also on subsequent sow longevity.

Both are good reasons to aim for an increase in the trait through selection, dependent on feed cost. That would lead to a breeding goal comprising three groups of traits:

- The grower-finisher traits as mentioned in the previous section.
- The usual reproduction traits such as litter size, piglet mortality rates and weaning-mating interval, and sow mortality rate.
- Litter weight gain, sow body tissue loss, and lactation feed intake.

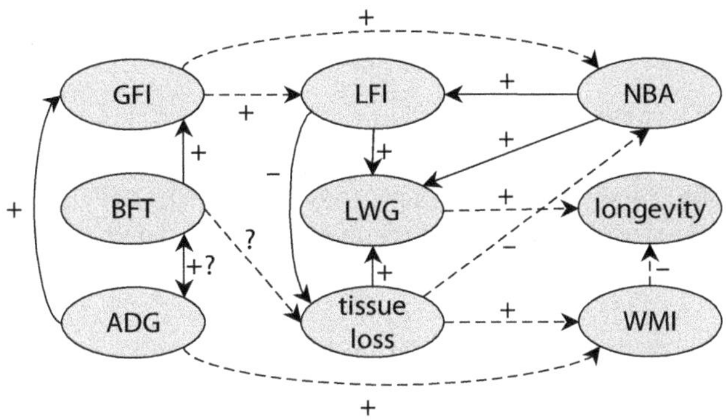

Figure 9. Summarised relationships among the entities described in this section. Symbols + and – denote the sign of a correlation between traits. Dotted arrows indicate effects on events in subsequent parities. ADG: grower-finisher growth rate; BFT: grower-finisher backfat depth; GFI: grower-finisher feed intake; LFI: sow lactation feed intake; LWG: litter weight gain; NBA: litter size (number born alive); WMI: weaning-mating interval.

Given proper data for breeding value estimation, this should allow for keeping each of those traits under genetic control and neutralising genetic antagonisms among them, where relevant.

Lactation feed intake, and sow body weight and backfat depth, have not been commonly recorded on a large-scale routine basis up to now. For a particular breeding program, the value of recording any of the traits of the third group above depends on the actual values of the above mentioned genetic parameters in that population (most notably the genetic correlations of lactation feed intake with grower-finisher feed intake and with litter weight gain), and on recording cost. Just as was worked out for partial recording of grower-finisher feed intake by Von Felde *et al.* (1996), Brisbane (2002), Grignola *et al.* (2002), Casey (2003) and Schulze *et al.* (2003), it may be feasible to restrict sow feed intake recording to just a part of the lactation period as shown by Hermesch (2007), who subdivided her data into four parts of five days each and concluded that recording in one of those periods only (during the second or third week of lactation) would provide sufficient information for breeding purposes.

As mentioned in the previous paragraph, the genetic correlation between grower-finisher feed intake and lactation feed intake is a critical parameter in this system – but it is currently very poorly documented. Until recently, the only published estimate (from Van Erp *et al.*, 1998) was based on a grossly inadequate number of observations so that its dubiously high value (+0.92±0.50) was paradoxically not significantly different from zero. Bunter *et al.* (2007) gave an equally confusing estimate of –0.26±0.33 which was recently updated for parity-1 records only, with preliminary (and much more credible) estimates at +0.3 to +0.6±0.14 (K.L. Bunter, personal communication, 2008) dependent on the statistical model and data subset used in the analysis. The 1990s Edinburgh-Wye selection experiment for lean growth component traits included divergent selection lines for grower-finisher feed intake. The lactation feed intake of parity-1 sows of the low and high lines was 94 kg and 102 kg, respectively (Cameron *et al.*, 2002); the difference of 8±5.2 kg is not significantly different from zero.

In mice, Archer *et al.* (1998) reported a genetic correlation estimate between pre-pubertal and lactation feed intake at +0.51, but did not give its standard error. In a mouse line selected for litter size, Rauw *et al.* (2009) reported phenotypic correlations of +0.32 (P<0.05) and +0.50 (P<0.01) dependent on the litter size standardisation regime; in the associated unselected control line, the estimates were not significantly different from zero.

Bunter *et al.* (2008) conclude from their own analyses that 'there are no strong antagonistic genetic correlations between sow body size or condition, lactation feed intake and litter size'. Any such antagonisms as indicated in Figure 9, can be neutralised by proper selection strategies; Knap (2008: Figures 17.2 and 17.4) gives examples of this, similar to the patterns shown for grower-finisher traits in Figure 5. The final graph in Figure 5 shows ten-year genetic trends of litter weight gain in four of the PIC lines. Comparison with the associated patterns of the other traits shows again that there is no simple rule to explain line differences in one trait from differences in another trait without knowledge of the line's breeding goal. Perhaps more importantly, this graph also shows that (given proper focus on the trait in the breeding goal, and given adequate selection methods) litter weight gain can be improved genetically just as readily as lean tissue growth rate. Given the genetic parameter structure reviewed above, this will lead to a correlated increase in lactation feed intake, particularly when the trend for grower-finisher feed intake is positive – such as for lines A and E in Figure 5.

More or less anticipating the evidence, dam line breeding goals in many of today's pig breeding programs aim at an increase of voluntary feed intake – not as measured in sows but as measured in growing pigs, because those records are widely available now. Which means effectively that lactation feed intake has become a dam line breeding goal trait with a poorly quantified (but implicitly positive) economic value, and with grower-finisher feed intake as its correlated selection trait, linked through equally poorly quantified genetic correlation guesstimates. But that is just a matter of nuance: the detail will be worked out eventually, so that the selection systems can be fine-tuned. The shift of paradigm of the late 1980s that was illustrated in Figure 1 worked the same way – for example, by 1990 very few pig breeders had any idea about the genetic correlation between voluntary feed intake and growth rate in their own populations, particularly in group housing. But Figure 2 shows very clearly that a large part of the breeding industry moved into the same direction nevertheless.

Feeding strategies for sows will have to be adapted as well: when the genotype changes, management must follow. It is unlikely that genetic improvement of lactation feed intake alone will be able to support the increasing litter size levels illustrated in Figure 12.1 of Knap and Rauw (2008). Eissen (2000) noticed that an additional piglet to rear increases the sow's lactation feed intake requirement by about 450 g/d (Verstegen *et al.*, 1985; Koketsu *et al.*, 1996); with an annual genetic trend in litter size of 0.2 piglets this requires a simultaneous increase in lactation feed intake of 90 g/d per year, i.e. about 2% of the mean level. As Eissen concluded, this 'will be difficult to realise genetically' even with dedicated selection on the trait, so nutritional and animal management strategies will have to be developed at the same time. Of course, the 'lactation efficiency' concept of Bergsma *et al.* (2008b, see above) becomes all the more interesting against this background.

Whittemore (1998) made the obvious comment that if the required nutrient intake cannot be realised during lactation, feeding strategies during pregnancy will have to be adapted, and Bunter *et al.* (2008) drew the same conclusion from their own data analyses: 'under current gestational feeding systems, sows are unable to adapt intake to meet both their own and litter requirements on an individual as-needs basis'.

Acknowledgements

Help from Kim Bunter, David Casey, Steve Chadd, Jaco Eissen, Susanne Hermesch, Andreas Hofer, Egbert Kanis and Wendy Rauw is gratefully acknowledged.

References

Andersen, I.L., Naevdal, E., Bakken, M. and Bøe, K.E., 2004. Aggression and group size in domesticated pigs, *Sus scrofa*: 'when the winner takes it all and the loser is standing small'. Animal Behaviour 68: 965-975.

Archer, J.A., Pitchford, W.S., Hughes, T.E. and Parnell, P.F., 1998. Genetic and phenotypic relationships between food intake, growth, efficiency and body composition of mice post weaning and at maturity. Animal Science 67: 171-182.

Bergsma, R., Kanis, E., Knol, E.F. and Bijma, P., 2008a. The contribution of social effects to heritable variation in finishing traits of domestic pigs (*Sus scrofa*). Genetics 178: 1559-1570.

Bergsma, R., Kanis, E., Verstegen, M.W.A. and Knol, E.F., 2008b. Genetic parameters and predicted selection results for maternal traits related to lactation efficiency in sows. Journal of Animal Science 86: 1067-1080.

Black, J.L, 2009. Models to predict feed intake. In: Torrallardona, D. and Roura, E. (eds.) Voluntary feed intake in pigs. Wageningen Academic Publishers, the Netherlands, pp. 323-351.

Brisbane, J.R., 2002. Prediction of within-herd differences in total feed intake between growing pigs. Canadian Journal of Animal Science 82: 283-293.

Bunter, K.L., Luxford, B.G. and Hermesch, S., 2007. Associations between feed intake of growing gilts, lactating sows and other reproductive or performance traits. Proceedings of the Association for Advancement of Animal Breeding and Genetics 17: 57-60.

Bunter, K.L., Smits, R., Luxford, B. and Hermesch, S., 2008. Sow body composition and its associations with reproductive and litter growth performance in the primiparous sow. Proceedings of AGBU Pig genetics workshop, pp. 67-81. Available at: http: //agbu.une.edu.au/pigs/pigblup/pdf/2008/Paper9_KB_Bodycomposition.pdf. Accessed 17 November 2008.

Cameron, N.D. and Curran, M.K., 1994. Selection for components of efficient lean growth rate in pigs. 4: Genetic and phenotypic parameter estimates and correlated responses in performance test traits with ad-libitum feeding. Animal Production 59: 281-291.

Cameron, N.D., Kerr, J.C., Garth, G.B., Fenty, R. and Peacock, A., 2002. Genetic and nutritional effects on lactational performance of gilts selected for components of efficient lean growth. Animal Science 74: 25-38.

Canario, L., Bergsma, R., D'eath, R.B., Lawrence, A.B., Roehe, R., Lundeheim, N., Rydhmer, L., Knol, E.F. and Turner, S.P., 2008. Genetic relations between the group effect for average daily gain, and post-mixing aggression and skin lesions in Swedish pigs. Abstracts of the 59th Annual meeting of the European Association for Animal Production, Vilnius, Lithuania. Paper 25.3.

Casey, D.S. 2003. The use of electronic feeders in genetic improvement programs for swine. PhD thesis, Iowa State University, Ames IO, USA.

Chen, C.Y., Johnson, R.K., Newman, S. and Van Vleck, L.D., 2007. A general review of competition genetic effects with an emphasis on swine breeding. Genetics and Molecular Research 6: 594-606.

Chen, C.Y., Kachman, S.D., Johnson, R.K., Newman, S. and Van Vleck, L.D., 2008. Estimation of genetic parameters for average daily gain using models with competition effects. Journal of Animal Science 86: 2525-2530.

Clowes, E.J., Aherne, F.X., Foxcroft, G.R. and Baracos, V.E., 2003. Selective protein loss in lactating sows is associated with reduced litter growth and ovarian function. Journal of Animal Science 81: 753-764.

De Haer, L.C.M., 1992. Relevance of eating pattern for selection of growing pigs. PhD thesis, Wageningen Agricultural University, Wageningen, The Netherlands.

De Haer, L.C.M and Merks, J.W.M., 1992. Patterns of daily feed intake in growing pigs. Animal Production 54: 95-104.

De Haer, L.C.M., Luiting, P. and Aarts, H.L.M., 1993. Relations among individual (residual) feed intake, growth performance and feed intake pattern of growing pigs in group housing. Livestock Production Science 36: 233-253.

De Haer, L.C.M. and De Vries, A.G., 1993. Feed intake patterns of and feed digestibility in growing pigs housed individually or in groups. Livestock Production Science 33: 277-292.

De Lange, C.F.M., Birkett, S.H. and Morel, P.C.H., 2001. Protein, fat and bone tissue growth in swine. In: Lewis, A.J. and Southern, L.L. (eds.) Swine nutrition 2nd edition. CRC Press, Boca Raton FL, USA, pp. 65-81.

Dourmad, J.Y., 1988. Ingestion spontanée d'aliment chez la truie en lactation: de nombreux facteurs de variation. INRA Production Animale 1: 141-146.

Duthie, C.A., Simm, G., Doeschl-Wilson, A.B., Kalm, E., Knap, P.W. and Roehe, R., 2008. Quantitative trait loci for chemical body composition traits in pigs and their positional associations with body tissues, growth and feed intake. Animal Genetics 39: 130-140.

Eissen, J.J., 2000. Breeding for feed intake capacity in pigs. PhD thesis, Wageningen University, Wageningen, the Netherlands.

Eissen J.J., Apeldoorn, E.J., Kanis, E., Verstegen, M.W.A. and De Greef, K.H., 2003. The importance of a high feed intake during lactation of primiparous sows nursing large litters. Journal of Animal Science 81: 594-603.

Ellis, M. and Augspurger, N., 2001. Feed intake in growing-finishing pigs. In: Lewis, A.J. and Southern, L.L. (eds.) Swine nutrition 2nd edition. CRC Press, Boca Raton FL, USA, pp. 447-467.

Emmans, G.C. 1988. Genetic components of potential and actual growth. In: Land, R.B., Bulfield, G. and Hill, W.G. (eds.) Animal breeding opportunities. British Society of Animal Production, Penicuik, UK. Occasional Publication no. 12, pp. 153-181.

Emmans, G.C. 1997. A method to predict the food intake of domestic animals from birth to maturity as a function of time. Journal of Theoretical Biology 186: 189-199.

Emmans, G.C. and Fisher, C. 1986. Problems in nutritional theory. In: Fisher, C. and Boorman, K.N. (eds.) Nutrient requirements of poultry and nutritional research. Butterworths, London, UK, pp. 9-39.

Fowler, V., Bichard, M. and Pease, A., 1976. Objectives in pig breeding. Animal Production 23: 365-387.

Grandinson, K., Rydhmer, L., Strandberg, E. and Solanes, F.X., 2005. Genetic analysis of body condition in the sow during lactation, and its relation to piglet survival and growth. Animal Science 80: 33-40.

Grignola, F., Clutter, A.C., Casey, D.S., Dekkers J.C.M. and Liu, X., 2002. Effects of feeding strategies using electronic feeders on data quality, feed intake, growth and feed efficiency of swine. Proceedings of the 7th World congress on genetics applied to livestock production, Montpellier, France, communication 10-12.

Hall, A.D., Hill, W.G., Bampton, P.R. and Webb, A.J., 1999. Predicted responses to selection from indices incorporating feeding pattern traits of pigs using electronic feeders. Animal Science 68: 407-412.

Henry, Y. 1985. Dietary factors involved in feed intake regulation in growing pig: a review. Livestock Production Science 12: 339-354.

Hermesch, S., 2004. Genetic improvement of lean meat growth and feed efficiency in pigs. Australian Journal of Experimental Agriculture 44: 1-9.

Hermesch, S., 2007. Genetic analysis of feed intake in lactating sows. Proceedings of the Association for Advancement of Animal Breeding and Genetics 17: 61-64.

Hermesch, S., Kanis, E. and Eissen, J.J., 2003. Economic weights for feed intake in the growing pig derived from a growth model and an economic model. Journal of Animal Science 81: 895-903.

Hermesch, S., Jones, R. and Bunter, K.L., 2008. Feed intake of sows during lactation has genetic relationships with growth and lifetime performance of sows. Proceedings of AGBU Pig genetics workshop, pp. 55-65. Available at: http: //agbu.une.edu.au/pigs/pigblup/pdf/2008/Paper 8_SH_Feed intake of sows.pdf. Accessed 17 November 2008.

Holm, B., Bakken, M., Klemetsdal, G. and Vangen, O., 2004. Genetic correlations between reproduction and production traits in swine. Journal of Animal Science 82: 3458-3464.

Hyun, Y., Ellis, M., McKeith, F.K. and Wilson, E.R., 1997. Feed intake pattern of group-housed growing-finishing pigs monitored using a computerized feed intake recording system. Journal of Animal Science 75: 1443-1451.

Kanis, E., 1990. Effect of food intake capacity on production traits in growing pigs with restricted feeding. Animal Production 50: 333-341.

Kanis, E. and De Vries, A.G., 1992. Optimization of selection for food intake capacity in pigs. Animal Production 55: 247-255.

Karsten, S., Roehe, R., Schulze, V., Looft, H., and Kalm, E., 2000. Genetische Beziehungen zwischen individueller Futteraufname während der Eigenleistungsprüfung und Fruchtbarkeitsmerkmalen beim Schwein. Archiv für Tierzucht 43: 451–461.

Kerr, J.C. and Cameron, N., 1996. Responses in gilt post-farrowing traits and pre-weaning piglet growth to divergent selection for components of efficient lean growth rate. Animal Science 63: 523-531.

Knap, P.W., 2000. Stochastic simulation of growth in pigs: relations between body composition and maintenance requirements as mediated through protein turnover and thermoregulation. Animal Science 71: 11-30.

Knap, P.W., 2005. Breeding robust pigs. Australian Journal of Experimental Agriculture 45: 763-774.

Knap, P.W., 2008. Robustness. In: Rauw, W.M. (ed.) Resource allocation theory applied to farm animal production. CABI, Wallingford, UK, pp. 288-301.

Knap, P.W. and Rauw, W.M., 2008. Selection for high performance in pigs. In: Rauw, W.M. (ed.) Resource allocation theory applied to farm animal production. CABI, Wallingford, UK, pp. 210-229.

Koketsu, Y. and Dial, G.D., 1997. Factors influencing the postweaning reproductive performance of sows on commercial farms. Theriogenology 47: 1445-1461.

Koketsu, Y., Dial, G.D., Pettigrew, J.E., March, W.E. and King, V.L., 1996. Characterization of feed intake patterns during lactation in commercial swine herds. Journal of Animal Science 74: 1202-1210.

Labroue, F., 1995. Facteurs de variation génétiques de la prise alimentaire chez le porc en croissance: le point des connaissances. INRA Productions Animales 8: 239-250.

Labroue, F., Sellier, P., Guéblez, R. and Meunier-Salaün, M.C., 1996. Estimation des paramètres génétiques pour les critères de comportement alimentaire dans les races Large White et Landrace Français. Journées de Recherche Porcine en France 28: 23-30.

Lorenzo Bermejo, J., Roehe, R., Schulze, V., Looft, H. and Kalm, E., 2003. Genetic change of feed intake curves in growing pigs using non-linear two-stage genetic analysis and linear random regression models. Journal of Animal Breeding and Genetics 120: 217-227.

Lynch, P.B., 1989. Voluntary food intake of sows and gilts. In: Forbes, J.M., Varley, M.A. and Lawrence, T.L.J. (eds.) The voluntary food intake of pigs. British Society of Animal Production, Penicuik, UK. Occasional Publication no. 13, pp. 71-77.

Manteca, X. and Edwards, S., 2009. Feeding behaviour and social influences on feed intake. In: Torrallardona, D. and Roura, E. (eds.) Voluntary feed intake in pigs. Wageningen Academic Publishers, the Netherlands, pp. 301-306.

McCracken, K.J., McEvoy, J., McAllister, A., Lilley, J. and Urquhart, R., 1994. Effects of overfeeding on protein/energy metabolism and body composition of high genetic potential boars. In: J.F. Aguilera (ed.) Energy metabolism of farm animals. CSIC, Madrid, Spain. EAAP publication no. 76, pp. 217-220.

Mohrmann, M., Roehe, R., Knap, P.W., Looft, H., Plastow, G.S. and Kalm, E., 2006. Quantitative trait loci associated with AutoFOM grading characteristics, carcass cuts and chemical body composition during growth of *Sus scrofa*. Animal Genetics 37: 435-443.

Muir, W.M. and Schinckel, A.P., 2002. Incorporation of competitive effects in breeding programs to improve productivity and animal well being. Proceedings of the 7th world congress of genetics applied to livestock production, Montpellier, France, communication 14-07.

Rauw, W.M., Soler, J., Tibau,, J., Reixach, J. and Gomez Raya, L., 2006. Feeding time and feeding rate and its relationship with feed intake, feed efficiency, growth rate and rate of fat deposition in growing Duroc barrows. Journal of Animal Science 84: 3404-3409.

Rauw, W.M., Hermesch, S., Bunter, K.L. and Gomez-Raya, L., 2009. The relationship between growth, mature and lactation food intake in a mouse model. Livestock Science, in press.

Rydhmer, L., Johansson, K., Stern, S. and L. Eliasson-Selling, L., 1992. A genetic study of pubertal age, litter traits, weight loss during lactation and relations to growth and leanness in gilts. Acta Agriculturae Scandinavica Section A 42: 211-219.

Schnyder, U., Hofer, A., Labroue, F. and Künzi, N., 2001. Genetic parameters of a random regression model for daily feed intake of performance tested French Landrace and Large White growing pigs. Genetics Selection Evolution 33: 635-658.

Schulze, V., Roehe, R., Looft, H. and Kalm, E., 2003. Effects of continuous and periodic feeding by electronic feeders on accuracy of measuring feed intake information and their genetic association with growth performances. Journal of Animal Breeding and Genetics 118: 403-416.

Schulze, V., Roehe, R., Lorenzo Bermejo J., Looft H. and Kalm, E., 2003. The influence of feeding behaviour on feed intake curve parameters and performance traits of station-tested boars. Livestock Production Science 82: 105-116.

SUISAG, 2007. Zahlen und Projekte 2007. SUISAG, Sempach, Switzerland. Available at: http: //www.suisag.ch/uploads/zahlen_und_projekte_2007_d.pdf. Accessed 21 October 2008.

Ten Napel, J. and Johnson, R.K., 1997. Genetic relationships among production traits and rebreeding performance. Journal of Animal Science 75: 51-60.

Van Erp, A.J.M., Molendijk, R.J.F., Eissen, J.J and Merks, J.W.M., 1998. Relation between ad libitum feed intake of gilts during rearing and feed intake capacity of lactating gilts. Proceedings of the 49th Annual meeting of the European Association for Animal Production, Warsaw, Poland. Paper G5.10.

Verstegen, M.W.A., Mesu, J., Van Kempen, G.J.M. and Geerse, C., 1985. Energy balances of lactating sows in relation to feeding level and stage of lactation. Journal of Animal Science 60: 731-740.

Von Felde, A., 1996. Genetische Analyse der Futteraufnahme-Informationen von Jungebern aus Gruppenprüfung mit automatischen Fütterungsanlagen. PhD thesis no. 90, Christian-Albrechts University, Kiel, Germany.

Von Felde, A., Roehe, R., Looft, H. and Kalm, E., 1996. Genetic association between feed intake and feed intake behaviour at different stages of growth of group-housed boars. Livestock Production Science 47: 11-22.

Webb, A.J., 1989. Genetics of food intake in the pig. In: Forbes, J.M., Varley, M.A. and Lawrence, T.L.J. (eds.) The voluntary food intake of pigs. British Society of Animal Production, Penicuik, UK. Occasional Publication no. 13, pp. 41-50.

Wellock, I.J., Emmans, G.C. and Kyriazakis, I., 2003. Predicting the consequences of social stressors on pig food intake and performance. Journal of Animal Science 81: 2995-3007.

Whittemore, C.T., 1998. Influence of pregnancy feeding on lactation performance. In: Verstegen, M.W.A., Moughan, P. J. and Schrama, J.W. (eds.) The lactating sow. Wageningen Pers, Wageningen, the Netherlands, pp. 183-200.

Whittemore, C.T. and Fawcett, R.H., 1976. Theoretical aspects of a flexible model to simulate protein and lipid growth in pigs. Animal Production 22: 87-96.

Chapter 2. Voluntary feed intake in piglets: potential impact of early experience with flavours derived from the maternal diet

J.E. Bolhuis[1], M. Oostindjer[1], H. Van den Brand[1], W.J.J. Gerrits[2] and B. Kemp[1]
[1]Adaptation Physiology Group, Wageningen Institute of Animal Sciences, Wageningen University, P.O. Box 338, 6700 AH Wageningen, the Netherlands; Liesbeth.Bolhuis@wur.nl
[2]Animal Nutrition Group, Wageningen Institute of Animal Sciences, Wageningen University, P.O. Box 338, 6700 AH Wageningen, the Netherlands

Abstract

The transition from suckling milk to the independent ingestion of solid feed is a challenge for piglets in commercial pig husbandry, which are usually weaned abruptly at a young age. A sufficient intake of nutrients after weaning is crucial for maintaining health and performance, but weanling piglets appear to be poorly adapted and motivated to ingest solid feed. The weaning transition therefore frequently results in a period of underfeeding, leading to health and welfare problems. To ensure a successful adaptation to solid feed, piglets should ideally learn what to eat and how to forage prior to weaning. In this period, they may benefit from the experience and skills of other individuals, i.e. social learning. In young mammals with intensive maternal care, mothers are likely the best models for social information transmission. There are several ways in which information transmission from mother to offspring, i.e. vertical transmission, may facilitate the suckling's transition to independent feeding. This chapter focuses on the effects of early exposure to flavours from the maternal diet on feed acceptance in later life. Transfer of dietary information from mother to offspring may start already before birth, as flavours from the maternal diet may be transmitted to the foetus through the amniotic fluid or foetal blood stream. In addition, flavours in mother's milk may be a source of chemosensory information about the maternal diet. Studies in other species suggest that perinatal exposure to flavours from the maternal diet may enhance the acceptance of similarly flavoured feeds during weaning in later life. In pigs, however, little is known regarding the impact of vertical flavour transmission, either pre- or postnatally, on feed intake around weaning and postweaning performance. We argue that adaptation of piglets to solid feed may be enhanced by facilitating the opportunities for vertical information transfer, which are currently limited and unexploited in commercial pig husbandry.

Keywords: weaning, vertical information transmission, prenatal perception, chemosensory learning, development, adaptation

Introduction and outline

Weaning is regarded as one of the most critical periods in pig production, imposing major limitations to its efficiency worldwide (King and Pluske, 2003). Timely and sufficient intake of nutrients during the immediate postweaning period is crucial for the performance, health and welfare of piglets. The transition from mother's milk to independent solid feeding, however, along with an array of other environmental changes, often causes a problematic drop in voluntary feed intake in the weanling pig. This chapter focuses on the potential effects of early exposure to flavours from the maternal diet on feed acceptance around weaning. After an introduction on voluntary feed intake in weanling pigs, the role of social information transmission in feed selection and acceptance will be briefly described. Then, the effects of early exposure, either prenatally or postnatally, to flavours from the maternal diet on feed preference and acceptance will be reviewed, using literature on several mammalian species. Subsequently, the scarce data available for pigs will be summarised. Finally, the potential impact of vertical social learning for performance of weanling pigs and implications for pig industry will be discussed.

The low voluntary feed intake around weaning and its consequences

For pigs under (semi)natural conditions, the weaning process comprises a gradual transition from total reliance on sow milk early in life to independent foraging at 9-22 weeks of age (e.g. Jensen, 1995; Jensen and Recén, 1989; Newberry and Wood-Gush, 1985, see Held and Mendl, 2001; Nielsen *et al.*, 2006 for review). In contrast, in modern pig husbandry, the lactation period is relatively short, 3-4 weeks in Europe and even shorter in the USA, and ends abruptly. Weaning is therefore considered to be a major stressor for commercially held piglets. In addition to alterations in their social and physical environment (see Weary *et al.*, 2008 for review), piglets have to change from a diet consisting predominantly of milk to solid feed. This abrupt transition is associated with a critical period of underfeeding: some piglets refrain from eating for over 50 h (Brooks and Tsourgiannis, 2003; Bruininx *et al.*, 2002). The voluntary feed intake in the immediate postweaning period varies largely between individual piglets (Appleby *et al.*, 1991; Pajor *et al.*, 1991; Pluske *et al.*, 1995), but is, on average, until days 3-4 after weaning too low to meet the metabolisable energy required for maintenance (Le Dividich and Sève, 2001).

The low nutrient intake during the first days after weaning is accompanied by a low and variable growth rate or an actual weight loss (Bruininx *et al.*, 2001; Pajor *et al.*, 1991). Moreover, the weaning transition may cause impaired intestinal functioning and an increased susceptibility to enteric pathogens, resulting in the occurrence of digestive disorders (King, 2003, King and Pluske, 2003; Madec *et al.*, 1998; Vente-Spreeuwenberg and Beynen, 2003). This is partly due to the relatively high feed intake following the period of underfeeding, which may exceed the digestive or absorptive capacity and cause an overload of undigested nutrients triggering undesirable microbial activity (Makkink, 1993). However, the low level of early postweaning feed intake, rather than the composition of the diet or the subsequent relative overeating, appears to be the primary and major determinant of reduced intestinal integrity (McCracken *et al.*, 1995, 1999; Spreeuwenberg *et al.*, 2001). Hence, performance of piglets at weaning depends primarily on their feed intake, providing an adequate diet is given (Leibbrandt *et al.*, 1975). Moreover, it has been suggested that the undernutrition and 'growth check' at weaning can also have a major impact on subsequent performance (Geary and Brooks, 1998; Le Dividich and Sève, 2001; Miller *et al.*, 1999; Williams, 2003), predisposing pigs to a future loss of growth potential. Several studies have therefore emphasised the importance of maintaining continuity of nutrient intake after weaning (Leibbrandt *et al.*, 1975; Brooks and Tsourgiannis, 2003).

Effects of preweaning experiences on feed intake around weaning

Preweaning experiences may affect the ability to cope with the weaning process and improve performance of piglets after weaning. The physiological and neural control of ingestive behaviour is, at least initially, different from that of suckling (Hall and Williams, 1983). Hence, for a successful transition from milk to solids, mechanisms promoting and regulating feed intake must have reached a certain level of maturity prior to weaning. Under (semi)natural conditions, wild and domestic piglets start exploring and ingesting solids early in life, long before the weaning process is fully terminated. For instance, piglets were reported to follow their mothers to feeding sites (Gundlach, 1968; Newberry and Wood-Gush, 1986; 1988) and to ingest acorns and corn (Meynhardt, 1980) at around 10-14 days of age. Although feed intake at this age is still low, the sampling of food and non-food substrates may prepare piglets, both behaviourally and physiologically, for the postweaning diet (see Brooks and Tsourgiannis, 2003; Labayen *et al.*, 1994; Petersen, 1994). For instance, exposure to solid feed influences the structure of the gastrointestinal tract and the activity of digestive enzymes (Makkink, 1993; Nabuurs *et al.*, 1993). In an attempt to adapt piglets in pig husbandry to solid feed prior to weaning, they are therefore frequently offered a highly palatable and easily digestible creep feed diet. Several studies show that creep feed intake of individual pigs during the suckling period is indeed positively correlated with their early postweaning feed intake or performance (Berkeveld *et al.*, 2007; Bruininx *et al.*, 2002, 2004; Kuller *et al.*, 2004), although a causal relationship has been questioned (Appleby *et al.*, 1991; Pajor *et al.*, 1991). The ingestion of creep feed during the lactation period is, however, often very low and varies considerably between suckling piglets (Appleby *et al.*, 1992; Bøe and Jensen, 1995; Pajor *et al.*, 1991; Pluske *et al.*, 1995).

Motivational aspects are thought to be critical factors limiting feed intake in young piglets (Pajor *et al.*, 1991). Young animals are innately reluctant to ingest a novel feed, i.e. neophobia. Neophobia may be adaptive under natural conditions as it prevents toxicosis while animals are learning from the post-ingestive consequences of eating a novel and potentially toxic feed (Rozin, 1976; Rozin and Vollmecke, 1986). It may, however, also hamper feed acceptance in suckling and weanling piglets. An abundance of studies have attempted to overcome neophobia and increase feed acceptance in suckling and weanling piglets by improving the palatability of the feed offered, enhancing its quality, or increasing its attractiveness by altering the way it is presented, with variable and mostly only minor effects (e.g. Appleby *et al.*, 1991; Edge *et al.*, 2004; see King and Pluske, 2003; Pajor *et al.*, 2002). Apparently, other strategies are required to stimulate and motivate piglets to start eating, for example, by manipulating social influences. Social aspects of the development of independent feeding behaviour in piglets have so far received little attention, both in research and in pig industry. In particular, little is known about the importance of the sow in transferring foraging information and skills to her inexperienced offspring. Studies in other species reveal, however, that social learning mechanisms may facilitate the young omnivores' transition from exclusive milk feeding to consuming solid feeds.

The role of social information transmission in the ontogeny of feeding

The social environment may influence or could even be crucial for an uncomplicated development of independent feed intake in piglets. Pigs are omnivorous animals that, under (semi)natural conditions, can exploit an enormous variety of potential food sources, such as nuts, grasses, fruits, twigs, herbs, fungi, agricultural crops, insect larvae, earthworms, snails and slugs, but also amphibians, reptiles, fish, small mammals, birds, and large mammals as carrion (Schley and Roper, 2003). For omnivores, the selection and ingestion of an appropriate diet is a complex challenge that requires the integration of information from a wide variety of sources. Omnivores appear to have developed a few genetically based predispositions, such as a congenital preference for sweet tasting substances and an aversion of bitter tasting ones, providing only a rough means of discriminating between nutritive feeds and toxic items (see Birch, 1998, 1999; Galef, 1996a; Rozin and Vollmecke, 1986, for review). Apart from these congenital responses to gustatory stimuli, learning, which enables flexibility in diet selection, is important for developing feed preferences (Galef, 1996a; Provenza and Balph, 1987; Rozin, 1976; Visalberghi *et al.*, 2003).

Animals acquire information about food items through a trial-and-error learning process of sampling different food items and monitoring the consequences of ingestion. In doing so, they use, amongst other characteristics, flavours to identify foods. The perception of flavours while ingesting feed involves at least three different chemosensory systems: olfaction, taste and trigeminal-mediated sensation. Retronasal olfaction, i.e. the stimulation of olfactory receptors by odours that travel from the mouth via the retronasal pharyngeal opening to the nasal cavity during suckling, swallowing and chewing, contributes probably most significantly to the complexity of flavours by inducing multimodal sensory integration (Bult *et al.*, 2007). Animals relate the flavours they identify in feeds to the positive and negative postingestive consequences of these feeds, and subsequently, through complex conditioning processes, may come to accept or avoid these flavours. In this way, they learn to distinguish safe and nutritious food items from toxic ones (see Galef, 1996a; Mehiel and Bolles, 1984, 1988; Provenza and Villalba, 2006).

Young animals may develop the appropriate responses to food through their own experience. There is considerable evidence that experiences early in life during sensitive periods produce enduring effects on later flavour preferences and food acceptance (e.g. Mennella *et al.*, 2004; Nolte and Provenza, 1992; Simitzis *et al.*, 2008). This individual food-related learning can, however, be time-consuming and risky. For inexperienced foragers, therefore, it may be safer and more efficient to benefit from the expertise of other individuals, concerning how, where, when and what to eat (Galef, 1996b; Galef and Giraldeau, 2001), i.e. social learning. Social learning refers to learning that is influenced by observation of or interaction with another animal or its products (see Nicol, 1995 for review). Social learning might particularly be crucial during the transition from suckling mother's milk to independent feeding (Galef, 1971; Galef and Clark, 1971, 1972; Valsecchi *et al.*, 1993). For young mammals, mothers are likely the best models for social information transfer (Nicol and Pope, 1994; Provenza and Balph, 1987; Thorhallsdottir *et al.*, 1990), i.e. vertical transfer. Dietary learning from mothers may be more powerful than learning from other experienced animals (Saint-Dizier *et al.*, 2007; Thorhallsdottir *et al.*, 1990). Chances that mothers, as transmitters of information, experience the same environment as their offspring are high, which enhances the adaptiveness of social learning relative to individual learning (see Sibly, 1999). Furthermore, because mother and offspring are genetically related, they will likely respond similarly to particular food items (Chapple and Lynch, 1986; Laland *et al.*, 1993). In addition, the transmission of feeding information from mother to offspring may start in an early stage of development, even prenatally. Although there are several other ways in which information transmission from mother to offspring (for review, see Nicol, 1995; Galef and Giraldeau, 2001) may facilitate the suckling's transition to independent feeding at weaning, this chapter focuses on the effects of early exposure to flavours from the maternal diet.

Perinatal exposure to flavours derived from the gestating and lactating mother's diet

Prenatal flavour experiences

Transfer of dietary information from mother to offspring may start already before birth. Even though the embryonic membranes provide a series of barriers that protect the mammalian foetus from most external disturbances, the foetus is prenatally exposed to a variety of chemosensory stimuli. Volatile compounds may enter the foetal blood stream after crossing the placental barrier, subsequently diffuse out of the nasal blood capillaries and come into contact with foetal olfactory receptors (Hepper, 1988). In addition, flavours may be transmitted to the amniotic fluid, and stimulate olfactory receptors and taste buds (Hepper, 1988, 1990; Mennella and Beauchamp, 1998) after inhalation or swallowing by the foetus. Swallowing of amniotic fluid during the latter part of gestation likely occurs in all mammals with significant foetal fluid production (Ross and Nijland, 1998), including pigs (Marrable, 1971; Sangild *et al.*, 2002, 2003).

Anatomical, behavioural and physiological evidence is accumulating that during the last part of gestation, the mammalian foetus is able to process and retain chemosensory information (for review, see Schaal and Orgeur, 1992). Exposure to flavours in the prenatal environment may subsequently influence behaviour after birth. For instance, neonates from several species exhibit attraction to the odour of amniotic fluid (humans: Soussignan *et al.*, 1997; Varendi *et al.*, 1996; rats: Hepper, 1987; sheep: Schaal *et al.*, 1995b; pigs: Rohde Parfet and Gonyou, 1991) and, moreover, can recognise their own amniotic fluid (Hepper, 1987; Schaal *et al.*, 1995a). Also prenatal experiences with artificial odours introduced through intra-amniotic injection have been demonstrated to influence postnatal behaviour (Hepper, 1990; Smotherman, 1982a,b; Schaal and Orgeur, 1992). For example, intra-amniotic injection of orange flavour induced a preference for orange odour in rat pups tested 12 days after birth (Hepper, 1990).

Prenatal chemosensory learning has not only been reported in mammals (humans: Schaal *et al.*, 2000; rabbits: Semke *et al.*, 1995; rats: Hepper, 1987; Smotherman, 1982a; sheep: Schaal *et al.*, 1995b), but also, *in ovo*, in birds (Sneddon *et al.*, 1998) and amphibians (Hepper and Waldman, 1992). This widespread occurrence of prenatal chemosensory learning suggests that it may be important in facilitating the transition from the prenatal environment to postnatal life. Apart from affecting nest localisation, maternal and kin recognition, nipple localisation and initiation of suckling in some species (Coureaud *et al.*, 2002; Hepper, 1987), prenatal flavour learning has been suggested to play an important role in establishing later dietary preferences and the acquisition of independent feeding at weaning (Altbäcker *et al.*, 1995; Bilkó *et al.*, 1994; Hepper, 1988; Hudson and Distel, 1999; Mennella *et al.*, 2004).

Effects of prenatal exposure to flavours from the maternal diet on feed acceptance

A factor that can affect the prenatal chemosensory environment is the maternal diet during gestation. Several studies have shown that flavours ingested by the mother appear in the foetal environment. For instance, it has been documented that the near-term human foetus and the amniotic fluid can adopt the odour of spices ingested by the mother, such as curry, cumin or fenugreek (Hauser *et al.*, 1985). In sheep, garlic odour was detected in the foetal blood and amniotic fluid at 50 and 100 min after ingestion by the ewe (Nolte *et al.*, 1992). Similarly, 45 min after garlic ingestion in humans its odour is present in the amniotic fluid (Mennella *et al.*, 1995).

Prenatal exposure to flavours from the maternal diet generally increases postnatal olfactory preference for these flavours, or reduces aversion; Table 1 illustrates the variety of flavours and species studied. Most of the studies cited have focussed on the short-term recall by neonates of flavours experienced *in utero*. Others indicate, however, that prenatally acquired flavour

Table 1. Overview of experiments demonstrating that prenatal exposure to flavours from the maternal diet affects postnatal preferences. Species, type of flavour, (significant) postnatal effects and references are indicated.

Species	Flavour[1]	Postnatal effects[5]	References
Rats	garlic	↑ olfactory preference d12	Hepper, 1988
Rabbits	juniper[2]	↑ intake juniper at weaning d28	Bilkó *et al.*, 1994; Semke *et al.*, 1995
	cumin	↑ olfactory preference d0	Coureaud *et al.*, 2002
Sheep	citral[2]	↑ olfactory preference d0	Schaal *et al.*, 1995b
	oregano[3]	↑ intake oregano-flavoured feed at 3, 4.5, 6, 7.5 mo	Simitzis *et al.*, 2008
Dogs	anise (trans-anethole)	↑ olfactory preference d0	Wells and Hepper, 2006
		↑ intake anise-flavoured treats w10, only when exposed through milk also	Hepper and Wells, 2006
Humans	anise[2]	↑ olfactory preference d0 and d4	Schaal *et al.*, 2000
	garlic	↑ olfactory preference d0	Hepper, 1995[6]
	carrot[4]	↓ negative facial responses while eating carrot-flavoured cereal at 6 mo	Mennella *et al.*, 2001
		↑ enjoyment perceived by mother while eating carrot-flavoured cereal	

[1] Provided in the maternal diet throughout gestation unless indicated otherwise.

[2] Last two gestational weeks.

[3] Day 50-130 of gestation.

[4] Last three gestational weeks.

[5] Relative to (non-exposed) control groups, i.e. ↑ preference can also be reduced avoidance. d=day; w=weeks; mo=months.

[6] Cited by Mennella *et al.*, 2001.

preferences may be long-lasting and persist until weaning and beyond. It has been suggested that the acceptance of solid foods at weaning is influenced by their correspondence with flavours experienced prior to birth (Schaal *et al.*, 2000). The few studies that have investigated the effect of prenatal flavour exposure on actual ingestion of similarly flavoured feeds during weaning generally support this suggestion (see Table 1). For instance, the ingestion of juniper berries on the day of weaning was five times higher in rabbit pups from does fed juniper during gestation compared to controls (Bilkó *et al.*, 1994). In humans, regular consumption of carrot juice by pregnant mothers increased the acceptance and enjoyment of a carrot-flavoured cereal meal in their 6-month old infants during the weaning stage (Mennella *et al.*, 2001). Moreover, prenatally exposed infants were reported to consume three times more carrot-flavoured cereal than plain cereal, whereas non-exposed infants did not show such a preference. Similarly, maternal ingestion of oregano during gestation increased the intake of oregano-supplemented feed in lambs at several ages, ranging from 3 to 7.5 months (Simitzis *et al.*, 2008). These findings are supported by data from studies in which flavours were artificially introduced into the prenatal environment. Exposure to apple juice by intra-amniotic injection on day 20 of gestation enhanced the preference for apple-flavoured solution in adult rats at 60 days of age (Smotherman, 1982a). Also intra-amniotic injection with ethanol increased ingestive preference for this solution in 9-day old rats (Chotro and Molina, 1990). Similarly, application of strawberry flavour on the eggshell during incubation resulted in an eightfold increase in the intake of strawberry-flavoured water by 4-day old chickens (Sneddon *et al.*, 1998).

Taken together, these findings indicate that prenatal learning plays a role in postnatal feeding behaviour and development of dietary habits. Prenatal exposure to flavours derived from the maternal diet may facilitate the acceptance of similarly flavoured feeds at weaning and in later life. However, also postnatally, during the lactation period, young mammals have the opportunity to become familiar with the flavours of feed eaten by their mother. The transmission of flavours from the lactating female's diet to her offspring and its consequences for food selection and acceptance are described below.

Effects of exposure to flavours from the maternal diet during lactation on feed acceptance

Studies in rats and mice have shown that for their first meals of solid feed, weanlings actively seek and preferentially ingest the feed their mother has eaten during lactation, even if that diet is relatively unpalatable (Galef and Henderson, 1972; Valsecchi *et al.*, 1989). Also in other species, the maternal diet during lactation profoundly influences the feed preferences of young animals around weaning and beyond (see Table 2 for summary of studies). For instance, mouse pups consumed three times more fennel at weaning when their mother had ingested fennel during lactation (Mainardi *et al.*, 1989). Moreover, the lactating mother's diet may influence the initial dietary choice at weaning (Bilkó *et al.*, 1994; Mainardi *et al.*, 1989).

Most of the studies reported in Table 2 used two- or three-way food choice tests to assess the influence of the maternal diet on feed preferences of the offspring, without paying much attention to feed intake per se. The maternal diet during lactation may, however, also affect feed intake at weaning when only a single diet is offered, which is often the case in pig husbandry. The first independent feed intake of weanling rats was relatively high if that feed was identical to

Table 2. Overview of experiments demonstrating that exposure to flavours from the diet of the lactating mother affects postweaning preferences. Species, type of flavour, postweaning effect and references are indicated.

Species	Flavour[1]	Testing age[8]	Postweaning effects[9]	References
Rats	lab chow[2]	d21 (w)	↑ intake lab chow maternal diet	Galef and Sherry, 1973
		d19-23 (w)	↑ intake lab chow maternal diet	Galef and Henderson, 1972
	lab chow[3]	d23 (w)	↑ intake in 30-min test if weanling diet matched with maternal diet	Bronstein *et al.*, 1975
	garlic[4]	d29-40	↑ intake of garlic-flavoured water	Capretta and Rawls, 1974
	onion	d21, d23 (w)	↑ intake of onion-flavoured diet	Wuensch, 1978
		d25	onion preference diminished	
Mice	fennel	d22 (w)	↑ intake of fennel	Mainardi *et al.*, 1989
Rabbits	juniper	d28 (w)	↑ intake of juniper	Bilkó *et al.*, 1994
Goats	rice straw[5]	6 mo	↓ latency to first ingestion ↑ intake of rice straw	Van Tien, 2002
Dogs	anise[6]	10 w	↑ intake anise-flavoured treats, only when exposed prenatally also	Hepper and Wells, 2006
Humans	carrot[7]	6 mo (w)	↓ negative facial responses while eating carrot-flavoured cereal	Mennella *et al.*, 2001
			↓ intake of carrot-flavoured cereal shortly after exposure to carrot-flavoured mother's milk	Mennella and Beauchamp, 1999

[1] Provided in the maternal diet throughout the lactation period unless indicated otherwise.

[2] Day 5-weaning.

[3] Day 2-weaning.

[4] Three days before parturition-weaning at day 22.

[5] Last three weeks before weaning at 3 months.

[6] First 20 days of lactation.

[7] First two months of lactation, for four days per week.

[8] (w) indicates that animals were tested at weaning or were not fully weaned at the time of testing. d=day; w=weeks; mo=months.

[9] Relative to (non-exposed) control groups, i.e. ↑ preference can also be reduced avoidance, and usually relative preference at the expense of a control feed.

the diet their mother had consumed during lactation, 64% and 87% higher compared to non-exposed controls, for two different feeds tested. Conversely, weanling rats that encountered a 'mismatch' between their mother's and their own diet showed a lower feed intake in the immediate postweaning period (Bronstein *et al.*, 1975; see Figure 1). Also beyond the immediate postweaning period, exposure to the maternal lactational diet may affect the acceptance of feed. For instance, the intake of rice straw by lactating goats markedly enhanced the acceptance of this

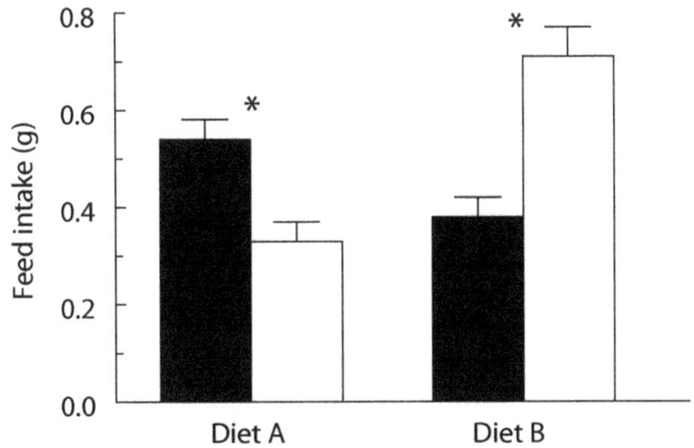

*Figure 1. Effects of the maternal diet during lactation (diet A=solid bars; diet B=open bars) on feed intake in rat pups offered a single diet, either A or B, during the first 30 min postweaning, n=8 litters per treatment. *P<0.05. Adapted from Bronstein* et al., *1975.*

feed type in their offspring at six months of age, three months after weaning (Van Tien, 2002). Young goats exposed through their lactating mother's diet started to ingest rice straw much sooner than control goats (on the fourth vs. 11th day of testing), and also consumed more rice straw until day 19 of testing. This indicates that preweaning exposure to feed flavours from the maternal diet can reduce food neophobia and promote the intake of an otherwise unfamiliar feed in the immediate postweaning period and beyond.

For suckling animals, exposure to mother's milk is likely the most salient route for information transfer concerning the lactating mother's diet. Volatiles can be transferred to the mammary glands via vascular routes from the digestive or respiratory system (Dougherty *et al.*, 1962; Shipe *et al.*, 1962). For instance, the flavour of human milk is altered by maternal consumption of vanilla (Mennella and Beauchamp, 1994), garlic (Mennella and Beauchamp, 1991), mint (Hausner *et al.*, 2008) and cheese (Mennella, 1995). In response to maternal ingestion of garlic, anise or vanilla, infants changed their suckling behaviour and milk intake, which implies that changes in milk flavour on the short-term are also detected by sucklings (Mennella and Beauchamp, 1991, 1993, 1996). Sucklings may even show short-term satiety for the flavours they experience through mother's milk: within hours after ingestion of carrot-flavoured milk, infants showed a reduced intake of carrot-flavoured cereal (Mennella and Beauchamp, 1999). Hence, although flavour exposure through mother's milk may on the long term facilitate the acceptance of a similarly flavoured feed at weaning, it may inhibit its intake on the very short term.

In most studies concerning the influence of the lactating female's diet on feed acceptance of her offspring, flavour transmission through milk was not experimentally separated from transmission through alternative routes. Even if the maternal diet is provided in the absence of the sucklings, thus preventing direct observational learning, flavours derived from the maternal diet may also be transmitted, for example, by the smell of the mother's body, breath, feed particles clinging to her skin, fur or teeth, or faecal droppings. For instance, after smelling the breath of their dam which had eaten an unfamiliar diet, weanling spiny mice doubled the number of nose-to-mouth and nose-to-head interactions with their dam, and ate more of this novel feed in a subsequent choice test than pups whose mother had not eaten it (McFadyen-Ketchum and Porter, 1989). It has been suggested, however, that in many species mother's milk is likely the most potent medium for transferring dietary information from mother to offspring during lactation. The mouth and swallowing movements made during suckling facilitate retronasal perception of the flavours present in milk, thereby providing both olfactory and gustatory stimulation (Mennella and Beauchamp, 1991). Indeed, in rodent studies controlling for exposure to the maternal diet in alternative ways, only pups that had suckled mother's milk preferentially ingested the feed eaten by their mother (Galef and Henderson, 1972; Wuensch, 1978), whereas exposure to the maternal diet through body odours or faeces was not effective (Galef and Sherry, 1973). Moreover, ingestion of mother's milk without the additional ways of transmission was sufficient to influence the dietary preference of rabbit and rat pups at weaning (Bilkó *et al.*, 1994; Capretta and Rawls, 1974; Galef and Sherry, 1973) and young goats in later life (Van Tien, 2002).

The effect of preweaning flavour exposure on feed preferences and acceptance in later life appears not to be restricted to mother's milk. Exposure to flavours in formula milk or milk replacer may also enhance the acceptance of similarly flavoured foods in later life (Mennella *et al.*, 2004; Mennella and Beauchamp, 2002; Nolte and Provenza, 1991), although it has been suggested that sucklings may form stronger preferences for flavours if they suckle from their mother (Nolte and Provenza, 1991, see next section).

Why do animals show enhanced acceptance of flavours from the maternal diet?

Evidence presented in the previous sections indicates that exposure to flavours derived from the maternal diet, either prenatally or postnatally through mother's milk, may enhance acceptance of a similarly flavoured diet at weaning, even if this diet is relatively unpalatable (e.g. Bilkó *et al.*, 1994; Mainardi *et al.*, 1989) or sour tasting (London *et al.*, 1979). This appears to hold for a wide range of flavours, with, to our knowledge, only one exception: it has been reported that maternal intake of quinine, a bitter tasting solution, during pregnancy and lactation slightly reduced its preference in young rats (London *et al.*, 1979). How these generally positive reactions to flavours from the maternal diet are established is largely unknown, but several different mechanisms may be involved. The simplest mechanism may be an effect of familiarity, the so-called ′mere exposure′ effect (Hill, 1978): due to repeated exposure to a flavour, this flavour may become increasingly acceptable and in this way neophobia is eliminated or attenuated (e.g. Hudson, 1993). If so, exposure to the maternal diet may simply overlap with a sensitive period during early development in which experiences with flavours produce enduring effects (Alberts and Gubernick, 1984). Indeed, evidence has been presented for a sensitive period for chemosensory learning (Blais *et al.*, 2006; Rochefort *et al.*, 2002; Rosselini-Austin and Williams, 1990). Exposure to a wide variety of odours or flavours during early life has been shown to enhance acceptance of both similarly flavoured and novel food items in later life (Capretta *et al.*, 1975; Gerrish and Mennella, 2001; Hennessy *et al.*, 1977).

However, it should be noted that animals are usually exposed to flavours from the maternal diet under highly reinforcing conditions, which may further strengthen their preference for and acceptance of these flavours. For instance, in the nursing situation sucklings are not only exposed to flavours, but simultaneously experience alleviation of hunger and thirst, achievement of warmth, and tactile social interaction, any of which might serve as a reinforcer. This might also explain to some extent why exposure per se is less effective than exposure in a social context, e.g. during suckling (e.g. Capretta and Rawls, 1974; London *et al.*, 1979; King, 1979; Patris *et al.*, 2008). In addition, it has been suggested that both amniotic fluid and mother's milk may act as unconditioned stimuli in the acquisition of flavour preferences, i.e. the foetus or suckling animal may associate the flavours transmitted to amniotic fluid and milk, respectively, with the reinforcing properties of these fluids (Arias and Chotro, 2007). Moreover, in rabbits, a mammary pheromone present in milk has been shown to act as a primary reinforcer by promoting rapid early learning, in only one exposure, of otherwise neutral odours or flavours (Coureaud *et al.*, 2006, 2008). This rapid acquisition of flavours may, as another form of flavour-flavour learning, contribute to feed selection and acceptance at weaning (Coureaud *et al.*, 2006, 2008; Shah, 2006).

It is largely unknown from what developmental stage young animals are capable of learning an association between flavours in food items and their postingestive nutritive consequences. However, recent studies indicate that at least in rats, mechanisms for flavour-nutrient conditioning appear to be functional before weaning (Myers *et al.*, 2005). It has been suggested that this conditioning may already play a role in the enhanced preference and acceptance of flavours experienced in mother's milk (Myers and Scaflani, 2006).

Impact of different routes for transfer of flavours derived from the maternal diet

Only few studies have compared the effectiveness of prenatal chemosensory learning through the maternal diet as opposed to postnatal learning through milk or other forms of social learning in enhancing feed acceptance at weaning. Combinations of flavour exposure through the maternal diet with other forms of social learning might be more effective than flavour conditioning alone. In goats, both 'observational learning' alone, i.e. feed offered to the mothers in the presence of their offspring, but no exposure through mother's milk, or exposure through mother's milk alone enhanced the acceptance of rice straw as compared with non-exposed controls (Van Tien, 2002). A combination of both learning strategies, however, was proven to be most effective: 6-month old goats trained in the preweaning period through both observational learning and mother's milk ingested more rice straw than groups exposed through mother's milk or observational learning alone (Van Tien, 2002). Rabbit pups showed a similar increase in preference for juniper berries at weaning when exposed to juniper flavour either prenatally via their mother's diet, or postnatally, or both, as compared with non-exposed control pups (Bilkó *et al.*, 1994). It has been suggested that the importance of any flavour transmission route is relatively high in rabbits, because opportunities for other forms of social learning (e.g. observation) before weaning are limited as the doe only returns for about 5 min per day to her offspring to nurse (Bilkó *et al.*, 1994). In other species, however, prolonged exposure to flavour cues from the mother's diet has been shown to strengthen its effect on feed preferences (Capretta and Rawls, 1974; Galef and Sherry, 1973). In dogs, exposure to anise flavour either prenatally or postnatally was not sufficient to enhance the ingestion of anise-flavoured treats in later life, but a clear preference was established when prenatal and postnatal exposure were combined (Hepper and Wells, 2005). It was suggested that this larger effect of perinatal exposure may not simply reflect a prolonged exposure to the flavour, but also a priming effect on the chemosensory system by prenatal experience, resulting in enhanced responsiveness to subsequent postnatal exposure (Hepper and Wells, 2005). In support of this, research in rabbits demonstrated that the introduction of a flavour *in utero* enhanced the sensitivity of the olfactory system to that same flavour in neonates (Semke *et al.*, 1995). Thus, the combined pre- and postnatal transmission of flavours from the maternal diet to amniotic fluid, colostrum and milk may provide a 'transnatal continuity' of chemosensory stimuli across the perinatal period (Coureaud *et al.*, 2002; Winberg and Porter, 1998). In this way, transitions from intra-uterine life to extra-uterine life and from suckling to independent feeding may be facilitated.

Role of social information transfer in feeding behaviour of pigs

The pig's natural behaviour of foraging in family groups (see Held and Mendl, 2001 for review) and its highly developed social cognitive skills (Held and Mendl, 2001; Held *et al.*, 2000, 2001, 2002) suggests ample opportunities for food-related social learning. Pigs can indeed acquire information from conspecifics concerning where (Held *et al.*, 2000; Nicol and Pope, 1994) and what (Nicol and Pope, 1994) to eat. It is also very likely that the development of their foraging behaviour and feed preferences are affected by what other group members ingest. For instance, there are indications that feed intake is stimulated when inexperienced weanling pigs are housed with experienced piglets of the same age (Morgan *et al.*, 2001). In support of this, the opportunity to mingle with other litters in a communal creep area increased solid feed intake and weight gain of piglets after weaning (Weary *et al.*, 1999, 2002) and, in sow-controlled get-away-systems, also increased the creep feed intake during the suckling period (Weary *et al.*, 2002), probably through both social facilitation and social learning. Social learning might be far more advantageous, however, if piglets could obtain information from an experienced conspecific, particularly their mother, rather than from piglets of the same age (Nicol and Pope, 1994), but studies on vertical information transfer in pigs are scarce.

Perinatal exposure to flavours from the maternal diet

Pigs are known for their well-developed sense of smell (Günterschulze, 1979; Morrow-Tesch and McGlone, 1990b) and probably also have high taste acuity (Hellekant and Danilova, 1999). Little is known about the ability of the porcine foetus to process and retain chemosensory information. Anatomical studies suggest, however, that chemosensory systems may be functional *in utero* in the pig (Holubcova *et al.*, 1997; Salazar *et al.*, 2004), like in other mammalian species. In support of this, piglets appear to have a well-developed olfactory system at birth (Horrell and Eaton, 1984; Morrow-Tesch and McGlone, 1990a). For instance, neonatal piglets were able to distinguish their mother's faecal and skin odours from those of other sows and to recognise shavings from their home pen (Horrell and Eaton, 1984; Horrell and Hodgson, 1992; Morrow-Tesch and McGlone, 1990b).

To our knowledge, only a small number of studies have investigated perinatal flavour learning in pigs, and its effect on postnatal feed acceptance. In a very small-scale study in Vietnam, it was shown that Large White piglets more readily ingested a novel feed, duckweed (fifth vs. eighth testing day), and also ingested more over a 28-day testing period, if this feed had been included in their mother's diet during pregnancy (Van Tien, 2002). This prenatal learning was ineffective in piglets from the Mong Cai breed, which, also unexposed, were found to have a higher acceptance of duckweed than Large White pigs.

Also in pigs, flavours from the maternal diet might be transferred to mother's milk and affect flavour preferences in later life (Madsen, 1977). Performance of weanling pigs was improved when they were fed a diet with a flavoured substance (called Firanor no. 3) that had also been present in their mother's diet during lactation (Campbell, 1976). During the first three weeks after weaning at 30 days of age, piglets that were re-exposed to the flavour from their lactating mother's diet ingested 12% more feed than the non-exposed piglets (818 vs. 729 g/d), and they gained 24% more weight (450 vs. 363 g/d), resulting in a weight difference of 1.8 kg over three weeks. Pigs that were exposed to the flavour through their lactating mother's diet or weaner diet only, did not differ in performance from the non-exposed controls, which demonstrates that it was the recognition of a flavour from the maternal diet resulting in improved performance rather than the flavour per se. The effect of vertical flavour transfer on feed intake was most pronounced in the first two postweaning weeks (Figure 2).

Similar results were obtained in a small-scale study by King (1979) who added a flavour called Firanor no. 24 to the sow's diet. Again, piglets exposed to the flavour through their mother's milk and subsequently offered feed with a similar flavour at weaning (day 27) ingested significantly more feed (10% more than non-exposed piglets or piglets exposed through the maternal diet or weaner diet only) and also tended to grow more (10 %) over 31 days after weaning. Flavour exposure through mother's milk did not increase the intake of similarly flavoured creep feed before weaning (King, 1979). Others described, however, that addition of both garlic and anise

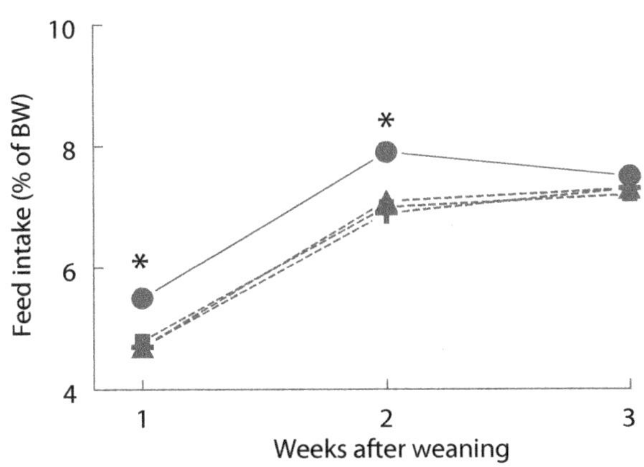

*Figure 2. Stimulating effect of flavouring the lactating mother's diet on intake of similarly flavoured feed in weanling piglets, as a percentage of their body weight. Piglets were either exposed to flavour both through their lactating mother's diet and weaner diet (solid line, circles), exposed through their lactating mother's diet only (triangles), weaner diet only (plus signs), or non-exposed (squares), n=6 per treatment. *Indicates difference between the doubly exposed group and all other groups, P<0.05. Adapted from Campbell (1976).*

to the maternal diet during gestation and lactation enhanced intake of similarly flavoured creep feed, but the effect became apparent only at 6 weeks of age (Langendijk *et al.*, 2007).

We recently conducted a 2×2 experiment to investigate the effects of prenatal and postnatal exposure to anise flavour via the maternal diet on performance and behaviour of piglets during the first two postweaning weeks (M. Oostindjer *et al.*, unpublished data). At weaning, piglets were both offered anise-flavoured feed and identical control feed without the flavour. Prenatally exposed piglets, irrespective of their postnatal treatment, tended to gain more weight during the postweaning period, and ingested more feed between 2 and 5 days after weaning, even though no clear preference for anise-flavoured feed was found. Prenatal flavour exposure also decreased the occurrence of diarrhoea and prolonged the latency to start fighting after weaning. In addition, the combination of pre- and postnatal experience reduced the occurrence of oral manipulation of pen mates, a maladaptive behaviour. Hence, prenatal flavour conditioning improved performance and welfare of piglets after weaning. Although we found no compelling evidence that prolonged exposure, i.e. also via mother's milk, would be more effective than prenatal exposure alone, it should be noted that numerically, but not statistically significant, piglets that were exposed to anise both prenatally and during lactation performed more favourably than all other treatment groups for all variables measured.

We can only speculate how prenatal flavour conditioning facilitated the adaptation of piglets to the postweaning situation, even without establishing a long-lasting preference for the flavoured postweaning feed. Firstly, prenatal experience may have reduced neophobia with respect to the anise-flavoured feed. It is possible that, subsequently, the prenatally exposed animals generalised their acceptance of anise-flavoured feed to the control feed, which was similar to the anise-flavoured feed offered, through flavour-flavour associations. Such generalisation of feed acceptance based on common flavour cues has also been reported in other species (Van Tien, 2006; Villalba and Provenza, 2000). For instance, in rabbits, juniper exposure *in utero* and/or through mother's milk increased acceptance of both juniper and thyme, which have properties in common, at weaning (Altbäcker *et al.*, 1995; Bilkó *et al.*, 1994). Hence, prenatal exposure to anise may have favourably affected feed intake and as a result growth and intestinal health through its reduction of neophobia (see Figure 3).

Alternatively, the mere presence of anise flavour (or odour) in the postweaning environment may have had a more general stress-reducing effect on the prenatally exposed piglets, which in turn may have affected a range of variables measured (see Figure 3). In support of this, we found that prenatally and pre- plus postnatally exposed piglets, when introduced individually to a novel environment at 16 days of age, displayed less escape attempts and vocalisations, respectively, if this environment was scented with anise odour. Also rats exposed *in utero* to apple juice not only showed a preference for this solution when tested at 60 days, but also had a lower corticosterone response compared to control animals during exposure (Smotherman, 1982a). During exposure to a novel solution, no such effect was found, suggesting that the effect of the *in utero* apple exposure was not a general reduction in the neophobic tendencies to all flavours. Likely, a 'match' between flavours or odours in the prenatal or suckling period and the postweaning period attenuates both food neophobia and more general stress responses in piglets. However, (olfactory) exposure per se in the preweaning period does not seem to be sufficient in reducing either food neophobia or

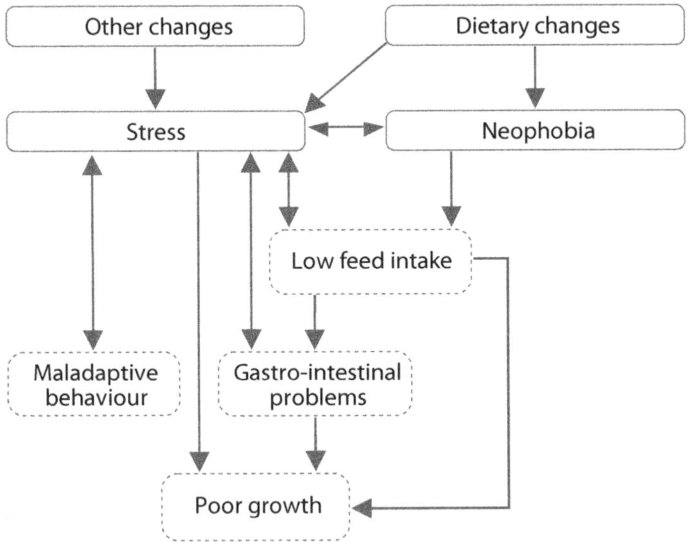

Figure 3. Schematic overview showing hypothetical links between neophobia, stress and performance at weaning. At weaning, piglets are subjected to dietary and other changes, which may lead to food neophobia and stress. These, in turn, may negatively affect feed intake, gastrointestinal health, growth rate and behaviour. Dotted lines indicate variables which were found to be affected by prenatal or pre- and postnatal flavour conditioning (M. Oostindjer et al., unpublished data).

stress of piglets at weaning. Adding a flavour to the preweaning creep feed diet was ineffective in enhancing flavoured feed intake or other performance parameters at weaning (King, 1979; Millet *et al.*, 2008). In addition, re-exposure to an artifical odourant (Ambi-Pur) that had been present in their farrowing pen failed to affect stress-related behaviour of piglets in a novel environment (Jones *et al.*, 2000). In contrast, exposure to a synthetic analogue of the putative porcine maternal pheromone has been demonstrated to reduce agonistic behaviours at weaning and improve weight gain (McGlone and Anderson, 2002). This suggests that re-exposure to chemosensory stimuli experienced in the prenatal environment and mother's milk, which holds likely both for the putative maternal pheromone (Guiraudie-Capraz *et al.*, 2005) and the volatile compounds of the maternal diet, is most effective in reducing postweaning neophobia and stress in piglets.

Concluding remarks and implications

In mammalian species, several ways of vertical information transmission have developed to ensure that weanlings can benefit from their mother's experience regarding dietary choices. The exposure to flavours from the maternal diet, both prenatally and postnatally, provides the young animal with information about the feed its mother has ingested, and has in a number of species been shown to increase the acceptance of similarly flavoured feed in later life. The natural behaviour and social structure of pigs suggest that information transfer from sow to offspring could contribute substantially to the development of independent feeding in piglets. Results of small-scale and preliminary studies indicate that offering a diet at weaning that is similar in flavour as their mother's diet during gestation and lactation may reduce food neophobia and stress in piglets. Thus, providing piglets with preweaning experiences with flavours from the sow's diet, starting even before birth, may facilitate adaptation to a similarly flavoured postweaning diet, and in this way enhance their performance, health and welfare.

In commercial pig husbandry, however, suckling piglets have little or no opportunity to observe the feeding behaviour of their mother, as sows are mostly confined in crates during lactation and fed in troughs that are usually inaccessible for piglets. Moreover, in contrast with (semi) natural conditions, piglets are mostly offered a diet that differs in composition and flavour from their mother's diet, and therefore the possibilities for learning through prenatal or postnatal exposure to flavours from the maternal diet are limited. The potential impact of information transmission from sow to piglets – including more direct forms of social learning - on feed acceptance and performance around weaning requires more attention in the light of the current weaning problems. If future research indicates that vertical social learning is effective in adapting piglets to solid feed, it may be worthwhile to develop management and feeding strategies that optimise the possibilities for vertical information transfer in pig husbandry.

References

Alberts, J.R. and Gubernick, D.J., 1984. Early learning as ontogenetic adaptation for ingestion by rats. Learning and Motivation 15: 334-359.

Appleby, M.C., Pajor, E.A. and Fraser, D., 1991. Effects of management options on creep feeding by piglets. Animal Production 53: 361-366.

Appleby, M.C., Pajor, E.A. and Fraser, D., 1992. Individual variation in feeding and growth of piglets: effects of increased access to creep food. Animal Production 55: 147-152.

Arias, C. and Chotro, M.G., 2007. Amniotic fluid can act as an appetitive unconditioned stimulus in preweanling rats. Developmental Psychobiology 49: 139-149.

Altbäcker, V., Hudson, R. and Bilkó, Á., 1995. Rabbit-mothers' diet influences pups' later food choice. Ethology 99: 107-116.

Berkeveld, M., Langendijk, P., Van Beers-Schreurs, H.G.M., Koets, A.P., Taverne, M.A.M., and Verheijden, J.H.M., 2007. Postweaning growth check in pigs is markedly reduced by intermittent suckling and extended lactation. Journal of Animal Science 85: 258-266.

Bilkó, Á., Altbäcker, V. and Hudson, R., 1994. Transmission of food preference in the rabbit: the means of information transfer. Physiology & Behavior 56: 07-912.

Birch, L.L., 1998. Development of food acceptance patterns in the first years of life. Proceedings of the Nutrition Society 57: 617-624.

Birch, L.L., 1999. Development of food preferences. Annual Review of Nutrition 19: 41-62.

Blais, I., Terkel, J. and Goldblatt, A., 2006. Long-term impact of early olfactory experience on later olfactory conditioning. Developmental Psychobiology 48: 501-507.

Bøe, K. and Jensen, P., 1995. Individual differences in suckling and solid food intake by piglets. Applied Animal Behaviour Science 42: 183-192.

Bronstein, P.M, Levine, M.J. and Marcus, M., 1975. A rat's first bite: the non-genetic, cross-generational transfer of information. Journal of Comparative and Physiological Psychology 80: 295-298.

Brooks, P.H. and Tsourgiannis, C.A., 2003. Factors affecting the voluntary feed intake of the weaned pig. In: Pluske, J.R., Le Dividich, J. and Verstegen, M.W.A. (eds). Weaning the pig. Concepts and consequences. Wageningen Academic Publishers, Wageningen, the Netherlands, pp. 81-116.

Bruininx, E.M.A.M., Van der Peet-Schwering, C.M.C. and Schrama J.W., 2001. Individual feed intake of group-housed weaned pigs and health status. In: Varley, M.A. and Wiseman, J. (eds). The weaner pig: nutrition and management. CAB International, Wallingford, United Kingdom, pp. 113-122.

Bruininx, E.M.A.M., Binnendijk, G.P., Van der Peet-Schwering, C.M.C., Schrama, J.W., Den Hartog, L.A., Evers, H. and Beynen, A.C., 2002. Effects of creep feed consumption on individual feed intake characteristics and performance of group-housed pigs. Journal of Animal Science 80: 1413-1418.

Bruininx, E.M.A.M., Schellingerhout, A.B., Binnendijk, G.P., Van der Peet-Schwering, C.M.C., Schrama, J.W., Den Hartog, L.A., Everts, H. and Beynen, A.C., 2004. Individually assessed creep food consumption by suckled piglets: influence on postweaning food intake characteristics and indicators of gut structure and hind-gut fermentation. Animal Science 78: 67-75.

Bult, J.H.F., De Wijk, R.A. and Hummel, T., 2007. Investigations on multimodal sensory integration: Texture, taste, and ortho- and retronasal olfactory stimuli in concert. Neuroscience Letters 411: 6-10.

Campbell, R.G., 1976. A note on the use of a feed flavour to stimulate the feed intake of weaner pigs. Animal Production 23: 417-419.

Capretta, P.J. and Rawls L.H., 1974. Establishment of a flavour preference in rats: importance of nursing and weaning experience. Journal of Comparative and Physiological Psychology 86: 670-673.

Capretta, P.J., Petersik, J.T. and Stewart, D.J., 1975. Acceptance of novel flavours is increased after early experience of diverse tastes. Nature 254: 689-691.

Chapple, R.S. and Lynch, J.J., 1986. Behavioural factors modifying the acceptance of supplementary foods by sheep. Research and Development in Agriculture 3: 113-120.

Chotro, M.G. and Molina, J.C., 1990. Acute ethanol contamination of the amniotic fluid during gestational day 21: Postnatal alcohol responsiveness in rats. Developmental Psychobiology 23: 535-547.

Coureaud, G., Schaal, B., Hudson, R., Orgeur, P. and Coudert, P., 2002. Transnatal olfactory continuity in the rabbit: behavioral evidence and short-term consequence of its disruption. Developmental Psychobiology 40: 372-390.

Coureaud, G., Moncomble, A.S., Motigny, D., Dewas, M., Perrier, G. and Schaal, B., 2006. A pheromone that rapidly promotes learning in the newborn. Current Biology 16: 1956-1961.

Coureaud, G., Fortun-Lamonthe, L., Rödel, H.G., Monclús, R. and Schaal, B., 2008. Development of social and feeding behaviour in young rabbits. Proceedings of the 9th World Rabbit Congress, June 10-13, Verona, Italy, pp. 1131-1146.

Doughertly, R.W., Shipe, W.F., Gudnason, G.V., Ledford, R.A., Peterson, R.D. and Scarpellino, R., 1962. Physiological mechanisms involved in transmitting flavors and odors to milk. I. Contribution of eructated gases to milk flavour. Journal of Dairy Science 45: 477-480.

Edge, H.L., Varley, M.A. and Rowlinson, P., 2004. The effect of trough colour on the voluntary food intake of postweaned pigs – a short communication. Livestock Production Science 88: 193-197.

Galef, B.G., 1971. Social factors in the weaning of domestic rat pups. Journal of Comparative Physiology and Psychology 75: 358-362.

Galef, B.G., 1996a. Food selection: problems in understanding how we choose foods to eat. Neuroscience and Biobehavioural Reviews 20: 67-73.

Galef, B.G., 1996b. Social influences on food preferences and feeding behaviors of vertebrates. In: Capaldi, E.D. (ed.). Why we eat what we eat. The psychology of eating. American Psychological Association, Washington D.C., USA, pp. 207-231.

Galef, B.G. and Clark, M.M., 1971. Parent-offspring interactions determine time and place of first ingestion of solid food by wild rat pups. Psychonomic Science Section on Animal and Physiological Psychology 25: 15-16.

Galef, B.G. and Clark, M.M., 1972. Mother's milk and adult presence: two factors determining initial dietary selection by weanling rats. Journal of Comparative Physiology and Psychology 78: 220-225.

Galef, B.G. and Giraldeau, L.A., 2001. Social influences on foraging in vertebrates: causal mechanisms and adaptive functions. Animal Behaviour 61: 3-15.

Galef, B.G. and Henderson, P.W., 1972. Mother's milk: a determinant of the feeding preferences of weaning rat pups. Journal of Comparative Physiology and Psychology 78: 213-219.

Galef, B.G. and Sherry, D.F., 1973. Mother's milk: a medium for transmission of cues reflecting the flavor of mother's diet. Journal of Comparative Physiology and Psychology 83: 374-378.

Geary, T.M. and Brooks, P.H., 1998. The effect of weaning weight and age on the postweaning growth performance of pigs fed fermented liquid diets. The Pig Journal 42: 10-23.

Gerrish, C.J. and Mennella, J.A., 2001. Flavor variety enhances food acceptance in formula-fed infants. American Journal of Clinical Nutrition 73: 1080-1085.

Guiraudie-Capraz, G., Slomianny, M.C., Pageat, P., Malosse, C., Cain, A.H., Orgeur, P. and Nagnan-Le Meillour, P., 2005. Biochemical and chemical supports for a transnatal olfactory continuity through sow maternal fluids. Chemical Senses 30: 241-251.

Gundlach, H., 1968. Brutfürsorge, Brutpflege, Verhaltensontogenese und Tagesperiodik beim Europäischen Wildschwein (*Sus scrofa L.*). Zeitschrift für Tierpsychologie 25: 955-995.

Günterschulze, J., 1979. Studies about the regio olfactoria of the wild boar (*Sus scrofa*) and of the domestic pig (*Sus scrofa domestica*). Zoologischer Anzeiger 202: 256-279.

Hall, W.G. and Williams, C.L., 1983. Suckling isn't feeding, or is it? A search for developmental continuities. Advances in the Study of Behavior 13: 219-255.

Hauser, G.J., Chitayat, D., Berns, L., Brauer, D. and Muhlbauer, B. 1985. Peculiar odours in newborns and maternal prenatal ingestion of spicy food. European Journal of Pediatrics 144: 403.

Hausner, H., Bredie, W.L.P., Mølgaard, C., Petersen, M.A. and Møller, P., 2008. Differential transfer of dietary flavour compounds into human breast milk. Physiology & Behavior 95: 118-124.

Held, S. and Mendl, M., 2001. Behaviour of the young weaner pig. In: Varley, M.A. and Wiseman, J. (eds). The weaner pig: nutrition and management. CAB International, UK, pp. 273-297.

Held, S., Mendl, M., Devereux, C. and Byrne, R.W., 2000. Social tactics of pigs in a competitive foraging task: The 'informed forager' paradigm. Animal Behaviour 59: 569-576.

Held, S,, Mendl, M., Devereux, C. and Byrne, R.W., 2001. Studies in social cognition: from primates to pigs. Animal Welfare 10: S209-217.

Held, S., Mendl, M., Laughlin, K. and Byrne, R.W., 2002. Cognition studies with pigs: Livestock cognition and its implication for production. Journal of Animal Science 80(E Suppl. 1): E10-E17.

Hellekant, G. and Danilova, V., 1999. Taste in domestic pig, *Sus scrofa*. Journal of Animal Physiology and Animal Nutrition 82: 8-24.

Hennessy, M.B., Smotherman, W.P. and Levine, S., 1977. Early olfactory enrichment enhances later consumption of novel substances. Physiology & Behavior 19: 481-483.

Hepper, P.G., 1987. The amniotic fluid: an important priming role in kin cognition. Animal Behaviour 35: 1343-1346.

Hepper, P.G., 1988. Adaptive foetal learning: prenatal exposure to garlic affects postnatal preferences. Animal Behaviour 36: 935-936.

Hepper, P.G., 1990. Foetal olfaction. In: Macdonald, D.W., Müller-Schwarze, D. and Natynczuk, S.E. (eds). Chemical signals in vertebrates 5. Oxford University Press, Oxford, United Kingdom, pp. 282-288.

Hepper, P.G., 1995. Human fetal 'olfactory' learning. International Journal of Prenatal and Perinatal Psychology and Medicine 7: 147-151.

Hepper, P.G. and Waldman, B., 1992. Embryonic olfactory learning in frogs. The Quarterly Journal of Experimental Psychology 44B: 179-197.

Hepper, P.G. and Wells, D.L., 2006. Perinatal olfactory learning in the domestic dog. Chemical Senses 31: 207-212.

Hill, W.F., 1978. Effects of mere exposure on preference in nonhuman mammals. Psychological Bulletin 85: 1177-1198.

Holubcova, S., Kocianova, I. and Tichy, F., 1997. Differentiation of the olfactory epithelium in swine in ontogenesis. Acta Veterinaria Brno 66: 135-139.

Horrell, R.I. and Eaton, M., 1984. Recognition of maternal environment in piglets: Effects of age and some discrete complex stimuli. Quarterly Journal of Experimental Psychology 36: 119-130.

Horrell, I. and Hodgson, J., 1992. The bases of sow-piglet identification. 2. Cues used by piglets to identify their dam and home pen. Applied Animal Behaviour Science 33: 329-343.

Hudson, R., 1993. Olfactory imprinting. Current Opinion in Neurobiology 3: 548-552.

Hudson, R. and Distel, H., 1999. The flavor of life: perinatal development of odor and taste preferences. Schweizerische Medizinische Wochenschrift 129: 176-181.

Jensen, P., 1995. The weaning process of free-ranging domestic pigs: within- and between-litter variations. Ethology 100: 14-25.

Jensen, P. and Recén, B., 1989. When to wean - Observations from free-ranging domestic pigs. Applied Animal Behaviour Science 23: 49-60.

Jones, J.B., Wathes, C.M., White, R.P. and Jones, R.B., 2002. Do pigs find a familiar odourant attractive in novel surroundings? Applied Animal Behaviour Science 70: 115-126.

King, D.E., 2003. Gastrointestinal disorders in neonatal pigs. In: Xu, R.J. and Cranwell, P.D. (eds). Gastrointestinal physiology and nutrition. Nottingham University Press, Nottingham, United Kingdom, pp. 275-308.

King, R.H., 1979. The effect of adding a feed flavour to the diets of young pigs before and after weaning. Australian Journal of Experimental Agriculture and Animal Husbandry 19: 695-697.

King, R.H. and Pluske, J.R., 2003. Nutritional management of the pig in preparation for weaning. In: Pluske, J.R., Le Dividich, J. and Verstegen, M.W.A. (eds). Weaning the pig. Concepts and consequences. Wageningen Academic Publishers, Wageningen, the Netherlands, pp. 37-52.

Kuller, W.I., Soede, N.M., Van Beers-Schreurs, H.M.G., Langendijk, P., Taverne, M.A.M., Verheijden, J.H.M. and Kemp, B., 2004. Intermittent suckling: effects on piglet and sow performance before and after weaning. Journal of Animal Science 82: 405-413.

Labayen, J.P., Batungbacal, M.R. and Lopez, P.L., 1994. The performance of suckling piglets given creep feed at seven and fourteen days of age. Philippine Journal of Veterinary Medicine 31: 38-41.

Laland, K.N, Richerson, P.J. and Boyd, R., 1993. Animal social learning: towards a new theoretical approach. In: Bateson, P.P.G. and Klopfer, P.H. (eds). Perspectives in Ethology: Behaviour and Evolution, Vol 10. Plenum Press, New York, USA, pp. 249-277.

Langendijk, P., Bolhuis, J.E. and Laurenssen, B.F.A., 2007. Effects of pre- and postnatal exposure to garlic and aniseed flavour on pre- and postweaning feed intake in pigs. Livestock Science 108: 284-287.

Le Dividich, J. and Sève, B., 2001. Energy requirements of the young pig. In: Varley, M.A. and Wiseman, J. (eds). The weaner pig: nutrition and management. CAB International, United Kingdom, pp. 17-44.

Leibbrandt, V.D., Ewan, R.C., Speer, V.C. and Zimmerman, D.R., 1975. Effect of weaning and age at weaning on baby pig performance. Journal of Animal Science 40: 1077-1080.

London, R.M., Snowdon, C.T. and Smithana, J.M., 1979. Early experience with sour and bitter solutions increase subsequent ingestion. Physiology & Behavior 22: 1149-1155.

Madec, F, Bridoux, N., Bounaix, S. and Jestin, A., 1998. Measurement of digestive disorders in the piglet at weaning and related risk factors. Preventive Veterinary Medicine 35: 53-72.

Madsen, F.C., 1977. Development of feed preference in young swine. Feedstuffs 49: 25.

Mainardi, M., Poli, M. and Valsecchi, P., 1989. Ontogeny of dietary selection in weaning mice: effects of early exposure and mother's milk. Biology of Behaviour 14: 185-194.

Makkink, C.A., 1993. Of piglets, proteins and pancreatic proteases. PhD thesis Wageningen University, the Netherlands.

Marrable, A.W., 1971. The embryonic pig. A chronological account. A. Wheaton & Co., Exeter, United Kingdom, 130 pp.

McCracken, B.A., Gaskins, H.R., Ruwe-Kaiser, P.J., Klasing, K.C. and Jewell, D.E., 1995. Diet-dependent and diet-independent metabolic responses underlie growth stasis of pigs at weaning. Journal of Nutrition 125: 2838-2845.

McCracken, B.A., Spurlock, M.E., Roos, M.A., Zuckermann, F.A. and Gaskins, H.R., 1999. Weaning anorexia may contribute to local inflammation in the piglet small intestine. Journal of Nutrition 129: 613-619.

McFadyen-Ketchum, S.A. and Porter, R.H., 1989. Transmission of food preferences in spiny mice *Acomys cahirinus* via nose-mouth interaction between mothers and weanlings. Behavioral Ecology and Sociobiology, 24: 59-62.

McGlone, J.J. and Anderson, D.L., 2002. Synthetic maternal pheromone stimulates feeding behavior and weight gain in weaned. Journal of Animal Science 80: 3179-3183.

Mehiel, R. and Bolles, R.C., 1984. Learned flavor preferences based on caloric outcome. Animal Learning and Behavior 12: 421-427.

Mehiel, R. and Bolles, R.C., 1988. Learned flavor preferences based on calories are independent of intial hedonic value. Animal Learning and Behavior 16: 383-387.

Mennella, J.A., 1995. Mother's milk: a medium for early flavor experiences. Journal of Human Lactation 11: 39-45.

Mennella, J.A. and Beauchamp, G.K., 1991. Maternal diet alters the sensory qualities of human milk and the nursling's behavior. Pediatrics 88: 737-744.

Mennella, J.A. and Beauchamp, G.K., 1993. The effects of repeated exposure to garlic-flavored milk on the nursling's behavior. Pediatric Research 34: 805-808.

Mennella, J.A. and Beauchamp, G.K., 1994. The infant's responses to flavoured milk. Infant Behavior and Development 17: 819.

Mennella, J.A. and Beauchamp, G.K., 1996. The human infants' responses to vanilla flavors in mother's milk and formula. Infant Behavior and Development 19: 13-19.

Mennella, J.A. and Beauchamp, G.K., 1998. Early flavor experiences: research update. Nutrition Reviews 56: 205-211.

Mennella, J.A. and Beauchamp, G.K., 1999. Experience with a flavor in mother's milk modifies the infant's acceptance of flavored cereal. Developmental Psychobiology 35: 197-203.

Mennella, J.A. and Beauchamp, G.K., 2002. Flavor experiences during formula feeding are related to preferences during childhood. Early Human Development 68: 71-82.

Mennella, J.A., Griffin, C.E. and Beauchamp, G.K., 2004. Flavor programming during infancy. Pediatrics 113: 840-845.

Mennella, J.A., Jagnow, C.P., and Beauchamp, G.K, 2001. Prenatal and postnatal flavor learning by human infants. Pediatrics 107: 88-94.

Mennella, J.A., Johnson, A. and Beauchamp, G.K, 1995. Garlic ingestion by pregnant women alters the odor of amniotic fluid. Chemical Senses 20: 207-209.

Meynhardt, H., 1980. Schwarzwildreport. Vier Jahre unter Wildschweinen. Verlag J Neumann-Neudamm, Melsungen.

Miller, H.M., Toplis, P. and Slade, R.D., 1999. Weaning weight and daily live weight gain in the week after weaning predict piglet performance. In: Cranwell, P.D. (ed.). Manipulating pig production, VII. Australasian Pig Science Association, Werribee, Australia, p. 130.

Millet, S., Aluwé, M, De Brabander, D.L. and Van Oeckel M.J., 2008. Effect of seven hours intermittent suckling and flavour recognition on piglet performance. Archives of Animal Nutrition 62: 1-9.

Morgan, C.A., Lawrence, A.B., Chirnside, J. and Deans, L.A., 2001. Can information about solid food be transmitted from one piglet to another? Animal Science 73: 471-478.

Morrow-Tesch, J. and McGlone, J.J., 1990a. Sensory systems and nipple attachment behavior in neonatal pigs. Physiology & Behavior 47: 1-4.

Morrow-Tesch, J. and McGlone, J.J., 1990b. Sources of maternal odors and the development of odor preferences in baby pigs. Journal of Animal Science 68: 3563-3571.

Myers, K.P., Ferris, J. and Sclafani, A., 2005. Flavor preferences conditioned by postingestive effects of nutrients in preweanling rats. Physiology & Behavior 84: 407-419.

Myers, K.P. and Sclafani, A., 2006. Development of learned flavour preferences. Developmental Psychobiology 48: 380-388.

Nabuurs, M.J.A., Van Zijderveld, F.G. and De Leeuw, P.W., 1993. Clinical and microbiological field studies in the Netherlands of diarrhoea in pigs at weaning. Research in Veterinary Science 55: 70-77.

Newberry, R.C. and Wood-Gush, D.G.M., 1985. The suckling behaviour of domestic pigs in a semi-natural environment. Behaviour 95: 11-25.

Newberry, R.C. and Wood-Gush, D.G.M., 1986. Social relationships of piglets, *Sus scrofa*, in a semi-natural environment. Animal Behaviour 34: 1311-1318.

Newberry, R.C. and Wood-Gush, D.G.M., 1988. Development of some behaviour patterns in piglets under semi-natural conditions. Animal Production 46: 103-110.

Nicol, C.J., 1995. The social transmission of information and behaviour. Applied Animal Behaviour Science 44: 79-98.

Nicol, C.J. and Pope, S.J., 1994. Social learning in sibling pigs. Applied Animal Behaviour Science 40: 31-43.

Nielsen, B.L., Thodberg, K., Dybkjaer, L. and Vestergaard E.M., 2006. Feeding behaviour in pigs. In: Bels, V. (ed.). Feeding in domestic vertebrates. From structure to behaviour. CABI Publishing, Wallingford, United Kingdom, pp. 156-178.

Nolte, D.L. and Provenza, F.D., 1991. Food preferences in lambs after exposure to flavours in milk. Applied Animal Behaviour Science 32: 381–389.

Nolte, D.L. and Provenza, F.D., 1992. Food preferences in lambs after exposure to flavors in solid foods. Applied Animal Behaviour Science 32: 337-347.

Nolte, D.L., Provenza, F.D., Callan, R. and Panter, K.E., 1992. Garlic in the ovine fetal environment. Physiology & Behavior 52: 1091-1093.

Pajor, E.A., Fraser, D. and Kramer, D.L., 1991. Consumption of solid food by suckling pigs: individual variation and relation to weight gain. Applied Animal Behaviour Science 32: 139-155.

Pajor, E.A., Weary, D.M., Caceres, C., Fraser, D. and Kramer, D.L., 2002. Alternative housing for sows and litters 1. Effects of piglet diet quality and sow-controlled housing on performance and behaviour. Applied Animal Behaviour Science 76: 267-277.

Patris, B., Perrier, G., Schaal, B. and Coureaud, G., 2008. Early development of filial preferences in the rabbit: implications of nursing- and pheromone-induced odour learning? Animal Behaviour 76: 305-314.

Petersen, V., 1994. The development of feeding and investigatory behaviour in free-ranging domestic pigs during their first 18 weeks of life. Applied Animal Behaviour Science 42: 87-98.

Pluske, J.R., Williams, I.H. and Aherne, F.X., 1995. Nutrition of the neonatal pig. In: Varley, M.A. (ed.). The neonatal pig. Development and survival. CAB International, Wallingford, United Kingdom, pp. 187-235.

Provenza, F.D. and Balph, D.F., 1987. Diet learning by domestic ruminants: theory, evidence and practical implications. Applied Animal Behaviour Science 18: 211-232.

Provenza, F.D. and Villalba, J.J., 2006. Foraging in domestic herbivores: linking the internal and external milieux. In: Bels, V. (ed.). Feeding in domestic vertebrates. From structure to behaviour. CABI Publishing, Wallingford, United Kingdom, pp. 210-240.

Rohde Parfet, K.A. and Gonyou, H.W., 1991. Attraction of newborn piglets to auditory, visual, olfactory and tactile stimuli. Journal of Animal Science 69: 125-133.

Rochefort, C., Gheusi, G., Vincent, J.D. and Lledo, P.M., 2002. Enriched odor exposure increases the number of newborn neurons in the adult olfactory bulb and improves odor memory. The Journal of Neuroscience 22: 2679-2689.

Ross, M.G. and Nijland, M.J.M., 1998. The development of ingestive behaviour. American Journal of Physiology 274: R879- R893.

Rosselini-Austin, L. and Williams. J., 1990. Enriched neonatal odor exposure leads to increased numbers of olfactory bulb mitral and granule cells. Developmental Brain Research 51: 135-137.

Rozin, P., 1976. The selection of foods by rats, humans and other animals. In: Rosenblatt, J., Hinde, R.A., Beerand, C. and Shaw, C.E. (eds). Advances in the study of behavior, vol. 6. Academic Press, New York, pp. 21-76.

Rozin, P., and Vollmecke, T.A., 1986. Food likes and dislikes. Annual Review of Nutrition 6: 433-456.

Saint-Dizier, H., Lévy, F. and Ferreira, G., 2007. Influence of the mother in the development of flavoured food-preference in lambs. Developmental Psychobiology 49: 98-106.

Salazar, I., Sánchez-Quinteiro, P., Lombardero, M., Aleman, N. and Fernández de Trocóniz, P., 2004. The prenatal maturity of the accessory bulb in pigs. Chemical Senses 29: 3-11.

Sangild, P.T., Schmidt, M., Elnif, J., Björnvad, C.R., Weström, B. and Buddington, R.K., 2002. Prenatal development of gastrointestingal function in the pig and effects of fetal esophageal obstruction. Pediatric Research 52: 416-424

Sangild, P.T., Hilsted, L., Fowden, A.L., and Silver, M., 2003. Effect of fluid ingestion on gastrin secretion in the fetal pig. Reproduction in Domestic Animals 28: 213-214.

Schaal, B., Marlier, L. and Soussignan R., 1995a. Responsiveness to the odour of amniotic fluid in the human neonate. Biology of the Neonate 67: 397-406.

Schaal, B. and Orgeur, P., 1992. Olfaction *in utero*: can the rodent model be generalized? The Quarterly Journal of Experimental Psychology 44B: 245-278.

Schaal, B., Orgeur, P. and Arnould, C., 1995b. Olfactory preferences in newborn lambs: possible influence of prenatal experience. Behaviour 132: 351-365.

Schaal, B., Marlier, L. and Soussignan R., 2000. Human foetuses learn odours from their pregnant mother's diet. Chemical Senses 25: 729-737.

Schley, L. and Roper, T.J., 2003. Diet of wild boar *Sus scrofa* in Western Europe, with particular reference to consumption of agricultural crops. Mammal Review 33: 43-56.

Semke, E., Distel, H. and Hudson, R., 1995. Specific enhancement of olfactory receptor sensivity associated with foetal learning of food odors in the rabbit. Naturwissenschaften 82: 148-149.

Shah, N.M., 2006. Nursing behaviour: remembrance of things past. Current Biology 16: R842-R844.

Shipe, W.F., Ledford, R.A., Peterson, R.D., Scanlan, R.A., Geerken, H.F., Dougherty, R.W. and Morgan, M.E., 1962. Physiological mechanisms involved in transmitting flavors and odors to milk. II. Transmission of some flavor components of silage. Journal of Dairy Science 45: 417-419.

Sibly, R.M., 1999. Evolutionary biology of skill and information transfer. In: Box, H.O. and Gibson, K.R. (eds). Mammalian social learning. Comparative and ecological perspectives. Zoological Society of London Symposium 73. London, United Kindom, pp. 57-71.

Simitzis, P.E., Deligeorgis, S.G., Bizelis, J.A. and Fegeros K., 2008. Feeding preferences in lambs influenced by prenatal flavour exposure. Physiology & Behavior 93: 529-536.

Smotherman, W.P., 1982a. *In utero* chemosensory experience alters taste preferences and corticosterone responsiveness. Behavioral and Neural Biology 36: 61-68.

Smotherman, W.P., 1982b. Odor aversion learning by the rat fetus. Physiology & Behavior 29: 769-771.

Sneddon, H., Hadden, R. and Hepper, P.G., 1998. Chemosensory learning in the chicken embryo. Physiology & Behavior 64: 133-139.

Soussignan, R., Schaal, B., Marlier, L. and Jiang, T., 1997. Facial and autonomic responses to biological and artificial olfactory stimuli in human neonates: re-examining early hedonic discrimination of odours. Physiology & Behavior 67: 745-758.

Spreeuwenberg, M.A.M., Verdonk, J.M.A.J., Gaskins, H.R. and Verstegen, M.A.W., 2001. Small intestine epithelial barrier function is compromised in pigs with low feed intake at weaning. Journal of Nutrition 131: 1520-1527.

Thorhallsdottir, A.G., Provenza, F.D. and Balph, D.F., 1990. Ability of lambs to learn about novel foods while observing or participating with social models. Applied Animal Behaviour Science 25: 25-33.

Valsecchi, P., Mainardi, M., Sgoifo, A. and Taticchi A., 1989. Maternal influences on food preferences in weanling mice. Behavioural Processes 19: 155-166.

Valsecchi, P., Moles, A. and Mainardi, M., 1993. Does mother's diet affect food selection of weanling wild mice? Animal Behaviour 46: 827-828.

Van Tien, D., 2002. Modifying ingestive behaviour to raise animal production in Central Vietnam. PhD thesis Utrecht University, the Netherlands.

Van Tien, D., 2006. Pigs familiar with duckweed readily accept another unfamiliar green feed, *Leucaena leucocephala*. In: Workshop-seminar, Workshop on Forages for Pigs and Rabbits, MEKARN-CelAgrid, 21-24 August 2006.

Varendi, H., Porter, R.H. and Winberg, J., 1996. Attractiveness of amniotic fluid odor: evidence of prenatal learning? Acta Pædiatrica 85: 1223-1227.

Vente-Spreeuwenberg, M.A.M. and Beynen, A.C., 2003. Diet-mediated modulation of small intestinal integrity in weaned piglets. In: Pluske, J.R., Le Dividich, J. and Verstegen, M.W.A. (eds). Weaning the pig. Concepts and consequences. Wageningen Academic Publishers, Wageningen, the Netherlands, pp. 145-198.

Villalba, J.J. and Provenza, F.D., 2000. Roles of novelty, generalization, and postingestive feedback in the recognition of foods by lambs. Journal of Animal Science 78: 3060-3069.

Visalberghi, E., Sabbatini, G., Stammati, M. and Adessi, E., 2003. Preferences towards novel foods in *Cebus apella*: the role of nutrients and social influences. Physiology & Behavior 80: 341-349.

Weary, D.M., Pajor, E.A., Bonenfant, M., Ross, S.K., Fraser, D. and Kramer, D.L., 1999. Alternative housing for sows and litters. Part 2. Effects of a communal piglet area on pre- and post-weaning behaviour and performance. Applied Animal Behaviour Science 65: 123-135.

Weary, D.M., Pajor, E.A., Bonenfant, M., Fraser, D. and Kramer, D.L., 2002. Alternative housing for sows and litters. Part 4. Effects of sow-controlled housing combined with a communal piglet area on pre- and post-weaning behaviour and performance. Applied Animal Behaviour Science 76: 279-290.

Weary, D.M., Jasper, J. and Hötzel M.J., 2008. Understanding weaning distress. Applied Animal Behaviour Science 110: 24-41.

Wells, D.L. and Hepper, P.G., 2006. Prenatal olfactory learning in the domestic dog. Animal Behaviour 72: 681-686.

Williams, I.H., 2003. Growth of the weaned pig. In: Pluske, J.R., Le Dividich, J. and Verstegen, M.W.A. (eds). Weaning the pig. Concepts and consequences. Wageningen Academic Publishers, Wageningen, the Netherlands, pp. 17-36.

Winberg, J. and Porter, R.H., 1998. Olfaction and human neonatal behaviour: clinical implications. Acta Pædiatrica 87: 6-10.

Wuensch, K.L., 1978. Exposure to onion taste in mother's milk leads to enhanced preference for onion diet among weanling rats. Journal of General Psychology 99: 163-167.

Chapter 3. Integration of pre- and post-absorptive factors in feed intake regulation and prediction with particular respect to the pig

J.M. Forbes
Institute of Integrative and Comparative Biology, University of Leeds, Leeds LS2 9JT, United Kingdom; j.m.forbes@leeds.ac.uk

Abstract

This chapter provides a broad overview of how food intake might be controlled and develops a theory to act as a framework for future research. Animals, including pigs, learn to associate the sensory properties of foods with the internal signals generated after consumption and use this learned information as a key part of determining intake of foods and selection between foods. There are well-described effects on voluntary food intake of infusions of metabolites into the gastrointestinal tract and hormones into the bloodstream which suggest feedback pathways, carried in the peripheral nervous system to influence the CNS. There is also considerable circumstantial evidence in the pig that distension of the stomach and intestines play a part in limiting intake under some circumstances. Because of the convergence of signals as they pass to the CNS, it is likely that it is the total of the strengths of these signals that is involved in intake control but that this is not sufficient to provide a full explanation. It is proposed that differences in the current rate of supply of a food resource (e.g. energy, protein, bulk) and the optimal supply of that resource for the animal in question generates discomfort which the animal seeks to minimise (other factors that might cause discomfort, such as space allowance, can also be incorporated). The error, as a proportion of the ideal, is squared and the square root of the total for all the resources under consideration is used as a signal of total discomfort. Total discomfort is calculated for a range of food intakes (and choices) and thus the intake (and choice) at which total discomfort is minimum, hence the Minimal Total Discomfort (MTD) framework. Animal and food(s) parameters can be changed and the new MTD is found, with its accompanying intake (and choice). The outcomes are realistic but considerable further work is required before there is a chance of validating the model for pigs and of predicting intake sufficiently well for use in practical conditions.

Keywords: diet selection, minimal total discomfort, learned aversion and preference, GIT receptors

Introduction

The title I have been invited to address, 'pre- and post-absorptive factors …', gives me freedom to discuss anything that affects and might be involved in the control of food intake, in any species of animal! Where possible I will use the pig for my examples, although there will be places in which relevant information is not available, or is compromised, and I will resort to results obtained with other species.

The pig has traditionally been thought of as a glutton, but genetic selection against fatness has resulted in growing animals that voluntarily eat amounts of food that broadly match their requirements; mature sows do, however, 'overeat' when given the chance, presumably to obey their genetic 'program' to store energy for use later in lactation. In this respect pigs are no different from other species of mammal and I will assume that results from one species can be applied to another, unless there are obvious anatomical or physiological reasons to the contrary. I will also include considerations of diet selection in my discussion as in nature most animals have a choice between two or more potential foods and the common agricultural situation, in which only one food is on offer, could be viewed as being a special case; anyway there has been considerable interest in choice feeding of farm animal, including pigs, in the recent past.

Feeding has traditionally been thought of as part of a simple negative feedback loop: the animal starts to eat; stomach fill and the concentration of metabolites increase and eventually the strength of the negative signals generated causes feeding to cease; while the stomach empties the concentration of metabolites initially continues to rise and to maintain satiety; eventually the strength of the various negative feedback signals falls as metabolites are utilised or stored; the feedback signals fall below a threshold and feeding starts again; and so on. However, it has become increasingly clear that learning plays a very important part in the relationship between the animal and its food, examples of which are described and discussed below.

Firstly we will provide evidence for many factors involved in the control of food intake and choice, from sensory properties of the food(s) to mechano- and chemo-receptors in the gut and deep body tissues. Secondly we will consider how the animal integrates the signals from these many factors (for surely intake is not controlled by any single factor acting on its own) and propose a semi-quantitative scheme to aid in understanding and predicting food intake and choice.

Many of the areas covered in this chapter are dealt with in more detail, and more fully referenced, in the book and review of Forbes (2007a,b).

Preingestional

Sensory factors

The chemistry and physics of food materials provides each one with a characteristic set of sensory properties (appearance, flavour, taste, texture) which enable the animal to identify it. The sum of these characteristics has often been called 'palatability' but, as we shall see, an animal's reaction to a food depends not only on its sensory properties but also on the animal's history and its experiences of foods, including the one in question. An extreme illustration of the relationship between the pig and its food is provided when Bitrex (the bitterest substance for the human palate – similar to that of the pig) is added to a normal food; the immediate effect is complete avoidance, followed by exploratory nibbling. As the pigs become hungrier, and as they learn that the 'new' food is not harmful, they accept it readily and after about three days are eating normally (Blair and Fitzsimons, 1970). Thus, an initially 'unpalatable' food has now become 'palatable'. There is good evidence that the milder differences in sensory properties between conventional foods are sufficient to enable conditioning to take place. There are relevant examples from pigs given a choice of two foods in which it has been observed to take several days to learn the nutritional differences between two foods (in protein content, for example) and relate these to the sensory differences between the foods. There is, however, little direct evidence of conditioning in the pig.

Animals of any type can use any of the senses to identify food and can even use memory of the placement of foods on previous occasions to recover hidden food. Baldwin and Cooper (1979) noted that anosmic pigs found apples buried in sand very quickly by burrowing around, as long as they had previously been able to learn that apples were likely to be buried in the test pen. When a novel food is offered the animal experiences the new appearance and flavour before eating the food and a new set of consequences after eating the food. From this we imagine they can learn to associate the two. However, in order to demonstrate unequivocally that the association is learned it is necessary to separate the two experimentally and for this it is necessary to turn to other species.

Birds generally use vision as the first sense in the characterisation of foods, whereas mammals normally use flavour. A comprehensive study with chicks of learned associations between the sensory properties of food and the internal consequences of eating that food was made by Covasa and Forbes (1994) whose aim it was to determine whether the reduction in food intake following an i.p. injection of cholecystokinin (CCK) was due to a pleasantly satiating effect or an unpleasantly nauseating effect of CCK. There were four phases to the experiment (Figure 1):

A. Birds were injected on some days with saline and then given a coloured normal food (*conditioned stimulus*, red or green) for 2 h and on other days with 100 µg CCK i.p. (*unconditioned stimulus*, sufficient to depress intake by about 20% over the next 2 h) and given the other colour of food for 2 h. When on other days they were given brief access to foods of both colours they progressively chose the saline-paired food more often.
B. For the next 11 days no injections were made but the preference test was repeated from time to time. The aversion was lost within a few days.

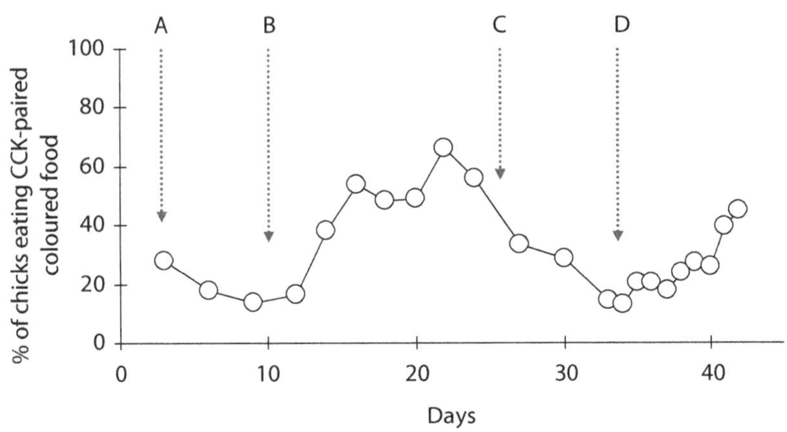

Figure 1. Percentage of chicks favouring the colour of food associated with CCK injection (Covasa and Forbes, 1994). See text for explanation.

C. The protocol of phase A was repeated but with treatments paired with the opposite colours to those used in phase A. Aversion rapidly developed against the new CCK-paired food colour.
D. No injections were made and again there was a return to no preference within a few days of stopping the injections.

Thus, chicks were made averse to the colour of food they had been given after an injection of CCK, irrespective of the colour. The conditioned stimuli, red and green food, can be switched and within a few days new conditioned responses are induced. Colour preference can be easily manipulated and the same is true for the association between food flavour and the consequences of eating that is more readily evident in mammals. Animals typically take several days to learn new associations – they do not immediately know how much of a new food to eat when it is first presented, nor do they know what proportion of two or more foods to eat when first presented with a choice.

Diet selection

When an animal given a choice of foods and chooses to eat predominantly one of them it is conventional to say that this preferred food is more 'palatable'. As we have just seen from the example with coloured foods for chicks, however, it is clear that the 'palatability' of a food depends not only on its sensory characteristics (red or green) but on the history that the chick has with the foods – the 'palatability' of the food can be changed at will by the experimenter by appropriate conditioning. The example of Bitrex in pig food teaches us the same lesson. There are numerous examples of diet selection in pigs, many of them from the research of Kyriazakis and his group in Edinburgh. From these we deduce that growing pigs offered a choice of two foods with different protein contents choose proportions of these foods that provide the optimum protein: energy ratio (P: E), as assessed by conventional dose: response experiments with single foods. Thus, as long as one food has a greater P: E and the other a lesser P: E than is optimal for the animals in question, they select a diet containing close to the optimal 210g CP/kg (Kyriazakis *et al.*, 1990). An example of diet selection pathways is shown in Figure 2. Newly-weaned piglets were offered a choice between foods of similar energy concentrations and crude protein concentrations of 291 and 155 g/kg for the high- (HP) and low-protein (LP) foods, respectively (Dalby *et al.*, 1995). The cumulative difference between the intakes of two foods (HP – LP) is plotted against the cumulative total food intake (TFI, HP + LP) of the foods, i.e. a horizontal line represents equal intakes of the two foods, a line that increases shows the animal eating more HP, while one declining shows a greater intake of LP (Kyriazakis *et al.*, 1990).

There are large differences between animals in the trajectories: one animal starts by eating almost all HP (pathway close to +45°) but after 6 days changes from mostly HP to approximately equal intakes of HP and LP; after about 14 days into the observation period it changes to eat predominantly LP. Three start by eating approximately equal amounts of the HP and LP (pathway close to horizontal) followed by a decline in HP: LP ratio. The remaining four eat mainly LP (pathway close to -45°) throughout. The downward trajectories eventually achieved by all but one animal after day 10 are approximately parallel and show a similar, low, proportion of HP eaten by these animals giving a diet containing 189 g CP/kg, declining to 182 g/kg by the end of the

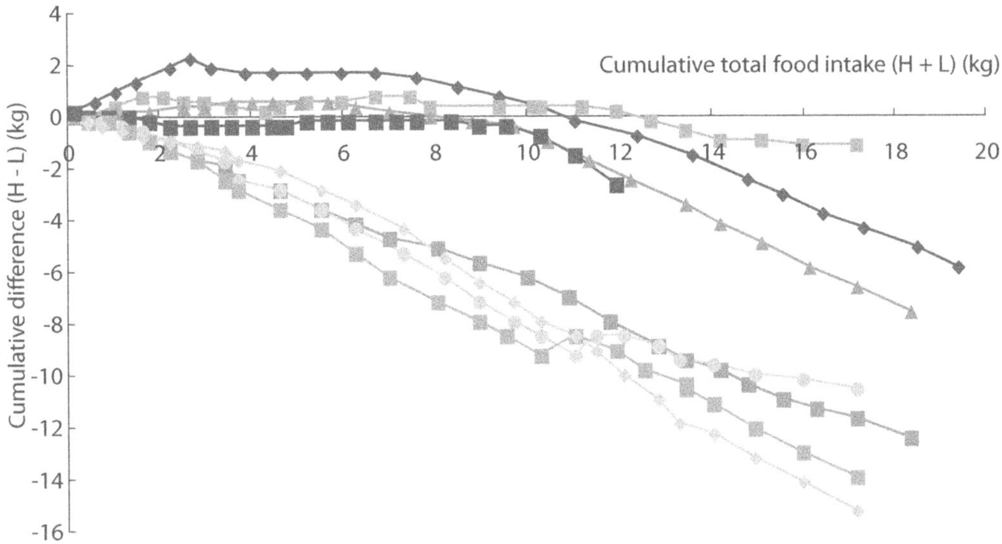

Figure 2. Diet selection pathways for 8 piglets offered a choice between foods containing 291(HP) and 155 (LP) g crude protein per kg (Dalby et al., *1995).*

observation period. The conclusion is that each pig learned that its needs were best met by taking a low proportion of HP, but different individuals learned this at different rates.

The conclusion to be drawn from these examples, and many other observations, is that animals learn to associate the sensory properties of foods with the metabolic consequences of eating those foods, and that this learning is an important factor in the control of food intake and selection.

Postingestional

Once food has been swallowed then the animal has to accept the metabolic consequences unless it vomits, which is highly aversive in most mammals. Between swallowing boluses and absorbing the products of their digestion there are numerous possibilities for food to influence intake before absorption. The first of these is the bulk of the food and the distension of the stomach and intestines; the second is the effect of the metabolites released by digestion on receptors in the wall of the digestive tract.

The small intestine is richly innervated by vagal and splanchnic branches whose sensory functions encompass mechano-, chemo- and osmo-reception. The following discussion of gastro-intestinal sensitivity in the context of food intake of pigs is necessarily brief and considerably more detail can be found in Gregory *et al.* (2002) 'Role of the intestine in regulation of food intake in growing pigs'.

Physical factors

There is circumstantial evidence of a physical limit to intake in the pig; when food is diluted with ingestible material then daily intake increases in compensation, but there comes a point beyond which daily intake of digestible nutrients is no longer sustained. This point comes sooner with smaller (younger) pigs than with larger pigs (Owen and Ridgman, 1967; Whittemore *et al.*, 2003). Even though direct evidence for effects of gastric distension on food intake is lacking (there are no reports of effects of balloon inflation in the stomach of pigs, for example) it is clear that there must be some limit to the extent to which the stomach wall can be stretched before the mechanoreceptors signal pain to the CNS. Whether this is part of the normal meal-by-meal regulation of intake is less certain. The question then arises as to whether there is a fixed limit to stomach capacity, or whether the animal will allow greater distension if it has a high demand for nutrients; this will be addressed later.

As well as the capacity of the stomach, the bulkiness of the food must be considered before any attempt can be made to predict whether physical factors are involved in the control of food intake in any given situation. The water holding capacity (WHC, g water held by 1 g of food, measured by centrifugation) of foods has been proposed as a predictor for their bulkiness in the stomach and the results of an experiment using various levels of inclusion of food ingredients of different fibre type and content, and WHC (Kyriazakis and Emmans, 1995) indicated a one unit increase in WHC to be associated with a decrease in food intake of about 6 g food DM per kg pig live weight, above a WHC of 4 g/g. The research of Whittemore *et al.* (2002) with bulky foods supports the concept of physical limitation of feeding by growing pigs: foods with moderate (MWHC, 3.5 kg/kg), high (HWHC, 4.6 kg/kg) and very high (VHWHC, 10.0 kg/kg) WHC were given to pigs whose individual feeding behaviour was automatically recorded. When offered HWHC or VHWHC pigs were unable to maintain food intake, compared to MWHC, and when time of access to the feeder was reduced from 24 h to 6 h/day there was no adaptive change in short-term feeding behaviour. Changing from bulky food to control food or vice versa caused a rapid change in feeding behaviour so that it became appropriate to the new food.

It has been suggested that the young pig cannot easily overcome such a limitation as intake of a bulky food was limited to about 50 g per day per kg live weight, even when energy requirements of the animals were increased by keeping in an environment of less than the animals' lower critical temperature (Whittemore *et al.*, 2001). It is, however, unlikely that stomach fill is ever the only factor affecting intake and we will suggest later in this chapter how signals from different types of receptor might be integrated.

The rate at which the stomach empties is obviously relevant to the duration of stomach distension, but also to the fill of the duodenum and small intestine, which possess stretch receptors. As far as the large intestine is concerned, no specific research has yielded firm evidence of involvement in intake control but distension would be expected to reduce intake.

Mechanoreceptors, responsive to distension and muscle contractions, are found in the stomach and intestines and are likely to be responsible for the sensation of fullness that plays a part in limiting food intake. Experimental treatments have often used unphysiological degrees of distension in the study of these receptors and it has been argued that the intake depression induced by such treatments is the result of discomfort, or even pain, rather than 'real satiety'. It has been suggested, however, that the satiety, subconscious discomfort or overt pain induced by increasing degrees of distension are a continuum of effect and that the mechanoreceptors play a part in the control of intake even with the mild degrees of distension associated with normal feeding (Forbes, 1996).

Gregory (2002) concludes that 'It seems likely that gastrointestinal distension is not normally a very important limiting factor in determining food intake in the growing pig. The pig is able to adjust its intake to compensate for quite wide variations in dietary energy concentrations, however there are physical limits, and eventually gastrointestinal fill limits intake'. The inhibitory feedback signals are probably located throughout the GI tract. In very young pigs, however 'Gastric distension is a major satiety stimulus'. Evidence, including that presented above, suggests that physical limitation might be of particular importance for foods with high WHC and for young animals.

Glucose

The major product of digestion of the high-carbohydrate foods normally given to pigs is glucose and it is therefore not surprising that many studies have focussed on glucose as a major satiating signal, first in the intestines and later in the body proper. Infusion of glucose solution into the duodenum of young pigs markedly reduced the amount eaten when given just after food was offered, but not when given 10 minutes before food was offered. This has been interpreted to mean that glucose is a very short-lived satiety signal (Gregory, 2002). However, it must be realised that the pigs had been fasted for almost 24 h before food was offered, in which case: (1) the glucose infused would be absorbed and metabolised very quickly; (2) the animals would be extremely hungry and thus of blunted susceptibility to satiating factors. The practice of long periods of fasting before experimental protocols were applied is a feature of much early work and was due to the fear of pigs grossly overeating if fed *ad libitum*. Now, however, following genetic selection against overeating and over-fatness, there should be no such fears.

Infusion into the duodenum of the glucose antimetabolite 2-deoxy-D-glucose (2-DG), which competes with glucose for uptake and phosphorylation without being itself metabolised, has no such effect, however, while prior administration of 2-DG blocks the response to glucose (Stephens, 1985). This suggests that glucose given into the duodenum only affects intake after absorption. There is, however, direct evidence for a direct duodenal action and this is likely to be largely due to its osmotic effect (see below).

The link between duodenum and the CNS is neural rather than humoral as prior bilateral thoracic vagotomy prevented the effect of glucose solution infused into the duodenum. The total effect of infusing glucose solutions into the duodenum, ileum or jejunum is to depress intake by about the same amount as the energy of the glucose infused, as long as the pigs are allowed free access to drinking water and glucose is infused at physiological rates (Gregory *et al.*, 1987). As detailed below, it is most likely that it is a combination of many factors acting at many sites that forms the contribution of the digestive tract to the control of intake (Houpt *et al.*, 1983).

There is, then, uncertainty as to how much of the effect of glucose in the duodenum and small intestine is due to osmolality, to specific sensitivity of receptors in the gut wall to glucose, or to post-absorptive effects of glucose, especially in the liver (see below). The presence of specific vagal duodenal glucoreceptors, which are unresponsive to distension, has been established in the cat while in the pig the situation is still uncertain, despite considerable research efforts. Problems of restricted feeding systems and withdrawal of access to water during infusions have rendered clear interpretation difficult.

Protein and amino acids

Infusion of protein into the stomach or protein hydrolysate into stomach, duodenum, jejunum or ileum all decrease intake approximately in proportion to the amount of energy infused, although the extent of the depression varies across different experiments from no effect whatsoever to a two-fold reduction compared to the energy content of the infusion (Gregory, 2002). Once again, some of this variation is due to the restricted nature of the feeding regime in some experiments.

It has been reported that some intestinal receptors that are not sensitive to glucose, osmolality or distension do respond to infusion of amino acids in laboratory animals; some of these respond only to certain specific amino acids while others are not selective. It is likely that these receptors are involved in intake control as vagal devervation prevents the suppression of intake caused by infusion of amino acids into the duodenum. In the only relevant research with the pig it was shown that local anaesthesia of the duodenum partly prevented the intake-depressing effects of protein hydrolysate infused into the duodenum (Anika *et al.*, 1981).

Fats

Intake is depressed in pigs by infusion of fat emulsions into stomach, duodenum or jejunum and, in the latter site at least, the response is to fatty acids rather than fat (Gregory, 2002). There is some evidence that only monoglycerides have the effect and other evidence suggests that the response depends on the degree of unsaturation of the fatty acids infused.

Specific receptors to short chain fatty acids and glycerol or to long chain fatty acids, but not both, have been demonstrated in the duodenum and jejunum of laboratory animals and these are not sensitive to mechanical, osmotic or other chemical stimuli. There are, however, some differences in sensitivity between species (cats and rats). As far as the pig is concerned, gastric motility and food intake are reduced by intestinal infusion of fat, an effect that in some cases has been blocked by local anaesthetic.

Osmolality

Effects of osmolality of rumen contents on food intake by sheep have been hotly debated, as a result of some experiments having been carried out with animals from which drinking water was withheld (Forbes, 2007b). Similar questions have arisen with pigs because, although infusion of hyper-osmotic solutions into the duodenum of pigs reduces food intake, osmolality of duodenal fluid does not exceed 300 mOsm/kg when free access to water is allowed. Higher duodenal osmolalities have been observed even when water was available, but only in the large meals following enforced fasts of several hours. With the small meals taken in the *ad libitum* situation the maximum osmolality occurring after a meal was only 308 mOsm/l, much lower than the levels achieved by most experimental infusions.

The significance of the effects of osmolality on intestinal receptors is open to debate as those apparently sensitive to osmolality are also sensitive to other stimuli. It may be that sensitivity to large changes in osmolality of chyme is a safety mechanism to prevent consumption of too much food. However, even a small effect of osmolality could contribute to satiety, alongside stimulation of gastrointestinal receptors by other stimuli (see below).

Interactions between stomach and intestinal effects on intake

There are complex interactions between the stomach and intestines whereby the presence of intake-limiting nutrients in the duodenum causes slowing of gastric emptying so as to conserve a constant rate of flow of digesta from stomach to duodenum. In some cases the control is such that it is the flow of energy from stomach to duodenum that is maintained at a constant rate during a meal of solid food in pigs. Whether gastric fill is best expressed by the volume, the increase in volume, the energy content, the tension or the pressure is a subject of debate (Gregory, 2002), as is the extent to which the stomach, and indeed other parts of the GIT, accommodate over time to changes in some or all of these parameters.

Taken together, then, the findings concerning the role of the stomach and intestines demonstrate that (1) there is considerable uncertainty about their importance in the normal control of intake; (2) there are numerous types of receptor throughout the length of the GIT; (3) we are not, therefore, in a position to use information on stimulation of these receptors in models of intake control.

On the other hand, signals from the GIT become integrated with each other, and with signals from other organs, including the liver, to generate a composite signal which could represent the overall consequences of eating and be used, along with learning, by the satiety/hunger complex of the CNS. These principles will be developed later in this chapter.

Postabsorptive

Intestinal hormones

In addition to direct effect on neural receptors, many intestinal stimulants result in the release of substances, commonly called hormones (carried in the bloodstream to affect distant organs and tissues), but including paracrine agents (affecting cells and tissues in the same organ from which they are secreted). As these are covered in detail by the chapters of Black *et al.* (2009) and Carroll and Allee (2009), and by Gregory (2002), their actions will only be summarised here. Of the many intestinal hormones and paracrines implicated in the control of food intake, the best known is undoubtedly cholecystokinin (CCK). Plasma CCK rises after a meal and with duodenal infusions of corn oil or starch or, to a lesser extent, of protein hydrolysate. Given intravenously in pigs it reduces food intake without necessarily affecting the rate of stomach emptying. Its site of action is most likely the upper small intestine but it may also have more direct affects on the CNS, carried there in the bloodstream. It has been suggested that exogenous CCK causes general malaise in pigs but it is very difficult to gauge the extent to which even low doses mimic the natural rates of release and concentrations in pigs post-feeding. The fact that immunisation against CCK stimulates food intake and weight gain in pigs provides strong evidence of a physiological role, as does research involving very low doses of CCK or its pharmacological antagonists. The degree to which CCK affects food intake directly via central mechanisms is as yet unclear.

Glucagon-like peptide-1 (GLP-1) is released by intestinal cells in response to carbohydrates and lipids and, as well as effects on insulin secretion and gastric motility, reduces food intake by central mechanisms.

Of the many other gastrointestinal hormones, apolipoprotein A-IV (Apo A-IV) is secreted by intestinal mucosal cells during absorption and transport of lipids and probably acts centrally to reduce food intake. Enterostatin reduces food intake when given through several routes and may synergise with the effect of CCK.

It can be seen, therefore, that there are numerous effects and interactions of gut hormones and paracrines in relation to food intake.

Liver

Once absorbed, metabolites are directed to the liver, except for triglycerides, which bypass the liver by transport in the lymphatic system. The liver takes up, metabolises, stores and releases several substances, depending on the animal's metabolic state. As glucose is the metabolite taken up in greatest quantities by non-ruminant animals the effects of infusing it into the liver have been studied. In several species glucose given into the hepatic portal vein (HPV), or its equivalent the coccygeo-mesenteric vein in the chicken, depress intake in a dose-dependant manner to a much greater extent than the same solution given into the jugular vein. The signal to the CNS is carried via both vagal and splanchnic afferents and is proportional to the rate at which substrates are being oxidised by the liver (Scharrer *et al.*, 1996). In ruminants, where the rate of glucose absorption is normally very low and the liver supplements the supply to the rest of the body by gluconeogenesis, infusion of glucose into the HPV has no effect while infusion of propionate (the major precursor of glucose) significantly depresses food intake (Anil and Forbes, 1987). This can be blocked permanently by denervation of the liver, or temporarily by local anaesthetisation of the splanchnic nerves (Figure 3).

Given that the role of the liver in the control of intake is widely accepted it is puzzling that intraportal injections of glucose or amino acids do not affect food intake in pigs (Houpt *et al.*, 1979; Stephens, 1980). There seems to be no *a priori* reason why the pig should be different to other mammals as its liver metabolism is similar, but the conditions of the experiments (following an overnight fast) would not necessarily allow an effect to be shown – severe hunger might far outweigh the effects of infused glucose, perhaps forcing reliance on stomach fill to terminate

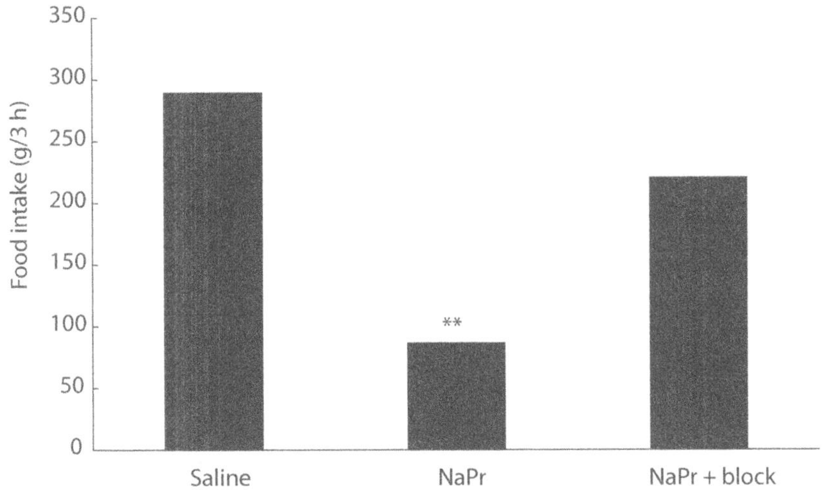

Figure 3. Effect of infusion of sodium propionate solution, with or without splanchnic nerve blockade by local anaesthetic, on voluntary intake of a complete pelleted food by castrated male sheep (Anil and Forbes, 1988).

feeding. It seems reasonable to assume that the liver plays a similar role in pigs as in other animals, in sensing rates of oxidation and signalling these to the CNS.

Metabolic hormones and adiposity

The ways in which insulin and leptin are involved in the CNS mechanisms of intake control, by providing signals concerning the adiposity of the animal, are covered in the chapters by Black *et al.* (2009) and Carroll and Allee (2009) to which the reader is directed for detailed information.

Protein and muscle

Millward (1995) highlights the important role of protein deposition in muscle in the growing animal in the control of food intake, proposing that there is strong evidence that it plays a more important role than adipose tissue in young animals. The importance of 'aminostatic' mechanisms in the control of food choice is clearly demonstrated by many choice-feeding studies in which growing pigs, and several other species, choose to eat proportions of high- and low-protein foods sufficient to support close to maximal growth of lean tissue. Not only do animals avoid protein deficiency by making appropriate choices, they equally avoid making for themselves diets with excessively high P: E ratios.

Offered a single food with a lower P: E ratio than optimal, animals tend to increase their food intake in an apparent effort to satisfy their need for more protein, but this involves excessive intakes of energy and so intake is limited by other factors, such as increased adiposity. Intake of foods with amino acid imbalances often depress food intake while alleviation of deficiency of an essential amino acid, either by addition to the diet or by infusion into the animal, can increase food intake to normal levels.

Central nervous system (CNS)

Although some parts of the CNS are sensitive to energy supply it is likely that this only comes into play in extremes of over-or under-supply of sources of energy, e.g. glucose. Most of the actions of the CNS are taken in response to neural and humoral messages from the periphery, as discussed above. Black *et al.* (2009) include considerable detail about the transmitters involved and how they relate to metabolism in the rest of the body.

Integration

So far we have presented a host of possible mechanisms in the control of food intake. Now we turn to how these might be integrated to provide overall regulation.

Convergence of neural and endocrine pathways; additivity

Most receptors in the gut wall respond to more than one type of stimulus (polymodal) and many of the same type of receptor are found in multiple sites along the tract (polytopic) (Forbes, 1996). Further, the satiating effects of the hormones glucagon, CCK and bombesin are additive at high doses and synergistic at low doses while leptin and CCK interact synergistically to reduce food intake in mice. Loading the stomach and administration of CCK, both at doses ineffective in depressing intake, significantly affect intake when given together. Infusion of acetate into the rumen of sheep with either propionate into the HPV or distension of the rumen have additive effects on food intake (Adams and Forbes, 1981). The effects of rumen distension, infusion of acetate and infusion of propionate, at doses too low to depress intake in cows significantly, generate a significant depression in the intake of silage when given in combination – the effect is approximately additive (Mbanya *et al.*, 1993).

Beyond local integration of the effects of different stimuli, there is also convergence of afferent fibres on the same dorsal root of the spinal cord and their messages to the brain are totally intermixed. There is also some evidence that integration occurs in the sympathetic ganglia. Further convergence takes place in the hind-, mid- and fore-brain. The signal eventually received by the intake-controlling circuits of the CNS is, therefore, rather non-specific and represents the total of various types of stimuli affecting the various classes of receptor in the abdomen.

Integration of pre- and post-absorptive factors

From the foregoing it has been assumed that the animal is a machine, responding automatically to (the total of) changes in satiating signals emanating from a host of sources. When some integral of these reaches a threshold then eating stops; when it falls below a threshold then eating starts again. On this basis it is possible to predict meal size and frequency from knowledge (or estimates) of stomach fill and rates of supply of nutrients from the digestive tract relative to rates of utilisation by tissues (e.g. Forbes, 1980). Models based on this premise give predictions of the same intake and meal patterns day after day, in contrast to observations of intakes and choices by real animals, where there is considerable daily variation (e.g. Figures 11 and 12). For a demonstration of the great flexibility in feeding behaviour consider what happened when birds had to peck at a button to obtain access to food and the number of pecks required was gradually increased, to a maximum of 160, and then slowly reduced again (Savory, 1989). Figure 4 shows that they took fewer but larger meals and ate more quickly, the greater the cost of getting to food; total daily food intake, however, remained unchanged apart from a small decrease at the highest fixed ratio of 160. If meal size in the free-feeding state was limited to about 3 g by the physiological factors discussed above then how is it that the chickens could manage meals of up to 15 g? Clearly feeding behaviour is flexible and any theory of how intake is controlled must encompass such flexibility.

The concept of a fixed limiting capacity of the stomach and fixed thresholds for blood glucose and/or other metabolites for feeding to start and stop seems, on the above evidence, unlikely. Given the continuum of response of receptors to increasing levels of stimulation, the greater the stimulation, the greater the contribution to reduce feeding, but without the fixed cut-offs envisaged by many theories of intake control.

It may be more realistic to envisage the effects of stimulation of GIT and liver being used as early-warning signals in order to prevent potentially dangerous situations arising. Thus, liver sensitivity to glucose (or rather the supply of oxidisable substrates) gives early warning of overload of the rest

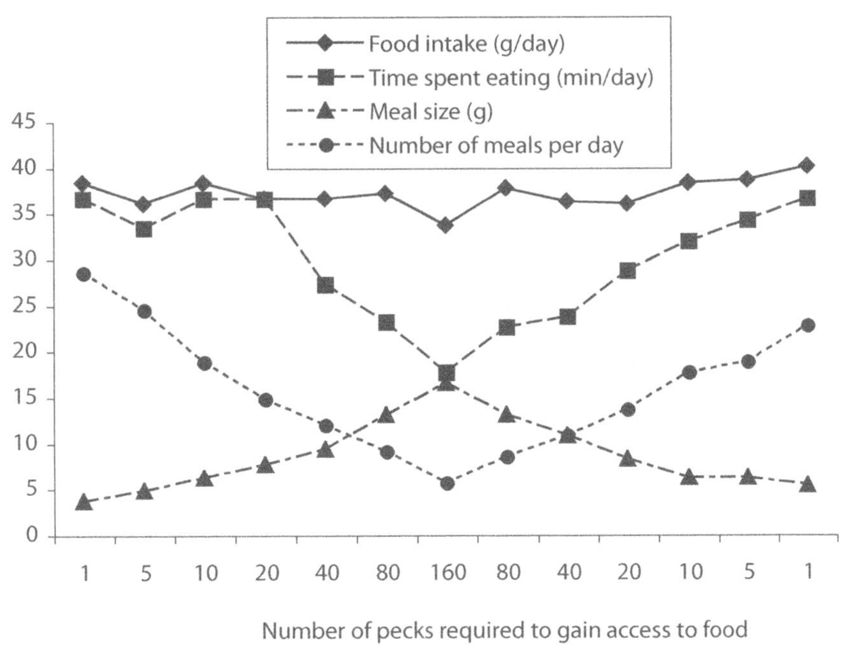

Figure 4. Food intake, meal size, time spent eating and number of meals by chickens made to deliver increasing, and then decreasing, numbers of pecks in order to obtain food (Savory, 1989).

of the body; lumen mechano- and chemo-stimulation provide warning of potential liver overload; taste and texture of food in the mouth give advanced information about events subsequent to swallowing; the senses of vision and smell provide even more opportunities for animals to learn how much food to eat in order to avoid later over- or under-stimulation of receptors.

Blundell and Halford (1994) envisaged a 'satiety cascade' in which the information available to the CNS about a particular food item becomes progressively more accurate as the food is seen or smelled in the environment; tasted and eaten; accumulates in the stomach; yields nutrients in the intestines; provides the body with nutrients via the liver; contributes to the body stores of energy and protein. Figure 5 is a representation of this concept.

The idea that stimulation of one part of the system is used to predict what will happen in other parts is consistent with the fact that animals become conditioned to expect certain discomforts if they eat foods with certain sensory properties, as outlined above. On first encountering a new food the animal has no experience to guide it and must rely only on post-ingestive consequences. Often a novel food will be eaten cautiously and daily intake will increase only slowly as the animal learns to balance the discomforts of eating too much of the food with the discomforts of eating too little. Once fully accustomed to a food the intake varies day-by-day about the optimum mean (see below), allowing the animal to experience the effects of eating too much and too little and to achieve new balances when, for example, the energy and protein requirements increase as a result of the onset of milk production.

Bearing in mind all these things, how can we make progress in understanding the controls of food intake and diet selection?

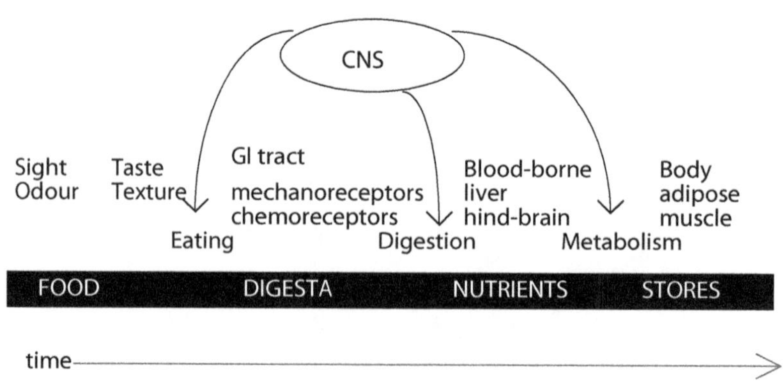

Figure 5. Satiety 'cascade'. Dotted arrows indicate flows of information to CNS as food goes from being identified in the environment to its nutrients being stored in the body (see text).

From physiological pathways to nutrient supply and demand

We have seen the potential for complexity in the control of food intake by negative feedback signals and cannot envisage including them individually in models of intake control. While the effects of distension, metabolites and hormones on the frequency of impulses transmitted to the CNS can be measured in anaesthetised animals it is not possible to use such information for predicting intake in the foreseeable future; if we cannot predict then can we claim to understand?

In order to proceed, therefore, we need to adopt some simplification. We have to admit that we cannot use the hour-by-hour changes in gut fill, nutrient flow or metabolism, which might be linked with short-term feeding behaviour, to predict or understand the longer-term control of intake – feeding behaviour is very flexible in the short term while at the same time providing relatively constant intakes of food over periods of a few days or more. If we are to provide a framework for understanding medium-term control of intake then instead of considering short-term changes in physiology we need to use currencies that are of practical relevance and that can be expressed numerically, such as: stomach capacity and food bulk (litres); energy requirements and DE concentration of food (MJ); protein requirements and concentration of available protein (g) in food. Of course, we should also be taking individual amino acids, minerals and vitamins into consideration, but let's learn to walk before we can run! In addition, it would be advantageous if environmental and social factors could be incorporated, as well as diet selection.

Minimal total discomfort (MTD)

It is clear from the discussion above that intake is controlled by a multitude of factors and it is a challenge to develop methods of integrating these factors in models of the control of intake. The most common assumption has been that whichever factor limits intake the most, then this is the factor in control, other factors being ignored. This concept has been applied to ruminant animals, the factors often being grouped as 'physical' and 'metabolic' (Forbes, 2007a) leading to the coining of the term 'Two phase hypothesis' (TPH, Pittroff and Kothmann, 1999). More complex models have been proposed (Poppi *et al.*, 1994), yet still relying on the 'first limiting factor' approach. The same principle has been used for pigs, and apparently validated by (Whittemore *et al.*, 2001, see above). However, this is not physiologically satisfying: it is difficult to imagine that at one moment the animal relies on one feedback signal (e.g. gastric distension) to control intake and at the next, after some small reduction in gut fill, suddenly changes to metabolic control. The problem is summed up by D'Eath *et al.*(2009): 'The main difficulty is that it is largely unclear how all the different processes are weighted and integrated …'.

An alternative to the TPH has been proposed and developed for ruminants and chickens (Forbes, 2007a,b) and will now be explored for pigs. It involves comparing the supply of resources from the food on offer with the optimum amounts of these resources for the animal in question, calculating the mismatch, integrating the mismatches and then progressively altering intake until the total mismatch is minimised. MTD cannot be called an hypothesis because to deserve that title methods would have to be available to allow it to be falsified; rather, view it as a framework to be used to stimulate discussion and experimentation.

Concept of discomfort

Before developing that line of thought further it is necessary to provide a method for expressing quantitatively, and in a common currency, the effects of a mismatch (excess or deficiency) of each of the factors (bulk, energy, protein, etc.) on the animal. If we offer an animal (or human being) a choice between two situations (foods, environments, etc.) and we observe that one is chosen over the other then we could say that the preferred one provides the individual with greater 'comfort'. Conversely, the other is avoided because experiencing it causes 'discomfort'. While discomfort might be more obviously associated with such physical stimuli as stomach distension, it can also be caused by metabolic factors such as excesses (however mild) of metabolites or toxins. Discomfort does not have to be sensed consciously in order to have physiological relevance; only severe discomfort is brought to the attention of the animal's consciousness. If we envisage that an animal seeks to maximise its comfort then we have a problem when it comes to quantification because we don't know how to quantify *maximum* comfort – is it infinity? By contrast, we can propose that the animal seeks to minimise 'discomfort' to zero, i.e. a finite level. We propose that animals seek to minimise their discomfort, whether given one food or more.

A common currency

It will have become clear that we propose to add together the strengths of the satiating signals coming from many sources, but how can we combine stomach fill (litres or kg) with quantities of nutrients (g) or energy (kJ)? If we can estimate an optimum intake of a particular resource then we can calculate by what proportion the current food intake over- or under-provides that resource. The difference between actual and desired is divided by the desired rate of supply of each resource to give the proportional error for each resource, which is the same currency for each. For the time being we will have to make assumptions that proportional under-supply of a particular resource will have the same effect as the same degree of over-supply, and that a given proportional over- or under-supply of one resource will have the same effect as the same proportional under- or over-supply of another resource.

Relationship between stimulus and response

Once we have calculated the 'stimuli' – the proportional deviations in the supply of a resource from the optimal – we can proceed to integrate them. However, what is the appropriate function relating the strength of the stimulus to the response in food intake? Examination of the literature provides a whole range of answers, including linearity.

It is unfortunate that there seem to be no usable dose-response experiments with pigs in the literature. Those few with more than two doses of stimulus either applied them in different experiments or the responses were too variable to enable conclusions about the dose-response curve to be made. Despite Gregory's (2002) assertion that in several experiments duodenal infusion of metabolites '... causes a linear, dose-dependent suppression of intake ...', examination of the data shows considerable variation which renders the conclusion unreliable, to say the least.

Additivity

Fisher and colleagues (1996) have integrated physical, energy, and protein factors in a multiplicative framework but this presents important conceptual difficulties including what happens when the strength of one of the signals being combined is zero (Forbes, 1995). In several cases cited earlier the effects of two satiating factors have been stated to be synergistic, i.e. the combined effect is greater than the sum of the separate effects; care needs to be taken in interpretation, however – it may be that one or both of the separate effects were not statistically significant even though the result of giving both together reached statistical significance. Is the evidence for synergy sufficiently strong and quantified for it to be used to further our understanding? In other cases the addition of a second satiating stimulus to the first has not been seen to have any additional effect; it seems that the first was having a maximal effect, giving the second no chance to show any additional effect.

Will it be sufficient for the present purposes to assume that signals affecting intake are integrated in an additive manner? Simple additivity of feedback signals is inadequate as it would predict a clear and simple point at which intake should settle, rather than the variable and flexible 'settling point' that would agree with reality.

A U-shaped response curve would fit the case better than a V-shaped one, but what transformation could be applied to provide this? Consider for a moment what the animal is trying to achieve: the best balance between providing the various essential nutrients in optimal amounts. For simplicity we will take the case of energy and protein. Figure 6 represents a 2-dimensional space with energy on the vertical axis and protein on the horizontal. The point marked 'current' is the animal's current point in this space and 'desired' is the point at which energy and protein intakes are optimised. The distance from 'current' to 'desired' can be calculated (Pythagorus) as the square root of the sum of the squares of the two separate deviations. In order to get from 'current'

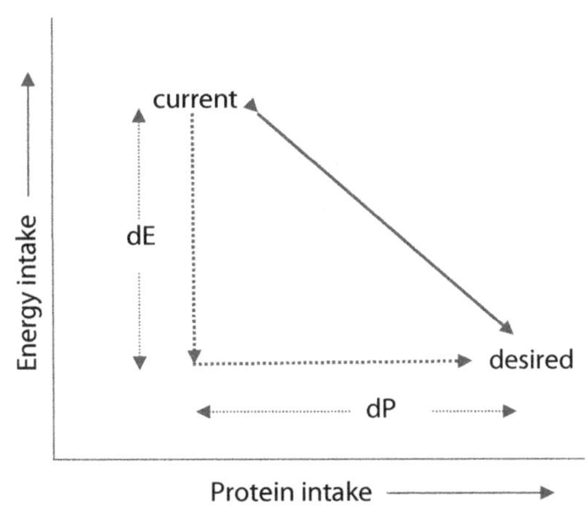

Figure 6. Notional pathways between an animal's current intakes of energy and protein, and its desired intakes.

(suboptimal) to 'desired' (optimal) we suggest that the animal 'experiments' with different rates of intake and learns as it goes, following approximately the direct the path towards optimal. Pythagorus works for any number of dimensions so we are not limited in how many resources we include in our considerations.

We have, therefore, decided to use square of the error for each resource. This gives a U-shaped response curve which, on the face of it, is at odds with the linear response seen in many experiments. This conundrum is discussed (and solved?) below.

An example of MTD in action

We must first of all decide what resources to include in our model; for the purposes of this example we will use energy, protein and bulk. A growing pig is arbitrarily specified as having optimum rates of intake of 20 MJ ME/day and 0.5 kg CP/day and a 'capacity' for WHC of 5 kg. Note that while under or over-provision of energy or protein generate discomforts, it is only an intake of WHC above the threshold that causes discomfort. The food provides 12 MJ ME/kg, 0.2 kg CP/kg and 4 kg WHC/kg. Thus this animal would satisfy its 'requirement' for energy by eating 20/12 = 1.83 kg of this food, but to optimise its protein intake it would have to eat 0.5/0.2 = 2.5 kg.

We then start to iterate, changing the daily intake for each iteration. Following Figure 7 and starting with an intake of 0.5 kg per day, we have an ME intake of 0.5 × 12 = 6 MJ ME/day, a difference of 20 - 6 = 14 MJ/day from optimal and a proportional shortfall of 14/20 = 0.7; squaring this gives a relative discomfort of 0.49. Similarly for protein and bulk. Then the relative discomforts are added and the square root taken: 1.06. Next we take a slightly higher (or lower) intake and repeat the calculations, moving in the direction of intake that reduces total discomfort, continuing until the MTD is reached. The intake at this point is taken as our prediction. Note that

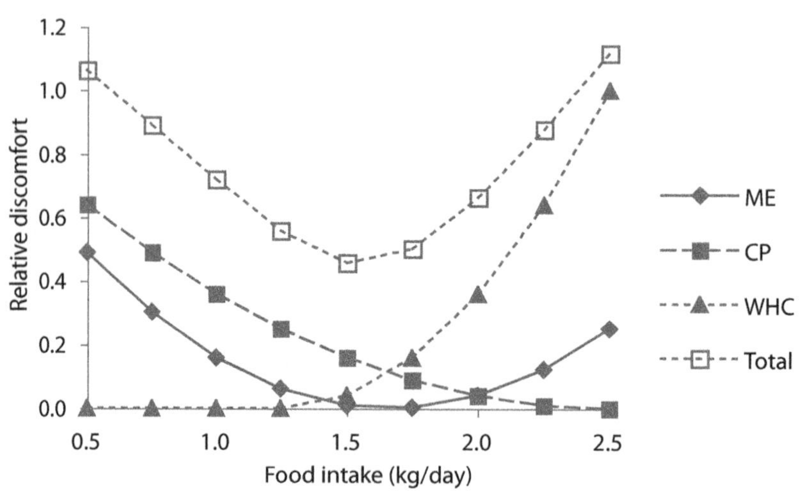

Figure 7. Calculations of discomfort due to deviations in supply of ME, CP, WHC from optimum, and total discomfort (see text for details).

food intake, rather than being placed on the y-axis as being a dependent variable, is on the x-axis as in this case it is the independent variable – it is the dependence of discomfort on food intake that is a major novel feature in the MTD approach.

The iteration involved in solving for MTD parallels the behaviour of the animal which, it is proposed, tries different rates of intake, settling on the one that provides the best compromise for providing optimal rates of intake of the various resources provided by the food(s). This proposed comparison of how the animal feels today compared with previous days implies learning.

The U-shape of the intake/discomfort curve means that there is a range of intakes at which discomfort is close to minimum and small deviations from optimal will have relatively small impacts on total discomfort. It should not be a surprise, therefore, to see day-to-day variability in intake (see below) which would provide the animals with the possibility to know whether the optimal intake has changed, due to changes in the nature of the food(s) available and/or its own requirements.

Further examples

One pig (B) is described as in the first example, the other as a smaller pig (A) with optimal intakes of 10 MJ ME/day, 0.25 kg CP/day and a threshold for WHC of 2.0 kg /day. Calculations were made of the voluntary intakes of foods with 12 MJ ME/kg and 0.2 kg CP/kg and WHC thresholds ranging from 1.5 to 6.0 g/g. Figure 8 shows the predicted food intakes and total discomforts and it will be noted that: (1) unsurprisingly, the big pig eats more than the little pig; (2) after a plateau at low WHC contents there is a progressive decline in intake with increasing food WHC content; (3) the rate of decline in intake is similar, in absolute terms, for the two pigs; (4) total discomfort rises with increasing WHC content; (5) at any given WHC total discomforts are similar for the two sizes of pig. Outcome (3) seems more realistic if the depression in intake is expressed relative

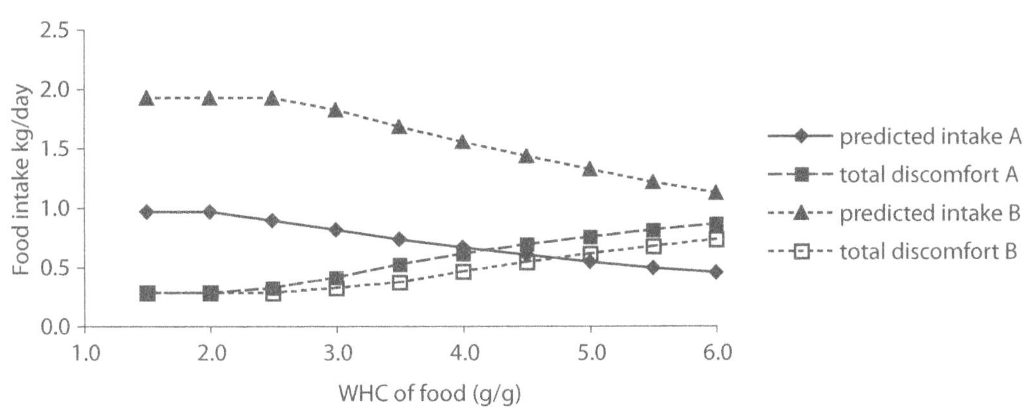

Figure 8. Outcome of running the MTD model with a range of food water holding capacity (WHC) for (A) small and (B) large pigs.

to animal weight, when the little pig is more severely affected by increasing WHC than the big pig, in line with expectations.

Note that this framework of ideas is not expected to replicate reality and that there has been deliberate avoidance of comparison of these outcomes with the results of observations on real pigs.

In a discussion of the TPH earlier it was noted that increasing pigs' energy requirements by lowering the environmental temperature below the critical temperature, increased intake to the same extent in pigs on high- and low-WHC foods. The MTD framework was computed with a range of energy requirements and foods with three levels of WHC (Figure 9). The simulation shows only small increases in intake as energy requirement increases with a food moderately high in WHC while with lower-WHC foods the increase is more marked. Would the conclusion of Whittemore *et al.* (2001) have been different had they used different foods?

The next example involves possible effects of balloon distension in the stomach of the pig, to explore the shape of the dose-response curve. Some experimental results with other species have shown a linear depression in intake with increasing distension, yet we have chosen a quadratic function for use in the MTD framework. Figure 10 shows the results of simulating gastric distension for foods of three WHC contents. The higher the WHC content the more linear the reduction in food intake in response to distension. How can a quadratic function give an almost linear response in some situations? The more is intake depressed by physical distension, the less stimulated are the other receptors that respond to chemical and osmotic stimulation in the intestines, and those that respond to metabolites in the body. Thus the total satiating signal is not elevated as much as the degree of distension might suggest at first sight, and the result is an almost

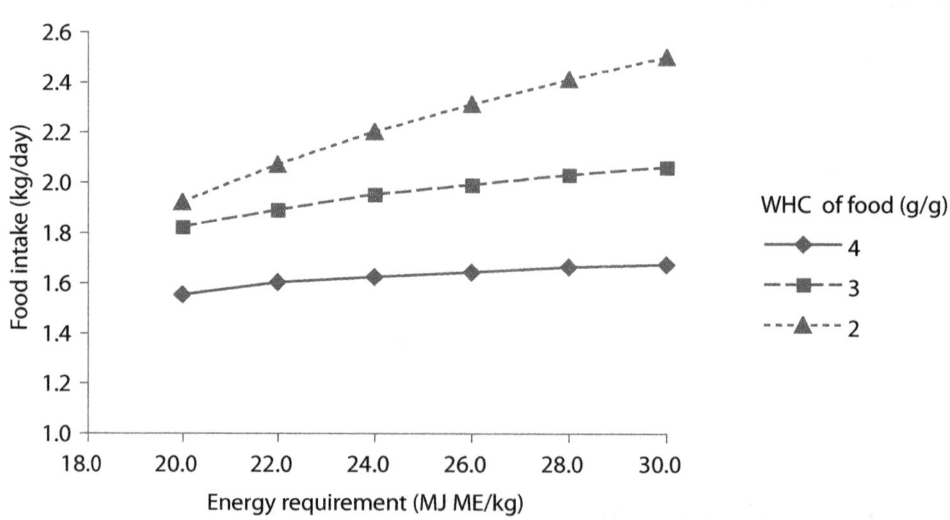

Figure 9. Outcome of running the MTD model with three levels of food WHC and a range of pig energy requirements.

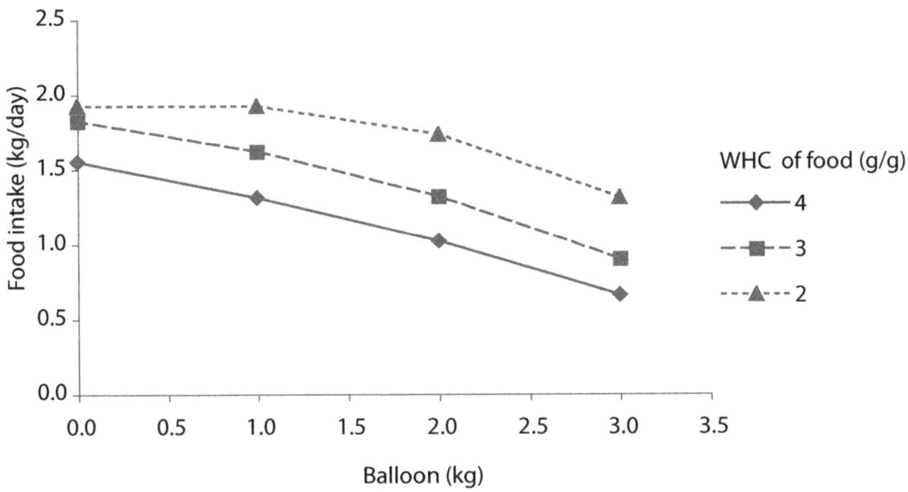

Figure 10. Outcome of funning the MTD model with increasing levels of distension of the GI tract with a balloon.

linear response. This demonstrates how a quadratic response to a single stimulus can generate a linear response when combined with other stimuli.

Finally in this section, we see how MTD fares in simulating the selection made by pigs given a choice of foods. Where it is possible to construct a balanced diet from the foods on offer then growing pigs succeed, by and large, and so does the MTD model. For example, HP and LP foods with similar DE contents are selected so as to produce a P: E ratio close to optimal. Where one food is more bulky than the other then MTD may be achieved with a mixture of the foods that does not provide the optimal P: E. If the LP food also has a lower DE content and a higher WHC than HP then MTD predicts selection of a mixture that provides an excess of protein and a deficiency of energy intake, because this is the optimal way to avoid an excessive intake of WHC. This area will not be followed further as it becomes complex and there is a paucity of experimental data with which to compare the predictions.

Shortcomings

As acknowledged above, MTD only provides an outline of a novel framework on which it should be possible to build a more complete scheme for predicting food intake and selection. If the general principles are accepted then major questions to be considered include:

- Is it appropriate to calculate the 'error' in the supply of each resource as a proportion of the optimal supply in order to achieve the same currency for each resource?
- Is squaring the appropriate function for relating the degree of discomfort to the extent of the proportional difference between supply and demand of each resource?
- Should the U-shaped response curve relating resource imbalance to discomfort be considered to be symmetrical?

- It is likely that different resources will require different weightings. How can these be derived?
- How can interactions between resources best be dealt with? For example, ME and CP are not independent as protein supplied in excess of requirements is used as a source of energy. A solution has been provided (Forbes, 2007a) but such considerations might still lead to violation of the claim of additivity and are a serious challenge to the implementation of the MTD framework.
- Can discomforting agents such as social and environmental stressors be included alongside food-related discomforts in the MTD approach, as suggested by Forbes (2006)?

Given these shortcomings and the fact that no attempt has been made to optimise the model, nor to validate it, does the MTD approach have any value? Indeed, there are no measurements of 'discomfort' with which to compare the predictions made above.

Variability in intake and selection

We have suggested above that variability in intake, both between animals of similar genotype, and in each animal from day to day, is the result of a control system that has imprecise outcomes. We have also speculated that such variability of intake in an individual provides the animal with information as to whether it should maintain, increase or decrease its mean intake. How would it know whether or not its current intake was optimal if it did not experience such variations day-to-day?

Figure 11 shows the daily intakes of three growing pigs selected at random from the records available for many hundred animals, monitored automatically in single-space feeders for groups

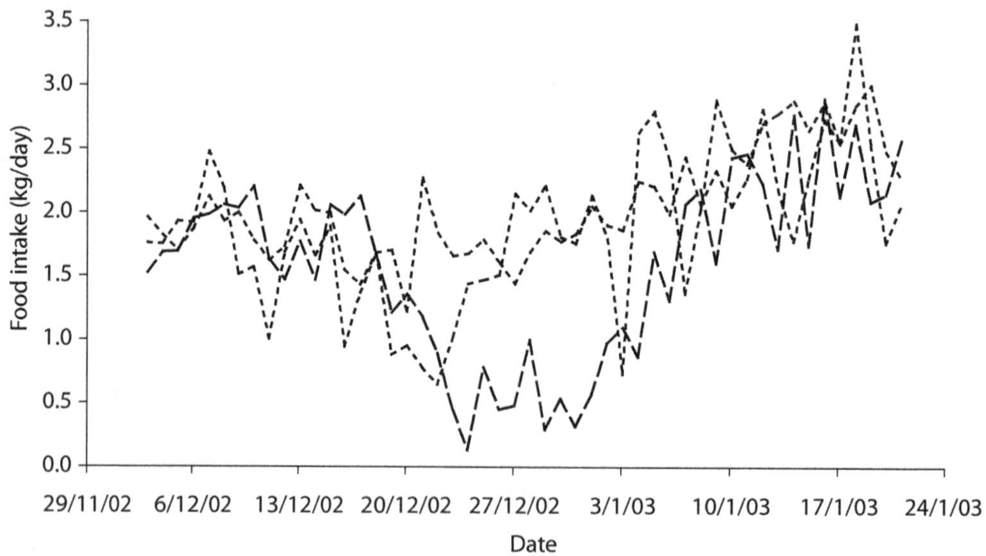

Figure 11. Daily food intake by three growing pigs over a period of 54 days (data provided by ACMC, East Yorkshire, UK).

of 15. On any given day there is wide variation between animals, and the within-pig day-to-day variation can clearly be seen. The longer the period considered, the more stable the intake, with the suggestion that control is over a period of at least several days.

What about variability in diet selection? In the absence of any data for pigs, we use an example from cattle. Daily intakes of silage by dairy cows vary to an extent similar to that shown above for growing pigs and Yeates *et al.* (2002) have examined variability in selection by cows between high (HP) and low (LP) protein foods. Figure 12 shows the proportion of HP chosen over periods of 8 h, 24 h and 7 d and it can clearly be seen that choice in consecutive 8 h periods is very variable and ranges from wholly HP to mostly LP. Cumulating these into 24 h periods greatly reduces, but certainly does not eliminate the variability, while 7 d means show great consistency, with a steady decline in selection for HP as milk yield declines in mid lactation. It can be concluded that the cows did not attempt to select a consistent diet in terms of P: E either within a single meal or in a short sequence of meals. The authors concluded 'Our present analysis does not suggest what the most relevant time scale is, except that is must be longer than a meal'.

This variability allows the animal to experience the effects of different rates of intake and it can be suggested that this enables it to determine whether it should, on average, be eating more or less than currently, or making different selection between foods. Adaptability is provided at the expense of stability.

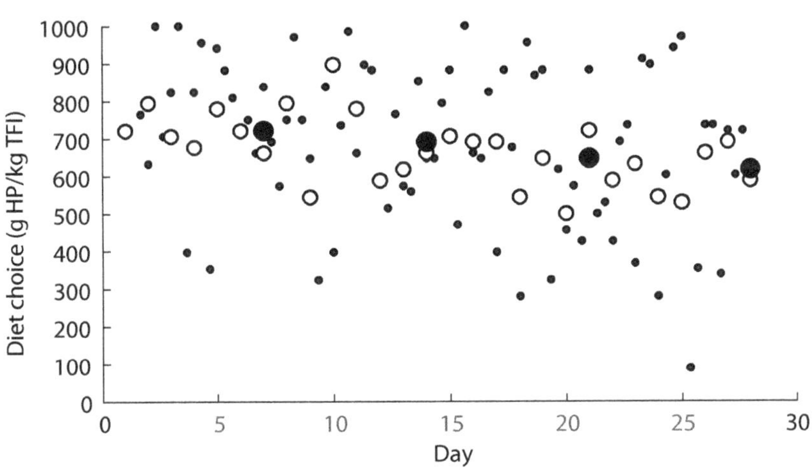

Figure 12. High protein (HP) food intake as a proportion of total food intake (TFI) by eight lactating cows given a choice of HP and LP over periods of 8 h (dots), 24 h (open circles) and 7 d (closed circles) (Yeates et al.*, 2002).*

Conclusions

It is absolutely clear that animals, including pigs, can learn to associate the sight, flavour and/or position of a food, with how they feel after eating that food. Learned preferences and aversions are then used to direct how readily a food is eaten and/or what proportion of that food is chosen if there is a choice.

Detailed knowledge of the effects of infusions of metabolites or hormones, or of the changes in levels of these during and between spontaneous meals, may contribute to our understanding of how feeding behaviour is controlled. However, there is considerable variation in meal size and frequency even when food composition and availability does not change and it is not possible to see, at the present time, how greater understanding of short-term changes in physiological parameters will lead to greater understanding of how intake is controlled over periods of days and weeks. Given that it is the latter that is of primary interest, at least to agricultural scientists, attempting to link physiology with short-term behaviour is not our most urgent concern. Rather, we are seeking a set of resources, required by animals and provided by food (and the environment), that can be used in outlining what happens when several such resources are considered simultaneously. The development of the MTD framework, while speculative and imperfect, does allow us to relate meaningful aspects of animal and food parameters in a novel manner. The author, in his retirement, looks forward to others taking up the challenge of developing this approach.

A spreadsheet on which the MTD model is programmed is available by contacting the author at *j.m.forbes@leeds.ac.uk*.

References

Adams, G.B. and Forbes, J.M., 1981. Additivity of effects of ruminal acetate and either portal propionate or rumen distension on food intake in sheep. Proceedings of the Nutrition Society 40: 44A.

Anika, S.M., Houpt, T.R. and Houpt, K.A., 1981. Cholecystokinin and satiety in pigs. American Journal of Physiology 240: R310-R318.

Anil, M.H. and Forbes, J.M., 1987. Neural control and sensory functions of the liver. Proceedings of the Nutrition Society 46: 125-133.

Anil, M.H. and Forbes, J.M., 1988. The roles of hepatic nerves in the reduction of food intake as a consequence of intraportal sodium propionate administration in sheep. Quarterly Journal of Experimental Physiology 73: 539-546.

Baldwin, B.A. and Cooper, T.R., 1979. The effects of olfactory bulbectomy on feeding behaviour in pigs. Applied Animal Ethology 5: 153-159.

Black, J.L., Williams, B.A. and Gidley, M.J., 2009. Metabolic regulation of feed intake in monogastric mammals. In: Torrallardona, D. and Roura, E. (eds.) Voluntary feed intake in pigs. Wageningen Academic Publishers, the Netherlands, pp. 189-213

Blair, R. and Fitzsimons, J., 1970. A note on the voluntary feed intake and growth of pigs given diets containing an extremely bitter compound. Animal Production 12: 529-530.

Blundell, J.E. and Halford, J.C.G., 1994. Regulation of nutrient supply: the brain and appetite control. Proceedings of the Nutrition Society 53: 407-418.

Carroll, J.A. and Allee, G.L., 2009. Hormonal control of feed intake in swine. In: Torrallardona, D. and Roura, E. (eds.) Voluntary feed intake in pigs. Wageningen Academic Publishers, the Netherlands, pp. 155-187

Covasa, M. and Forbes, J.M., 1994. Exogenous cholecystokinin octapeptide in broiler chickens: Satiety, conditioned colour aversion and vagal mediation. Physiology and Behavior 56: 39-49.

D'Eath, R.B., Tolkamp, B.J., Kyriazakis, I. and Lawrence, A.B., 2009. 'Freedom from hunger' and preventing obesity: the animal welfare implications of reducing food quantity or quality. Animal Behaviour doi: 10.1016/j.anbehav.2008.10.028.

Dalby, J.A., Forbes, J.M., Varley, M.A. and Jagger, S., 1995. The requirements of weaned piglets for a training period prior to a choice feeding regime. Animal Science 61: 311-319.

Fisher, D.S., 1996. Mathematical integration of distension and chemostatic feedbacks for regulation of ruminant intake in diets with varied dietary protein. Journal of Animal Science 74: 3076-3081.

Forbes, J.M., 1980. A model of the short-term control of feeding in the ruminant: effects of changing animal or feed characteristics. Appetite 1: 21-41.

Forbes, J.M., 1995. Voluntary Food Intake and Diet Selection in Farm Animals. CAB International, Wallingford, pp. 520.

Forbes, J.M., 1996. Integration of regulatory signals controlling forage intake in ruminants. Journal of Animal Science 74: 3029-3035.

Forbes, J.M., 2006. Conclusion and perspectives. In: Bels, V.L. (ed.) Food and Feeding in Domestic Vertebrates. A Comprehensive Approach from Structure to Behaviour. CABI, Wallingford, UK, pp. 349-358.

Forbes, J.M., 2007a. A personal view of how ruminant animals control their intake and choice of food: Minimal Total Discomfort. Nutrition Research Reviews 20: 132-146.

Forbes, J.M., 2007b. Voluntary Food Intake and Diet Selection in Farm Animals. CABI, Wallingford, UK, pp. 453.

Gregory, P.C., 2002. Role of the intestine in regulation of food intake in growing pigs. In: Zabielski, R., Gregory, P.C. and Weström, B. (ed.) Biology of the Intestine in Growing Animals. Elsevier Science B.V., the Netherlands, pp. 427-464.

Gregory, P.C., McFadyen, M. and Rayner, D.V., 1987. The influence of gastrointestinal infusions of glucose on regulation of food intake in pigs. Quarterly Journal of Experimental Physiology,72: 525-536.

Houpt, K.A., Houpt, T.R. and Pond, W.G., 1979. The pig as a model for the study of obesity and of control of food intake: A review. Yale Journal of Biology and Medicine 52: 307-329.

Houpt, T.R., Baldwin, B.A. and Houpt, K.A., 1983. Effects of duodenal osmotic loads on spontaneous meals in pigs. Physiology and Behavior 30: 787-796.

Kyriazakis, I. and Emmans, G.C., 1995. The voluntary feed intake of pigs given feeds based on wheat bran, dried citrus pulp and grass meal, in relation to measurements of feed bulk. British Journal of Nutrition 73: 191-207.

Kyriazakis, I., Emmans, G.C. and Whittemore, C.T., 1990. Diet selection in pigs: choices made by growing pigs given foods of different protein concentrations. Animal Production 51: 189-199.

Mbanya, J.N., Anil, M.H. and Forbes, J.M., 1993. The voluntary intake of hay and silage by lactating cows in response to ruminal infusion of acetate or propionate, or both, with and without distension of the rumen by a balloon. British Journal of Nutrition 69: 713-720.

Millward, D.J., 1995. A protein-stat mechanism for the regulation of growth and maintenance of the lean-body mass. Nutrition Research Reviews 8: 93-120.

Owen, J.B. and Ridgman, W.J., 1967. The effect of dietary energy content on the voluntary feed intake of pigs. Animal Production 9: 107-113.

Pittroff, W. and Kothmann, M.M., 1999. Regulation of intake and diet selection by herbivores. In: Jung, H. G. and Fahey, G. C. (ed.) Nutritional Ecology of Herbivores. American Society of Animal Science, Savoy, Illinois, USA, pp. 366-422.

Poppi, D.P., Gill, M. and France, J., 1994. Integration of theories of intake regulation in growing ruminants. Journal of Theoretical Biology 167: 129-145.

Savory, C.J., 1989. Responses of fowls to an operant feeding procedure and its potential use for reducing randomness in meal occurrence. Physiology and Behavior 45: 373-379.

Scharrer, E., Lutz, T.A. and Rossi, R., 1996. Coding of metabolic information by hepatic sensors controlling food intake. In: Shimazu, T. (ed.) Liver Innervation. John Libbey, Eastleigh, UK, pp. 381-388.

Stephens, D.B., 1980. The effects of alimentary infusions of glucose, amino acids, or neutral fat on meal size in hungry pigs. Journal of Physiology 299: 453-463.

Stephens, D.B., 1985. Influence of intraduodenal glucose on meal size and its modification by 2-deoxy-d-glucose or vagotomy in hungry pigs. Quarterly Journal of Experimental Physiology 70: 129-135.

Whittemore, E.C., Emmans, G.C. and Kyriazakis, I., 2003. The relationship between live weight and the intake of bulky foods in pigs. Animal Science 76: 89-100.

Whittemore, E.C., Kyriazakis, I., Emmans, G.C. and Tolkamp, B.T., 2001. Tests of two theories of food intake using growing pigs - 1. The effect of ambient temperature on the intake of foods of differing bulk content. Animal Science 72: 351-360.

Whittemore, E.C., Kyriazakis, I., Tolkamp, B.J. and Emmans, G.C., 2002. The short-term feeding behavior of growing pigs fed foods differing in bulk content. Physiology and Behavior 76: 131-141.

Yeates, M.P., Tolkamp, B.J. and Kyriazakis, I., 2002. The relationship between meal composition and long-term diet choice. Journal of Animal Science 80: 3165-3178.

Chapter 4. Taste receptor biology

K.C. Klasing[1] and B.D. Humphrey[2]
[1]University of California, Davis, CA, USA; [2]Dept. of Animal Science, California Polytechnic Institute, San Luis Obispo, CA, USA; kcklasing@ucdavis.edu

Abstract

Vertebrates need to procure foods that provide adequate levels of required nutrients. However many potential food items protect themselves with chemical defenses. Taste permits the sensing of nutrients and harmful substances in order to guide selection of nutritious foods and avoidance of toxins. There are at least six primary tastes for which receptors have been identified: sweet, umami, fat, salt, sour, and bitter. The sensory cells responsible for perceiving taste are bipolar epithelial cells found in clusters in each taste bud. Each taste cell is endowed with a specific set of receptors embedded in the apical membrane and serve as the 'molecular antennas' that listen into the chemical environment of the oral cavity. Each of the nutrient tastes (sweet, umami, fat, salt) is identified by only one or a few TRs, whereas the bitter tasting toxins are identified by a very large family of taste receptors. Recent research has identified a wide range of ligands for specific receptors. The nuances of signal transduction within the taste cells, the neurotransmitters released by these cells, and the complex neurobiology of taste perception have been explored. This cellular machinery of taste has been manipulated by gene knockout and transgenesis technologies to reveal important biological functions of taste. Within a species, genetic polymorphisms in taste receptors mediate individual differences in ligand binding, taste perception and food choices. The evolution of taste receptors illustrates the gain and loss of specific receptors as well as major changes in ligand binding preferences of a receptor that appear to be driven by dietary history. Understanding and manipulating taste receptor biology will be used in animal agriculture to discover taste enhancers and to improve intake of ingredients that contain bitter tasting secondary plant compounds.

Keywords: receptors, gustation, signal transduction, evolution, agriculture

Overview of taste biology

Vertebrates need to procure foods that provide adequate levels of required nutrients. However many potential food items protect themselves with chemical defenses. Taste permits the sensing of both nutrients and harmful substances in order to guide selection of nutritious foods and avoidance of toxins. Additional information about food items is obtained by tactile, cold and pain receptors in the oral cavity and olfactory receptors in the nasal epithelium. Species have evolved taste systems that are tuned differently toward different nutrients or potential toxins, depending upon their dietary niche (e.g. carnivore, herbivore, omnivore). Despite many decades of research by anatomists, physiologists and behaviorists, a unified synthesis of taste biology was beyond reach until molecular biologists identified the genes that code for membrane bound receptors (taste receptors; TRs) that recognise the sugars, amino acids, lipids, ions, and toxins responsible for the primary tastes. Discovery of the TRs as well as the signal transduction pathways that amplify the information for transmission to the brain has transformed the way we view the evolution of taste in vertebrates and heralds novel applications in animal nutrition.

Basic research on taste biology has utilised rats and mice. The sensory cells responsible for perceiving taste are bipolar epithelial cells found in clusters of about 100 in each taste bud. These epithelial cells have several characteristics of neural cells including the ability to form synapses and to depolarise. TRs extend microvilli to the epithelial surfaces where they are in contact with mucus and saliva (Ganchrow, 2000). The TRs are embedded in the membrane at the apical surface of the microvilli and serve as the 'molecular antennas' that sense into the chemical environment of the oral cavity. Each taste cell forms a synapse at its basal membrane with a neuron. Taste cells maintain an action potential and depolarise when they have identified a taste molecule. This depolarisation results in the release of neurotransmitters that stimulate their post-synaptic neuron to depolarise and transmit the information to the brain. The brain interprets the aggregate of incoming information as a specific taste quality (Uematsu, 1999; Drewnowski and Gomez-Carneros, 2000; Spector, 2000; Yamaguchi and Ninomiya, 2000). A preponderance of bitter or sour taste is typically interpreted as bad tasting and initiates a food rejection response. A preponderance of sweet, umami, or salt taste is typically interpreted as good and food consumption is stimulated. These interpretations can be modified by prior experience with a specific food item and by the animal's physiological state, including nutritional and disease status (Spector, 2000).

Individual taste cells typically express only a single type of taste receptor and it was originally thought that there were 'labeled lines' between the taste receptor cells and specific areas of the brain. Recent work demonstrates that taste receptor cells expressing a single receptor type secrete ATP that stimulates both presynaptic (type III) cells and sensory afferent nerve fibers (Finger *et al.*, 2005; Huang *et al.*, 2007). Once stimulated by taste receptor cells, the presynaptic cells release serotonin and stimulate afferent neurons that reach the brain (Kaya *et al.*, 2004). The convergence of taste receptor cells that are individually tuned to salt, sour, sweet, umami or bitter onto the same presynaptic cell results in more broadly tuned information reaching the afferent neurons (DeFazio *et al.*, 2006; Simon *et al.*, 2006). Thus, there are two types of signals that leave the tongue, labeled lines that keep taste information specific to the taste cell that detected it, and integrated signals that provide more generalised and integrated information.

In the lower brain, the nucleus of the solitary tact, parabrachial nucleus, and thalamic gustatory areas are organised into spatial patterns corresponding to each of the primary taste modalities (sweet, umami, bitter, sour, salty) (Simon *et al.*, 2006; Sugita, 2006; Accolla *et al.*, 2007). There are also specific areas for that integrate the perception of pleasant (sweet, umami, salty) taste signals and other areas that integrate unpleasant (bitter, sour) signals. There must be a specific 'hard wired' connection from a specific taste cell decorated with a specific type of receptor (sweet, umami, bitter) to specific areas of the brain that interpret the taste quality because switching the TR in an epithelial taste cell does not change the perceived taste. For examples changing the receptor of a cell that normally expresses a sweet receptor so that is expresses a bitter receptor results in a sweet taste when a bitter compound is consumed (Mueller *et al.*, 2005).

Our detailed knowledge of the cellular and molecular knowledge of taste in rodents and, to a lesser extent, humans explains a very wide variety of taste phenomenon ranging from specific 'after tastes', to individual differences in taste perceptions and preferences, to susceptibilities to alcoholism, obesity, and other diseases facilitated by food choices. More broadly, knowledge of the types of TRs and their ligand preferences across a wide variety of taxa begins to explain how taste co-evolves with and is molded by the foods consumed. Thus, the tuning of taste perception appears to match the habitual diet of a species.

Taste receptors

Confirmation that a specific receptor in the genome is a bona fide taste receptor requires three lines of evidence. First, it must be expressed robustly in taste cells and not most other cell types. Second, transfection of non-taste cells with the candidate receptor confers responsiveness to the taste stimuli. Third, knocking out the gene eliminates responsiveness to the taste stimuli. These criteria have been used to identify the T1R family as receptors for sweet and umami tastes and the T2R family as receptors for bitter. All members of the T1R and T2R families are guanine-coupled-nucleotide-binding protein-coupled receptors (GPCRs). The T1R, T2R receptor proteins are true receptors in that they bind their extracellular ligands (sugars, amino acids, toxins) and transmit this information into the cell through a signal transduction cascade to result in a response. The TRs for salt and sour are ion channels that selectively permit the entrance of sodium or hydrogen, respectively, and couple this entrance to the depolarisation of the cell membrane, which causes the response. The three criteria for identifying receptors have not been rigorously met for fat, sour and salt tastes. Table 1 summarises the TRs, their signaling pathways, and their nutritional implications.

Table 1. Receptors that detect the primary tastes.

Taste	Number of receptors	Type of receptors	Ligands	Signal transduction	Physiological response
sweet	1 heterodimer	T1R2/T1R3	sugars, d-amino acids	G-proteins	pleasant taste
umami	1 heterodimer	T1R1/T1R3	l-amino acids	G proteins	pleasant taste
fat	1?	CD36?	linoleate	Ca	pleasant taste
salty	1?	ENaC sodium channel	sodium	Na	pleasant taste
sour	several	PKD1L3 PKD2L1	hydrogen	ions	unpleasant taste
bitter	3-49 depending on species	T2R	>100 unrelated molecules	G proteins	unpleasant - repulsive taste

Sweet

Sweetness is the strongest pleasurable and hedonistic taste in humans, providing the most powerful positive experience of the primary tastes. Sugars including sucrose, glucose, fructose and lactose bind to T1R2/T1R3 (Li *et al.*, 2002). Some d-amino acids have a sweet taste (e.g. phenylalanine and tryptophan). These d-amino acids, but not their l-amino acid enantiomers, bind to T1R2/T1R3 (Nelson *et al.*, 2001; Li *et al.*, 2002; Nelson *et al.*, 2002). Additionally several proteins have a sweet taste and also bind to T1R2/T1R3.

The binding of artificial sweeteners to T1R2/T1R3 is species dependent. This is because there are species differences in the amino acid sequences of T1R2/T1R3. For example, aspartame is exceptionally sweet in humans and binds human T1R2/T1R3, but it has no taste in rodents and does not bind rodent T1R2/T1R3. It also has no taste in pigs (Danilova *et al.*, 1999) and presumably does not bind to their T1R2/T1R3. Therefore evolution has resulted in variations in the sequences of the TRs that do not affect the binding of authentic sugars, but markedly affects the binding of artificial ligands.

Sucrose has affinity for the extracellular domain of T1R2 but has even greater affinity for the extracellular domain of T1R3 (Nie *et al.*, 2006). Conversely, glucose has greater affinity for the extracellular domain of T1R2 than that of T1R3. Because sucrose is perceived to be sweeter than glucose, binding to T1R3 may give a more robust signal than binding to T1R2 (Nie *et al.*, 2006). In humans, T1R2 binds many of the artificial sweeteners (e.g. aspartame, glycyrrhizic acid, neotame, thaoumatin, brazzein, and monellin) (Xu *et al.*, 2004; Cui *et al.*, 2006; Walters and Hellekant, 2006). The amino acids in T1R2-T1-R3 receptor that interact with specific moieties of various sweet tastants have been identified and this knowledge can be used to engineer better artificial sweeteners (Esposito *et al.*, 2006).

Zinc sulphate inhibits the sweetness of 11 of 12 chemically diverse sweeteners in humans (Keast *et al.*, 2004). For most compounds the degree of inhibition is dose dependent. Given that zinc sulphate is often to pig diets, its role in affecting taste of this species should be investigated.

Umami

The umami taste of some amino acids, peptides and purine nucleotides provides a pleasant sensation. The primary umami receptor is a heterodimer of T1R1 and T1R3 proteins (Brand, 2000; Li *et al.*, 2002; Nelson *et al.*, 2002). The umami receptor uses the same T1R3 subunit as the sweet receptor, but combines it with different subunit (T1R1) that tunes it to a different set of molecules. The T1R1/T1R3 receptor recognises only L-amino acids and not their D-amino acid enantiomers or sugars. The mouse umami receptor is broadly tuned and detects most L-amino acids, whereas the human receptor is tuned much more narrowly with glutamate being the preferred ligand. Glutamate binds to the Venus flytrap domain of T1R1 (Zhang *et al.*, 2008). Fish have very broad amino acid perception, which is mediated by both T1R1 and T1R2 (see below).

Fat

Fat is now thought by many to be a primary taste. Several possible receptors have been identified. Evidence suggests that CD36 mediates food choices based on fatty acids (Laugerette *et al.*, 2007). CD36 is expressed in gustatory cells of the circumvallate and binds to long chain fatty acids, especially linoleic acid. Circumvallate papillae taste-bud cells are immersed in lingual lipase secreted from von Ebner's glands so they are strategically located for interception of dietary triglycerides. Transgenic mice lacking CD36 lack a preference for fatty acids (Laugerette *et al.*, 2005, 2007; Gaillard *et al.*, 2008a,b). Linoleic acid binding to CD36 on taste receptors cells results in an increase in intracellular calcium. Other candidate fatty acid receptors are being investigated and may be important in fatty acid taste. Ultimately, a hedonistic response to fat is induced by activation of the receptors as indicated by the release of beta-endorphin and dopamine in the brain and the involvement of the opioid mu receptor (Mizushige *et al.*, 2007).

Salt

Salt is a well known pleasant taste caused by ionic sodium and to a lesser extent by other cations. In rodents ENaC, the sodium-specific and amiloride-sensitive sodium channel, is an important mediator of the salt taste (Gilbertson *et al.*, 2000, 2001; Lindemann, 2001). ENaC is a hetero-oligomeric complex consisting of α, β and γ subunits. When sodium is present in the oral cavity, it can travel through the sodium channel formed by ENaC and trigger an action potential in the taste cell, followed by release of neurotransmitter. The evidence that this protein is the sole receptor for salt taste is not compelling. A role second mechanism that involves the transient receptor potential vanilloid receptor-1 (TRPV1) has been demonstrated for some regions of the tongue using knockout mice (Lyall *et al.*, 2004; Ruiz *et al.*, 2006; Treesukosol *et al.*, 2007). The ENaC receptor is highly selective towards sodium whereas theTRPV1 receptor is more broadly tuned and is responsible for rejection-responses to K^+, NH_4^+, and Ca^+ salts (Lyall *et al.*, 2004; DeSimone and Lyall, 2006; Ruiz *et al.*, 2006).

During a nutritional sodium deficiency, high levels of the renal hormone aldosterone induce at least one of the subunits of ENaC (Lin *et al.*, 1999). This upregulation of sodium receptors increases the taste sensitivity for sodium and aids in the specific appetite for this nutrient. Thus, the nutritional state tunes taste acuity to help maintain homeostasis.

Sour

Spoiled foods often taste sour and this taste often results in food rejection. Our current knowledge about the detection of sourness is least advanced of any of the taste senses. Taste cells that mediate a sour taste are sensitive to intracellular hydrogen ion concentration and respond when intracellular pH drops below a threshold (Huang *et al.*, 2008). Many candidate sour receptors have been reported from *in vitro* work, but none have been shown conclusively to be solely responsible *in vivo*. The best evidence is for PKD1L3 and PKD2L1 to function as the sour taste receptor. They are expressed in sour taste receptor cells and appear to form heterodimers at the cell membrane (Huang *et al.*, 2006; Ishimaru *et al.*, 2006). Knockout mice that lack PKD2L1 do not respond to sour stimuli indicating that PKD2L1 is required to taste acids (Huang *et al.*, 2006).

Bitter

A bitter taste is initiated by a very large number of unrelated organic molecules, many of which have a low toxicity threshold. Most bitter tasting compounds are produced by plants to defend themselves from herbivory. These molecularly diverse bitter compounds are identified by a large repertoire of receptors of the T2R family (Adler *et al.*, 2000; Chandrashekar *et al.*, 2000; Lindemann, 2001; Margolskee, 2002). Sequence identities among individual T2R receptors vary by 30-70%, indicating robust diversity among the individual receptors enabling them to respond to a staggering variety of natural bitter compounds. A single bitter taste receptor cell expresses an assortment of different T2R receptors and can respond to many different compounds (Adler *et al.*, 2000). Consequently a single bitter taste modality arises from many structurally unrelated but potentially toxic compounds.

Interestingly, the T2R receptors are not closely related to the T1R receptors but show homology to the vomeronasal receptors (V1R) of the olfactory system and are probably of common origin. The size of the T2R family is species dependent (Table 2) ranging from 49 in frogs to 3 in chickens and sticklebacks (Dong *et al.*, 2009). There are also a large number of pseudogenes in many species, suggesting a fast evolutionary history of this gene family (see below).

The spectrum of ligands recognised by each of the many T2Rs has yet to be fully characterised. Several of the receptors that have been examined in detail are tuned widely, having many ligands, and others are tuned narrowly (Andres-Barquin and Conte, 2004; Pronin *et al.*, 2004; Sainz *et al.*, 2007). All of the tastants found to bind to human T2Rs expressed *in vitro* have been found to have a bitter taste *in vivo*.

The human *TAS2R38* receptor has been studied in great detail (Floriano *et al.*, 2006; Miguet *et al.*, 2006). Its ligands include TPC, POP, and related compounds that contain a thiourea moiety including the glucosinolates. A recent study demonstrates that allelic variation in *TAS2R38* explains sensitivity to glucosinolate-containing foods such as broccoli, turnips and horseradish (Sandell and Breslin, 2006; Sacerdote *et al.*, 2007). *TAS2R43 and TAS2R44* receptors are responsible for the bitter aftertaste of saccharin and acesulfame (Kuhn *et al.*, 2004).

T2Rs of different species are tuned differently. For example, mice, rats, and golden hamsters have very high sensitivities to the antibiotic cycloheximide via their *Tas2r105* whereas humans and rabbits do not detect it (Hettinger *et al.*, 2007). There appear to be many species-specific ligand-receptor interactions reflecting the biologic need to detect different toxins in different nutritional niches.

It is not yet clear if mammals can resolve differing bitter tastes or if there is only one bitter taste for all compounds (Spector, 2000; Spector and Kopka, 2002; Sacerdote *et al.*, 2007). The observation that most bitter taste cells have many T2Rs, all of which use the same signal transduction system, suggests that there should be few distinct bitter tastes. Behavioral evidence supports the idea of a single bitter taste modality.

Table 2. Species variation in numbers of T1R and T2R genes in vertebrates.

	T1R family		T2R family			
Species[1]	Intact	Pseudogene	Intact	Partial + pseudogene	Total	Fraction Intact
Human	3	0	24	10	34	71
Macaque	3	0	26	12	38	68
Mouse	3	0	33	9	42	79
Rat	3	0	36	6	42	86
Dog	3	0	15	5	20	75
Cat	2	1	?	?	?	-
Pig	3	0?	?	?	?	-
Cow	3	0	18	15	33	55
Horse	?	?	19	36	55	35
Opossum	3	0	26	10	36	72
Platypus	?	?	4	2	6	67
Chicken	2	0	3	0	3	100
Lizard	?	?	37	12	49	76
Frog	0	0	49	17	66	74
Fugu	5	0	4	0	4	100
Puffer fish	6	0	6	0	6	100
Stickleback	10	2	3	0	3	100
Zebrafish	4	?	4	0	4	100

[1] Data from Dong *et al.*, 2009; Nei *et al.*, 2008; Hashiguchi *et al.*, 2007, and Holt and Klasing (unpublished). Fugu refers to *Takifugu rubripes* and puffer fish refers to *Tetraodon nigroviridis*.

Signal transduction

The T1R and T2R receptors reside in the cell membrane at the apical surface of the taste cell and signals through membrane-bound G-proteins to cause an influx of calcium into the cytoplasm, which initiates depolarisation of the taste cell and the release of neurotransmitters. Two G-protein families mediate signal transduction, α-gustducin and α-transducin. The specific G-protein utilised depends on the TR type (Sainz *et al.*, 2007) and the region of the oral cavity (Miura *et al.*, 2007). G-protein activation initiates a cascade of events through two parallel streams of intracellular events that contribute to the depolarisation of the taste receptor cell (Trubey *et al.*, 2006; Roper, 2007). First, the coupling of taste receptors to $G_{\alpha i}$ results in the activation of phosphodiesterases, which decreases cyclic nucleotides. This stimulates phosphokinase A, which phosphorylates potassium channels and causes their inhibition. Reduced potassium channel activity causes depolarisation and neurotransmitter release. Second, the coupling of taste receptors to $G_{\beta\gamma}$ activates phospholipase$_{\beta 2}$, which catalyses an increase in inositol triphosphate (IP_3). IP_3 stimulates the type-III IP_3 receptor, which mediates an increase in intracellular calcium. The increase in intracellular calcium arises from two sources: influx of extracellular calcium and release of calcium from intracellular stores (Narukawa *et al.*, 2006). Increased calcium activates the ion channel TRPM5, which causes sodium influx and depolarisation of the cell (Zhang *et al.*, 2007). Depolarisation of the taste receptor cell causes release of ATP through pannexin 1 hemichannels (Finger *et al.*, 2005; Huang *et al.*, 2007). Pannexin 1 serves as the neurotransmitter for activation of afferent gustatory nerves and presynaptic cells via their P2X and P2Y receptors (Kataoka *et al.*, 2006; Fedorov *et al.*, 2007).

Taste acuity increases with the temperature of the tastant. There are several thermosensitive molecules in the taste transduction pathway (Talavera *et al.*, 2005). TRPV, which is one of the receptors for salt, is temperature responsive. Heat activation of TRPM5, which mediates taste cell depolarisation, explains the enhanced perception of sweet flavors by warm temperatures.

Tissue expression

Taste receptors are expressed in the oral cavity, where they mediate taste, but they are also expressed in strategic areas along the gastrointestinal tract where they mediate important digestive processes (see Sbarbati *et al.*, 2009). In mice and rats, T1R1(umami TR) is expressed at high levels in fungiform and palate taste buds, lower levels in some foliate taste buds, and rarely in the circumvallate taste buds. It is also expressed in the palate (Bachmanov and Beauchamp, 2007; Stone *et al.*, 2007). T1R2 (sweet TR) is highly expressed in circumvallate and foliate taste buds, expressed at low levels in palate taste buds, and is rarely expressed in the fungiform taste buds. It is also expressed in the palate (Bachmanov and Beauchamp, 2007; Stone *et al.*, 2007). When T1R2 is expressed in a taste receptor cell, T1R3 is almost always expressed in that same cell. Similarly, cells that express T1R1 almost always express T1R3. Coexpression of T1R1 and T1R2 in the same taste receptor cell is rare (Kim *et al.*, 2003; Liao and Schultz, 2003; Bachmanov and Beauchamp, 2007; Tomchik *et al.*, 2007). In rats, all three subunits ENaC are expressed in fungiform papillae and vallate papillae.

T1R and T2R genes are rarely co-expressed in the same taste receptor cell. T2Rs are highly expressed in taste receptor cells of the circumvallate, foliate, palate, and epiglottis taste buds (Bufe *et al.*, 2002; Kuhn *et al.*, 2004; Bachmanov and Beauchamp, 2007). Fungiform taste buds have low levels of expression. A subset of the T2Rs is expressed in each bitter-type taste receptor cell, meaning that each receptor cell can survey a very wide, but not complete, spectrum of bitter compounds (Mueller *et al.*, 2005). This explains the phenomenon of poor discrimination between the very wide varieties of molecules that bind to T2Rs. But functional test confirm that there is discrimination between some bitter tastants (Brasser *et al.*, 2005; Geran and Travers, 2006).

Knockouts and transgenics

Mice have been engineered that lack the *mTas1r1* gene, the *mTas1r2* gene, the *mTas1r3* gene, and various combinations of the three genes (Damak *et al.*, 2003; Zhao *et al.*, 2003). Results examining the taste preferences of these mice confirm the general conclusions found previously using in vitro expression systems: *Tas1r1* knockouts are unable to taste most L-amino acids and other umami tastants. *Tas1r2* knockouts are unable to taste many sweet tastants and are compromised in the perception of others. *Tas1r3* knockouts are deficient in both sweet and umami recognition.

Transgenic mice have had their sweet taste 'humanised' by replacing the mouse *Tas1r2* with human *TAS1R2*. The humanised mice are able to taste the artificial sweetener, aspartame, but wild-type mice do not (Zhao *et al.*, 2003).

Several T2R knockout mouse strains have been generated. For example, the *Tas2R105* gene codes for a receptor with high affinity for cycloheximide in *in vitro* expression systems. *Tas2R105* knockout mice do not respond to cycloheximide but have normal responses to other bitter compounds (Mueller *et al.*, 2005).

β-glucopyranoside (BGP) is very bitter to humans, but not mice, and has been shown to bind to human bitter receptor *TAS2R16* (Bufe *et al.*, 2002). Mice are normally indifferent to BGP. Interestingly, when mice are made transgenic for the human *TAS2R16* driven by a promoter that causes expression in bitter-type taste receptor cells, they avoid BGP; but when the *TAS2R16* is driven by a promotor that drives expression in sweet-type taste receptor cells (*Tas1r2*), they show preference for BGP (Mueller *et al.*, 2005). Thus, a bitter receptor will confer a positive taste stimulus when it is expressed in a taste cell that normally detects sweet tastants.

Transgenic medaka fish expressing a dysfunctional $G\alpha_{i2}$, the G-protein that mediates signal transduction in both T1R and T2R receptors in fish, are taste-blind. These transgenic fish do not have a preference for amino acids as do wild type fish, nor do they have an aversion to bitter compounds (Aihara *et al.*, 2008).

Physiological tuning

Taste cells have receptors for leptin, a hormone secreted by adipocytes that regulates body mass. Leptin inhibits the depolarisation of taste cells containing sweet receptors and blunts sweet taste sensations (Ninomiya *et al.*, 2002; Shigemura *et al.*, 2004). Thus leptin may decrease food consumption, in part, by decreasing the pleasure of eating.

It is well known that taste acuity is diminished during bacterial or viral infections. Taste receptor cells express receptors for interferon γ and respond to this inflammatory cytokine with activation of the STAT1 signaling pathway leading to apoptosis (Wang *et al.*, 2007). Simulating infections with lipopolysaccharides or polyinosinic:plycytidylic acid activated STAT1 and markers of apoptosis. Thus, infection appears to dampen taste acuity by pro-inflammatory cytokine-medicated loss of taste receptor cells.

Evolution

The T1R and T2R taste receptor families are structurally similar (seven transmembrane G-protein coupled receptors) and functionally similar (binding aqueous tastants), but have very different evolutionary histories. Both T1R and T2R gene families diverged from the more ancient vomeronasal receptor family sometime around the separation between the lamprey and jawed vertebrates (Grus and Zhang, 2009). Vertebrates usually have three or fewer genes in the T1R gene family (T1R1, T1R2, T1R3) and these share significant sequence similarity with vomeronasal type 2 receptors but not with the other taste gene family, T2R. The number of genes in the T2R family is highly variable across species (Table 2) and genes in this family share significant sequence similarity with vomeronasal type 1 receptors. Pseudogenes, which are genes that have lost their protein coding ability, are extremely common in the T2R family but are rare in the T1R gene family (Nei *et al.*, 2008).

The T1R family is involved in the sensing of nutrients. Phylogenetic analysis of T1R genes shows that the T1R1, T1R2, and T1R3 genes diverged before the separation of fishes and tetrapods (Shi and Zhang, 2006; Hashiguchi *et al.*, 2007; Grus and Zhang, 2009).Although much work remains to be done, known prototypic members of the T1R family (i.e. those in fish) are involved in sensing animo acids and mammalian T1R2 are thought to have acquired the property of sensing sweet compounds later in evolution (Ishimaru, 2009). The very low rate of gene duplication and pseudogenisation in the T1R family relative to T2R or odor receptors is likely because there is little variability in the chemical forms of plant nutrients. All vertebrate species require amino acids and one broadly tuned receptor is sufficient to recognise these important nutrients. Interestingly, carnivorous cats consume a natural diet that is almost devoid of glucose and sucrose, so the evolutionary gain in function of T1R2 for detecting sweet became futile. Thus, it is not surprising that T1R2 became a pseudogene in cats (Li *et al.*, 2006).

The rapid evolution of T2R genes due to gene duplication and death (Kim *et al.*, 2004; Parry *et al.*, 2004) is likely due to the huge diversity of toxins produced by plants (and accumulated by insect herbivores), the variability of these toxins across food types and environments, and the continual change in toxin structures due to evolutionary arms race to prevent herbivory. Species comparisons of pseudogenes indicates a general lack of orthologous functional losses, indicating that pseudogenisation is species specific and likely driven by the differing biologically important taste needs of each species (Go *et al.*, 2005, 2006; Behrens and Meyerhof, 2006). The high number of pseudogenes in humans is due to relaxed selection and indicates a decreasing reliance on bitter taste perception during recent evolutionary time. From the limited information available, it appears that omnivores have the highest number of TR2 genes and numbers are diminished in carnivores and herbivores. The low number of T2R genes in cows may be due to the high detoxification capacity of the rumen, which would diminish the need to identify bitter plants that are potentially poisonous to non-ruminants (Shi and Zhang, 2006). Lower numbers of T2R genes in dogs may be related to their mostly carnivorous diet strategy because vertebrate tissue is usually devoid of toxins so the need to taste bitter would be minimised. The very small number of T2R genes in the chicken apparently is due to a major genome size reduction event that occurred early in the evolution of Aves and, interestingly, this gene family has not expanded in the ensuing 100s of millions of years (Go, 2006). This may be because most avian species are not herbivorous and their ability to fly resulted in a co-evolutionary strategy where plants supply them with low toxin foods in return for spreading seeds (Klasing, 1998). Speculatively, it is possible that chickens have recruited a different receptor family to detect bitter compounds.

Within a species, both T1R and T2R gene families show high rates of polymorphism. Variability in the T1R2 gene is among the highest 10% in the human genome (Kim *et al.*, 2006). In a survey of 100 people, 18 SNPs were found that resulted in 10 different coding amino acid sequences (Drayna, 2005). Most of the changes are amino acid substitutions in the N-terminus extracellular domain, which is responsible for ligand binding. Many of these polymorphisms result in changes in the perception of sweet compounds. The T1R1 gene shows less polymorphism in humans than the T1R2 gene and the T1R3 gene has even less (Kim *et al.*, 2006). This indicates that there should be more individual variation in taste perception of sweet than umami.

Human T2R genes have substantial polymorphisms that are likely responsible for the widespread individual differences in bitter taste perception (Drayna, 2005; Wooding, 2006; Wooding *et al.*, 2006). For example, 17 SNPs have been identified in the T2R44 gene that results in 11 different polymorphisms in amino acid sequences plus two premature stop codons (Drayna, 2005). Human T2R genes that code for TRs recognising cyanogenic glycosides found in many domesticated food crops have undergone evolutionarily recent positive selection (Soranzo *et al.*, 2005). However in recent evolutionary time, most human T2R genes have incurred loss of function mutations indicating that there has been little evolutionary pressure to maintain their genetic integrity (Wang *et al.*, 2004). This is likely due to use of fire in cooking, change in foods to those that have low levels of bitter compounds, and culturally learned avoidance of toxins. This loss in bitter tasting ability is likely occurring in domestic animals as well.

Differences in taste perception may also be due to different levels of expression of a specific taste gene, with individuals expressing high levels of the same gene having higher sensitivity than those expressing low levels. There are also phenotypic differences in the number of taste buds in the oral cavity. In human subjects, there is a strong positive correlation between the number of fungiform papillae and the intensity of sensation from a fixed concentration of sugar (Reed *et al.*, 2006). Also prior experience and conditioning to the tastant can interact with genetics to markedly affect central motivation systems that drive the strength of preferences (Sclafani, 2006).

Applications in animal agriculture

TRs in livestock poultry and fish

T1R1 and T1R3 genes are present in the chicken genome (Lagerstrom *et al.*, 2006; Shi and Zhang, 2006) and are expressed in the tongue (Humphrey and Roura, unpublished observations). Five lines of evidence suggest that T1R2 gene is actually missing and is not an artifact (Shi and Zhang, 2006; Holt, Humphrey, and Klasing, unpublished observations). First, fish T1R2 genes should be more difficult to detect than chicken T1R2 when mammalian genes are used as queries; yet fish genes were easily identifiable. Second, the chicken draft genome sequence has 6.63 coverage and the flanking genes of mouse T1R2 gene have been identified in the syntenic regions in the chicken genome. Thus, the region where the gene should occur was sequenced completely. Third, T1R2 is not represented in the chicken EST databases for mRNA. Forth, degenerative PCR using primers from highly conserved regions of mammalian T1R2 do not produce products with homology to known T1R2s. Finally, many historical studies indicate poor taste perception of sugars in chickens. Chickens have only three T2Rs (Lagerstrom *et al.*, 2006; Shi and Zhang, 2006; Dong *et al.*, 2009). However, historical studies indicate good perception of bitter compounds by chickens so it is possible that chickens have recruited a different receptor family to detect bitter compounds. Vomeronasal receptors are good candidates because they are the progenitor family of T2Rs. Based on representative sequences from human, mouse and rat receptors for salty

(ENaC), sour (MDEG-1), hot (TRPV-1) and cold (TRPM-8), BlastN searches have identified homologous taste receptors in the chicken genome (Humphrey, unpublished observations).

Pigs possess T1R1, T1R2 (Holt, Roura, and Klasing, unpublished observations) and T1R3 (Kiuchi *et al.*, 2006). A search for SNPs in nine pigs from various breeds resulted in eight in exons and five in introns, but none resulted in an amino acid substitutions (Kiuchi *et al.*, 2006). The cow genome contains sequences for T1R2, T1R2 and T1R3 genes and 15 functional T2Rs (Go, 2006; Dong *et al.*, 2009). Furthermore, type II taste cells in bovine circumvallate and fungiform papillae express gustducin protein (Tabata *et al.*, 2003; Tabata *et al.*, 2006). Although TRs have been characterised in Fugu, Puffer fish, Stickleback, Medaka and Zebrafish (Shi and Zhang, 2006; Hashiguchi *et al.*, 2007; Oike *et al.*, 2007; Dong *et al.*, 2009; Doving *et al.*, 2009; Grus and Zhang, 2009; Ishimaru, 2009; Yasuoka and Abe, 2009), they have yet to be characterised in economically important species.

Commercial applications

Research to date clearly indicates that anthropomorphising human experience is perilous for predicting animal taste preferences. Research using humans has been successful at finding artificial sweeteners and umami additives for human foods, but these molecules have not been very useful for animal feed applications because they are poor ligands for non-human TRs. Finding similarly cost effective feed additives for animals will require screening systems that are specific to the target species. Screening cannot be efficiently done in living animals because the time and money required has always been prohibitive. *In vitro* expression systems where reporter cells are transfected with the T1R genes has permitted the identification of preferred taste substances for mice and are now being used to screen combinatorial libraries for novel artificial sweeteners for humans. Many patents have resulted from in vitro screens using human TRs. Now that the many of the TRs have been identified in agriculturally important animals, screening for potent natural and artificial taste molecules can be done in vitro using cell culture instead of in live animals. Promising applications of taste biology to animal nutrition involve finding compounds that enhance sweet, umami, fat and sodium tastes. To be cost effective taste agonists should have exceptionally high potency; several artificial sweeteners are more than 1000 times sweeter than sucrose on a molar basis.

A second promising application to agriculture arising from our new understanding of taste biology involves masking bitter tastes. Many feed ingredients have phenols, flavonoids, isoflavones, terpenes, glucosinolates and other bitter compounds that stimulate T2Rs and cause a decrease in food intake. These bitter compounds are useful to plants because they repel pathogens and herbivores ranging from insects to vertebrates. A classic example is sorghum, which is an excellent feed source for chickens and pigs with the exception of their high tannin content. Low tannin sorghums are accepted well by chickens, however yields are low because of insect and wild bird predation. Conversely, high tannin sorghums provide good yields because of their natural pesticides but must be fed to cattle because chickens and pigs reject them. A very similar scenario exists for lupins, which are naturally high in bitter alkaloids, and sweet lupins, which are devoured by insects in the field. Additionally many byproducts of the human food industry are very high in bitter tasting compounds, which is why they are removed prior to human consumption. It is often the decrease in food intake due to bitter taste and not their inherent toxicity that limits the use of many potential feed ingredients. In vitro screening systems based on cell reporter systems expressing the three chicken T2Rs should prove useful for finding molecules that mask bitter tastes. Alternately, knock out or silencing the three T2Rs should result in a bitter-blind chicken. Proof of concept of a similar approach has been accomplished in Medaka fish (Aihara *et al.*, 2008). Such bitter-blind farm animals could be fed high levels of lupines, tannins and byproducts. In the future it might be possible to engineer the expression of high levels of pest-protective bitter compounds into corn, soy and other plants and feed them to bitter-blind poultry and livestock. This would permit GMO crops expressing natural pesticides to be grown for animal feed.

Bitter tastes are intrinsic to some proteins found in plant foods and animal meals. Exopeptidases and other process techniques can be used to mitigate these flavors (Raksakulthai and Haard, 2003). Screening systems based on TRs might be used to identify enzymes and other treatments that are most efficacious for process removal of bitter taste compounds in feeds.

Many medicines and growth promoters have a negative taste. Most drugs that have a flavour are bitter (Gilman, Goodman, and Gilman, 1980). Masking negative tastes of medicines might allow greater inclusion rates or simpler delivery systems. For example, the antibiotic tilmicosin is very efficacious in pigs but its use is limited by its bitter taste (Hellekant and Danilova, 1999). It usually must be given by gavage because pigs reject it when added to feed or water. Masking bitter flavors would provide added value to many pharmaceuticals. Clearly our new understanding of taste biology portends many developments in animal nutrition and health.

References

Accolla, R., Bathellier, B., Petersen, C.C., and Carleton, A., 2007. Differential spatial representation of taste modalities in the rat gustatory cortex. Journal of Neuroscience 27: 1396-1404.

Adler, E., Hoon, M.A., Mueller, K.L., Chandrashekar, J., Ryba, N.J., and Zuker, C.S., 2000. A novel family of mammalian taste receptors. Cell 100: 693-702.

Aihara, Y., Yasuoka, A., Iwamoto, S., Yoshida, Y., Misaka, T., and Abe, K., 2008. Construction of a taste-blind medaka fish and quantitative assay of its preference-aversion behavior. Genes, Brain and Behavior 7: 924-932.

Andres-Barquin, P.J., and Conte, C., 2004. Molecular Basis of Bitter Taste: The T2R Family of G Protein-Coupled Receptors. Cell Biochemistry and Biophysics 41: 99-112.

Bachmanov, A.A. and Beauchamp, G.K., 2007. Taste receptor genes. Annual Review of Nutrition 27: 389-414.

Behrens, M. and Meyerhof, W., 2006. Bitter taste receptors and human bitter taste perception. Cell and Molecular Life Sciences 63: 1501-1509.

Brand, J.G., 2000. Receptor and transduction processes for umami taste. J Nutr 130:942S-945S.

Brasser, S.M., Mozhui, K. and Smith, D.V., 2005. Differential covariation in taste responsiveness to bitter stimuli in rats. Chemical Senses 30: 793-799.

Bufe, B., Hofmann, T., Krautwurst, D., Raguse, J.D. and Meyerhof, W., 2002. The human TAS2R16 receptor mediates bitter taste in response to beta-glucopyranosides. Nature Genetics 32: 397-401.

Chandrashekar, J., Mueller, K.L., Hoon, M.A., Adler, E., Feng, L., Guo, W., Zuker, C.S. and Ryba, N. J., 2000. T2Rs function as bitter taste receptors. Cell 100: 703-711.

Cui, M., Jiang, P., Maillet, E., Max, M., Margolskee, R.F. and Osman, R., 2006. The heterodimeric sweet taste receptor has multiple potential ligand binding sites. Current Pharmaceutical Design 12:4591-4600.

Damak, S., Rong, M., Yasumatsu, K., Kokrashvili, Z., Varadarajan, V., Zou, S., Jiang, P., Ninomiya, Y. and Margolskee, R.F., 2003. Detection of sweet and umami taste in the absence of taste receptor T1r3. Science 301: 850-853.

Danilova, V., Roberts, T. and Hellekant, G., 1999. Responses of single taste fibers and whole chorda tympani and glossopharyngeal nerve in the domestic pig, Sus scrofa. Chemical Senses 24: 301-316.

DeFazio, R.A., Dvoryanchikov, G., Maruyama, Y., Kim, J.W., Pereira, E., Roper, S.D. and Chaudhari, N., 2006. Separate populations of receptor cells and presynaptic cells in mouse taste buds. Journal of Neuroscience 26: 3971-3980.

DeSimone, J.A. and Lyall, V., 2006. Taste receptors in the gastrointestinal tract III. Salty and sour taste: sensing of sodium and protons by the tongue. American Journal of Physiology - Gastrointestinal and Liver Physiology 291: G1005-1010.

Dong, D., Jones, G. and Zhang, S., 2009. Dynamic evolution of bitter taste receptor genes in vertebrates. BMC Evolutionary Biology 9: 12.

Doving, K.B., Sandvig, K. and Kasumyan, A., 2009. Ligand-specific induction of endocytosis in taste receptor cells. Journal of Experimental Biology 212: 42-49.

Drayna, D., 2005. Human taste genetics. Annual Review of Genomics and Human Genetics 6: 217-235.

Drewnowski, A., and Gomez-Carneros, C., 2000. Bitter taste, phytonutrients, and the consumer: A review. American Journal of Clinical Nutrition 72: 1424-1435.

Esposito, V., Gallucci, R., Picone, D., Saviano, G., Tancredi, T. and Temussi, P.A., 2006. The importance of electrostatic potential in the interaction of sweet proteins with the sweet taste receptor. Journal of Molecular Biology 360: 448-456.

Fedorov, I.V., Rogachevskaja, O.A. and Kolesnikov, S.S., 2007. Modeling P2Y receptor-Ca(2+) response coupling in taste cells. Biochimica et Biophysica Acta 1768: 1727-1740.

Finger, T.E., Danilova, V., Barrows, J., Bartel, D.L., Vigers, A.J., Stone, L., Hellekant, G. and Kinnamon, S.C., 2005. ATP signaling is crucial for communication from taste buds to gustatory nerves. Science 310: 1495-1499.

Floriano, W.B., Hall, S., Vaidehi, N., Kim, U., Drayna, D., and Goddard, W.A., 3rd., 2006. Modeling the human PTC bitter-taste receptor interactions with bitter tastants. Journal of Molecular Modeling 12: 931-941.

Gaillard, D., Laugerette, F., Darcel, N., El-Yassimi, A., Passilly-Degrace, P., Hichami, A., Khan, N.A., Montmayeur, J.P. and Besnard, P., 2008a. The gustatory pathway is involved in CD36-mediated orosensory perception of long-chain fatty acids in the mouse. FASEB Journal 22: 1458-1468.

Gaillard, D., Passilly-Degrace, P. and Besnard, P., 2008b. Molecular mechanisms of fat preference and overeating. Annals of the New York Acadamy of Science 1141: 163-175.

Ganchrow, J.R., 2000. Taste cell function: Structural and biochemical implications. Physiology & Behavior 69: 29-40.

Geran, L.C. and Travers, S.P., 2006. Single neurons in the nucleus of the solitary tract respond selectively to bitter taste stimuli. Journal of Neurophysiology 96: 2513-2527.

Gilbertson, T.A., Boughter, J.D., Jr., Zhang, H. and Smith, D.V., 2001. Distribution of gustatory sensitivities in rat taste cells: whole-cell responses to apical chemical stimulation. Journal of Neuroscience 21: 4931-4941.

Gilbertson, T.A., Damak, S., and Margolskee, R. F., 2000. The molecular physiology of taste transduction. Current Opinion in Neurobiology 10: 519-527.

Gilman, A.G., Goodman, L.S. and Gilman, A., 1980. The phamacological basis of therapeutics. Macmillan Press, New York.

Go, Y., 2006. Lineage-specific expansions and contractions of the bitter taste receptor gene repertoire in vertebrates. Molecular Biology and Evolution 23: 964-972.

Go, Y., Satta, Y., Takenaka, O. and Takahata, N., 2005. Lineage-specific loss of function of bitter taste receptor genes in humans and nonhuman primates. Genetics 170: 313-326.

Grus, W.E. and Zhang, J., 2009. Origin of the genetic components of the vomeronasal system in the common ancestor of all extant vertebrates. Molecular Biology and Evolution 26: 407-419.

Hashiguchi, Y., Furuta, Y., Kawahara, R. and Nishida, M., 2007. Diversification and adaptive evolution of putative sweet taste receptors in threespine stickleback. Gene 396: 170-179.

Hellekant, G. and Danilova, V., 1999. Taste in domestic pig, Sus scrofa. Journal of Animal Physiology and Animal Nutrition 82: 8-24.

Hettinger, T.P., Formaker, B.K. and Frank, M.E., 2007. Cycloheximide: no ordinary bitter stimulus. Behavioural Brain Research 180: 4-17.

Huang, A.L., Chen, X., Hoon, M.A., Chandrashekar, J., Guo, W., Trankner, D., Ryba, N.J. and Zuker, C.S., 2006. The cells and logic for mammalian sour taste detection. Nature 442: 934-938.

Huang, Y.A., Maruyama, Y., Stimac, R. and Roper, S.D., 2008. Presynaptic (Type III) cells in mouse taste buds sense sour (acid) taste. Journal of Physiology 586: 2903-2912.

Huang, Y.J., Maruyama, Y., Dvoryanchikov, G., Pereira, E., Chaudhari, N.,and Roper, S.D., 2007. The role of pannexin 1 hemichannels in ATP release and cell-cell communication in mouse taste buds. Proceedings of the National Academy of Sciences of the USA 104: 6436-6441.

Ishimaru, Y., 2009. Molecular mechanisms of taste transduction in vertebrates. Odontology 97: 1-7.

Ishimaru, Y., Inada, H., Kubota, M., Zhuang, H., Tominaga, M. and Matsunami, H., 2006. Transient receptor potential family members PKD1L3 and PKD2L1 form a candidate sour taste receptor. Proceedings of the National Academy of Sciences of the USA 103: 12569-12574.

Kataoka, S., Toyono, T., Seta, Y. and Toyoshima, K., 2006. Expression of ATP-gated P2X3 receptors in rat gustatory papillae and taste buds. Archives of Histology and Cytology 69: 281-288.

Kaya, N., Shen, T., Lu, S.G., Zhao, F.L. and Herness, S., 2004. A paracrine signaling role for serotonin in rat taste buds: expression and localization of serotonin receptor subtypes. American Journal of Physiology - Regulatory, Integrative and Comparative Physiology 286: R649-658.

Keast, R.S., Canty, T.M. and Breslin, P.A., 2004. Oral zinc sulfate solutions inhibit sweet taste perception. Chemical Senses 29: 513-521.

Kim, M.R., Kusakabe, Y., Miura, H., Shindo, Y., Ninomiya, Y. and Hino, A., 2003. Regional expression patterns of taste receptors and gustducin in the mouse tongue. Biochemical and Biophysical Research Communications 312: 500-506.

Kim, U.K., Breslin, P.A., Reed, D. and Drayna, D., 2004. Genetics of human taste perception. Journal of Dental Research 83: 448-453.

Kim, U.K., Wooding, S., Riaz, N., Jorde, L.B. and Drayna, D., 2006. Variation in the human TAS1R taste receptor genes. Chemical Senses 31: 599-611.

Kiuchi, S., Yamada, T., Kiyokawa, N., Saito, T., Fujimoto, J. and Yasue, H., 2006. Genomic structure of swine taste receptor family 1 member 3, TAS1R3, and its expression in tissues. Cytogenetic and Genome Research 115: 51-61.

Klasing, K.C., 1998. Comparative avian nutrition. CAB International, Wallingford, UK.

Kuhn, C., Bufe, B., Winnig, M., Hofmann, T., Frank, O., Behrens, M., Lewtschenko, T., Slack, J.P., Ward, C.D. and Meyerhof, W., 2004. Bitter taste receptors for saccharin and acesulfame K. Journal of Neuroscience 24: 10260-10265.

Lagerstrom, M.C., Hellstrom, A.R., Gloriam, D.E., Larsson, T.P., Schioth, H.B. and Fredriksson, R., 2006. The G protein-coupled receptor subset of the chicken genome. PLoS Computational Biology 2: e54.

Laugerette, F., Gaillard, D., Passilly-Degrace, P., Niot, I. and Besnard, P., 2007. Do we taste fat? Biochimie 89: 265-269.

Laugerette, F., Passilly-Degrace, P., Patris, B., Niot, I., Febbraio, M., Montmayeur, J.P. and Besnard, P., 2005. CD36 involvement in orosensory detection of dietary lipids, spontaneous fat preference, and digestive secretions. Journal of Clinical Investigation 115: 3177-3184.

Li, X., Li, W., Wang, H., Bayley, D.L., Cao, J., Reed, D.R., Bachmanov, A.A., Huang, L., Legrand-Defretin, V., Beauchamp, G.K. and Brand, J.G., 2006. Cats lack a sweet taste receptor. Journal of Nutrition 136: 1932S-1934S.

Li, X., Staszewski, L., Xu, H., Durick, K., Zoller, M. and Adler, E., 2002. Human receptors for sweet and umami taste. Proceedings of the National Academy of Sciences of the USA 99: 4692-4696.

Liao, J. and Schultz, P.G., 2003. Three sweet receptor genes are clustered in human chromosome 1. Mammalian Genome 14: 291-301.

Lin, W., Finger, T.E., Rossier, B.C. and Kinnamon, S.C., 1999. Epithelial Na+ channel subunits in rat taste cells: Localization and regulation by aldosterone. Journal of Comparative Neurology 405: 406-420.

Lindemann, B., 2001. Receptors and transduction in taste. Nature 413: 219-225.

Lyall, V., Heck, G.L., Vinnikova, A.K., Ghosh, S., Phan, T.H., Alam, R.I., Russell, O.F., Malik, S.A., Bigbee, J.W. and DeSimone, J.A., 2004. The mammalian amiloride-insensitive non-specific salt taste receptor is a vanilloid receptor-1 variant. Journal of Physiology 558: 147-159.

Margolskee, R.F., 2002. Molecular mechanisms of bitter and sweet taste transduction. Journal of Biological Chemistry 277: 1-4.

Miguet, L., Zhang, Z. and Grigorov, M.G., 2006. Computational studies of ligand-receptor interactions in bitter taste receptors. Journal of Receptor and Signal Transduction Research 26: 611-630.

Miura, H., Nakayama, A., Shindo, Y., Kusakabe, Y., Tomonari, H. and Harada, S., 2007. Expression of gustducin overlaps with that of type III IP3 receptor in taste buds of the rat soft palate. Chemical Senses 32: 689-696..

Mizushige, T., Inoue, K. and Fushiki, T., 2007. Why is fat so tasty? Chemical reception of fatty acid on the tongue. Journal of Nutritional Science & Vitaminology 53: 1-4.

Mueller, K.L., Hoon, M.A., Erlenbach, I., Chandrashekar, J., Zuker, C.S. and Ryba, N.J., 2005. The receptors and coding logic for bitter taste. Nature 434: 225-229.

Narukawa, M., Mori, T. and Hayashi, Y., 2006. Umami changes intracellular Ca2+ levels using intracellular and extracellular sources in mouse taste receptor cells. Bioscience, Biotechnology & Biochemistry 70: 2613-2619.

Nei, M., Niimura, Y. and Nozawa, M., 2008. The evolution of animal chemosensory receptor gene repertoires: roles of chance and necessity. Nature Review Genetics 9: 951-963.

Nelson, G., Chandrashekar, J., Hoon, M.A., Feng, L., Zhao, G., Ryba, N.J. and Zuker, C.S., 2002. An amino-acid taste receptor. Nature 416: 199-202.

Nelson, G., Hoon, M.A., Chandrashekar, J., Zhang, Y., Ryba, N.J. and Zuker, C.S., 2001. Mammalian sweet taste receptors. Cell 106: 381-390.

Nie, Y., Hobbs, J.R., Vigues, S., Olson, W.J., Conn, G.L. and Munger, S.D., 2006. Expression and purification of functional ligand-binding domains of T1R3 taste receptors. Chemical Senses 31: 505-513.

Ninomiya, Y., Shigemura, N., Yasumatsu, K., Ohta, R., Sugimoto, K., Nakashima, K. and Lindemann, B., 2002. Leptin and sweet taste. Vitamins & Hormones 64: 221-248.

Oike, H., Nagai, T., Furuyama, A., Okada, S., Aihara, Y., Ishimaru, Y., Marui, T., Matsumoto, I., Misaka, T. and Abe, K., 2007. Characterization of ligands for fish taste receptors. Journal of Neuroscience 27: 5584-5592.

Parry, C.M., Erkner, A. and Le Coutre, J., 2004. Divergence of T2R chemosensory receptor families in humans, bonobos, and chimpanzees. Proceedings of the National Academy of Sciences of the USA 101: 14830-14834.

Pronin, A. N., Tang, H., Connor, J., and Keung, W., 2004. Identification of ligands for two human bitter T2R receptors. Chemical Senses 29: 583-593.

Raksakulthai, R. and Haard, N.F., 2003. Exopeptidases and their application to reduce bitterness in food: a review. Critical Reviews in Food Science and Nutrition 43: 401-445.

Reed, D.R., Tanaka, T. and McDaniel, A.H., 2006. Diverse tastes: Genetics of sweet and bitter perception. Physiology and Behavior 88: 215-226.

Roper, S.D., 2007. Signal transduction and information processing in mammalian taste buds. Pflugers Archiv 454: 759-776.

Ruiz, C., Gutknecht, S., Delay, E. and Kinnamon, S., 2006. Detection of NaCl and KCl in TRPV1 knockout mice. Chemical Senses 31: 813-820.

Sacerdote, C., Guarrera, S., Smith, G.D., Grioni, S., Krogh, V., Masala, G., Mattiello, A., Palli, D., Panico, S., Tumino, R., Veglia, F., Matullo, G. and Vineis, P., 2007. Lactase persistence and bitter taste response: instrumental variables and mendelian randomization in epidemiologic studies of dietary factors and cancer risk. American Journal of Epidemiology 166: 576-581.

Sainz, E., Cavenagh, M.M., Gutierrez, J., Battey, J.F., Northup, J.K. and Sullivan, S.L., 2007. Functional characterization of human bitter taste receptors. Biochemistry Journal 403: 537-543.

Sandell, M.A. and Breslin, P.A., 2006. Variability in a taste-receptor gene determines whether we taste toxins in food. Current Biology 16: R792-794.

Sbarbati, A., Benati, D. and Merigo, F., 2009. The diffuse chemosensory system. In: Torrallardona, D. and Roura, E. (eds.) Voluntary feed intake in pigs. Wageningen Academic Publishers, the Netherlands, pp. 141-153.

Sclafani, A., 2006. Oral, post-oral and genetic interactions in sweet appetite. Physiology and Behavior 89: 525-530.

Shi, P. and Zhang, J., 2006. Contrasting modes of evolution between vertebrate sweet/umami receptor genes and bitter receptor genes. Molecular Biology and Evolution 23: 292-300.

Shigemura, N., Ohta, R., Kusakabe, Y., Miura, H., Hino, A., Koyano, K., Nakashima, K. and Ninomiya, Y., 2004. Leptin modulates behavioral responses to sweet substances by influencing peripheral taste structures. Endocrinology 145: 839-847.

Simon, S.A., De Araujo, I.E., Gutierrez, R. and Nicolelis, M.A., 2006. The neural mechanisms of gustation: a distributed processing code. Nature Review Neuroscience 7: 890-901.

Soranzo, N., Bufe, B., Sabeti, P.C., Wilson, J.F., Weale, M.E., Marguerie, R., Meyerhof, W. and Goldstein, D.B., 2005. Positive selection on a high-sensitivity allele of the human bitter-taste receptor TAS2R16. Current Biology 15: 1257-1265.

Spector, A.C., 2000. Linking gustatory neurobiology to behavior in vertebrates. Neuroscience & Biobehavioral Reviews 24: 391-416.

Spector, A.C. and Kopka, S.L., 2002. Rats fail to discriminate quinine from denatonium: implications for the neural coding of bitter-tasting compounds. Journal of Neuroscience 22: 1937-1941.

Stone, L.M., Barrows, J., Finger, T.E. and Kinnamon, S.C., 2007. Expression of T1Rs and gustducin in palatal taste buds of mice. Chemical Senses 32: 255-262.

Sugita, M., 2006. Taste perception and coding in the periphery. Cell and Molecular Life Sciences 63: 2000-2015.

Tabata, S., Kudo, K., Wada-Takemura, A., Nishimura, S. and Iwamoto, H., 2006. Structure of bovine fungiform taste buds and their immunoreactivity for gustducin. Journal of Veterinary Medical Science 68: 953-957.

Tabata, S., Wada, A., Kobayashi, T., Nishimura, S., Muguruma, M. and Iwamoto, H., 2003. Bovine circumvallate taste buds: taste cell structure and immunoreactivity to alpha-gustducin. The Anatomical Record. Part A, Discoveries in Molecular, Cellular, and Evolutionary Biology 271: 217-224.

Talavera, K., Yasumatsu, K., Voets, T., Droogmans, G., Shigemura, N., Ninomiya, Y., Margolskee, R.F. and Nilius, B., 2005. Heat activation of TRPM5 underlies thermal sensitivity of sweet taste. Nature 438: 1022-1025.

Tomchik, S.M., Berg, S., Kim, J.W., Chaudhari, N. and Roper, S.D., 2007. Breadth of tuning and taste coding in mammalian taste buds. Journal of Neuroscience 27: 10840-10848.

Treesukosol, Y., Lyall, V., Heck, G.L., DeSimone, J.A. and Spector, A.C., 2007. A psychophysical and electrophysiological analysis of salt taste in Trpv1 null mice. American Journal of Physiology - Regulatory, Integrative and Comparative Physiology 292: R1799-1809.

Trubey, K.R., Culpepper, S., Maruyama, Y., Kinnamon, S.C. and Chaudhari, N., 2006. Tastants evoke cAMP signal in taste buds that is independent of calcium signaling. American Journal of Physiology - Cell Physiology 291: C237-244.

Uematsu, H., 1999. Oral cavity and feeding. Journal of Medical and Dental Sciences 46: 105-110.

Walters, D.E. and Hellekant, G., 2006. Interactions of the sweet protein brazzein with the sweet taste receptor. Journal of Agriculture and Food Chemistry 54: 10129-10133.

Wang, H., Zhou, M., Brand, J. and Huang, L., 2007. Inflammation activates the interferon signaling pathways in taste bud cells. Journal of Neuroscience 27: 10703-10713.

Wang, X., Thomas, S.D. and Zhang, J., 2004. Relaxation of selective constraint and loss of function in the evolution of human bitter taste receptor genes. Human Molecular Genetics 13: 2671-2678.

Wooding, S., 2006. Phenylthiocarbamide: a 75-year adventure in genetics and natural selection. Genetics 172: 2015-2023.

Wooding, S., Bufe, B., Grassi, C., Howard, M. T., Stone, A.C., Vazquez, M., Dunn, D.M., Meyerhof, W., Weiss, R.B. and Bamshad, M.J., 2006. Independent evolution of bitter-taste sensitivity in humans and chimpanzees. Nature 440: 930-934.

Xu, H., Staszewski, L., Tang, H., Adler, E., Zoller, M. and Li, X., 2004. Different functional roles of T1R subunits in the heteromeric taste receptors. Proceedings of the National Academy of Sciences of the USA 101: 14258-14263.

Yamaguchi, S. and Ninomiya, K., 2000. Umami and food palatability. Journal of Nutrition 130: 921S-926S.

Yasuoka, A. and Abe, K., 2009. Gustation in fish: search for prototype of taste perception. Results and Problems in Cell Differentiation, in press.

Zhang, F., Klebansky, B., Fine, R.M., Xu, H., Pronin, A., Liu, H., Tachdjian, C. and Li, X., 2008. Molecular mechanism for the umami taste synergism. Proceedings of the National Academy of Sciences of the USA 105: 20930-20934.

Zhang, Z., Zhao, Z., Margolskee, R. and Liman, E., 2007. The transduction channel TRPM5 is gated by intracellular calcium in taste cells. Journal of Neuroscience 27: 5777-5786.

Zhao, G.Q., Zhang, Y., Hoon, M.A., Chandrashekar, J., Erlenbach, I., Ryba, N.J. and Zuker, C.S., 2003. The receptors for mammalian sweet and umami taste. Cell 115: 255-266.

Chapter 5. Feed appetence in pigs: an oronasal sensing perspective

E. Roura and G. Tedó
R+D Lucta SA, Ctra. Masnou a Granollers 12,4 km, 08170 Montornés del Vallès (Barcelona), Spain; eugeni.roura@lucta.es

Abstract

Peripheral chemosensing functions as a decoding system that interprets among other stimuli those relevant to the nutritional value of feedstuffs. Food ingestion simultaneously evokes odor, taste and thermo-mechanical (somatosensing) sensations that stimulates appetite for appropriate ingredients that ensure self-nourishment. This chapter reviews our current understanding of these three main oronasal sensing systems in pigs with emphasis in the physiological principles. Recent discoveries in the porcine smell, taste and somatosensing biology drives a comparative study with other animal species particularly human and farm animals. Deeper discussions relevant to our current knowledge on the swine taste receptor biology and structural functionalities are outlined including a description of the two main taste receptor families (T1R – for sweet and umami taste, T2R – for bitter taste) and other tastes sensed by pigs (salty, acid and new taste candidates). Finally the links between the sensing biology and the *in vivo* ingredient preference data in pigs are discussed.

Keywords: comparative chemosensing, smell, taste, somatic, receptors, pigs

Introduction

The sensing of the nutritional value of available food sources is key to successful self-nourishment mechanisms. These sensing mechanisms in mammals and birds are part of the *peripheral senses* and have a direct and short term impact on meal initiation and meal size (Hyde and Witherly, 1993; Rauw *et al.*, 2006). Peripheral senses particularly the smell, the taste and the oral somatosensing present in the oral and nasal cavities (from now on referred to as oronasal) interpret those stimuli relevant to feed location and nutritional value (Dulac, 2000; Forbes, 1998; Kuenzel, 1989; Roura *et al.*, 2008a). Furthermore, oronasal food related sensing plays a key role in signaling food arrival to the rest of the digestive system and as a consequence digestive secretions and gut motility are stimulated. These responses have been described as part of the 'cephalic phase' of feed intake (Katschinski, 2000; Laugerette *et al.*, 2005; Mattes, 2000; Ohara *et al.*, 1988; Pavlov, 1902; Teff, 2000). The objective of this chapter is to summarise our current understanding of oronasal sensing and the links that lead to feed appetence and voluntary feed intake in pigs.

Peripheral sensing in pigs

The peripheral nervous system (PNS) connects the central nervous system – CNS - (brain and spinal cord) with the periphery. The PNS consists of two types of neurons: afferent (carrying information from organ, skin or muscles into the CNS) and efferent (carrying information away from the CNS). Information from the sense organs is specifically collected through afferent neurons named after some of the XII cranial nerves (Table 1).

Table 1. Cranial nerves of the peripheral nervous system and functions.

Cranial nerve		Description of main functions: Sensory (S) or/and Motor (M)	
I[1]	Olfactory	S/M	Sense of *smell*
II	Optic	S	Vision
III	Oculomotor	M	Eye, pupil and eyelid movements
IV	Trochlear	M	Eye movements
V[1]	Trigeminal	S/M	*Somatic* sensations from face, mouth, cornea and muscles of mastication (chewing)
VI	Abducens	M	Abducens Nerve: eye movements
VII[1]	Facial	S/M	Facial expression, *taste* from anterior tongue (*Chorda Tympani* fibers), lachrymal (tears) and salivary glands.
VIII	Vestibulocochlear	S	Auditory functions (hearing) and sense of balance
IX[1]	Glossopharyngeal	S/M	Sensation from pharynx, *taste* from posterior tongue, carotid baroceptors
X[1]	Vagus	S/M	Autonomic functions of gut, *taste* and *somatic* sensation from pharynx, epiglottis, muscles of vocal cords, swallowing
XI	Spinal Accessory	M	Shoulder and neck muscles
XII	Hypoglossal	M	Movements of tongue

[1] Cranial nerves I, V, VII, IX and X are involved in the food-born 'flavour' sensing.

The entrance of nutrients to the digestive system follows a non-continuous pattern (i.e. meals) through the oral cavity that allows food discrimination before swallowing occurs. In the oronasal cavity food is scrutinised according to several chemical and physical parameters that will determine the nutritional value and safety of food (Forbes, 1998; Hyde and Witherly, 1993; Kuenzel, 1989). Food borne oronasal stimulation in mammals and birds has been generically referred to as food 'flavour' sensing or palatability and is mediated by the peripheral senses known as somatosensing (based on thermal or mechanical sensations), smell (detection of odorants and pheromones) and taste (Dulac, 2000; Rouquier and Giorgi, 2007; Roura *et al.*, 2008a; Shepherd, 2006). Thus, the 'flavour system' (Shepherd, 2006) consists of the PNS cranial nerves V (Trigeminial) and X (Vagus) for somatic sensing, cranial nerve I (Olfactory) for the sense of smell and cranial nerves VII (*Glossopharyngeal* – GP), IX (facial through the *Chorda tympani* – CT- fibers) and X (Vagus) for taste (Table 1).

In mammals, early in the embryonic development the skin evolves as the first sensory organ capable of detecting mechanic stimulus including pain and temperature. The other senses will emerge progressively as the embryo matures. The somatosensing properties of the skin are the baseline of the whole peripheral sensing system in embryos but also after birth. Grosmaitre *et al.* (2007) reported that all oronasal neurons responding to smell or taste are also responsive to mechanical stimuli. The development of the non-somatic peripheral sensing does not necessarily end with the mammalian embryonic development. In fact, hearing, taste and sight in mammals will experience their full development postnatal (Marlier *et al.*, 2007; Matsson *et al.*, 1978; Oakley,

1998). For example, newborn piglets did not respond to auditive (maternal vocalisations) stimuli during the first 24 hours and preferred darkness to light (Rohde Parfet and Gonyou, 1991).

In turn, similar to humans, the smell emerges as the main sense after birth in pigs. Nevertheless, the development and relevance of the other senses differs between the two species as individuals mature. For example, in primates the sight will quickly develop early in the life and become the main sense. The importance of the sight in human might be linked to the development of trichromatic vision. While most mammals including pigs, possess dichromatic vision (able to discern only between short and long wavelengths) primates are unique as they posses trichromatic vision (able to discern short, medium and long wavelengths) (Dulai and von Dornum, 1999). The relevance of the sense of smell in primates but not in pigs seems to decrease as individuals mature. Furthermore, Gilad *et al.* (2004) concluded that the deterioration of the olfactory gene repertoire in primates occurred concomitant with the acquisition of the trichromatic color vision. The development of sight acuity in pigs is also age dependent and follows many maturational similarities with humans (Matsson *et al.*, 1978). Nevertheless, in pigs visual development does not seem to be deterring the importance of smell. Pigs have a well developed olfactory system with extremely high sensitivity compared to primates and other animal species according to published evidences already at birth (Guiraudie-Capraz *et al.*, 2005; Morrow-Tesch and McGlone, 1990; Rohde Parfet and Gonyou, 1991; Roura *et al.*, 2008a). In contrast, taste in mammals may undergo further development after birth (Oakley, 1998). Therefore, pre and peri natal taste exposures may not determine preferences later in life to the same extend than smell exposures do.

The sensorial process of food-evoked stimuli or food palatability is in a continuous age-dependent evolution (Marlier *et al.*, 2007). Basic or innate taste, smell and somatic cues are maintained through life but both the sensitivity of each of the senses, the reward perception they produce and the interactions between senses will change with age.

Somatosensing

Oral-born somatic sensing has been defined as the non-taste and non-odorous sensations evoked by the presence of food in the mouth. Such sensing follows thermal and mechanical but also some chemical stimulations (Hyde and Witherly, 1993; Miller and Teates, 1984; Roura *et al.*, 2008a). Szczesniak (2002) described them according to the following categories: (1) viscosity (thin, thick and viscous); (2) feel on soft tissue surfaces (smooth, pulpy and creamy); (3) carbonation-related terms (bubbly, tingly and foamy); (4) body-related terms (heavy, watery and light); (5) chemical effect (astringent, burning, sharp); (6) coating of oral cavity (mouth coating, clinging, fatty, oil); (7) resistance to tongue movement (slimy, syrupy, pasty and sticky); (8) afterfeel mouth (clean, drying and lingering); (9) afterfeel physiological (refreshing, warming, thirst-quenching and filling); (10) temperature related (cold or hot); (11) wetness related (wet or dry).

Biological principles of somatic sensing

In mammals the somatic sensing of the oronasal cavity is linked to the cranial nerve V (Figure 1), the Trigeminal (Jones and Roper, 1997; Miller and Teates, 1984; Witt *et al.*, 2003). Sensory neurons of the trigeminal nerve are part of the pain pathway and are involved in the detection

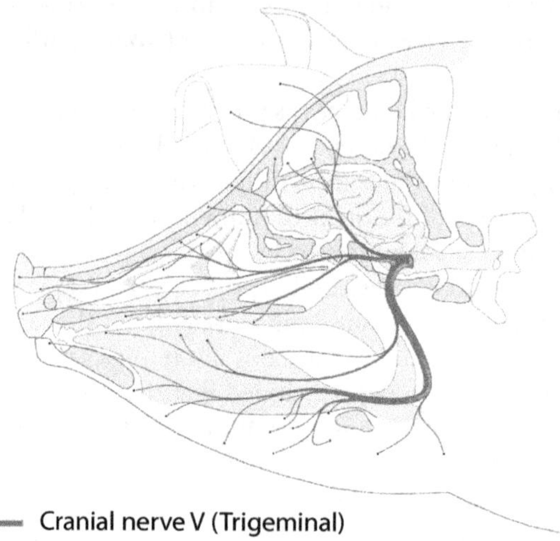

Figure 1. Somatic sensing of the oronasal cavity is driven mainly through two of the branches of the bilateral cranial nerve V (Trigeminal). (drawing by Joaquim Roura).

of noxious stimuli such as mechanical (e.g. viscosity, sticky, hardness), thermal (low or high temperatures) and pungent (e.g. some volatile acids and spices) (Hyde and Witherly, 1993; Miller and Teates, 1984; Roura *et al.*, 2008a; Tominaga and Julius, 2000). Transmembrane ion channel receptors of the transient receptor potential (TRP) family seem to be responsible for the detection of high (TRPVs) and low (TRPMs) temperatures and pungency (TRPAs) (Caterina, 2007; Dhaka *et al.*, 2006; Julius and Basbaum, 2001). Noxious hot temperatures (higher than 43 °C) activate the receptor identified as TRPV1 that also responds to chemical entities such as capsaicin (and other pungent vanilloid compounds) that, in turn, will be sensed as hot material (Caterina *et al.*, 1997; Jordt *et al.*, 2003). Noxious cold (lower than 4 °C) activates the receptor TRPM8 (Dhaka *et al.*, 2006) but so do menthol and derivatives perceived as cooling agents (Bautista *et al.*, 2007; Jordt *et al.*, 2003).

Somatosensing and feed appetence in pigs

Physical characteristics of feed and feed ingredients have an impact on feed palatability through the stimulation of the somatosensing system. According to Szczesniak (2002) there are at least 6 categories of mechanical stimuli that may influence palatability. Solà-Oriol *et al.* (2007) studied the correlation between the preference of pigs for diets with 60% of rice, barley, sorghum or oats (47.1, 22.8, 11.2, 3.0% of preference, respectively) and their physical properties. Rheological characteristics such as feed viscosity and water retention capacity showed no significant correlation with the preference values. On the other hand, feed hardness and chewing effort were negatively correlated with preference so that the higher the hardness and effort the lower the preference. Particle size was also studied and found to be related to preferences. Both the biggest ($P<0.10$) and numerically ($P>0.10$) the smallest particle sizes were negatively correlated with preferences while intermediate values showed a positive correlation ($P<0.05$).

Noxious high and low temperatures as well as pungent compounds may damage the mucosa of the gastrointestinal tract. The trigeminal stimulation consequently leads to an alarm response characterised by feed avoidance, a strong stimulation of digestive secretions and an increase in the intestinal motility aimed at protecting the digestive epithelium (Jamroz *et al.*, 2006; Platel and Srinivasan, 2004; Westendarp, 2005).

There has been a lot of research on the use of herbs and spices (often referred in the market also as botanicals, phytogenic or essential oils) to improve health and growth in pigs particularly after the banning of feed antibiotic growth promoters implemented in Europe (Langhout, 2000; McIntosh *et al.*, 2003; Stammati *et al.*, 1999; Westendarp, 2005; Windisch *et al.*, 2008). In contrast, many herbs and spices are popular food condiments. Essential oils and oleoresins of garlic and capsicum as well as cinnamic aldehyde, carvacrol and pipperine (from black pepper), among others, have for long been used due to their food flavor enhancement properties. Together with some characteristic smell cues, their action involves the activation of the oral somatosensing system (Platel and Srinivasan, 2004; Srinivasan, 2007; Tominaga and Julius, 2000). Moderate seasoning stimulates digestive secretions (salivary, gastric, bile, pancreatic and intestinal mucosa secretions – Platel and Srinivasan, 2001 and, 2004; Westendarp, 2005) while heavily spiced food such as with capsaicin might be perceived as noxious stimuli and trigger a pain-mediated alarm response (Tominaga and Julius, 2000) causing food rejection (Tewksbury and Nabhan, 2001).

In pigs, there are no clear evidences that herbs and spices improve palatability and feed intake according to two recent reviews on the topic (Wenk, 2006; Windisch *et al.*, 2008). Similarly, essential oils from fennel and caraway seed resulted in decrease feed intake in weaned piglets (Schöne *et al.*, 2006). Nevertheless, some researchers have found that many herbs and botanicals are able to improve growth rate through increased feed intake (Wenk, 2006). Herbs and spices is a very heterogeneous group of ingredients and that may contribute to inconsistency of results and conclusions. Particular interest from our group has been focused on the effect of specific spices on the performance of piglets. Hypothetically the pungency of some spices will trigger a somatic

Table 2. Effect of several spices and herbs on feed intake in piglets from 0 to 28 days post-weaning (from Bikker et al.*, 2003).*

	Feed intake (g/day)					
Product	control	Spice 1[a] 500 ppm	Spice 2[a] 40 ppm	Spice 3[a] 30 ppm	Spice 4[a] 410 ppm	LSD[b] (α<0.05)
week 1-2	332[d]	300[c]	311[cd]	300[cd]	318[cd]	33
week 3-4	832[d]	799[cd]	795[cd]	767[c]	821[cd]	61
week 1-4	574[d]	541[cd]	545[cd]	526[c]	561[cd]	42

[a] Spice 1: *Cinnamaldehyde*; Spice 2: *Carvacrol*; Spice 3: *Salvia dalmacia*; Spice 4: *Cinnamaldehyde* + *Capsicum*

[b] LSD: least significant differences.

[c,d] Means with different superidices within the same file are significantly (*P*<0.05) different.

response. We have shown that feed supplemented with moderate levels of cinnamaldehyde, carvacrol or capsicum oleoresin resulted in decreased feed intake (Bikker *et al.*, 2003). Daily feed intake of piglets fed the supplemented diets was between 2 and 10% lower than the control group (Table 2). In a later study by the same authors, feed intake depression after addition of several capsicum based additives was prevented with the use of a cherry-honey flavour (Roura *et al.*, 2004) proving that palatability is an important issue when using this type of spices in pig diets.

Low molecular weight acids such as formic, acetic and propionic among others, are highly volatile and known to be pungent as well. Therefore, they may stimulate the somatic sensing system through TRPA receptors (Caterina, 2007). The effect of organic acids on feed intake is controversial (Lallès *et al.*, 2009). Nevertheless in pigs, formic acid and combinations of formic and propionic acids have consistently been reported to decrease feed intake (Eisemann and van Heugten, 2007; Ettle *et al.*, 2004; Partanen *et al.*, 2007). In a choice feeding piglets refuse diets acidified with 1.2% formic acid (14% and 86% preference values for formic acid supplemented diet and control respectively) presumably because of palatability (Ettle *et al.*, 2004). In a performance trial, Eisemann and van Heugten (2007) showed significant ($P<0.05$) decreases of feed intake after 1.2% inclusion of formic acid during the nursery phase. Partanen *et al.* (2007) tested formic acid 'as is' (F), formic acid absorbed (FAb) in a diatomaceous earth carrier and two combination of formic acid and propionic acids (with potassium sorbate – FPS - and with sodium benzoate – FPB). Up to 60 days of age of the piglets, F and FPB decreased total feed consumed in 8%, FPS by 2% while pigs supplemented with the absorbed form (FAb) where able to maintain the consumption compared to the control group. Although impact on feed intake was not statistically significant, F and FPB showed a significant ($P<0.05$) decrease in feed to gain ratio driven mainly by the lower feed consumption. For a general overview on the effect of organic acids in pig performance including feed intake the reader is referred to other more specific reviews (Diebold and Eidelsburger, 2006; Lallès *et al.*, 2009; Partanen, 2001).

Olfaction/smell

Olfaction is a chemosensing mechanism present across the whole animal kingdom to find food, detect potential predators and preys and mark territory (Firestein, 2001; Jones and Roper, 1997; Niimura and Nei, 2007; Wicher *et al.*, 2008). The sense of smell emerges as the main sense after birth in pigs and most mammals. The relevance of the sense of smell in primates seems to decrease as individuals mature a phenomenon that might be related to the evolution of the trichromatic vision in primates (Dulai and Von Dornum, 1999; Gilad *et al.*, 2004). In turn the olfactory sensitivity in pigs have been shown to be extremely high compared to primates and other animal species according to published evidences (Morrow-Tesch and McGlone, 1990; Rohde Parfet and Gonyou, 1991; Roura *et al.*, 2008a).

Olfactory perception is classified into two categories (Table 3) depending on the source and direction of the air flow through the nasal cavity: orthonasal (external air is inspired) and retronasal (air expired from the oral cavity). The two types of olfaction will be part of two different cognitive pathways that involve integration of other peripheral sensing mechanisms. The orthonasal olfactory perception integrates the olfactory pathway with other external stimuli, particularly the vomeronasal (pheromonal), sight, sound and touch (somatic). In contrast the retronasal

Table 3. The dual olfactory perception system (adapted from Shepherd, 2006).

Olfaction type	Stimulation route / air flow	Stimuli	Integrated pathways
Orthonasal	*air inspired* through the external nares into the nasal cavity and to the pharynx and larynx	floral scents; perfumes; smoke; food aromas; prey/ predator smells; social odours; pheromones; MHC molecules	olfactory (visual and auditory may also take place)
Retronasal	from the back of the oral cavity up through the nasopharynx into the nasal cavity and *air expired* through the nares	food volatiles	somatic, gustatory and olfactory: 'the flavour system'

olfactory perception involves only food volatiles integrated with taste and oral somatosensing cues. The 'flavour system' arises from the retronasal olfaction and integrates somatic, gustatory and olfactory stimuli (Shepherd, 2006).

Identification of food sources through smell cues requires establishing a link between odour stimulating molecules and the nutritional value of the food. Essential nutrients such as amino acids and fatty acids, among others, are precursors of volatile compounds produced by plants. Other precursors of plant volatiles that may stimulate consumption are health-promoting compounds (Du Toit *et al.*, 1991; Goff and Klee, 2006; Jones and Roper, 1997; Pass and Foley, 2000). Tomato volatile compounds identified by the human olfactory system are derived from long chain – cis - polyunsaturated fatty acids, leucine, carotenoids or isoleucine, supporting the relationship between volatile/odorant perception and nutrient value (Goff and Klee, 2006). On the other side, plant volatile secondary metabolites that may discourage consumption include those derived from antinutritional factors (ANFs) and potentially toxicant compounds (Du Toit *et al.*, 1991; Pass and Foley, 2000).

The nutritional value of foods is species dependent. The olfactory system has evolved as a highly adaptable physiological system that reads and interprets external smell cues but modulated by the central monitoring of the nutritional and health status of the animal. The integration of peripheral and central signalling will lead to feeding behaviours adapting to the varying nutritional needs and digestive capabilities of the different animal species (Roura *et al.*, 2008a).

Biological principles of olfactory sensing

The afferent sensory fibre of the PNS cranial nerve I (the Olfactory) is devoted to the sense of smell (Table 1, Figure 2). The olfactory nerve has a dual functionality (sensory and motor) since responds as well to mechanical stimuli (Grosmaitre *et al.*, 2007). In pigs and most other mammals, excluding humans and other higher apes, the anatomical features of the nasal cavity reflect its primary function as odorant detection (Reznik, 1990). Odour is generated mostly by

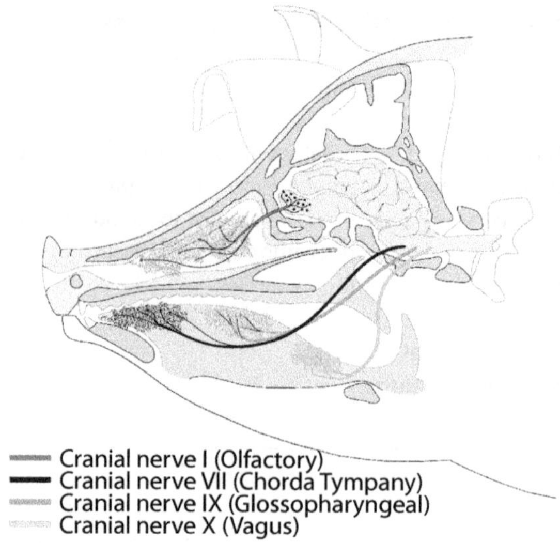

Figure 2. Smell is sensed through cranial nerve I (Olfactory) while full taste sensing requires bilateral cranial nerves VII (Facial, through its branch Chorda Tympany*), IX (*Glossopharingeal*) and X (Vagus). (drawing by Joaquim Roura).*

volatile compounds that reach the olfactory neuro-epithelium (OE) located in the upper wall of the nasal cavity. The ratio of OE to respiratory nasal epithelium determines the classification of mammals as macrosmatic (dogs, cats, rabbits, pigs and lab rodents) or microsmatic (sheep and primates). Macrosmates have a large proportion of surface of OE (Reznik, 1990).

The OE consists of the turbinates, an anatomical structure that increases the epithelial surface area, and the neuroepithelium composed of only three cell types (Figure 3): the olfactory sensory neurons (OSN), the supporting cells and a basal layer of stem-cells from which new OSN are continuously being generated (Firestein, 2001; Reznik, 1990). The OSN are bipolar neurons with a single dendrite that reaches up to the surface of the OE projecting some, 20–30 cilia that lie in the surface of the epithelium allowing it to come in contact with odorants dissolved in the thin nasal mucus. Olfactory receptors (OR) on the plasma membrane of the cilia are responsible for ligand recognition (Firestein, 2001; Shepherd, 2006).

Two classes of OR have been characterised: class I related to fish-like receptors since they bind water-soluble odorants, and class II receptors binding to air-born molecules (Rouquier and Giorgi, 2007). Class I OR represent only around 10 to, 20% of the OR family (Niimura and Nei, 2007). In, 1991, 2004 Nobel Prize winners Linda Buck and Richard Axel (Buck and Axel, 1991) discovered the multigene family of mammalian OR. The family has 636 members in humans (Malnic *et al.*, 2004) but over 1000 in the other non-primate mammals studied to date (Roura *et al.*, 2008a). This indicates that roughly 1-5% of the genes in the genome are devoted to detection of odorants what makes the olfactory gene repertoire the largest family in the entire genome of mammals (Buck, 2005). As expected the OR were shown to belong to a superfamily of receptors

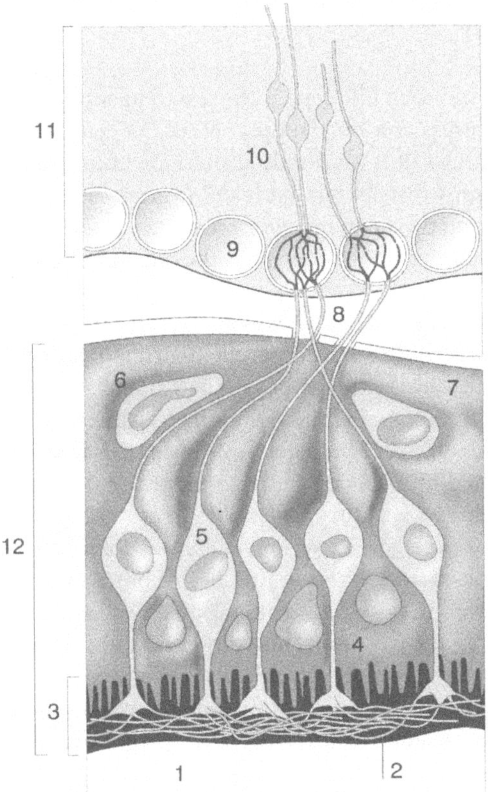

Figure 3. The Olfactory Epithelium (OE) and Olfactory Bulb (OB). Olfactory sensory neurons (5) reach the lumen of the nasal cavity through the mucosa (3) by setting a net of olfactory cilia (2). The axons (8) of the neurons reach the glomerulus (9) in the olfactory bulb (11) where mitral cells (10) will take the signal to the olfactory cortex (not shown). (drawing by Joaquim Roura).

known as GPCR (Guanine-nucleotide-binding (G) protein-coupled receptors). The GPCR superfamily includes other chemosensory receptors such as the taste (TR), the vomeronasal (VR), neuroreceptors, photoreceptors (rhodopsin has been stablished as a well-known model for GPCR) and many other receptors involved in cellular metabolism such as the adrenergic receptors or the metabotropic glutamate receptor family (Firestein, 2001). Vertebrate OR have a structure of seven α-helical transmembrane domains connected by intracellular and extracellular loops of variable lengths including an unusually long, characteristic to OR, second extracellular loop (Mombaerts, 1999). Within the family of ORs there is a region of hypervariability in the third, fourth and fifth transmembrane regions that are thought to face each other and form a pocket that has been determined as the probable binding site for ligands (Firestein, 2001; Mombaerts, 1999). A significant portion of the OR genes in mammals are considered to be inactive, or pseudogenes. Pseudogenes have been defined as genes with sequences that have premature stop codons or lack intact seven-transmembrane domains (Shi and Zhang, 2006). The number of OR pseudogenes differs among species with humans showing one of the highest pseudogenisation rates among mammals with 46% (Roura *et al.*, 2008a).

Each OSN expresses only one OR gene and each OR gene is expressed in about 1/1000 neurons. The OE has distinct spatial zones that express non-overlapping sets of OR genes. That means that each group of neurons expressing the same OR is randomly scattered within one zone ensuring that the input from one type of OR is highly distributed in the OE (Malnic *et al.*, 1999; Ressler *et al.*, 1993). Eventhough the OSN expressing the same receptor are scattered within one epithelial zone, nevertheless their axons converge in specific glomeruli in the olfactory bulb.

In the review by Buck (2005) the sequence of an odoriferous event starts in the nose by excitation of several OR. That is a segregation of the signal in different neurons belonging to several neuronal subsets. Each neuron communicates with the brain via a single axon that it extends to the olfactory bulb. The axons belonging to OSN of the same subset, expressing the same OR, converge in the same 2-5 glomeruli of the olfactory bulb determining an odorant-related surface map. Each odorant, therefore, is identified by a specific activity pattern in the glomerular layer of the olfactory bulb. Every glomerulus in the olfactory bulb relays to around 50 mitral cells that extend their axons to the olfactory cortex (a large part ventro-lateral section of the brain). In the olfactory cortex, information is reorganised and sent to a number of other brain areas such as higher cortical areas thought to be involved in odour discrimination as well as deep limbic areas of the brain, which are thought to mediate the emotional and physiological effects of odours.

There is a wide variety of chemical entities that may stimulate olfaction (Pichersky *et al.*, 2006). Detecting such variety of odorants requires therefore a large number of receptors for molecular recognition and a complex transduction and further processing of the olfactory stimuli (Duchamp-Viret *et al.*, 1999; Dudai, 1999). Four main features characterise the olfactory system:

1. A high number of OR (around 1000 intact OR genes in non-primate mammalian species studied to date; Roura *et al.*, 2008a).
2. Each OR can recognise multiple odorants (Zhao *et al.*, 1998).
3. Each odorant binds to multiple different OR (Firestein, 2001).
4. The OR are used combinatorially to encode odour identities (Malnic *et al.*, 1999).

The combinatorial coding scheme of 1000 different OR allows for the discrimination of an almost unlimited number of odorants. For example, if each odorant would stimulate only three ORs, this scheme would generate almost one billion different odour codes (Buck, 2005).

Olfactory cues and feed appetence in pigs

The sense of smell seems to be the most developed of all the senses in non-primate mammals (Reznik, 1990; Young and Trask, 2002) particularly in pigs (Morrow-Tesch and McGlone, 1990). The smell emerges as the main sense after birth both in pigs and humans but the development of sight acuity, age dependent, does not seem to be deterring the importance of smell in pigs such as in primates. Pigs show an olfactory system with extremely high sensitivity compared to other animal species according to published evidences. Pigs have the largest OE reported to date with a surface area of close to 300 cm^2 while the human OE is around only 5 cm^2. Similarly the largest number of OSN in the OE has been found in pigs with 576 million neurons when in humans has been estimated around only 10 million (Roura *et al.*, 2008a).

The availability of 'in vivo' data specifically addressing the relevance of smell sensing in pigs is still limited but offers compelling evidences. Piglets at birth discern the odour signature of the mother and show significant preferences for odours associated with maternal feces and skin secretions (Morrow-Tesch and McGlone, 1990). Other smell driven behaviours in young pigs have been shown related to placental fluids and sow milk (Rohde Parfet and Gonyou, 1991) and a simulated udder-odour by (Lay *et al.*, 1999). Thus, smell cues early in the life of the pig seem to play a critical role in assuring appropriate nourishment and safety as a way to secure survivability.

Odorants have been tested by several authors to condition the weaning process in mammals. Hepper (1988) showed that selected aniseed and garlic flavours contain odorants that cross into both amniotic and allantoic fluids and milk. In his early work on that matter, Campbell (1976) tried to reduce the weaning stress in piglets through flavour conditioning. A flavour was added to the sow's gestation and lactation diets expecting that some of the flavouring components would further incorporate into the sow's milk and get to the suckling piglets. Furthermore the same flavour was incorporated in the weaner diet of the treated piglets. The trial resulted in that the pigs fed a flavoured diet and weaned from sows offered a flavoured diet ate significantly more feed and grew significantly faster than the other pigs. Similarly Langendijk *et al.* (2007) used prenatal and postnatal exposure to garlic and aniseed flavours aiming at enhancing postweaning feed intake in piglets. Sows received the treated diets during the last month of gestation and during lactation. Litters of which the mother received the flavoured diet, had a higher feed intake of a flavoured creep feed and of a pre-starter (postweaning) diet than litters of sows without the flavour.

Double choice models have been widely used in determining flavour preferences in pigs. McLaughlin *et al.* (1983) studied 73 taste-free flavors and 23 sweet flavours divided in 7 basic groups by olfactory notes: buttery, cheesy, fatty, fruity, green, meaty and musty. Pigs showed positive preferences (preference index >55%) for 25 taste-free flavours over the 73 flavours tested. Highest rate of preferred over non-preferred flavours were recorded for buttery (50%), green (40%) and meaty (38%) notes.

The palatability of feed and feed ingredients will also largely depend on the odorants they release. For example, olfactory cues associated with lipid oxidation (rancidity and peroxide values) may compromise palatability and explain a depression on early feed intake in young animals (DeRouchey *et al.*, 2004). Valerianate and aldehyde C-6 are known volatile compounds that contribute to give rancid notes to an ingredient and have been identified in samples of dried distillers grain with solubles (DDGS). The quality of DDGS may vary widely depending on the industrial processing (Linneen *et al.*, 2008). Such differences may affect the nutritional value but also their chemosensing characterisation that seem to have an impact on palatability (Hastad, 2005; Seabolt, 2008). After performing headspace gas chromatography analysis significant differences in several volatile compounds released by different samples of DDGS were identified. Seabolt (2008) found valerianate and aldehyde C-6 to be responsible for the rancid notes of some of the samples that had the lowest preference and the highest impact in decreasing feed intake (Figure 4). On the contrary, furfural (an odorant known for its smoky and almond notes) was

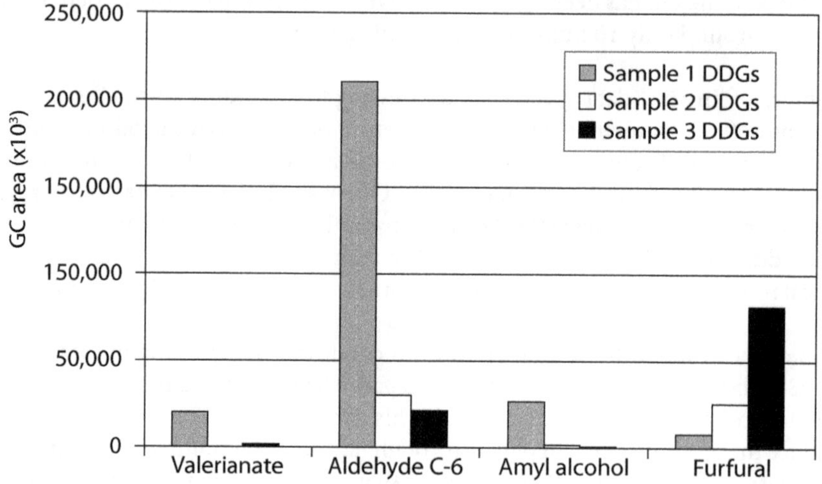

Figure 4. Identification of some volatile compounds from different DDGs samples (Seabolt, 2008).

higher in the DDGS samples with better acceptance. Nevertheless, Hastad (2005) was unable to reproduce any significant impact on palatability when either Furfural or Phenyl-ethyl-alcohol or γ-Butyrolactone were added to a grower pig diet. So, based on sensorial analysis, it seems that rancid notes present in pig feed decrease palatability and feed acceptance, while smoky-almond notes may increase palatability but further studies based on isolated plant derived odorants need to be performed.

Taste/gustation

The taste system is defined as the oral chemosensory system that recognises a diverse repertoire of non-volatile compounds. Not less than five different tastes have been defined and are widely accepted: sweet, umami, salty, sour and bitter. Sweet taste is related to carbohydrates such as sugars. Umami to protein born nutrients such as amino acids (aa) and oligopeptides. Salty and sour are responsive to salts and acids. While bitter taste relates to toxic or antinutritional chemical entities. Currently, other taste candidates are under study such as fat, calcium (Ca^{+2}) or water tastes. All in all, the objective of the gustatory system is to provide valuable information about the quality of food (Bachmanov and Beauchamp, 2007; Dulac, 2000; Grigorov *et al.*, 2003; Margolskee, 2002; Matsunami and Amrein, 2003; Nelson *et al.*, 2002; Roura *et al.*, 2008a).

Taste perception in pigs, similar to other mammals, may not be fully develop at birth (Oakley, 1998). Thus, pre-natal taste exposures may not be determining the post-natal preferences to the same extend than smell does. Nevertheless, taste will quickly develop after birth since taste perception is intimately related to the quality of food and to diet selection and varies among species. Taste perception diversity across species reflects their nutritional adaptation to different ecological niches (Forbes, 1998; Goff and Klee, 2006; Li *et al.*, 2005; Roura *et al.*, 2008a; Shi and Zhang, 2006).

Biological principles of taste sensing

The field of taste biology has undergone a large increase in research activity particularly over the past decade since the discovery of taste receptors in the early, 2000s. The taste system in mammals (including pigs) is mainly located in the oral papillae (there are 3 types of papillae: fungiform, cirumvallate and foliate) that are primarily located on the tongue, epiglottis and soft palate epitheliums (Figure 2). Papillae contain high numbers of a garlic-clove like organelle named taste bud (Figure 5). Type and number of papillae and the number of taste buds in each papillae differ among animal species (Chandrashekar *et al.*, 2006; Dulac, 2000; Mombaerts, 2000; Roper, 1992; Roura *et al.*, 2008a). The cranial nerve IX (*Glossopharingeal* - GP) innervates mainly circumvallate and foliate papillae located in the back and laterals of the tongue respectively. In turn the tip of the tongue allocates mainly fungiform papillae sensed by the cranial nerve VII (*Chorda tympani* – CT). Furthermore, cranial nerve X (Vagus) also senses taste stimulations from taste buds located in the pharynx and epiglottis (Table 1, Figure 2) (Shepherd, 2006; Witt *et al.*, 2003).

The taste buds consist of groups of 50 to 120 sensory cells that project a number of microvilli that reach the mucus layer of the tongue (Figure 5) (Chandrashekar *et al.*, 2006; Dulac, 2000; Matsunami and Amrein, 2003; Roper, 1992). At least four different types of cell constitute a taste bud; three taste-type cells (I, II and III) and one basal-type cell believed to be a progenitor of the other three (DeFazio *et al.*, 2006; Romanov and Kolesnikov, 2006). Taste cells type I are sour-sensing, type II are sweet, umami and bitter sensing and type III cells play an intermediate signaling role between the true taste cells (type I and II) and the sensory neurons (DeFazio *et al.*, 2006; Palmer, 2007; Romanov and Kolesnikov, 2006).

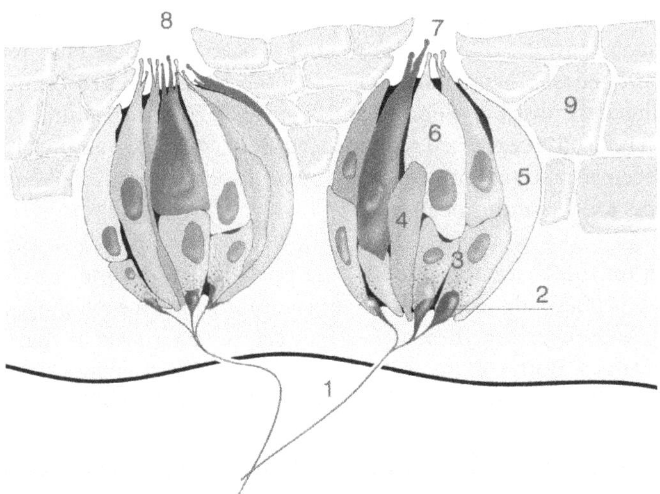

1 Nerve fibers
2 Synapses
3 Type III sensory cell
4 Basal cell
5 Supporting cell
6 Type II sensory cell
7 Microvilli
8 Outer taste pore
9 Epithelial cell

Figure 5. Lingual epithelium showing the main structural and functional cells and other characteristic features associated with taste buds (drawing by Joaquim Roura).

Stimulation of taste cells is mediated through transmembrane taste receptors (TR). Each taste bud seems to recognise all basic tastes nevertheless any single cell expresses only one family of TR (Dulac, 2000; Huang, 2006; Mombaerts, 2000). The TR for sweet, umami, bitter, salty and sour tastes have been already described particularly in humans and laboratory rodents while other candidate TR such as fat, calcium (Ca^{+2}) or water are still under investigation (Bachmanov and Beauchamp, 2007; Roura *et al.*, 2008a). The TR will recognise specific soluble taste ligands present in the oral cavity and trigger an intracellular metabolic cascade followed by an intercellular communication between the taste-sensing cell and the type III cells (Bachmanov and Beauchamp, 2007; Firestein, 2001; Huang, 2006; Matsunami and Amrein, 2003; Mombaerts, 2000; Roper, 2006). Type III cells form synapses (Figure 5) with sensory neurons that transmit the nerve impulse through the nucleus of the solitary tract of the medulla and to the somatosensorial brain cortex (Bachmanov and Beauchamp, 2007; Palmer, 2007; Romanov and Kolesnikov, 2006).

Similar to the taste biology found in other mammals the pig follows a similar taste system based on umami (Humphrey *et al.*, 2009; Roura *et al.*, 2008b) and presumably sweet and bitter (unpublished results) TR that belong to the superfamily of GPCR (same superfamily than the OR – see the previous section of this chapter). Taste GPCR have been divided into two classes: T1R (related to umami and sweet tastes) and bitter related T2R (Bachmanov and Beauchamp, 2007; Chandrashekar *et al.*, 2000; Matsunami and Amrein, 2003; Mombaerts, 2000). Porcine umami and sweet are perceived through heterodimers that include one specific umami receptor (pT1R1) and one specific sweet receptor (pT1R2) with a common unit for the two tastes (pT1R3) (Humphrey *et al.*, 2009; Roura *et al.*, 2008a,b). A second receptor may also be involved in the umami taste of monosodium glutamate (Chaudhari *et al.*, 2000; Toyono *et al.*, 2002). This receptor is a metabotropic GPCR called mGluR4, that recognises only glutamate and its sensitivity is enhanced by inositol monophosphate (IMP).

The size of the T2R family (bitter TR) differs among animal species and may be constituted by up to 35 different members such as in the mouse (Roura *et al.*, 2008a). A single taste receptor cell expresses all or most *TAS2R* genes (Chandrashekar *et al.*, 2000) what makes discrimination between different bitter stimuli barely possible (Behrens and Meyerhof, 2006). The number of pig T2R is still currently unknown but some expressed sequence tags for candidate porcine bitter TR (pTas2r7 and pTas2r8) have been recently identified in The Institute for Genomic Research pig EST database (B. Humphrey personal communication).

Our group has been working on the characterisation of the porcine heterodimer umami TR (pT1R1/pT1R3). Roura *et al.* (2008b) determined the nucleotide sequence of the pTas1r1 and pTas1r3 genes in the pig and expressed these genes in a cell reporter system that allows quantification of receptor activation. Both pT1R1 and pT1R3 amino acid sequences begin with methionine which is required for all mammalian proteins. The pT1R1 protein contains 844 amino acid residues and the pT1R3 contains 855 (Table 4). Similar to other known mammalian species, our current findings show that pig T1R1/T1R3 receptors are located in the plasma membrane and their sequence contains a ligand binding domain from amino acid 1 to 571 in T1R1 and 1 to 565 in T1R3 and 7 transmembrane domains (Humphrey *et al.*, 2009).

Table 4. Identified porcine taste receptors (full or partial sequences – Expressed Sequence Tags) and putative ligands.

Sensorial response	Receptor		Amino acid sequences		Porcine ligands
	Name	Type	Full length	Main extracel.	Preferred
Umami[1]	T1R1	GPCR/dimer	844	571	L-amino acids, IMP, GMP
Sweet[2]	T1R2	GPCR/dimer	n/a	n/a	carbohydrates, D-amino acids
Umami[1]/sweet	T1R3	GPCR/dimer	855	565	-
Bitter[3]	T2R7	GPCR	n/a	n/a	unknown
Bitter[3]	T2R8	GPCR	n/a	n/a	unknown
Bitter	several unknown	GPCR	n/a	n/a	denatonium, quinine, caffeine, tilmicosin
Salty[3]	ENAC-γ	LGIC/tetramer	n/a	n/a	Na^+ salts

Abbreviations: T1R1,2,3: taste receptor family 1 types 1, 2 and 3; GPCR: guanine-nucleotide-binding protein-coupled receptors; ENaC: epithelial amiloride-sensitive ion channel; LGIC: ligan-gated ion channels; n/a: means the full gene sequence is not available, only partial sequences found in gene data banks (Expressed Sequence Tags).

[1] Source Humphrey *et al.* (2009).

[2]Holt, Roura and Klasing, unpublished observations.

[3]Humphrey and Roura, unpublished observations.

Other taste stimuli such as sodium (salty) and protons (sour) may penetrate cell membranes through ion or ligand gated protein channels or other mechanisms that include potential intracellular targets (Bachmanov and Beauchamp, 2007; Palmer, 2007). Salt is a pleasant taste caused by ionic sodium and to a lesser extent by other cations. The epithelial sodium channel receptor in rodents is called ENaC (Lin *et al.*, 1999) but a second mechanism involving the transient receptor potential vanilloid receptor-1 (TRPV1) has been also described (Lyall *et al.*, 2004; Ruiz *et al.*, 2006). The ENaC is a trimeric receptor, composed of three subunits (α, β, γ), highly selective towards sodium whereas the TRPV1 receptor is more broadly tuned and is responsible for rejection-responses to K^+, NH_4^+, and Ca^+ salts (DeSimone and Lyall, 2006). Candidate porcine TRPV1 and ENAC-α-β-γ receptors (Table 4) have been recently identified in The Institute for Genomic Research pig EST database (B. Humphrey personal communication).

Sour taste is known for acid perception, nevertheless, only a moderate correlation exists between the hydrogen ion concentration of a food and its perceived sourness. This is partly because some molecules, such as weak organic acids, are able to penetrate the cell membrane in an undissociated form and then dissociate inside the taste cell. Hydrogen gated channels are also transmembrane protein complexes that serve as ion channels for the selective entrance of hydrogen ions into the taste cell. Protein complexes HCN (named after hyperpolarisation-activated cyclic nucleotide-gated channel) and PKD1L3 and PKD2L1 (named after polycystic kidney disease gene family) have been proposed as responsible for sourness perception (Huang *et al.*, 2006; Ishimaru *et al.*, 2006; Lindemann, 2001).

Our knowledge on the biology of the sour and other candidate taste families in the pig has not been developed yet. The availability of the pig genome in the near future will allow a much faster and thorough advance in this area.

Taste agonists in pigs

Sweetness is a pleasurable taste in pigs. First work reported on sugar preferences was published in the mid 50s (Salmon-Legagneur and Fevrier, 1956). The authors showed increases in feed consumption after dietary inclusion of 3 or 5% of sucrose. In the same direction, Kennedy and Baldwin (1972) studied taste preferences to sweet solutions. Animals responded to sucrose, glucose and sodium saccharin in water. The threshold preference ranged from 0.005 to 0.01M for sucrose and from 0.01 to 0.03M for glucose. For both substances animals showed consistent preferences (higher than 90%). Results for sodium saccharin were more variable. The threshold preference was between 0.005 to 0.01M but pigs never showed preferences higher than 90% and even between 0.1 to 1 M animals started to reject the solution.

Danilova *et al.* (1999) performed electrophysiological studies in pigs to observe the responses of the CT and the GP nerves to several hypothetical tastants. Both nerves exhibited large responses to carbohydrate sweeteners such as fructose, sucrose, lactose, maltose, glucose and galactose among other mono and oligosaccharides. These data is consistent with the pig preference values for sweeteners observed by Glaser et al (2000). Sucrose showed the highest intensity of all the carbohydrates tested and their responses were very similar to sugar sensory tests in humans (Table 5). Furthermore, pigs seem to be responsive to most D-AA and a few L-AA (Alanine, Serine and Threonine) and glycine (Table 6) known to be sweet to humans. Glycine, nevertheless,

Table 5. Sweetener potencies relative to sucrose (on a molar basis) in humans and pigs (adapted from Glaser et al.*, 2000).*

		Potency relative to sucrose	
		Humans	Pigs [a]
Carbohydrates	sucrose	1.00	1.00
	D-fructose	0.50	0.50
	lactose	0.33	0.15
	D/l-glucose	0.25	0.13
Non-carbohydrates	aspartame	155	<1.00[b]
	cyclamate (Na^+ salt)	17.6	<0.15[b]
	nehoesperidin (dihydrocalcone)	3,600	<151[b]
	thaumatin	2,000	<1,622[b]
	saccharin	215	3.34

[a] Sweetener potencies in pigs were measured based on double choice preference tests.

[b] Equivalent to the lowest dose tested that gave no response. The exact potency was not determined.

Table 6. Gustatory responsiveness of pigs to L and D isomers of amino acids and predominant hedonic response in humans (adapted from Tinti et al.*, 2000).*

Amino acid	L-isomer		D-isomer	
	Human taste[1]	Pig response	Human taste[1]	Pig response
Alanine	sweet	yes	sweet	yes
Asparagine	complex	yes	sweet	yes
Glutamine	complex	yes	sweet	yes
Glycine[2]	-	-	-	-
Histidine	bitter	no	sweet	no
Hydroxyproline	sweet	yes	complex	no
Isoleucine	bitter	no	sweet	no
Leucine	bitter	no	sweet	no
Phenylalanine	bitter	no	sweet	yes
Proline	complex	no	bitter	no
Serine	sweet	yes	sweet	yes
Threonine	sweet	yes	sweet	yes
Tryptophan	bitter	no	sweet	yes
Valine	bitter	no	sweet	no

[1] The predominant human taste was qualified by the authors in three categories: sweet, bitter or complex. Complex includes hedonic descriptions by the authors such as: flat, meaty, bitter and sweet.
[2] Glycine is a non chiral amino acid. Glycine is perceived sweet by humans and elicits pig responses.

might have a complex taste for pigs including stimulation of putative umami, bitter and sour fibers together with the sweet ones (Danilova *et al.*, 1999; Glaser *et al.*, 2000; Hellekant and Danilova, 1999; Tinti *et al.*, 2000).

On the other hand, only minor or no responses to other non-carbohydrate substances known to be sweet to humans have been reported in pigs. For example, pigs did not respond to human-based doses of aspartame, neohesperidin, thaumatin or saccharin (Table 5), four well-known high intensity sweetners (HIS) in humans (Glaser *et al.*, 2000; Matsunami *et al.*, 2000). In the sweet clusters of both taste nerves (GP and CT) studied by Danilova *et al.* (1999), acesulfame-k and stevioside resulted in higher responses than any of the other HIS tested including combinations of saccharin with neohesperidin or thaumatin. Sodium saccharin, a sweetener widely used in pig diets, elicited a response only three times above that of sucrose (Glaser *et al.*, 2000).

Umami taste is mainly related to protein-born chemicals such as L-AA and peptides but also to inosine monophosphate (IMP) and other nucleotides (Nelson *et al.*, 2002; Ninomiya, 2002). The umami taste was first discovered by Ikeda (1909) who isolated glutamate in a Japanese broth. Similarly some L-AA and nucleotides are also effective in stimulating pig taste (Danilova *et al.*, 1999; Roura *et al.*, 2008b; Tinti *et al.*, 2000). Danilova *et al.* (1999) found strong responses to umami compounds (monosodium glutamate alone or in combination with nucleotides) in both taste nerves (GP and CT) suggesting that umami taste is important in pigs. Tinti *et al.* (2000) also identified several L-AA giving an hedonic response, presumably umami, in pigs: Glutamine, Alanine, Asparagine, Hyrdorxyproline, Serine and Threonine (Table 6). On a five minute double choice test a diet supplemented with L-Tryptophane (Trp) was preferred over a Trp-deficient diet (Ettle *et al.*, 2004) by pigs. Nevertheless, L-Trp showed no preference according to Tinti *et al.* (2000).

Danilova *et al.* (1999) reported as well that coupling between sweet and umami fibers may be occurring in response to umami tastants in pigs. Monosodium glutamate alone or in combination with nucleotides was able to stimulate fibers identified as umami but also those identified as sweet. These coupled responses may lead to a stronger positive hedonic responses for umami compared to sweet tastants.

Many potentially toxicant compounds present in feeds including some antinutritional factors such as glucosinolates, are known to taste bitter to animals and cause feed avoidance (Mawson *et al.*, 1993). Pigs seem to respond more vigorously to bitter taste aversions than to hedonic preferences (Tedó *et al.*, 2007). Plant extracts (Schöne *et al.*, 2006; Windisch *et al.*, 2008) and pharmaceutical compounds such as antibiotics are often added to pig feeds and may further elicit bitter taste driven diet rejections (Nelson and Sanregret, 1997). For example, tilmicosin has been found to elicit a strong electrophysiological response in pig taste fibers. All the ten pharmaceuticals tested by Nelson and Sanregret (1997) resulted in aversion behaviours in pigs some of which were found to be dose dependent. Similarly some chemical compounds such as quinine HCl, denatonium, caffeine and sucrose octaacetate also elicited avoidance responses in pigs (Danilova *et al.*, 1999; Nelson and Sanregret, 1997). Bitrex (the harmless but bitterest compound known to humans) addition abruptly reduced feed intake in pigs. Nevertheless, the consumption was back to normal after only a few meals (Blair and FitzSimons, 1970).

Pigs have a high sensitivity for acids such as citric acid particularly at a young age. Danilova *et al.* (1999) found citric but also ascorbic acids to elicit the largest CT electrophysiological responses of all the 30 potential tastants they tested. Acids such as citric, malic, tartaric or fumaric are present in several potential dietary sources for pigs such as a variety of fruits and tubers among other plant products. Acetic, formic and lactic may also be found in grapes or bananas while lactic and butyric acids are commonly found in fermented foods (e.g. fermented milk). It has been proposed by several authors that some organic acids may improve performance in weanling pigs thanks to lower dietary pH and buffer capacity and a decrease in intestinal bacterial growth (Partanen, 2001; Roth and Kirchgessner, 1998). Nevertheless unpublished results from our lab showed that 0.5% inclusion in piglet feed of citric, acetic, malic, lactic or butyric acids, among others, failed to lower acid binding capacity of feeds at a simulated gastric environment with a pH set at 3. Furthermore, citric, acetic or butyric proved also relatively low anti-microbial capacities (e.g. against *E. coli*) showing instead high taste and olfactory stimulatory effects. The chemosensing properties and palatability implications of such dietary acids in piglets merits further consideration.

Finally, the response to salty tastants such as chloride salts, where also significant but les pronounced than for umami or acid agonists (Danilova *et al.*, 1999). Nevertheless, weanling pigs are known to respond to the addition of sodium chloride to complete diets that were already relatively high in sodium (Mahan *et al.*, 1996).

Saliva and taste sensitivity

Saliva is a generic term that refers to the watery secretions present in the oral cavity produced by several glands mainly the parotid, the submandibular, the sublingual and several hundreds of minor glands (including the lingual Von Ebner's glands – VEG) located throughout the oral cavity within the lamina propria of the oral mucosa. The secretions are mainly mucous in nature except for the VEG products, and contain electrolytes, mucins, antibacterial compounds, enzymes and other functional proteins. The VEG are found in the clefts of the circumvallate and foliate papillae (Matsuo, 2000) and secrete a serous fluid rich in lipase activity and carrier proteins. Further to the generic function to coat the oral cavity and facilitate swallowing of alimentary boluses the salivary secretions may play an important role related to peripheral sensing.

The secretion of saliva is induced by oral food-born cues. As the principal fluid component of the extracellular environment of taste buds saliva plays a major role in modulating taste sensitivity (Matsuo, 2000; Spielman, 1990). Its main functions linked to chemosensing start with the transport and solving of taste substances. The content in bicarbonate and sodium-chloride buffers free hydrogen ions (sour taste) and increases the threshold for salt taste. Endogenous salivary glutamate levels may contribute and enhance the umami taste. Also, some salivary constituent proteins of the VEG secretions may bind to insoluble material such as some bitter (which are often hydrophobic molecules; Matsuo, 2000) or lipophilic (Schmale *et al.*, 1990) compounds.

Little is known regarding specific features of pig saliva. A pheromonal steroid-binding protein from miniature wild boar saliva was isolated by Booth and White (1988). In our laboratory we tested the '*in vitro*' solubility of several carbohydrate (sucrose, fructose and dextrose) and non-carbohydrate (stevioside and saccharine) sweeteners in distilled water compared to an artificial saliva containing porcine mucin (Table 7). After thirty six seconds all sweeteners showed a high rate of solution in distilled water. Dextrose and fructose dissolved completely and sucrose up to 92%. Similarly stevioside also dissolved completely but not saccharin (92%). The change of solvent from distilled water to artificial porcine saliva dramatically changed the scenario. Carbohydrate sweeteners hardly dissolved in the saliva with sucrose showing now the highest rate and dextrose the lowest at both time points (eighteen or thirty six seconds). The HIS tested also decreased their solubility rates in saliva compared to distilled water, but to a lesser extend than carbohydrate sweeteners. Stevioside maintained a higher solubility rate than saccharin after eighteen seconds (41 vs. 30% respectively) although the latter tended to equalise after thirty six seconds (47 vs. 45% for stevioside and saccharin respectively).

After eighteen seconds, the carbohydrate sweetener solubility in saliva, but not in distilled water, correlates with the 'in vivo' data shown in Table 5 from Glaser *et al.* (2000). For example, relative potency of sucrose in pigs was twice as much as the potency of fructose and more than five times that of dextrose. Nevertheless, our 'in vitro' data contrasts with results reported by Matsuo *et al.*

Table 7. Cumulative sweetener recovery (%) after 18 or 36 seconds of distilled water or porcine artificial saliva[1] flows.

	Cumulative sweetener recovery from solution (%)				
	Distilled water		Porcine artificial saliva		
	18 sec.	36 sec.	18 sec.	36 sec.	Ratio to sucrose [2]
Sucrose	53.94	92.33	1.41	3.03	1.00
Fructose	71.38	100	0.76	2.16	0.54
Glucose (dextrose)	93.78	100	0.26	1.28	0.19
Stevioside	79.62	100	40.97	46.75	29.10
Saccharin	74.59	92.13	29.98	45.42	21.29

[1] A commercial artificial saliva with a pH of 7.2 and containing porcine mucin (AS Saliva Orthana® from AS Pharma) was used in these studies.

[2] Ratio of sweetener solution recovery in porcine artificial saliva after 18 seconds relative to sucrose.

(2000) in rats where sucrose gave a larger response with artificial saliva than with distilled water. The composition of artificial saliva (e.g. pH, electrolytes and viscosity among other features) seems to be essential to understand the interactions between saliva and taste sensitivity. Further research in this area should be granted.

The comparative oronasal sensing biology of the pig

The biodiversity in nature is a result of the adaptation of living species to the variety of ecological niches on earth. That principle is particularly relevant when discussing food sources. Animal species have adapted their feeding habits, anatomical and physiological systems as well as their metabolism to the dietary niches. For example, the adaptations of the beak of the finch for feeding purposes were a main contributor to Charles Darwin's initial theories on evolution (Klasing, 1998). Peripheral senses are key to interpret those stimuli relevant to nutritional value of available food sources and are bound to be highly adaptive mechanisms to species specific dietary needs (Dulac, 2000; Forbes, 1998; Kuenzel, 1989; Roura *et al.*, 2008a).

Comparative anatomical, electrophysiological (most with invasive techniques) and feeding behaviour studies have fulfilled a gap with an integrated view on how dietary perception may differ among animal species. Nevertheless, oronasal smell, taste and somatic sensing stimulation in animal species have not been measured properly until animal genomics have appeared in the horizon. Bioinformatic tools provide the ability to search the farm and companion animal genomes available through public gene data banks and identify putative sensing receptor homologues (OR, TR and TRP among others). Unfortunately, the pig genome is not yet available what limits our means for a better understanding of the pig peripheral sensing system. The data presented in this section from literature (Kiuchi *et al.*, 2006; Li *et al.*, 2005; Roura *et al.*, 2008a; Shi and Zhang,

2006) or from our own research group on pig comparative peripheral sensing receptor biology (Humphrey *et al.*, 2009; Roura *et al.*, 2008a,b) have been obtained by bioinformatic tools that may require confirmation by direct sequencing.

The somatosensing system is part of the pain pathway. Differences among mammals have not been reported and singularities of the pig somatic sensing are yet to be uncovered. The identification from available gene bank data bases of the porcine genes coding for TRPV1,2,3,4 (hot and spicy) and the TRPM8 (cold) have been recently accomplished (Humphrey, personal communication). Birds might be more tolerant than mammals to TRPV1 mediated responses including some spices (Roura *et al.*, 2008a). The effect of specific spices on the performance of piglets supports the hypothesis that pungent/hot spices will trigger a somatic response. We have shown that feed supplemented with moderate levels of cinnamaldehyde, carvacrol or capsicum oleoresin resulted in decreased feed intake in piglets (Bikker *et al.*, 2003) but not in chickens (unpublished data). A sigmoid type response of the TRPV1 receptor to capsaicin has been reported for both mammals and chickens although the responsive concentration range to capsaicin in lab rodents is in the nM (Tóth *et al.*, 2004) while in chickens is in the μM and does not result in a full neuronal response such as in lab rodents (Kirifides *et al.*, 2004). Our data is not supportive to the use of common spices in piglets.

Smell seems to be the most developed sense in pigs (Guiraudie-Capraz *et al.*, 2005; Morrow-Tesch and McGlone, 1990; Rohde Parfet and Gonyou, 1991; Roura *et al.*, 2008a). The olfactory to respiratory nasal areas ratio has been used to classify mammals as macrosmatic (dogs, cats, rabbits, pigs and lab rodents) or microsmatic (sheep and primates) (Reznik, 1990). Macrosmates have a large proportion of surface of olfactory epithelium, larger brain structures involved in the olfactory function, particularly the olfactory bulb, and a higher number of OR (Rouquier and Giorgi, 2007). The pig and the dog have larger OE surfaces and number of OSN than humans and other mammals studied to date (Table 8). The number of OR in the pig have not been described yet but the other macrosmate mammalians with known genomes such as dogs and mice have around 1,100-1,200 functional OR, the largest number of OR known to date (Dryer, 2000; Godfrey *et al.*, 2004; Quignon *et al.*, 2005; Young and Trask, 2002).

Part of the OR genes in mammalians and chickens are inactive, or pseudogenes (Table 9). The number of OR pseudogenes differs among species and have been estimated to be around, 20% in macrosmates such as the dog (Quignon *et al.*, 2005) and the mouse (Godfrey *et al.*, 2004). Humans and cows have a much higher rate of pseudogenisation both around 46% (Malnic *et al.*, 2004; Niimura and Nei, 2007). It has been hypothesised that the increase in the number of olfactory pseudogenes may reflect a decrease in the animal reliance on the sense of smell. For example, Gilad *et al.* (2004) concluded that the deterioration of the olfactory repertoire in primates occurred concomitant with the acquisition of the full trichromatic colour vision.

The comparative taste biology shows important differences among mammalian species. Pigs and cows with roughly, 20,000 taste buds in the oral cavity have the highest known number of these organelles (Roura *et al.*, 2008a). Pigs have taste buds in all the three different types of papillae as opposed to other mammals such as cats and cattle that have no foliate papillae or no buds in the

Table 8. Differences in the main olfactory epithelium (OE) area, number of olfactory sensory neurons (OSN) and the olfactory receptor genes in several vertebrate species (reproduced with permission from Roura et al.*, 2008a).*

Animal species	OE area (cm²)	OSNs ×10⁶	Olfactory receptor genes			
			Total	Intact	Pseudogenes	References
Pig[c]	288[a]	576[a]	n/a	n/a	n/a	-
Rabbit	8.32[a]	16[a]	n/a	n/a	n/a	-
Cow	n/a	n/a	2,129	1,152	977	Niimura and Nei, 2007
Dog (German Shep.)	139[a]	278[a]	~1,094	~872	~222	Quignon *et al.*, 2005
Mouse	1.67[d]	3[a]	1,209	913	296	Godfrey *et al.*, 2004
Human	5[a]	10[a]	636	339	297	Malnic *et al.*, 2004
Chicken	5[b]	10-20[b]	~383	283	~100	International Chicken Genome Sequencing Consortium, 2004

Abbreviations: OE=olfactory epithelium; OSN= olfactory sensory neurons; n/a=not available

[a] Adapted from Leibetseder (1980).

[b] Reviewed by Pihlström *et al.* (2005).

[c] Extrapolated from data corresponding to the wildboar (*Sus escrofa*) by Pihlström *et al.* (2005).

[d] Extrapolated from data corresponding to the 'deer mouse' (*Peromyscus maniculatus*) by Adams (2005).

foliate papillae, respectively (Dasgupta *et al.*, 1990; Robinson and Winkles, 1990). Pigs have 3 to 4 times more taste buds than humans.

All mammals studied, except cats (Beauchamp *et al.*, 1977; Li *et al.*, 2005), seem to share significant similarities in the nature of their five primary taste senses: sweet, umami, salty, sour and bitter (Roura *et al.*, 2008a). Pigs, similar to the rest of non-feline mammals, have the three T1R related to sweet and umami tastes (Table 9). Cats, like chickens, lack a functional sweet T1R2. Not much is known related to the pig T2R (bitter) family except that a few gene candidates have been identified (Table 5).

Calves (Hard *et al.*, 1989a,b; Hellekant *et al.*, 1994) and sheep (Goatcher and Church, 1970) showed preferences for carbohydrate sweeteners to a similar extent than pigs do but cats and chickens are unable to taste sweets (Balog and Millar, 1989; Beauchamp *et al.*, 1977; Li *et al.*, 2005). On the other hand, the binding of non-carbohydrate HIS to the T1R2/T1R3 dimer appears to be species dependent. Artificial and non-carbohydrate HIS have been developed after human sweet perception and may not be sweet to other animal species such as the pig (Danilova *et al.*, 1999; Glaser *et al.*, 2000; Kennedy and Baldwin, 1972).

Little is known in pigs and other farm animals regarding umami tastants. Nevertheless, on top of the pig, humans and lab rodents, umami T1R1/T1R3 receptors have also been identified in cow,

Table 9. Number of functional[1] or pseudogenes[2] in several farm and companion animals, mouse and human belonging to the Tas1r *and* Tas2r *gene families (adapted from Roura* et al.*, 2008a).*

Animal species	Tas1r		Tas2r		Total taste GPCR genes
	Functional	Pseudogenes	Functional	Pseudogenes	
Human[3]	3	0	25	11	39
Cow[4]	3	0	19	15	37
Pig[5]	3	0	n/a	n/a	n/a
Dog[3]	3	0	16	5	24
Cat[6]	2	1	n/a	n/a	n/a
Horse[4]	3	0	n/a	n/a	n/a
Mouse[3]	3	0	35	6	44
Chicken[3]	2	0	3	0	5

Abbreviations: *Tas1-2r*=taste receptor genes, type 1 or 2.

[1] Functional genes include genes with 7 transmembrane domains and complete open reading frame (ORF).

[2] Pseudogenes include genes with missing transmembrane domains or part of the ORF sequence.

[3] Data extracted from the review by Shi and Zhang (2006).

[4] Cow and horse (*Tas1r1*, *Tas1r2* and *Tas1r3)* genes have been identified (Humphrey, personal communication).

[5] Pig genome is not fully available. Full pig *Tas1r1* and *Tas1r3* genes have been identified and published by Roura *et al.* (2008b).

[6] Cat *caTas1r2* has been identified as a pseudogene (Li *et al.*, 2005).

dog, cat and several fish species (Table 9). Such a broad distribution across vertebrate species makes the umami taste a true universal taste. Umami compounds known to elicit a hedonic response in pigs include monosodium glutamate and some L-AA (Glutamic, Alanine, Asparagine, Hyrdorxyproline, Serine and Threonine) and nucleotides (Danilova *et al.*, 1999, Roura *et al.*, 2008b; Tinti *et al.*, 2000). Nevertheless, only L-Glutamic and L-Glutamine have been identified as being umami in humans (Table 6). In turn, the umami T1R1/T1R3 receptor of the mouse is more responsive to cysteine, methionine, arginine and threonine than to glutamate (Li *et al.*, 2002; Nelson *et al.*, 2002). Chickens are also responsive to dietary supplementation of essential amino acids although may or may not be peripheral sensing driven (Picard *et al.*, 1993). Danilova *et al.* (1999) have reported as well that umami tastants in pigs may elicit coupled responses that included stimulatory effects on both umami but also sweet fibers. These results are in contrast with studies in other species were umami compounds coupled better with salty than with sweet compounds (Ninomiya and Fuyakoshi, 1989, Yamamoto *et al.*, 1991).

Based on umami porcine *pTas1r1/ pTas1r3* nucleotide homologies, pig shares the highest identities with cat (Humphrey *et al.*, 2009). Cow, horse and dog also appear to have very similar umami receptor genes to swine. It is interesting to note that, among mammals, Humphrey and co-workers (2009) found that human and mouse umami systems show the lowest resemblance to the pig. Roura *et al.* (2008b) examined *Tas1r* based phylogenetic relationships among several mammalian species and the *Gallus domesticus*. Again, the pig is closer to the carnivorous dog and cat than to the primates. Based on that relationship, the human umami hTAS1R1 seems to be half the way between the rodents (omnivores) and the dietary highly adapted herbivores (cow) and carnivores (cat and dog). Therefore, based on umami taste, pigs and primates are not a good model the one for the other.

In contrast to the sweet and umami, the bitter receptor (*Tas2r*) is a large mammalian gene family ranging from 21 (dog) to 41 (mouse) members (Table 9). The mammalian bitter gene family has been found to show a strong adaptive response through gene inactivation or pseudogenisation (Roura *et al.*, 2008b). Chickens uniquely appear to have a very small 3 member *Tas2r* family (Shi and Zhang, 2006). Unfortunately very little information is known about the bitter taste system in pigs. Within the known mammalian species a large variation of T2R numbers and functionality has been described. Presumably linked to diet profile and digestive strategies the degree of bitter gene pseudogenisation varies greatly among mammalians with ruminants having the largest proportion of non-functional genes. In pigs two sequence tags belonging to two bitter tastes receptors have been identified (Table 4) from the limited material available in porcine gene data banks (B. Humprey personal communicaton).

Pigs elicit avoidance responses to antibiotics and several chemicals such as quinine HCl, denatonium, caffeine, sucrose octaacetate and bitrex (Blair and FitzSimons, 1970; Danilova *et al.*, 1999; Nelson and Sanregret, 1997). The sensitivity to denatonium seems to be species dependent with humans responding to nanomolar while pigs showing a much lower sensitivity and responding to milimolar levels (Danilova *et al.*, 1999). Responses to quinine HCl have also been reported in calves and chickens (Roura *et al.*, 2008b) but Tripathi and Mishra (2007) described that ruminants had a 4 times higher tolerance level to dietary glucosinolates from rapeseed meal than pigs. In general, ruminants seem to be less sensitive than non-ruminants to antinutritional factors (ANF) due to the ability to digest or inactivate them in the rumen (Du Toit *et al.*, 1991, Hill and Tamminga, 1998).

Pigs seem to respond to salts and acids (Danilova *et al.*, 1999) to a similar extend that other mammalian species such as calves (Hard *et al.*, 1989a,b), sheep (Goatcher and Church, 1970) and chickens. Piglets were highly responsive to the putative salty taste of several 0.1M chloride salts solutions have and to several acids according to electrophysiological studies. Unfortunately, no further comparative studies have been developed to described species differences on these two tastes.

The concept of palatability in pigs - feed appetence and feed intake - practical applications in swine nutrition

Palatability is the perception of the quality of the food present in the oral cavity through oronasal sensing mechanisms. The perception is related to the food characteristics (quantity and nature of volatile – i.e. odorants - and non-volatile – i.e. tastants - compounds released, nutrient density, physical features such as texture, particle size and water holding capacity among others) but also to the physiological and genetic capacity of the individual and its metabolic status.

Optimisation of feed intake in farm animals has been a major area of agricultural research over the last decades. Appetite measured as meal length and eating rate has the highest phenotypic correlation with rates of growth and fat deposition in fattening pigs when compared with other feed intake measures (Rauw *et al.*, 2006). The short-term event of meal initiation is determined in part by feed appetence resulting from the processing of peripheral sensing (i.e. somatic, olfactory and gustatory stimuli).

Feed intake (i.e. a meal) is a result of integrating the short-term events with long-term signals that monitor the nutritional status of the animal so that the drive to eat decreases progressively as feed is consumed until a state of 'satiation' arrives and meal termination occurs (Cummings and Overduin, 2007; Forbes, 1998). Peripheral chemosensing seems to be subject to regulation by satiation signals. Smell acuity increased in fasting rats while decreased during fed states (Baly *et al.*, 2007; Aimé *et al.*, 2007). Receptors for anorectic (leptin and cholecistokinine – CCK) and orexigenic hormones have been identified in the olfactory epithelium and taste buds (Hoppe *et al.*, 2006;). These receptors mediate the smell and taste sensitivity depending on the animal nutritional status. Therefore feed appetence is modulated by the physiological feeding state of animals through alterations in the sensitivity of the chemosensory systems. In pigs, both gastric and intra-duodenal injections of glucose but not elevated plasma glucose levels significantly decreased appetite for sugar (Baldwin, 1996).

The physiological capacity of peripheral sensing is species dependent an adapted to the recognition of the food available in a given ecological neach. Pigs (*Sus scrofa domesticus*) are ungulate mammalians of the order of artiodactyla. Other farm or domestic ungulate animals include the horse, donkey, cattle/bison, sheep and goat. All domestic ungulate animals are considered ruminant (cattle, sheep and goat) or non-ruminant (horse and donkey) herbivores (i.e. an animal that is adapted to eat primarily plants and not meat) except pigs. Uniquely among domestic ungulates, hogs in the wild have evolved to an omnivorous diet eating grass, leaves, roots, grains (e.g. cereals) but also insects, worms, frogs, toads, mice, eggs, chicks and animal remains when found.

Many of the innate based feeding behaviours will be determined by the taste and smell perception mechanisms. Animals react with caution in front of novel food sources. Neophobia (defined as initial reluctance or avoidance to ingest a novel feed or food) seems to be quite universally conserved across species from humans to chickens including lambs, cattle, dog and pigs (Roura *et al.*, 2008a). Novel ingredients evoking bitter taste elicit a strong neophobic feeding behaviour (Mawson *et al.*, 1993). Bitter taste has been related to naturally occurring compounds mainly of plant origin and mediates the behavioural aversions to feed (Du Toit *et al.*, 1991; Forbes, 1998; Mote *et al.*, 2007). Several are carbon-based (such as terpenes and phenolics – e.g. lignin and tannins) while a few others are nitrogen-based (alkaloids – e.g. caffeine - and cyanogens).

Feed ingredients used in pig diets have been selected for a high nutritional value what implies a low ANF content in most cases. Nevertheless, market prices of traditional feed ingredients, such as corn, wheat, rice, milk whey, soybean and protein concentrates, may fluctuate and reach costs that may force animal nutritionists to consider the utilisation of alternative protein and energy sources, such as rapeseed, peas, DDGS, sunflower and sorghum, in which palatability is likely to play a limiting role and trigger strong neophobic responses. According to our double choice feed model in piglets, responses to bitter taste aversions are more vigorous than hedonic preferences (Tedó *et al.*, 2007). Although not necessarily related only to bitter taste, pigs have shown depressed feed intake and performance to plant derived compounds present in wheat (Cadogan and Choct, 2003), soybean (Palacios *et al.*, 2004) and soybean concentrates (Lenehan *et al.*, 2007), sorghum (Roese *et al.*, 2003), rapeseed and canola meals (Lee and Hill, 1983; Mawson *et al.*, 1993; Owusu-Asiedu *et al.*, 2003; Roura *et al.*, 2007), peas (Heng *et al.*, 2006; Mathé *et al.*, 2004), sunflower meal (Roura *et al.*, 2007), DDGS (Hastad, 2005; Linneen *et al.*, 2008; Seabolt, 2008) and plant extracts (Schöne *et al.*, 2006; Windisch *et al.*, 2008) as well as to other non-plant derived feed additives such as antibiotics growth promoters and other pharmaceutical compounds (Nelson and Sanregret, 1997).

Preferences for cereals and protein sources in piglets have been recently reported by Solà-Oriol *et al.* (2009) and Solà-Oriol (2008a). Among the 15 protein sources tested, potato protein, rapeseed and sunflower meals showed the lowest values of acceptance presumably related to their content in alkaloids, glucosinolates and lignin among other ANF. Strong acids such as sulphuric and chloride present in the acid milk whey could have caused an unpleasant stimulus as well. On the other hand, ingredients releasing stimulating umami (free AA), sweet (high lactose levels) or salt (high Na levels) tastes in absence of putative bitter compounds showed the highest preference values (Roura *et al.*, 2007, 2008a).

Complex diets result from feed formulation using ingredients of high nutritional value, such as processed cereals and highly digestible protein sources, and are widely being used in young animals mainly to facilitate the weaning process by maximising nutrient intake. Significant increases in feed intake together with improved growth have been reported in complex diets compared to simple diets (Mahan *et al.*, 2004; Mavromichalis *et al.*, 2001; McCracken *et al.*, 1995; Wolter *et al.*, 2003). We observed from several of these studies that the increases of feed intake could explain most of the improved performance and we believe that feed palatability may play a central role in increasing meal size and frequency particularly in early weaned young animals. For example, rice was highly preferred to corn in nursing pigs (Solà-Oriol *et al.*, 2009) and weanling pigs fed a corn based diet had significantly poorer feed intake and growth than those fed a rice based diet (Hongtrakul *et al.*, 1998; Vicente *et al.*, 2008). A positive significant Pearson's correlation ($r=0.48$; $P=0.015$) was found between the preference and the rate of glucose release of 25 feed ingredients in piglets (Solà-Oriol *et al.*, 2008). Thus, piglets seem to prefer cereal sources that have rapid glucose availability (such as rice) in contrast with the post-ingestive down regulating mechanisms described by Baldwin (1996).

In the review by Tokach *et al.* (2003), a few ingredients are identified as having a direct positive impact on feed intake in piglets: dried whey and to a lesser extend other lactose and carbohydrate sources, whey protein concentrate, spray-dried animal plasma and blood meals, dried porcine solubles and high quality fish meal. Based on these studies, some of the most palatable ingredients in pig diets have been found to share several features in common: (1) they are animal derived; (2) they do not have known ANF or other toxic compounds that may stimulate bitterness or pungency; (3) they are highly digestible energy or protein sources, (4) they all contain significant fractions of putative salty or sweet and/or umami tastants (Roura *et al.*, 2008a).

Substitution of high quality and highly preferred ingredients in piglet diets has been addressed by several authors. High quality lactose sources (dried whey) and spray-dried animal plasma may only partially be substituted by other good quality ingredients in piglets (Tokach *et al.*, 2003). Our research group and partners have been heavily involved in the development and application of new flavours to improve palatability and feed intake in piglets. We have been successful in proving the effectiveness of selected flavours in masking some low palatability ingredients such as rapeseed and sunflower meals by preventing strong neophobic responses (Roura *et al.*, 2007). Furthermore, milk derivatives (dried whey) and cooked cereals where totally or partially substituted by non-cooked cereals and soybean and rapeseed meals without an impact on piglet performance only when a selected flavour was added (Tedó *et al.*, 2008).

There are several circumstances under modern pig husbandry were feed consumption is well below the animal's maximum capacity: lactating sow, weanling piglets and diet changes (neophobia) particularly when including alternative protein sources likely to elicit bitter cues (ANFs). Other circumstances that involve feed intake regulation mechanisms include environmental effects (heat stress) that might be particularly alarming in lactating sows or excessive energy intake (and fat deposition) such as in fattening-finishing pigs. The review of oronasal sensing mechanisms and effects on swine feeding behaviour leads us to believe that the manipulation of the sensorial properties of feed is a viable strategy to help optimise the feed intake patterns under commercial conditions. Plant derived ANFs do not always pose a true potential threat for animal health (Pass and Foley, 2000) what opens a door to the use of masking sensory feed additives (Launchbaugh *et al.*, 1997; Roura *et al.*, 2007). Innate feed preferences (linked with peripheral sensing) are particularly relevant and may determine feeding rates in young animals. Both sweet and umami and possibly salty tastes trigger pleasant stimuli, therefore the discovery of high intensity tastants and pleasant odorants should enhance feed appetence and presumably lead to increases in feed intake in young animals. The opposite may also be adequate, that is stimulation of bitter, pungent and deterrent flavours (such as rancid notes) to modulate neophobic responses and limit feed intake. Other learnings from the peripheral sensing mechanisms include the potential use of cooling agents to help prevent heat stress.

In summary, we propose several peripheral-based strategies that may help improve feed intake in swine:
1. Piglet chemosensing conditioning by early maternally driven exposure.
2. Hull removal and lowering ANF content of feeds and feed ingredients.
3. Ingredient technology (cooking, enzymatic treatments…) that facilitates either oral nutrient release such as glucose or some savoury amino acids or inactivates ingredient ANFs.
4. Use of high preference flavours to enhance early feed intake in the piglet or to mask dietary changes and use of alternative feed ingredients.
5. Development of new high intensity tastants or bitter blockers based on the pig chemosensing sensitivity.
6. Use of dietary cooling agents to help prevent heat stress conditions particularly in sows.
7. Use of our understanding of the pig chemosensing biology to prevent excess energy intakes in adult animals.

References

Adams, D.R., 2005. Olfactory and non-olfactory epithelia in the nasal cavity of the mouse, *Peromyscus*. American Journal of Anatomy 133: 37-49.

Aimé, P. Duchamp-Viret, P., Chaput, M.A., Savinger, A., Mahfouz, M. and Julliard, A.K., 2007. Hunger increases and satiation decreases rat olfactory detection performances for a neutral odor. Behavioural Brain Research 179: 258-264.

Bachmanov, A.A. and Beauchamp, G.K., 2007. Taste receptor genes. Annual Review of Nutrition 27: 387-412.

Baldwin, B.A., 1996. Postingestive factors influencing operant sugar intake in pigs. Physiology & Behavioural 60: 283-286.

Balog, J.M. and Millar, R.I., 1989. Influence of the sense of taste on broiler chick feed consumption. Poultry Science 68: 1519-1526.

Baly, C., Aioun, J., Badonnel, K., Lacroix M.C., Durieux, D. and Schlegel, C., 2007. Leptin and its receptors are present in the rat olfactory mucosa and modulated by the nutritional status. Brain Research 1129: 130-141.

Bautista, D.M., Siemens, J., Glazer, J.M., Tsurudal, P.R., Basbaum, A.I., Stucky, C.L., Jordt, S.E. and Julius, D., 2007. The menthol receptor TRPM8 is the principal detector of environmental cold. Nature 448: 204-208.

Beauchamp, G.K., Maller, O. and Roger jr., J.G., 1977. Flavor preferences in cats (Felis catus and Panthera sp.). Journal of Comparative Physiology 91: 1118-1127.

Behrens, M. and Meyerhof, W., 2006. Bitter taste receptors and human bitter taste perception. Cellular and Molecular Life Sciences 63: 1501-1509.

Bikker, P., Fontanillas, R. and Roura, E., 2003. Dietary supplementation with botanical compounds depresses piglet feed intake while faecal E. Coli counts remain unchanged. Journal of Animal Science 81 (suppl.1): 203.

Blair, R. and FitzSimons, J., 1970. A note on the voluntary feed intake and growth of pigs diets containing an extremely bitter compound. Animal Production 12: 529-530.

Booth, W.D. and White, C.A., 1988. The isolation, purification and some properties of pheromaxein, the pheromonal steroid-binding protein, in porcine submaxillary glands and saliva. Journal of Endocrinology 118: 47-57.

Buck, L. and Axel, R., 1991. A novel multigene family may encode odorant receptors - a molecular-basis for odor recognition. Cell 65: 175-187.

Buck, L., 2005. Unraveling the sense of smell (Nobel lecture). Angewandte Chemie (International Edition in English) 44: 6128-6140.

Cadogan, D.J. and Choct, M., 2003. Wheat carbohydrate constituents and voluntary feed intake in pigs. Recent Advances in Animal Nutrition in Australia 14: 181-192.

Campbell, R.G., 1976. A note on the use of a feed flavour to stimulate the feed intake of weaner pigs. Animal Production 23: 417-419.

Caterina, M.J., Schumacher, M.A., Tominaga, M., Rosen, T.A., Levine, J.D. and Julius, D., 1997. The capsaicin receptor: a heat-activated ion channel in the pain pathway. Nature 389: 816-824.

Caterina, M.J., 2007. Sticky spices. Nature 445: 491-492.

Chandrashekar, J., Mueller, K.L., Hoon, M.A., Adler, E., Feng, L., Guo, W. and Zuker, C.S. and Ryba, N.J., 2000. T2Rs function as bitter taste receptors. Cell 100: 703-711.

Chandrashekar, J., Hoon, M.A., Ryba, N.J.P. and Zuker, C.S., 2006. The receptors and cells for mammalian taste. Nature 444: 288-294.

Chaudhari, N., Landin, A.M and Roper, S.D., 2000. A metabotropic glutamate receptor variant functions as a taste receptor. Nature Neuroscience 3: 113-119.

Cummings, D.E. and Overduin, J., 2007. Gastrointestinal regulation of food intake. The Journal of Clinical Investigation 117: 13-23.

Danilova, V., Roberts, T. and Hellekant, G., 1999. Responses of single taste fibers and whole chorda tympani and glossopharyngeal nerve in the domestic pig, sus scrofa. Chemical Senses 24: 301-316.

Dasgupta, K., Singh, A. and Ireland, W.P., 1990. Taste bud density in circumvallate and fungiform papillae of the bovine tongue. Histology and Histopathology 5: 169-172.

DeFazio, R.A., Dvoryanchikov, G., Maruyama, Y., Kim, J.W., Pereira, E., Roper, S.D. and Chaudhari, N., 2006. Separate populations of receptor cells and presynaptic cells in mouse taste buds. The Journal of Neuroscience 26: 3971-3980.

DeRouchey, J.M., Hancock, J.D., Hines, R.H., Maloney, C.A., Lee, D.J., Cao, H., Dean, D.W. and Park, S., 2004. Effects of rancidity and free fatty acids in choice white grease on growth performance and nutrient digestibility in weanling pigs. Journal Animal Science 82: 2937-2944.

DeSimone J.A. and Lyall, V., 2006. Taste receptors in the gastrointestinal tract III. Salty and sour taste: sensing of sodium and protons by the tongue. American Journal Physiology. Gastrointestinal and Liver Physiology 291: G1005-10.

Dhaka, A., Viswanath, V. and Patapoutian, A., 2006. Trp ion channels and temperature sensation. Annual Review of Neuroscience 29: 135-161.

Diebold, G. and Eidelsburger, U., 2006. Acidification of diets as an alternative to antibiotic growth promoters. In: Barug, D., De Jong, J., Kies, A.K. and Verstegen, M.W.A. (eds.) Antimicrobial growth promoters. Wageningen Academic Publishers, the Netherlands, pp. 311-327.

Dryer, L., 2000. Evolution of odorant receptors. BioEssays 22: 803-810.

Du Toit, J.T., Provenza, F.D. and Nastis, A., 1991. Conditioned taste aversions: how sick must a ruminant get before it learns about toxicity in foods? Applied Animal Behaviour Science 30: 35-46.

Duchamp-Viret, P., Chaput, M.A. and Duchamp, A., 1999. Odor response properties of rat olfactory receptor neurons. Science 284: 2171-2174.

Dudai, Y., 1999. The smell of representations. Neuron 23: 633-635.

Dulac, C., 2000. The physiology of taste, vintage 2000. Cell 100: 607-610.

Dulai, K.S. and von Dornum, M., 1999. The evolution of trichromatic color vision by opsin gene duplication of new world and old world primates. Genome Research 9: 629-638.

Eisemann, J.H. and van Heugten, E., 2007. Response of pigs to dietary inclusion of formic acid and ammonium formate. Journal of Animal Science 85: 1530-1539.

Ettle, T., Mentschel, K. and Roth, F.X., 2004. Dietary self-selection for organic acids by the piglet. Archives of Animal Nutrition 58: 379-386.

Firestein, S., 2001. How the olfactory system makes sense of scents. Nature l413: 211-218.

Forbes, J.M., 1998. Dietary awareness. Applied Animal Behaviour Science 57: 287-297.

Gilad, Y., Wiebe, V., Przeworski, M., Lancet, D. and Pääbo, S., 2004. Loss of olfactory receptor genes coincides with the acquisition of full trichromatic vision in primates. Public Library of Science Biology. 2: 120-124.

Glaser, D., Wanner, M., Tinti, J.M. and Nofre, C., 2000. Gustatory responses of pigs to various natural and artificial compounds known to be sweet in man. Food Chemistry 68: 375-385.

Goatcher, W.D. and Church, D.C., 1970. Taste responses in ruminants. I. Reactions of sheep to sugars, saccharin, ethanol and salt. Journal of Animal Science 30: 777-783.

Godfrey, P.A., Malnic, B. and Buck, L.B., 2004. The mouse olfactory receptor gene family. Proceedings of the National Academy of Science of the USA 101: 2156-2161.

Goff, S. A. and Klee, H.J., 2006. Plant volatile compounds: Sensory cues for health and nutritional value? Science 311: 815-819.

Grigorov, M.G., Schlichtherle-Cerny, H., Affolter, M. and Kochhar, S., 2003. Design of Virtual Libraries of Umami-Tasting Molecules. Journal of Chemical Information and Modeling 43: 1248-1258.

Grosmaitre, X., Santarelli, L.C., Tan, J., Luo, M. and Ma, M., 2007. Dual functions of mammalian olfactory sensory neurons as odor detectors and mechanical sensors. Nature Neuroscience 10: 348-354.

Guiraudie-Capraz, G., Slomianny, M.C., Pageat, P., Malosse, C., Cain, A.H., Orgeur, P. and Nagnan-Le Meillour, P., 2005. Biochemical and chemical supports for a transnatal olfactory continuity through sow maternal fluids. Chemical Senses 30: 241-251.

Hard, C., Segerstad, A.F. and Hellekant, G., 1989a. The sweet taste in the calf. I. Chorda Tympani proper nerve responses to taste stimulation of the tongue. Physiology & Behavior 45: 633-638.

Hard, C., Segerstad, A.F. and Hellekant, G., 1989b. The sweet taste in the calf. II. Glossopharyngeal nerve responses to taste stimulation of the tongue. Physiology & Behavior 45: 1043-1047.

Hastad, C., 2005. The use of dried distillers grains with solubles in swine diets. PhD Dissertation. Kansas State University, Manhattan, Ka, USA.

Hellekant, G. and Danilova, V., 1999. Taste in domestic pig, sus scrofa. Journal of Animal Physiology and Animal Nutrition 82: 8-24.

Hellekant, G., Hard, C., Segerstad, A.F. and Roberts, T.W., 1994. The sweet taste in the calf. III. Behavioral responses to sweeteners. Physiology & Behavior 56: 555-562.

Hepper, P.G., 1988. Adaptive foetal learning: prenatal exposure to garlic affects postnatal preferences. Animal Behaviour 36: 935-936.

Heng, L., Vincken, J.P., Van Koningsveld, G., Legger, A., Gruppen, H., Van Boekel, T., Roozen, J. and Voragen, F., 2006. Bitterness of saponins and their content in dry peas. Journal of the Science of Food and Agriculture 86: 1225-1231.

Hill, G.D. and Tamminga, S., 1998. The effects of antinutritional factors in legume seed and rapeseed on ruminant nutrition. In: Jansman, A.J.M., Hill, G.D., Huisman J. and Van der Poel, A.F.B. (eds.) Recent advances of research in anti-nutritional factors in legume seeds and rapeseed. Wageningen Academic Publishers, the Netherlands, pp. 157-172.

Hongtrakul, K., Goodband, R.D., Behnke, K.C., Nelssen, J.L., Tokach, M.D., Bergström, J.R., Nessmith Jr., W.B. and Kim, I.H., 1998. The effects of extrusion processing of carbohydrate sources on weanling pig performance. Journal of Animal Science 76: 3034-3042.

Hoppe, R., Schwarzenbacher, K., Hass, N., Kaiser, T., Horenburg, S., and Breer, H., 2006. Gut hormones in chemosensory systems. A chemosensory system in the gut? Chemical Senses 31: 64 (abstract).

Huang, A.L., Chen, X., Hoon, M.A., Chadrashekar, J., Guo, W., Trankner, D., Ryba, N.J.P. and Zuker C.S., 2006. The cells and logic for mammalian sour taste detection. Nature 442: 934-938.

Humphrey, B., Tedó, G., Klasing, K.C. and Roura, E., 2009. Caractérisation des polypeptides récepteurs umami porcins (pT1r1 et pT1r3). Journées de la Recherche Porcine 41: 165-166.

Hyde, R.J. and Witherly, S.A., 1993. Dynamic contrast: A sensory contribution to palatability. Appetite 21: 1-16.

Ikeda, K., 1909. On a new seasoning. Journal of the Tokyo Chemical Society 30: 820-83.

International Chicken Genome Sequencing Consortium., 2004. Sequence and comparative analysis of the chicken genome provide unique perspectives on vertebrate evolution. Nature 432: 695-716.

Ishimaru, Y., Inada, H., Kubota, M., Zhuang, H., Tominaga, M. and Matsunami, H., 2006. Transient receptor potential family members PKD1L3 and PKD2L1 form a candidate sour taste receptor. Proceedings of the National Academy of Science of the USA 103: 12569-12574.

Jamroz, D., Wertelecki, T., Houszka, M. and Kamel, C., 2006. Influence of diet type on the inclusion of plant origin active substances on morphological and histochemical characteristics of the stomach and jejunum walls in chicken. Journal of Animal Physiology and Animal Nutrition 90: 255-268.

Jones, R.B. and Roper, T.J., 1997. Olfaction in the domestic fowl: a critical review. Physiology & Behavior 62: 1009-1018.

Jordt, S.E., McKemy, D.D. and Julius, D., 2003. Lessons from peppers and peppermint: the molecular logic of thermosensation. Current Opinion in Neurobiology 13: 487-492.

Julius, D. and Basbaum, A.I., 2001. Molecular mechanisms of nociception. Nature 413: 203-210.

Katschinski, M., 2000. Nutritional implications of cephalic phase gastrointestinal responses. Appetite 34: 189-196.

Kennedy, J.M. and Baldwin, B.A. 1972., Taste preferences in pigs for nutritive and non-nutritive sweet solutions. Animal Behaviour 20: 706-718.

Kirifides, M.L., Kurnellas, M.P., Clark, L. and Bryant, B.P., 2004. Calcium responses of chicken trigeminal ganglion neurons to methyl anthranilate and capsaicin. Journal of Experimental Biology 207: 715-722.

Kiuchi, S., Yamada, T., Kiyokawa, N., Saito, T., Fujimoto, J. and Yasue, H. 2006. Genomic structure of swine taste receptor family 1 member 3, TAS1R3, and its expression in tissues. Cytogenetic and Genome Research 115: 51-61.

Klasing, K.C., 1998. Comparative avian nutrition. CAB International, Wallingford, United Kingdom, Preface vi pp.

Kuenzel, W.J., 1989. Neuroanatomical substrates involved in the control of food intake. Poultry Science 68: 926-937.

Lallès, J.P., Bosi, P., Janczyk, P., Koopmans, S.J. and Torrallardona, D. 2009. Impact of bioactive substances on the gastrointestinal tract and performance of weaned piglets: a review. Animal, in press. doi:10.1017/S175173110900398X

Langendijk, P., Bolhuis, J.E. and Laurenssen, B.F.A., 2007. Effects of pre- and postnatal exposure to garlic and aniseed flavour on pre- and postweaning feed intake in pigs. Livestock Science 108: 284-287.

Langhout, P., 2000. Part 2: New additives for broiler chickens. World Poultry 16: 22-27.

Laugerette, F., Passilly-Degrace, P., Patris, B., Niot, I., Febbraio, M., Montmayeur, J.P. and Besnard, P., 2005. CD36 involvement in orosensory detection of dietary lipids, spontaneous fat preference, and digestive secretions. The Journal of Clinical Investigation 115: 3177-3184.

Launchbaugh, K. L., Provenza, F.D. and Werkmeister, M.J., 1997. Overcoming feed neophobia by domestic ruminants through addition of a familiar flavor and repeated exposure to novel foods. Applied Animal Behaviour Science 54: 327-334.

Lay, D.C., Haussman, M.F., Buchanan, H.S. and Daniels, M.J., 1999. Danger to pigs due to crushing can be reduced by the use of a simulated udder. Journal of Animal Science 77: 2060-2064.

Lee, P.A. and Hill, R., 1983. Voluntary feed intake of growing pigs given diets containing rapeseed meal, from different types and varieties of rape, as the only protein supplement. Journal of Nutrition 50: 661-671.

Leibetseder, J., 1980. Introduction to animal olfactory and taste physiology. Fortschritte in der Tierphysiologie und Tierernährung 11: 11-20.

Lenehan, N.A., DeRouchey, J.M., Goodband, R.D., Tokach, M.D., Dritz, S.S., Nelssen, J.L., Groesbeck, C.N. and Lawrence, K.R., 2007. Evaluation of soy protein concentrates in nursery diets. Journal of Animal Science 85: 3013-3021.

Li, X., Staszewski, L., Xu, H., Durick, K., Zoller, M. and Adler, E. 2002. Human receptors for sweet and umami taste. The Proceedings of the National Academy of Science of the United States of America 99: 4692-4696.

Li, X., Li, W., Wang, H., Cao, J., Maehashi, K., Huang, L., Bachmanov, A.A., Reed, D.R., Legrand-Defretin, V., Bauchamp, G.K. and Brand, J.G., 2005. Pseudogenization of a sweet-receptor gene accounts for cats' indifference toward sugar. Public Library of Science Genetics 1: 27-35.

Lin, W., Finger, T.E., Rossier, B.C. and Kinnamon, S.C., 1999. Epithelial Na+ channel subunits in rat taste cells: Localization and regulation by aldosterone. Journal of Computational Neuroscience 405: 406-420.

Lindemann, B., 2001. Receptors and transduction in taste. Nature 413: 219-225.

Linneen, S.K., DeRouchey, J.M., Dritz,S.S., Goodband, R.D., Tokach, M.D. and Nelssen, J.L., 2008. Effects of dried distillers grains with solubles on growing and finishing pig performance in a commercial environment. Journal of Animal Science 86: 1579-1587.

Lyall V., Heck G.L., Vinnikova, A.K., Ghosh, S., Phan, T.T., Alam, R.I., Russell, O.F., Malik, S.A., Bigbee, J.W. and DeSimone, J.A., 2004. The mammalian amiloride-insensitive non-specific salt taste receptor is a vanilloid receptor-1 variant. The Journal of Physiology 558: 147-159.

Mahan, D.C., Newton, E.A. and Cera, K.R., 1996. Effect of supplemental sodium chloride, sodium phosphate, or hydrochloric acid in starter pig diets containing dried whey. Journal of Animal Science 74: 1217-1222.

Mahan, D.C., Fastinger, N.D. and Peters, J.C., 2004. Effect of dietary complexity and dietary lactose levels during three starter phases on postweaning pig performance. Journal of Animal Science 82: 2790-2797.

Malnic, B., Hirono, J., Sato, T., and Buck, L.B., 1999. Combinatorial receptor codes for odors. Cell 96: 713-723.

Malnic, B., Godfrey, P.A. and Buck, L.B., 2004. The human olfactory receptor gene family. Proceedings of the National Academy of Sciences of the United States of America 101: 2584-2589.

Margolskee, R.F., 2002. Molecular mechanisms of bitter and sweet taste transduction. Journal of Biological Chemistry 277: 1-4.

Marlier, L., Gaugler, C., Astruc, D. and Messer, J., 2007. The olfactory sensitivity of the premature newborn. Archives of Pediatrics 14: 45-53.

Mathé, D., Monéger, R. and Guillou, D., 2004. Effet du pois protéagineux sur les performances et le comportement du porc lors des transitions alimentaires. Journées de la Recherche Porcine 35: 127-132.

Matsson, J.L., Fry, W.N., Boward, C.A. and Miller, E., 1978. Maturation of the visual evoked response in newborn miniature pigs. American Journal of Veterinary Research 39: 1279-1281.

Matsunami, H., Montmayeur, J.P. and Buck, L.B., 2000. A family of candidate taste receptors in human and mouse. Nature 404: 601-604.

Matsunami, H. and Amrein, H., 2003. Taste and pheromone perception in mammals and flies. Genome Biology 4: 220.

Matsuo, R., 2000. Role of saliva in the maintenance of taste sensitivity. Critical Reviews on Oral Biology and Medicine 11: 216-229.

Mattes, R.D., 2000. Nutritional implications of the cephalic-phase salivary response. Appetite 34: 177-183.

Mavromichalis, I., Hancock, J.D., Hines, R.H., Senne, B.W. and Cao, H., 2001. Lactose, sucrose, and molasses in simple and complex diets for nursery pigs. Animal Feed Science and Technology 93: 127-135.

Mawson, R., Heaney, R.K., Zdunczyk, Z. and Kozlowska, H., 1993. Rapeseed meal-glucosinolates and their antinutritional effects. Part II. Flavour and palatability. Die Nahrung 37: 336-344.

McCracken, B.A., Gaskins, H.R., Ruwe-Kaiser, P.J., Klasing, K.C. and Jewell, D.E., 1995. Diet-dependent and diet-independent metabolic responses underlie growth stasis of pigs at weaning. Journal of Nutrition 125: 2838-2845.

McIntosh, F.M., Williams, P., Losa, R., Wallace, R.J., Beever, A. and Newbold, C.J., 2003. Effects of essential oils on ruminal microorganisms and their protein metabolism. Applied and Environmental Microbiology 69: 5011-5014.

McLaughlin, C.L., Baile, C.A., Buckholtz, L.L. and Freeman, S.K., 1983. Preferred flavours and performance of weaning pigs. Journal of Animal Science 56: 1287-1293.

Miller, M.G. and Teates, J.F., 1984. Oral somatosensory factors in dietary self-selection after food deprivation and supplementation. Behavioral Neuroscience 98: 424-434.

Mombaerts, P., 1999. Seven-transmembrane proteins as odorant and chemosensory receptors. Science 286: 707-711.

Mombaerts, P., 2000. Better taste through chemistry. Nature Genetics 25: 130-132.

Morrow-Tesch, J. and McGlone, J.J., 1990. Sources of maternal odors and the development of odor preferences in baby pigs. Journal of Animal Science 68: 3563-3571.

Mote, T.E., Villalba, J.J. and Provenza, F.D., 2007. Relative availability of tannin- and terpene-containing foods affects food intake and preference in lambs. Journal of Chemical Ecology 33: 1197-1206.

Nelson, S. L. and Sanregret, J. D., 1997. Response of pigs to bitter-tasting compounds. Chemical Senses 22: 129-132.

Nelson, G., Chandrashekar, J., Hoon, M.A., Feng, L., Zhao, G., Ryba, N.J.P. and Zuker, C.S., 2002. An amino-acid taste receptor. Nature 416: 199-202.

Niimura, Y. and Nei, M., 2007. Extensive gains and losses of olfactory receptor genes in mammalian evolution. Public Library of Science ONE 2(8): e708. doi:10.1371/journal.pone.0000708.

Ninomiya, K., 2002. Umami: a universal taste. Food Reviews International 18: 23-38.

Ninomiya, Y. and Fuyakoshi, M., 1989. Peripheral neural basis for behavioural discrimination between glutamate and the four basic taste substances in mice. Comparative Biochemistry and Physiology 93: 371-376.

Oakley, B., 1998. Taste neurons have multiple inductive roles in mammalian gustatory development. Annals of the New York Academy of Sciences. 855: 50-57.

Ohara, I., Otsuka, S.I. and Yugari Y., 1988. Cephalic phase response of pancreatic exocrine secretion in conscious dogs. American Journal of Physiology. Gastrointestinal and Liver Physiology 17: G424-G428.

Owusu-Asiedu, A., Ekpe, E.D., Zijlstra, R.T., Patience, J.F., Classen, H.L. and Simmins, P.H., 2003. Effect of dietary fibre on feed intake and growth performance in weaned pigs. Canadian Journal of Animal Science 83: 628 (abstract).

Palacios, M.F., Easter, R.A., Soltwedel, K.T., Parsons, C.M., Douglas, M.W., Hymowitz, T. and Pettigrew, J.E., 2004. Effect of soybean variety and processing on growth performance of young chicks and pigs. Journal of Animal Science 82: 1108-1114.

Palmer, R.K., 2007. The pharmacology and signaling of bitter, sweet, and umami taste sensing. Molecular Interventions 7: 87-98.

Partanen, K., 2001. Organic acids. Their efficacy and modes of action in pigs. In: Piva, A., Bach Knudsen K.E. and Lindberg, J.E. (eds.). Gut environment of pigs. Nottingham University Press, Nottingham, UK, pp. 201-218.

Partanen, K., Siljander-Rasi, H., Pentikäinen, J., Pelkonen, S. and Fossi, M., 2007. Effects of weaning age and formic acid-based feed additives on pigs from weaning to slaughter. Archives of Animal Nutrition 61: 336-356.

Pass, G.J. and Foley, W.J., 2000. Plant secondary metabolites as mammalian feeding deterrents: separating the effects of the taste of salicin from its post-ingestive consequences in the common brushtail possum (*Trichosurus vulpecula*). Journal of Comparative Physiology 170: 185-192.

Pavlov, I.P., 1902. The work of the digestive glands. Charles Griffin, London, UK.

Picard, M.L., Uzu, G., Dunnington, E.A. and Siegel, P.B., 1993. Food intake adjustments of chicks: short term reactions to deficiencies in lysine, methionine and tryptophan. British Poultry Science 34: 737-746.

Pihlström, H., Fortelius, M., Hemilä, S., Forsman, R. and Reuter, T., 2005. Scaling of mammalian ethmoid bones can predict olfactory organ size and performance. Proceedings of the Royal Society B 272: 957-962.

Pichersky, E., Noel, J.P. and Dudareva, N., 2006. Biosynthesis of plant volatiles: Nature's diversity and inigenuity. Science 311: 808-811.

Platel, K. and Srinivasan, K., 2001. A study of the digestive stimulant action of select spices in experimental rats. Journal of Food Science & Technology 38: 358-361.

Platel, K. and Srinivasan, K., 2004. Digestive stimulant action of spices: A myth or reality? Indian Journal of Medical Research 119: 167-179.

Quignon. P., Giraud, M., Rimbault, M., Lavigne, P., Tacher, S., Morin, E., Retout, E., Valin, A.S., Lindblad-Toh, K., Nicolas, J. and Galibert, F., 2005. The dog and rat olfactory repertoires. Genome Biology 6: R83.

Rauw, W.M., Soler, J., Tibau, J., Reixach, J. and Gomez Raya, L., 2006. Feeding time and feeding rate and its relationship with feed intake, feed efficiency, growth rate, and rate of fat deposition in growing Duroc barrows. Journal of Animal Science 84: 3404-3409.

Ressler, K.J., Sullivan, S.L., and Buck, L.B., 1993. A zonal organization of odorant receptor gene expression in the olfactory epithelium. Cell 73: 597-609.

Reznik, G.K., 1990. Comparative anatomy, physiology, and function of the upper respiratory tract. Environmental Health Perspectives 85: 171-176.

Robinson, P.P. and Winkles, P.A., 1990. Quantitative study of fungiform papillae and taste buds on the cat's tongue. Anatomical Record 225: 108-111.

Roese, G.J., Barker, L.J., Collins, D.P., Brewster, C.J., Holland, J.F. and Giles, L.R., 2003. Sorghum or sorghum kernel can replace wheat in weaner pig diets. In: Paterson, J.E. (ed.) Manipulating Pig Production IX, Australian Pig Science Association (Inc.), Werribee, Australia, pp. 161.

Rohde Parfet, K.A. and Gonyou, H.W. 1991. Attraction of newborn piglets to auditory, visual, olfactory and tactile stimuli. Journal of Animal Science 69: 125-133.

Romanov R.A. and Kolesnikov S.S., 2006. Electrophysiologically identified subpopulations of taste bud cells. Neuroscience Letters 395: 249-254.

Roper, S.D., 1992. The microphysiology of peripheral taste organs. The Journal of Neuroscience 12: 1127-1134.

Roper, S.D., 2006. Cell communication in taste buds. Cellular and Molecular Life Sciences 63: 1494-1500.

Roth, F.X. and Kirchgessner, M., 1998. Organic acids as feed additives for young pigs: nutritional and gastrointestinal effects. Journal of Animal and Feed Sciences 7: 25-33.

Rouquier, S. and Giorgi, D., 2007. Olfactory receptor gene repertoires in mammals. Mutation Research 616: 95-102.

Roura, E., Fontanillas, R. and Bikker, P., 2004. Botanical additives did not affect feed intake in weanling pigs when masked by a flavor but growth or faecal consistency were not affected either. Journal of Animal Science 82 (Suppl. 1): 27.

Roura, E., Solà-Oriol, D., Mallo, J.J., van Hees, H., Tedó, G. and Torrallardona, D., 2007. Utilisation de tourteau de colza et de tourteau de tournesol à doses élevées dans les aliments porcs. Evaluation des préférences alimentaires et de la consommation volontaire d'aliment. Journées de la Recherche Porcine 39: 163-166.

Roura, E., Holt, R. and Klasing, K.C., 2008b. Identification of the porcine umami taste receptor dimer responsible for the taste of amino acids. Journal of Animal Science 81 (Suppl. 1): 203.

Roura, E., Humphrey, B., Tedó, G. and Ipharragerre, I., 2008a. Unfolding the codes of short-term feed appetence in farm and companion animals. A comparative oronasal nutrient sensing biology review. Canadian Journal of Animal Science 88: 535-558.

Ruiz, C., Gutknecht, S., Delay, E. and Kinnamon, S., 2006. Detection of NaCl and KCl in TRPV1 knockout mice. Chemical Senses 31: 813-820.

Salmon-Legagneur, E. and Fevrier, R., 1956. Les préférences alimentaires du porcelet II. – le sucre dans les aliments pour porcelets. Annales de Zootechnie 5: 73-79.

Schöne, F., Vetter, A., Hartung, H., Bergmann, H., Biertümpfel, A., Richter, G., Müller, S. and Breitschuh, G., 2006. Effects of essential oils from fenner (Foeniculi aetheroleum) and caraway (Carvi aetheroleum) in pigs. Journal of Animal Physiology and Animal Nutrition 90: 500-510.

Seabolt, B.S., 2008. Palatability of feed ingredients in nursery pigs. Master Thesis. Animal Science/Nutrition, North Carolina State University, Raleigh, NC, USA, pp. 64-83.

Shepherd, G.M., 2006. Smell images and the flavour system in the human brain. Nature 444: 316-321.

Shi, P. and Zhang, J., 2006. Contrasting modes of evolution between vertebrate sweet/umami receptor genes and bitter receptor genes. Molecular Biology and Evolution 23: 292-300.

Schmale, H., Holtgreve-Grez, H. And Christiansen, H., 1990. Possible role of salivary gland protein in taste perception indicated by homology to lipophilic-ligand carrier proteins. Nature 343: 366-369.

Solà-Oriol, D., Roura, E. and Torrallardona, D., 2007. Pig preference for cereal based diets, relationship with their digestibility and physical properties. Livestock Science 108: 190-193.

Solà-Oriol, D., 2008. Quantitative evaluation of the palatability of feed ingredients in swine. PhD Thesis, Animal Nutrition, IRTA, Centre Mas de Bover, Constantí, Tarragona, Spain, pp. 130-152.

Solà-Oriol, D., Pujol, S., van Kempen, T., Roura, E. and Torrallardona, D., 2008. Glucose release correlates with piglet's preference for cereal based feeds. Journal of Animal Science 86 (E-suppl. 3): 96.

Solà-Oriol, D., Roura, E. and Torrallardona, D., 2009. Feed preferentes in pigs: Effect of cereal sources at different inclusion rates. Journal of Animal Science, in press.

Spielman, A.I., 1990. Interaction of saliva and taste. Journal of Dental Research 69: 838-843.

Srinivasan, K., 2007. Black pepper and its pungent principle-piperine: a review of diverse physiological effects. Critical Reviews in Food Science and Nutrition 47: 735-748.

Stammati, A., Bonsi, P., Zucco, F., Moezelaar, R., Alakomi, H.L. and Von Wright, A., 1999. Toxicity of selected plant volatiles in microbial and mammalian short-term assays. Food and Chemical Toxicology 37: 813-823.

Szczesniak, A. S., 2002. Texture is a sensory property. Food Quality and Preference 13: 215-225.

Tedó, G., Puigvert, X., Manteca, X. and Roura, E., 2007. An '*in vivo*' 48-hour model to study feed ingredients preferences in post-weaned pigs. In: Van der Honing, Y. (ed.), Book of abstracts of the 58th EAAP Annual Meeting, Dublin, Ireland. Wageningen Academic Publishers, the Netherlands, p. 178.

Tedó, G., Bourry, C., Planchenault, D. and Roura, E., 2008. Effet de l'aromatisation et de la stratégie de formulation sur les performances zootechniques des porcelets en post sevrage. Journées de la Recherche Porcine 40: 223-224.

Teff, K., 2000. Nutritional implications of the cephalic-phase reflexes: endocrine responses. Appetite 34: 206-213.

Tewksbury, J.J. and Nabhan, G.P., 2001. Seed dispersal: directed deterrence by capsaicin in chillies. Nature 412: 403-404.

Tinti, J.M., Glaser, D., Wanner, M. and Nofre, C., 2000. Comparison of gustatory responses to amino acids in pigs and in humans. Lebensmittel Wisssenschaft und Technologie 33:578-583.

Tokach, M.D., Dritz, S.S., Goodband, R.D. and Nelssen, J.S., 2003. Nutritional requirements of the weaned pig. In: Pluske, J.R., Le Dividich, J. and Verstegen M.W.A. (eds.) Weaning the pig. Concepts and consequences. Wageningen Academic Publishers, Wageningen, the Netherlands, pp. 259-299.

Tominaga, M. and Julius, D., 2000. Capsaicin receptor in the pain pathway. The Japanese Journal of Pharmacology 83: 20-24.

Toyono, T., Seta, Y., Kataoka, S., Harada, H., Morotomi, T., Kawano, S., Shigemoto, R. and Toyosima, K., 2002. Expression of the metabotropic glutamate receptor, mGluR4a, in the taste hairs of taste buds in rat gustatory papillae. Archives of Histology and Cytology 65: 91-96.

Toth, A., Blumberg, P.M., Chen, Z. and Kozikowski, A.P., 2004. Design of a high affinity campetitive antagonist of the vanilloid receptor selective for the calcium entry-linked receptor population. Molecular Pharmacology 65: 282-291.

Tripathi, M.K. and Mishra, A.S., 2007. Glucosinolates in animal nutrition: A review. Animal Feed Science and Technology 132: 1-27.

Vicente, B., Valencia, D.G., Pérez-Serrano, M., Lázaro, R. and Mateos, G.G., 2008. The effects of feeding rice in substitution of corn and the degree of starch gelatinization of rice on the digestibility of dietary components and productive performance of young pigs. Journal of Animal Science 86: 119-126.

Wenk, C., 2006. Are herbs, botanicals and other related substances adequate replacements for antimicrobial growth promoters? In: Barug, D., Jong, J., Kies, A.K. and Verstegen, M.W.A. (eds.) Antimicrobial growth promoters. Wageningen Academic Publishers, the Netherlands, pp. 329-340.

Westendarp H., 2005. Essential oils for the nutrition of poultry, swine and ruminants. Deutsche Tierarztliche Wochenschrift 112: 375-380.

Wicher, D., Schäfer, R., Bauernfeind, R., STensmyr, M.C., Heller, R., Heinemann, S.H. and Hansson, B.S., 2008. Drosophila odorant receptors are both ligand-gated and cyclic-nucleotide-activated cation channels. Nature 452: 1007-1011.

Windisch, W., Schedle, K., Plitzner, C. and Kroismayr, A., 2008. Use of phytogenic products as feed additives for swine and poultry. Journal of Animal Science 86: E140-E148, doi:10.2527/jas.2007-0459.

Witt, M., Reutter, K. and Miller, I.J., 2003. Morphology of peripheral taste system. In: Doty, R.L. (ed.) Handbook of olfaction and gustation. Marcel Dekker, New York, USA, pp. 651-677.

Wolter, B.F., Ellis, M., Corrigan, B.P., DeDecker, J.M., Curtis, S.E., Parr, E.N. and Webel, D.M., 2003. Impact of early postweaning growth rate as affected by diet complexity and space allocation on subsequent growth performance of pigs in a wean-to-finish production system. Journal of Animal Science 81: 353-359.

Yamamoto, T., Matsuo, R., Fujimoto, Y., Fukunaga, I., Miyasaka, A. and Imoto, T., 1991. Electrophysiological and behavioural studies on the taste of umami substances in the rat. Physiology & Behavior 49: 919-925.

Young J.M. and Trask B.J., 2002. The sense of smell: Genomics of vertebrate odorant receptors. Human Molecular Genetics 11: 1153-1160.

Zhao, H., Ivic, L., Otaki, J. M., Hashimoto, M., Mikoshiba, K. and Firestein, S., 1998. Functional expression of a mammalian odorant receptor. Science 279: 237-242.

Chapter 6. The diffuse chemosensory system

A. Sbarbati, D. Benati and F. Merigo
Morphological-Biomedical Sciences Department, Human Anatomy and Histology Section, University of Verona, Strada Le Grazie 8, 37134 Verona, Italy; andrea.sbarbati@univr.it

Abstract

The diffuse chemosensory system (DCS) of the digestive apparatus presents a multiplicity of morphological aspects. Recent research provided a description of these elements, which contain molecules of the chemosensory intracellular machinery. However, the functional roles of the different cytotypes belonging to the DCS are not well known. The description of the functional roles of the DCS might be a first step towards the discovery of nutritional approaches targeted on chemosensory mechanisms. The data recently obtained on the DCS suggest that in the future it could be the target of strategic actions aimed to modify the voluntary feed intake in mammals.

Keywords: chemoreception, taste receptor, solitary chemosensory cell, gustducin

Introduction

In the apparatuses of endodermic origin (i.e. the digestive and respiratory apparatuses), a diffuse chemosensory system (DCS) composed by solitary chemosensory cells (SCCs, also called solitary chemoreceptor cells), which have analogies with the taste cells but are not aggregated in buds is present (Sbarbati and Osculati, 2003, 2005b; Finger *et al.*, 2003).

Recent studies demonstrate that the SCCs are densely distributed in strategic areas of the body and suggest that chemoreceptors are involved in important physiologic processes both in respiratory (Gulbransen *et al.*, 2008a) and digestive apparatus (Rozengurt, 2006; Salmon *et al.*, 2007; Mace *et al.*, 2007). Recent studies have linked the chemosensation of the intraluminal content not only to the central nervous system activity (Gulbransen *et al.*, 2008b) but also to local control of the absorptive and secretory processes (Salmon *et al.*, 2007; Mace *et al.*, 2007; Sclafani, 2007, Kellett *et al.*, 2008). The control of the microbial population (Sbarbati and Osculati, 2006; Merigo *et al.*, 2008) and the detection of irritants seem to be other possible functions of the DCS (Finger *et al.*, 2003; Gulbransen *et al.*, 2008a; Osculati *et al.*, 2007).

The SCCs were first described in aquatic vertebrates (Kotrschal, 1996) where they mediate functions linked to external cues such as feeding and predator avoidance (Whitear, 1992). Until recently, the SCCs have been considered typical of the aquatic vertebrates and absent in mammals. However, a wide series of data accumulated in the last twenty years led to the concept that they are present and play an important role in mammals too (Sbarbati and Osculati 2003). Since the mid 80's, the availability of an important marker of chemoreceptive mechanism (i.e. the G protein subunit alpha-gustducin) (McLaughlin *et al.*, 1992) made possible the discovery of elements with putative chemosensory role in several organs (Hofer *et al.*, 1996). Gustducin was first described in the brush cells of the digestive apparatus (Hofer and Drenckhahn, 1998). Such elements were in the past indicated with different names (for a recent review see Sbarbati and

Osculati, 2005a). Then SCCs have been described in the ontogenesis of the rat taste buds (Sbarbati *et al.*, 1998) and the presence of gustducin in these elements was clearly demonstrated (Sbarbati *et al.*, 1999c). A similar role for the SCCs seems to be played in the ontogenesis of the fish taste buds (Hansen *et al.*, 2002). The presence in the oral SCCs of the G protein gustducin strongly supported the concept of an analogy between these elements and the taste cells. The presence of gustducin-immunoreactive elements was then signalled in the nasal cavity (Zancanaro *et al.*, 1999) and in the palate (El Sharaby *et al.*, 2001).

The SCCs are characterised by genes coding for olfactory and gustatory receptors (Mombaerts, 1999; Sullivan, 2002). At the beginning, these receptors were considered to be specific for the olfactory and oro-pharyngeal mucosae but later it appeared evident that they are widely expressed in several organs (Feldmesser *et al.*, 2006). The individuation of these heterotopic expressions raised a series of questions about their role. In the airway, the presence of chemoreceptorial molecules was suggested to be linked to the control of secretion (Merigo *et al.*, 2007) and microbial population (Sbarbati and Osculati, 2005b, 2006). The evidences are not limited to rodents but data were also obtained in large size mammals Tizzano *et al.*, 2006).

In the digestive system, the presence of these molecules has been demonstrated to control absorptive mechanisms (Salmon *et al.*, 2007; Mace *et al.*, 2007) or endocrine secretion (Rozengurt, 2006). It is now widely accepted that, in addition to the 'classical' taste receptor cells (TRCs) organised in taste buds, chemosensory cell clusters or SCCs are found in non-lingual epithelia of endodermal origin (i.e. respiratory and digestive). This system of chemosensory cells has been termed DCS and shares signal transduction components typical for classical TRCs.

The

diffuse chemosensory system

It is well known that changes in the luminal contents of the gastrointestinal tract modulate gastrointestinal functions, including absorption of nutrients, food intake, and protection against harmful substances. Several good reviews have recently dealt with these aspects (see for instance Rozengurt, 2006 or Sternini, 2007). Apart the local reflexes, evidence from the intestinal taste aversion paradigm also suggests that nutrient tastes detected in the intestines can be recognised centrally (Tracy *et al.*, 2004).

It is well known that post-ingestive effects can influence food preference independently of palatability and recent findings suggest that calorie-rich nutrients can directly influence brain reward circuits that control food intake independently of palatability or functional taste transduction (De Araujo *et al.*, 2008).

In the past, the idea that chemoreceptorial cells are present in the digestive apparatus was advanced on the basis of ultrastructural data, which clearly showed brush cells with a morphology suggesting detection of intraluminal chemicals (Luciano *et al.*, 2003). When immunological markers for 'taste cells' (i.e. the G-protein gustducin) became available, the chemosensory role for these cells appeared more probable and recently, methods of molecular biology definitively confirmed the presence of these markers in wide areas of the apparatus (Hofer *et al.*, 1996).

Regions enriched with elements of the DCS are located in particular at interface among different microenvironments such as the glandular ducts and the cardiac region of the stomach (Luciano and Reale, 1992).

In the digestive apparatus, the DCS co-exists with the peripheral nervous system and diffuse endocrine system. Therefore, the digestive system appears to be a main site to study the interactions among the DCS and these structures. To date, the interest of the alimentary industry to develop functional foods represent a strong impulse in study the chemosensory structure of the digestive apparatus. First studies performed in the respiratory system suggested that the expression of molecules of the chemoreceptorial cascade is not restricted to a single cell type (Merigo *et al.*, 2007). These molecules are highly expressed in elements with morphological features of SCCs but are also present in elements with different morphology (in particular in cells expressing a secretory phenotype) (Merigo *et al.*, 2007).

In the bowel about 60% of the elements expressing chemoreceptorial markers are 'pure chemoreceptors' (i.e. SCCs) the remaining 40% seems to be composed of other elements (Sutherland *et al.*, 2007). The composition of this enormous number of cells is in function of the site (e.g. in the colon they are in large part entero-endocrine elements) (Rozengurt, 2006).

Transient receptor potential (TRP) proteins form a large gene family of ion channels characterised by distinct activation mechanisms and biophysical properties. A growing number of TRP channels have been identified as key players in the sensation of smell, temperature, mechanical forces and taste. TRPM5 is known to be abundantly expressed in taste receptor cells where it participates in sweet, amino acid and bitter perception. Kaske *et al.* (2007) found solitary TRPM5-enriched epithelial cells in all parts of the respiratory and gastrointestinal tract. Based on their tissue distribution, the low cell density, morphological features and co-immunostaining with different epithelial markers, they identified these cells as brush cells (also known as tuft, fibrillovesicular, multivesicular or caveolated cells).

Other data give more ground to the existence of a diffuse taste system in the GI by showing that several taste-signaling proteins are coexpressed in solitary epithelial cells of the intestine. Bezencon *et al.* (2007) show that T1r1, T1r3, Trpm5, and α-gustducin are also expressed in the human GI and found colocalisation of Trpm5 and alpha-gustducin in tufted cells at the surface epithelium of the colon, but these cells did not express T1r3 or PLCbeta2. In the duodenal glands, 43%, 33%, and 38% of Trpm5-expressing cells also express PLCbeta2, T1r3, or alpha-gustducin, respectively. The duodenal gland cells that coexpress PLCbeta2 and Trpm5 morphologically resemble enteroendocrine cells. They found a large degree of colocalisation of Trpm5, alpha-gustducin, T1r1, and T1r3 in tufted cells of the duodenal villi. In the duodenal villi, tufted cells coexpress T1r1, T1r3, α-gustducin, and Trpm5, suggesting that they are implicated in L-amino acid detection. Unlike TRCs, most of these cells do not express PLCß2, suggesting that although they use the same receptors and G-protein as TRCs, they may use a different downstream transduction pathway.

In the study of Mace *et al.* (2007) western blotting and immunocytochemistry revealed that all T1R members are expressed in rat jejunum at strategic locations including Paneth cells, SCCs or the apical membrane of enterocytes; T1Rs are colocalised with each other and with alpha-gustducin, transducin or PLC beta2 to different extents (Mace *et al.*, 2008).

In the same year, Sutherland *et al.* (2007) showed that gustducin is expressed in solitary epithelial cells of the mid to upper villous, which are distributed in a regional manner with most occurring within the midjejunum. Here, 27% of α-gustducin cells colabeled for 5-HT and 15% for GLP-1; 57% of -gustducin cells colabeled UEA-1, with no triple labelling. α-Gustducin cells that colabeled for 5-HT or GLP-1 were of distinct morphology and exhibited a different α-gustducin immunolabeling pattern to those colabeled with UEA-1. The authors concluded that subsets of enteroendocrine cells in the midjejunum and brush cells (more generally distributed) are equipped to utilise α-gustducin signalling in mice.

Both in the respiratory and in the digestive system the cells that express molecules of the chemoreceptive cascade can be inserted in supracellular aggregates that may have the morphology of specialised epithelia or cell clusters).

Hass *et al.* (2007) have investigated a large cluster of gustducin-positive cells which is located exactly at the boundary between the fundic and the oxyntic mucosa of the mouse stomach, at the so-called 'limiting ridge'. This location was previously studied by Luciano and Reale (1992), which found an accumulation of brush cells. In close association with the candidate chemosensory cluster, they found two populations of enteroendocrine cells: one population containing the satiety regulating hormone ghrelin, the other population comprising serotonin-secreting enterochromaffin cells. The particular arrangement of gustducin-expressing cells and enteroendocrine cells at the limiting ridge suggested a direct interplay between these cell types with immediate implications, not only for digestive processes in the stomach, but also for parameters controlling the satiety status (Hass *et al.*, 2007).

In the past, ultrastructural studies of the rat ileum have demonstrated the presence within the mucosa of 'clear' cells with apical cytoplasmic processes into the gut lumen and basal cytoplasmic processes extending underneath adjacent epithelial cells. The 'clear' cells were devoid of secretory granules, as e.g. endocrine cells, and had a morphological resemblance to taste cells. Occasionally the 'clear' cells made contacts with submucosal nerve terminals (Newson *et al.*, 1982).

The ultrastructure of human intestinal brush cells were recently studied by Morroni *et al.* (2007). The surface brush cells were characterised by a narrow apical pole from which emerged microvilli that were longer and thicker than those of enterocytes. The filamentous core extended far into the cell body without forming the terminal web. Caveolae were abundant. Filaments were in the form of microfilaments and intermediate filaments. Cytoplasmic projections containing filaments were found on the basolateral surface of brush cells. Axons containing vesicles and dense core granules were in close contact both with the basal and the lateral surface of the cell. The observation of nerve fibres closely associated with brush cell ends supported the hypothesis of a receptor role for these cells.

Considerations about the role of the DCS

Molecular sensing by gastrointestinal cells plays a critical role in the control of multiple fundamental functions in digestion and also initiates hormonal and/or neural pathways leading to the regulation of caloric intake, pancreatic insulin secretion, and metabolism. Molecular sensing in the GI tract is also responsible for the detection of ingested harmful drugs and toxins, thereby initiating responses critical for survival (Rosengurt, 2006).

The most interesting aspects are surely those linked to the sugar-absorption, however, the role of taste receptors does not seem strictly linked to the transport of sugar.

Mace *et al.* (2008) investigated whether non-sugar nutrients are regulated by T1R taste receptors using perfused rat jejunum in vivo. They concluded that a taste receptor-coordinated transport network incorporating nutrients and/or stimuli and capable of control energy supply exists.

Food evaluation

The diffuse presence of 'taste markers' in the SCCs is the witness of a commune signature for these elements and the gustatory epithelium. The bipolar morphology and ultrastructural aspects are also common features. In mammals, both in the tongue and the palate, the taste buds develop after the SCCs and could be considered big clusters of SCCs. The same ontogenesis was described in fish. These commonalities justify the term taste cell related DCS that was used. Among the gustatory structures and the SCCs of the airway, the laryngeal chemosensory clusters seem to represent a transitional form. In the digestive apparatus, the anatomical continuity is interrupted by the oesophagus, which is lined by a pluristratified epithelium in which chemosensory cells were not described until now. A main difference between the taste buds and the SCCs seems to be in the innervation (Finger *et al.*, 2005). The difference in innervation is also quantitative being largely more abundant in the taste buds. It is evident that a structure like the vallate papilla presents a degree of complexity much more elevated that a simple SCC. This is probably in relation with execution of a complex analysis of the food, whose data are rapidly transmitted to the brain with the scope to elicit reflex behaviours. On the other face, despite their simple organisation, the widespread distribution of the SCCs makes them well suited to monitor the wide surfaces of the digestive system.

Despite the differences, the taste buds and the SCCs present common features that suggest the inclusion of both in the DCS.

From an anatomical point of view, the organisation of the DCS in the digestive apparatus recalls that of the endocrine structures. In the digestive apparatus, this latter may be in form of isolated elements or to be organised in *insulae* such as in the pancreas.

Salivary secretion

The relationship among the stimulation of the taste buds and the salivary secretion are well known and the discussion of this topic goes beyond the present review. Part of this reflex secretion is mediated by central reflex involving the brain.

However, it was underlined as the vallate papilla-vonEbner gland complex if perhaps the most representative example of senso-effector secretory organ in the body (Sbarbati *et al.*, 1999a,b). This structure is basically composed of a sensory portion (the gustatory epithelium), an integrative centre (the Remak intrinsic ganglion) and a secretory portion (the vonEbner gland) (Sbarbati *et al.*, 2001). The relation among the different parties is so strict that the taste buds can be considered embedded in the excretory duct of the vonEbner gland (Sbarbati *et al.*, 2002). SCCs are also present in the ducts of other salivary glands, therefore is possible that they have a role in evaluation on the produced saliva.

Gastric emptying

It is well known that the gastric emptying is linked to the chemical composition of the intraluminal material. In particular it is slow in presence of lipids and sugars at high concentration. In rodents, the gastric SCCs are mainly localised in the cardiac portion, which is a ridge between the squamous lined-forestomach and the glandular portion (Luciano and Reale, 1992; Hass *et al.*, 2007). The prevalence of these elements at interfaces between different microenvironments seems to be confirmed.

Enteric, biliary and pancreatic secretion

As previously described, the presence of SCCs in glands located in different parts of the body suggests a function of the DCS is the evaluation of the glandular secretion. This concept seems to be supported by the notion that, as in the salivary, biliary and pancreatic ductal systems, in the enteric glands, the SCCs are mainly located in the proximal portion. Another example of the link between the DCS and the secretory events is the described presence of brush cells in the biliary tree. The expression of gustducin in these elements was demonstrated (Toyono *et al.*, 2007). There are no functional studies on these elements but the possibility to monitor the composition of the bile before its entrance in the duodenum is a suggestive possibility. The importance of this function may be demonstrated by the several diseases (e.g. the cystic fibrosis) that are related to incorrect biliary secretion. Obviously other hypotheses cannot be rejected without specific studies. As an example, the possible evaluation of bacterial colonisation of the biliary tree could be object of research.

Toyono *et al.* (2007) studied the expression of chemosensory markers in the biliary tree. Immunohistochemistry showed that T1R2 and T1R3 were expressed in the intrahepatic bile duct. Promoter analysis of Tas1r3 was performed in the Tas1r3-expressing cell line, HuCCT1. The 226-bp region upstream of the ATG had promoter activity, and C/EBPbeta activated the Tas1r3 promoter activity in HuCCT1 cells. These results showed that T1R2 and T1R3 receptors, in addition to their role in taste perception, may also have a role in intrahepatic cholangiocytes.

C/EBPbeta was identified as the transcription factor regulating Tas1r3 promoter activity in HuCCT1 cells. Similar consideration can be made for the SCCs located in the pancreatic ducts (Hofer and Drenckhahan, 1998). Around these ducts also specific structures of the peripheral nervous system are present. Also for chemosensors of the pancreatic ductal system a role in pathology must be evaluated.

The microbic control

The lumen of the enteric tube hosts a microenvironment in which an impressive number of microorganisms is present. Here the microbic population is for some aspects more 'dangerous' than in the respiratory apparatus for its density, plurality and fecaloid nature of several species. The different portions of the apparatus host different microenvironments with a different flora. These microenvironments must be separated in a dynamic way without stopping the progression of the aliments. It is interesting to note that specific points, which represent interfaces among the different microenvironments are equipped with a battery of SCCs. Therefore, these areas seem to dispose of specific check-points that, at least in principle could be able to perform a real time analysis of internal environment. The hypothesis that the DCS has a role in innate immunity and in the control of the microbic flora has been advanced (Sbarbati and Osculati, 2006) and has recently obtained (Gulbransen *et al.*, 2008b).

Endocrine control

There are important similarities between the chemosensory machinery elucidated in specialised neuroepithelial taste receptor cells of the lingual epithelium and the molecular transducers localised recently in enteroendocrine open gastro-intestinal cells that sense the chemical composition of the luminal contents of the gut (Sternini, 2007).

The possible interaction with endocrine elements in the digestive system was recently explored (Sternini, 2007). This association seems to be particularly strong in the colon where the expression of taste molecules is abundant.

The current notion is that mucosal enteroendocrine cells act as primary chemoreceptors by releasing signalling molecules in response to changes in the luminal environment, which in turn activate nerve terminals.

The recent discovery that taste receptors and G protein subunits alpha-gustducin and alpha-transducin, involved in gustatory signal transduction, are expressed in the gastrointestinal mucosa supports the concept of a chemosensory machinery in the gastrointestinal tract.

An understanding of luminal sensing processes responsible for the generation of the appropriate functional response to specific nutrients and nonnutrients is of clinical importance since aberrant or unsteady responses to changes in luminal contents might result in disease states ranging from intoxication to feeding disorders and inflammation.

The discussion about the functional implications of taste signalling molecules in the gastrointestinal tract and their relationship with enteroendocrine pathways responsible for the generation of specific responses to luminal contents has been based on a series of experimental works.

Chen *et al.* (2006) demonstrated the expression of bitter taste receptors of the type 2 family (T2R) and the alpha-subunits of the G protein gustducin in the rodent gastrointestinal tract and in gastrointestinal endocrine cells. Then they characterised mechanisms of calcium fluxes induced by two distinct T2R ligands: denatonium benzoate and phenylthiocarbamide, in mouse enteroendocrine cell line STC-1. Both these substances induced a marked increase in intracellular calcium in a dose- and time-dependent manner. Collectively, their results demonstrated that bitter tastants increase cholecystokinin release through calcium influx in enteroendocrine STC-1 cells.

Intestinal glucose sensing and regulation of intestinal glucose absorption

The role of the sweet taste receptors in sugar absorption in the intestine have recently been reviewed by Kellett *et al.* (2008). A growing body of evidence demonstrates that chemoreceptory mechanisms similar to the lingual ones are involved in intestinal glucose sensing and regulation of intestinal glucose absorption (for a recent review see also Dyer *et al.*, 2007). This action seems to operate at different levels and modify the absorption acting on transporters such as SGLT1 (Salmon *et al.*, 2007) or GLUT2 (Mace *et al.*, 2008).

The composition of the intestinal luminal content varies considerably with diet. It is important therefore that the intestinal epithelium senses and responds to these significant changes and regulates its functions accordingly. A prototype example is the modulation in the capacity of the gut to absorb monosaccharides via the intestinal luminal membrane Na(+)/glucose cotransporter, SGLT1. Experimental evidence suggests that luminal sugar is sensed by a glucose sensor residing on the luminal membrane of the gut epithelium and linked to a G-protein-coupled receptor, cAMP/PKA (protein kinase A) pathway, resulting ultimately in modulation of intestinal monosaccharide absorption. Salmon *et al.* (2007) report the expression, at mRNA and protein levels, of members of the T1R sweet taste receptors, and the alpha-subunit of the gustducin, in the small intestine and the enteroendocrine cell line, STC-1. In the small intestine, there is a highly coordinated expression of sweet taste receptors and gustducin. The potential involvement of these receptors in sugar sensing in the intestine will facilitate our understanding of intestinal nutrient sensing, with implications for better nutrition and health maintenance (Salmon *et al.*, 2007).

Mace *et al.* (2007) also provided functional evidence for an intestinal sensing system based on lingual taste receptors. Intestinal glucose absorption consists of two components: one is classical active Na+-glucose cotransport, the other is the diffusive apical GLUT2 pathway. Artificial sweeteners increase glucose in parallel with their ability to increase intracellular calcium concentration. Stimulation occurs within minutes by an increase in apical GLUT2, which correlates with reciprocal regulation of T1R2, T1R3 and alpha-gustducin versus T1R1, transducin and phospholipase C beta2. This observation demonstrates that artificial sweeteners are nutritionally active, because they can signal to a functional taste reception system to increase sugar absorption during a meal.

Sugar sensing and regulation of SGLT1

In this action the sodium-dependent glucose transporter isoform 1 (SGLT1) is involved (Margolskee *et al.*, 2007).

Dietary sugars are transported from the intestinal lumen into absorptive enterocytes by the sodium-dependent glucose transporter isoform 1 (SGLT1). Regulation of this protein is important for the provision of glucose to the body and avoidance of intestinal malabsorption. Although expression of SGLT1 is regulated by luminal monosaccharides, the luminal glucose sensor mediating this process is unknown. Margolskee *et al.* (2007) showed that the sweet taste receptor subunit T1R3 and the taste G protein gustducin, expressed in enteroendocrine cells, underlie intestinal sugar sensing and regulation of SGLT1 mRNA and protein. Dietary sugar and artificial sweeteners increased SGLT1 mRNA and protein expression, and glucose absorptive capacity in wild-type mice, but not in knockout mice lacking T1R3 or alpha-gustducin. Artificial sweeteners, acting on sweet taste receptors expressed on enteroendocrine GLUTag cells, stimulated secretion of gut hormones implicated in SGLT1 up-regulation. Gut-expressed taste signalling elements involved in regulating SGLT1 expression could provide novel therapeutic targets for modulating the gut's capacity to absorb sugars, with implications for the prevention and/or treatment of malabsorption syndromes and diet-related disorders including diabetes and obesity.

Modulation of glucagon-like peptide-1 (GLP-1) in duodenal L cells

Glucagon-like peptide-1 (GLP-1), released from gut endocrine L cells in response to glucose, regulates appetite, insulin secretion, and gut motility. How glucose given orally, but not systemically, induces GLP-1 secretion is unknown. Jang *et al.* (2007) showed that human duodenal L cells express sweet taste receptors, the taste G protein gustducin, and several other taste transduction elements. Mouse intestinal L cells also express alpha-gustducin. Ingestion of glucose by alpha-gustducin null mice revealed deficiencies in secretion of GLP-1 and the regulation of plasma insulin and glucose. Isolated small bowel and intestinal villi from alpha-gustducin null mice showed markedly defective GLP-1 secretion in response to glucose. The human L cell line NCI-H716 expresses alpha-gustducin, taste receptors, and several other taste signalling elements. GLP-1 release from NCI-H716 cells was promoted by sugars and the noncaloric sweetener sucralose, and blocked by the sweet receptor antagonist lactisole or siRNA for alpha-gustducin. The authors concluded that L cells of the gut 'taste' glucose through the same mechanisms used by taste cells of the tongue. These data suggest that modulating GLP-1 secretion in gut 'taste cells' may provide an important treatment for obesity, diabetes and abnormal gut motility (Jang *et al.* 2007).

Conclusions

In the last years, the concept of a *second brain* composed by neurones located in the digestive apparatus is now becoming popular (Gershon, 1999). According to this concept, the autonomy of the digestive apparatus from the central nervous system is now considered to involve integrative functions. It is now clear that the internal organs also posses *sense organs* by which they test their internal microenvironment. It was noted that after the new researches, the chemosensory structures appear to be disposed in an iceberg-like configuration with a small top located cranially end a larger submerged portion located in the internal viscera. The functions of the DCS are still in large part unknown. The links with both the immune cells and with the peripheral nerve system suggest that it could have a role in the *neurobiology of the innate immunity*. It is probable that new research may have a strong impact in both immunology and neurobiology.

The common feature of all the elements of the DCS is the presence of molecules of the chemoreceptorial cascade. It is now evident that these molecules have other roles in addition to evaluating the food. In the present work, we have listed a wide series of possible functions. All them are sustained by some data but none is clearly documented. Despite these limitations due to a still small number of studies a concept appears to be clear: in the internal viscera, sensory inputs originate from specialised epithelial cells and not only from free nerve endings. It is well known for a long time that these inputs and in particular the vagal and trigeminal ones are really important in the control of the peripheral viscera and can modify the central nervous system activity.

References

Bezencon, C., Le Coutre, J. and Damak, S., 2007. Taste-signaling proteins are coexpressed in solitary intestinal epithelial cells. Chemical Senses 32: 41-49.

Chen, M.C., Wu, S.V. and Reeve, J.R. Jr., 2006. Rozengurt E. Bitter stimuli induce Ca2+ signaling and CCK release in enteroendocrine STC-1 cells: role of L-type voltage-sensitive Ca2+ channels. American Journal of Physiology - Cell Physiology 291: C726-C739.

De Araujo, I.E. Oliveira-Maia, A.J., Sotnikova, T.D., Gainetdinov, R.R., Caron, M.G., Nicolelis, M.A. and Simon, S.A., 2008. Food reward in the absence of taste receptor signaling. Neuron 57: 930-941.

Dyer, J., Daly, K., Salmon, K.S., Arora, D.K., Kokrashvili, Z., Margolskee, R.F. and Shirazi-Beechey, S.P., 2007. Intestinal glucose sensing and regulation of intestinal glucose absorption. Biochemical Society Transactions 35: 1191-1194.

El-Sharaby, A., Ueda, K. and Wakisaka, S., 2001. Differentiation of the lingual and palatal gustatory epithelium of the rat as revealed by immunohistochemistry of alpha-gustducin. Archives of Histology and Cytology 64: 401-409.

Feldmesser, E., Olender, T., Khen, M., Yanai, I., Ophir, R. and Lancet, D., 2006. Widespread ectopic expression of olfactory receptor genes. BMC Genomics 7: 121.

Finger, T.E., Bottger, B., Hansen, A., Anderson, K.T., Alimohammadi, H. and Silver, W.L., 2003. Solitary chemoreceptor cells in the nasal cavity serve as sentinels of respiration. Proceedings of the National Academy of Sciences of the USA 100: 8981-8986.

Finger, T.E., Danilova, V., Barrows, J., Bartel, D.L., Vigers, A.J., Stone, L., Hellekant, G. and Kinnamon, S.C., 2005. ATP signaling is crucial for communication from taste buds to gustatory nerves. Science 310: 1495-1499.

Gershon, M.D., 1999. The enteric nervous system: a second brain. Hospital Practice 34: 31-42.

Gulbransen, B., Silver, W. and Finger, T.E., 2008b. Solitary chemoreceptor cell survival is independent of intact trigeminal innervation. J Comp Neurol. 508: 62-71.

Gulbransen, B.D., Clapp, T.R., Finger, T.E. and Kinnamon, S.C., 2008a. Nasal solitary chemoreceptor cell responses to bitter and trigeminal stimulants in vitro. Journal of Neurophysiology 99: 2929-2937.

Hansen, A., Reutter, K. and Zeiske, E., 2002. Taste bud development in the zebrafish, *Danio rerio*. Developmental Dynamics 223: 483-496.

Hass, N., Schwarzenbacher, K. and Breer, H., 2007. A cluster of gustducin-expressing cells in the mouse stomach associated with two distinct populations of enteroendocrine cells. Histochemistry and Cell Biology 128: 457-471.

Hofer, D. and Drenckhahn, D., 1998. Identification of the taste cell G-protein, alpha-gustducin, in brush cells of the rat pancreatic duct system. Histochemistry and Cell Biology 110: 303–309.

Hofer, D., Puschel, B. and Drenckhahn, D., 1996 Taste receptor-like cells in the rat gut identified by expression of alpha-gustducin. Proc Natl Acad Sci USA 93: 6631-6634.

Jang, H.J., Kokrashvili, Z., Theodorakis, M.J., Carlson, O.D., Kim, B.J., Zhou, J., Kim, H.H., Xu, X., Chan, S.L., Juhaszova, M., Bernier, M., Mosinger, B., Margolskee, R.F. and Egan, J.M., 2007. Gut-expressed gustducin and taste receptors regulate secretion of glucagon-like peptide-1. Proceedings of the National Academy of Sciences of the USA 104: 15069-15074.

Kaske, S., Krasteva, G., Konig, P., Kummer, W., Hofmann, T., Gudermann, T. and Chubanov, V., 2007. TRPM5, a taste-signaling transient receptor potential ion-channel, is a ubiquitous signaling component in chemosensory cells. BMC Neuroscience 8: 49.

Kellett, G.L., Brot-Laroche, E., Mace, O.J. and Leturque, A., 2008. Sugar Absorption in the Intestine: The Role of GLUT2. Annual Review of Nutrition 28: 35-54.

Kotrschal, K., 1996. Solitary chemosensory cells: why do primary aquatic vertebrates need another taste system? Trends in Ecology and Evolution 11: 110-113.

Luciano, L. and Reale, E., 1992. The 'limiting ridge' of the rat stomach. Archives of Histology and Cytology 55 (Suppl): 131-138.

Luciano, L., Groos, S. and Reale, E., 2003. Brush cells of rodent gallbladder and stomach epithelia express neurofilaments. Archives of Histology and Cytology 51: 187-198.

Mace, O.J., Affleck, J., Patel, N. and Kellett, G.L., 2007. Sweet taste receptors in rat small intestine stimulate glucose absorption through apical GLUT2. Journal of Physiology 582: 379-392.

Mace, O.J., Lister, N., Morgan, E.L., West, E.J., Affleck, J., Helliwell, P.A., Bronk, J.R., Kellett, G.L., Meredith, D., Boyd, C.A., Myrtani, P., Bailey, P.D., Pettecrew, R. and Foley, D., 2008. An energy supply network of nutrient absorption coordinated by calcium and T1R taste receptors in rat small intestine. Journal of Physiology 587: 195-210.

Margolskee, R.F., Dyer, J., Kokrashvili, Z., Salmon, K.S., Ilegems, E., Daly, K., Maillet, E.L., Ninomiya, Y., Mosinger, B. and Shirazi-Beechey, S.P., 2007. T1R3 and gustducin in gut sense sugars to regulate expression of Na+-glucose cotransporter 1. Proceedings of the National Academy of Sciences of the USA 104: 15075-15080.

McLaughlin, S.K., McKinnon, P.J. and Margolskee, R.F., 1992. Gustducin is a taste-cell-specific G protein closely related to the transducin. Nature 357: 563-569.

Merigo, F., Benati, D., Di Chio, M., Osculati, F. and Sbarbati, A., 2007. Secretory cells of the airway express molecules of the chemoreceptive cascade. Cell and Tissue Research 327: 231-247.

Merigo, F., Benati, D., Galiè, M., Crescimanno, C., Osculati, F. and Sbarbati, A., 2008. Immunohistochemical localization of cystic fibrosis transmembrane regulator and clara cell secretory protein in taste receptor cells of rat circumvallate papillae. Chemical Senses 33: 231-241.

Mombaerts, P., 1999. Seven-transmembrane proteins as odorant and chemosensory receptors. Science 286: 707-711.

Morroni, M., Cangiotti, A.M. and Cinti, S., 2007. Brush cells in the human duodenojejunal junction: an ultrastructural study. Journal of Anatomy 211: 125-131.

Newson, B., Ahlman, H., Dahlström, A. and Nyhus, L.M., 1982. Ultrastructural observations in the rat ileal mucosa of possible epithelial 'taste cells' and submucosal sensory neurons. Acta Physiology Scandinavia 114: 161-164.

Osculati, F., Bentivoglio, M., Castellucci, M., Cinti, S., Zancanaro, C. and Sbarbati, A., 2007. The solitary chemosensory cells and the diffuse chemosensory system of the airway. European Journal of Histochemistry 51 (Suppl 1): 65-72.

Rozengurt, E., 2006. Taste receptors in the gastrointestinal tract. I. Bitter taste receptors and alpha-gustducin in the mammalian gut. American Journal of Physiology - Gastrointestinal and Liver Physiology 291: G171-G177.

Salmon, K.S., Arora, D.K., Kokrashvili, Z., Margolskee, R.F. and Shirazi-Beechey, S.P., 2007. Intestinal glucose sensing and regulation of intestinal glucose absorption. Biochemical Society Transactions 35: 1191-1194.

Sbarbati, A. and Osculati, F., 2003. Solitary chemosensory cells in mammals? Cells Tissues Organs 175: 51-55.

Sbarbati, A. and Osculati, F., 2005a. A new fate for old cells: brush cells and related elements. Journal of Anatomy 206: 349-358.

Sbarbati, A. and Osculati, F., 2005b. The taste cell-related diffuse chemosensory system. Progress in Neurobiology 75: 295-307.

Sbarbati, A. and Osculati, F., 2006. Allelochemical communication in vertebrates: kairomones, allomones and synomones. Cells Tissues Organs 183: 206-219.

Sbarbati, A., Crescimanno, C. and Osculati, F., 1999a. The anatomy and functional role of the circumvallate papilla/von Ebner gland complex. Medical Hypotheses 53: 40-44.

Sbarbati, A., Crescimanno, C., Benati, D. and Osculati, F., 1998. Solitary chemosensory cells in the developing chemoreceptorial epithelium of the vallate papilla. Journal of Neurocytology 27: 631-635.

Sbarbati, A., Crescimanno, C., Bernardi, P. and Osculati, F., 1999c. Alpha-gustducin-immunoreactive solitary chemosensory cells in the developing chemoreceptorial epithelium of the rat vallate papilla. Chemical Senses 24: 469-472.

Sbarbati, A., Crescimanno, C., De Rossi, V., Bernardi, P. and Osculati, F., 1999b. NADPH-diaphorase and NOS-1 positive ganglion cells are found in the rat vallate papilla/von Ebner gland complex. Histochemical Journal 31: 417-424.

Sbarbati, A., Crescimanno, C., Merigo, F., Benati, D., Bernardi, P., Bertini, M. and Osculati, F., 2001. A brief survey of the modifications in sensory-secretory organs of the neonatal rat tongue. Biology of the Neonate 80: 1-6.

Sbarbati, A., Merigo, F., Bernardi, P., Crescimanno, C., Benati, D. and Osculati, F., 2002. Ganglion cells and topographically related nerves in the vallate papilla/von Ebner gland complex. Journal of Histochemistry and Cytochemistry 50: 709-718.

Sclafani, A., 2007. Sweet taste signaling in the gut. Proceedings of the National Academy of Sciences of the USA 104: 14887-14888.

Sternini, C., 2007. Taste receptors in the gastrointestinal tract. IV. Functional implications of bitter taste receptors in gastrointestinal chemosensing. American Journal of Physiology - Gastrointestinal and Liver Physiology 292: G457-G461.

Sullivan, S.L., 2002. Mammalian chemosensory receptors. Neuroreport 13: A9-A17.

Sutherland, K., Young, R.L., Cooper, N.J., Horowitz, M. and Blackshaw, L.A., 2007. Phenotypic characterization of taste cells of the mouse small intestine. American Journal of Physiology - Gastrointestinal and Liver Physiology 292: G1420-G1428.

Tizzano, M., Merigo, F. and Sbarbati, A., 2006. Evidence of solitary chemosensory cells in a large mammal: the diffuse chemosensory system in Bos taurus airways. Journal of Anatomy 209: 333-337.

Toyono, T., Seta, Y., Kataoka, S. and Toyoshima, K., 2007. CCAAT/Enhancer-binding protein beta regulates expression of human T1R3 taste receptor gene in the bile duct carcinoma cell line, HuCCT1. Biochimica et Biophysica Acta 1769: 641-648.

Tracy, L., Phillips, R.J., Chi, M.M., Powley, T.L. and Davidson T.L., 2004. The gastrointestinal tract 'tastes' nutrients: evidence from the intestinal taste aversion paradigm American Journal of Physiology - Regulatory, Integrative and Comparative Physiology 287: R1086-R1100.

Whitear, M., 1992. Solitary chemoreceptor cells. In: Hara, T.J. (ed.), Chemoreception in fishes. Chapman and Hall, New York, USA, pp. 103-125.

Zancanaro, C., Caretta, C.M., Merigo, F., Cavaggioni, A. and Osculati, F., 1999. Alpha-gustducin expression in the vomeronasal organ of the mouse. European Journal of Neuroscience 11: 4473-4475.

Chapter 7. Hormonal control of feed intake in swine

J.A. Carroll[1] and G.L. Allee[2]
[1]USDA-ARS Livestock Issues Research Unit, Lubbock, TX 79403, USA; Jeff.Carroll@ars.usda.gov
[2]Animal Sciences Division, University of Missouri, Columbia, MO 65211, USA

Abstract

Voluntary feed intake is controlled by a plethora of factors including, but not limited to, day length, social interactions, environmental conditions, oronasal sensory cues (i.e. taste, smell, texture), gastrointestinal fill, health status, metabolic status, dietary composition, drug interactions, exercise (physical activity), mental status (i.e. depression) and gender. However, most, if not all, of these factors can mediate the hormonal milieu within the body that can ultimately control feeding behaviour. Therefore, the focus of this chapter will be on hormonal control of voluntary feed intake in swine. For detailed information pertaining to other factors that affect voluntary feed intake in swine such as perinatal flavour programming, oronasal chemosensory cues, metabolism, antinutritional factors, social interactions, environmental conditions and pathogen exposure, the reader is referred to the accompanying chapters within this book. With regard to hormonal regulation, voluntary feed intake is controlled by intricate and diverse afferent and efferent communication pathways that transmit hormonal and neural messages throughout the body that ultimately reflect the status of the body's energy balance to the hypothalamus. Collectively, these signals are responsible for coordinating a delicate balance between energy utilisation and energy intake in order to maintain energy homeostasis. Some of these biological messages such as glucose and cholecystokinin tend to be more reflective of short-term energy balances associated with food consumption, whereas others, such as insulin and leptin tend to be more reflective of long-term energy stores within the body. While numerous hormones have been shown to be involved in feed intake, this chapter will focus on hormones associated with the control of feed intake specifically in swine with supporting information generated in other species. Specifically, information will be provided on the relationship between feed intake and the somatotrophic axis, the appetite stimulating affects of ghrelin, orexin, neuropeptide Y (NPY) and agouti-related protein (AGRP), and the appetite suppressive effects of leptin and urocortin. Additionally, information will be presented related to the potential effects of glucocorticoids on appetite stimulation in swine.

Keywords: appetite, somatotrophic axis, neuropeptides, glucocorticoids

Introduction

Under normal circumstances, the regulation of feed intake is a complex, albeit finely orchestrated, biological process that involves multiple biochemical pathways, physiological processes and physical constraints within the body. Even minor disruptions to this finely tuned process can manifest themselves as phenotypic characteristics such as those associated with obesity and anorexia in humans. Even in an ideal nutritional environment, complications associated with adequate food consumption and appropriate nutrient intake becomes a complicated process regulated by psychological, sociological, physical, environmental and physiological inputs. In livestock production, however, feed intake is largely controlled through management practices that strive to optimise the economic balance between food intake and overall animal health and productivity. For the most part, maintaining this balance in mature animals is less complex than that associated with young, growing animals as mature animals are typically retained in livestock production settings for reproductive and lactation purposes. Therefore, the nutritional requirements of these mature animals are associated with the maintenance energy required for a specific biological process and are typically easy to accomplish through proper nutritional management in the absence of extenuating circumstances that lead to undernourishment such as fluctuations in environmental conditions, social interactions, hormonal imbalances and disease challenges. Conversely, in the young, growing animal, nutritional requirements are changing rapidly as the animal grows and as they transition through the various stages of their production cycle. This is especially relevant in young pigs during the weaning process when they must transition from a liquid milk diet to a diet consisting of solid food and water at a time of rapid growth and during a time when their immune system has not fully developed. Complicating the situation even more for the young pig is the limited energy reserves to draw upon during this critical time.

In the absence of swine management intervention strategies, pigs gradually undergo the weaning process by consuming a solid diet while they are still suckling from the sow. However, in the majority of modern, intensive swine production systems, young pigs are weaned from the sow and moved to a nursery facility where they must be transitioned to a solid diet rapidly in order to survive and thrive in their new environment. During this transition, weaned pigs routinely experience a 1 to 3 day period of inappetence or voluntary feed deprivation and weight loss resulting in a period of growth stasis that can be detrimental to their health, productivity and overall well-being. While the precise mechanisms associated with this period of inappetence have not been fully elucidated, it is routinely attributed to several factors including maternal separation, transportation stress, stress associated with new environmental and social surroundings, possible exposure to new pathogens and the associated immunological challenges, and the transition from a liquid to solid feed diet (Forbes, 1995; Riley, 1989). The ability to successfully transition from a liquid milk to a solid food diet at weaning can be influenced by several factors including prior access to solid food while the pig is still suckling (Funderburke and Seerley, 1990), the establishment of social hierarchy within a group of weaned pigs, the availability of adequate feeder space, and the size of the pig at weaning as lighter weight pigs eat less and tend to have a lower daily weight gain in comparison to larger pigs (Georgsson and Svendsen, 2002).

Unravelling the controlling mechanisms underneath this common phenomenon becomes even more challenging given the individual piglet variation in behavior that exists and the differential responses to the weaning experience that are ultimately linked to post-weaning feed intake and growth rate (Giroux *et al.*, 2000). Fortunately, the majority of pigs adjust fairly rapidly to the various stressors imposed upon them at weaning including the transition to solid feed. However, even a successful transition at weaning does not typically occur without an associated biological and economic cost associated with weight loss, the potential for increased susceptibility to pathogens due to an energy deficit, and expenditure of an already scarce energy reserve in order to maintain homeostasis while contending with numerous stress-related events. Therefore, one of the utmost challenges, and potentially one of the most lucrative opportunities, in swine production is the ability to prevent or decrease the amount of time that young pigs spend in this weaning-induce period of inappetence, and the associated detrimental effects on health and productivity brought about by inadequate nutrition, by being able to stimulate feed intake in the newly weaned pig. However, to be able to capitalise on this opportunity by devising intervention strategies that enhance post-weaning feed intake and growth performance, a better understanding of the impact of weaning on the hormones that regulate growth and feed intake during the post-weaning period would be beneficial, if not obligatory. To this end, the intent of this chapter is to provide the reader with a broad overview of published literature pertaining to the association between feed intake and regulation of the somatotrophic axis in swine, and the use of exogenous hormonal treatments as a means to regulate feed intake in swine.

Feed intake and the somatotrophic axis

Brief overview of growth hormone regulation

The synthesis and secretion of growth hormone (GH) from somatotrophs within the anterior pituitary gland is regulated by a complex interaction among hormonal signals and intracellular signal transduction pathways. Classically, the primary stimulator of GH synthesis and secretion is GH-releasing hormone (GHRH) and the primary inhibitor of GHRH-induced GH secretion is somatostatin or somatrotropin-release inhibiting factor (SRIF) (Plotsky and Vale, 1985) which inhibits GHRH-induced GH secretion; however, it does not inhibit the synthesis of GH (Harvey, 1995). Both GHRH and SRIF are neurohormones synthesised in hypothalamic neurons that enter the anterior pituitary via the hypothalamo-hypophyseal portal system. While GHRH is synthesised in the arcuate and ventromedial nuclei of the hypothalamus, SRIF is predominantly synthesised in the periventricular nuclei. Both of these neuropeptides are transported via axons to nerve ending in the external layer of the median eminence where they are released into the hypothalamo-hypophyseal portal system. The pulsatile pattern of GH secretion observed *in vivo* may be explained primarily by the pattern of GHRH and SRIF release from the hypothalamus into the hypophysial portal system (Muller, 1987). Plotsky and Vale (1985) demonstrated that these two neurohormones are released in episodic fashions which are 180° out of phase with each other. A rapid release of GH is stimulated by a hypothalamic pulse of GHRH which results in a subsequent hypothalamic pulse of SRIF that will terminate further GH release from somatotrophs (Berelowitz *et al.*, 1981; Molitch and Hlivyak, 1980).

Biological action of growth hormone

Postnatal growth in the weaned pig is strongly influenced by production, secretion, and activity of somatotrophic hormones (Carroll *et al.*, 2000) and insulin-like growth factors. Growth hormone is considered to be a major endocrine factor in regulating postnatal muscle growth with direct influences on muscle metabolism (Novakofski and McCusker, 1993) and is considered to be the key homeorrhetic controller of partitioning nutrients for growth (Harrell *et al.*, 1999). Growth hormone, also known as somatotropin, regulates numerous biological activities within the body including activation of hyperplasia, enhancement of amino acid uptake and nitrogen retention, increasing mRNA for protein synthesis, decreasing lipogenesis while increasing fatty acid oxidation and fatty acid release, increasing plasma glucose, triggering insulin secretion, and enhancing bone mineralisation. With regard to the development of skeletal muscle, GH elicits direct actions through amino acid uptake and nitrogen retention, increased mRNA for protein synthesis, and glucose sparing, and elicits indirect actions through insulin-like growth factor-1 (IGF-1) and/or cell proliferation. Growth hormone stimulates IGF-1 release from the liver as well as from target tissues such as muscle and bone. Insulin-like growth factor-1 exhibits both paracrine and endocrine actions with the major target tissues being muscle, cartilage, bone, liver, kidney, nerves, skin, and lungs. It promotes the growth of muscle and bone, and tissue regeneration of the other more biologically active target organs in part through regulation of cell growth by moving cells from a resting phase (G0) to an active phase of the cell cycle and by increasing a cell's ability to complete DNA synthesis.

Growth hormone received considerable attention over the years in growth physiology research as a primary mediator of animal growth. Several studies have demonstrated that exogenous GH enhances growth rate, feed efficiency, bone mineralisation and lean tissue deposition while reducing carcass fatness in livestock (McLaughlin *et al.*, 1991; Preston *et al.*, 1995). However, the biological action of growth hormone within the body cannot be simply assigned to the endocrine, neuroendocrine or immune system due to its broad and diverse biological actions. Even the digestive and reproductive systems do not escape functional dependency from GH. It has been suggested however, that the biological actions of GH could be broadly classified as either somatogenic or metabolic (Underwood, 1995). The somatogenic roles include any GH effects which result in cell proliferation which is mediated either directly or indirectly by the somatomedins. The metabolic role of GH includes direct actions on a variety of tissues in addition to the metabolism of protein, carbohydrates, lipids and minerals.

With regard to carcass composition, GH exhibits both anabolic and catabolic actions. Its anabolic action results from an increase in muscle and intestinal protein synthesis (Eisemann *et al.*, 1989) without affecting protein degradation (Hart and Johnsson, 1986) which in turn results in a net increase in protein accretion in the muscle and liver. The increase in whole body protein synthesis is associated with the ability of GH to increase nitrogen retention that results from a decrease in urinary nitrogen excretion (Eisemann *et al.*, 1986). In cattle, exogenous GH increases nitrogen retention by 20 to 30% (Crooker *et al.*, 1990; Eisemann *et al.*, 1989). This anabolic action of GH on muscle tissue is coupled with the catabolic action of GH on adipose tissue (Eisemann *et al.*, 1989; Moseley *et al.*, 1992; Phillips, 1986; Schwarz *et al.*, 1993).

The lipolytic action of GH becomes especially important during times of feed restriction. In fasting cattle, GH response to a stimulus such as intravenous injection of arginine is increased as compared to fed cattle (McAtee and Trenkle, 1971). This increase in GH during periods of fasting results in an increase of free fatty acids and glycerol from the hydrolysis of triglycerides. Harvey (1995) suggested that the lipolytic action of GH may be the result of one or more of the following actions: (1) the action of GH may be indirect by altering the secretion of endogenous hormones that mediate lipolysis. For example, GH may increase the secretion of lipolytic hormones such as epinephrine, norepinephrine and glucagon, or perhaps GH inhibits the secretion of anti-lipolytic hormones (e.g. insulin); (2) GH could possibly effect the sensitivity of adipose tissue to these lipolytic or anti-lipolytic hormones; or (3) GH may interact with its receptor on adipose tissue to directly increase lipolysis. The overall net result of the anabolic and catabolic actions is an increase in lean tissue accretion and a subsequent decrease in lipid deposition which produces a desirable carcass.

In addition to the somatogenic and metabolic actions in the body, GH is now being recognised as an important mediator of the immune system. Interestingly, it has been demonstrated that GH is not only synthesised by somatotrophs in the anterior pituitary gland, but is also produced by cells of the immune system. Several research groups have demonstrated that there are similarities, as well as differences, that exist in the regulation of anterior pituitary synthesis and secretion of GH as compared to leukocyte synthesis and secretion of GH (Hattori *et al.*, 1990; Kato *et al.*, 1989; Weigent *et al.*, 1988). Additionally, it has been reported that immune system peptides are capable of mediating the synthesis and secretion of anterior pituitary GH (Weigent and Blalock, 1990). This apparent bi-directional communication between the immune system and the endocrine system is further supported by evidence that demonstrated that GH is capable of altering the function and activity of all the major immune system cell types including T-cells, B-cells, natural killer cells (NK) and macrophages (Kelley, 1989). Establishing this bi-directional communication between the neuroendocrine, endocrine and immune systems has lead to the use of the term 'neuroimmunoendocrine system' by endocrinologists and physiologists.

A discussion pertaining to the biological action of GH would be incomplete without at least a brief introduction of The Dual Effector Theory (Green *et al.*, 1985). The basis for this theory originates from the observation that IGF-1 stimulates *in vitro* amino acid uptake into the muscle but does not necessarily promote *in vivo* growth. Thus, GH is thought to control circulating as well as local tissue production of IGF-1. The theory proposes that GH not only controls IGF-1 production, but also controls differentiation of cells as mentioned above so that they can respond to the mitogenic effect of IGF-1. The Dual Effector Theory really comes into play during periods when there is uncoupling of the GH/IGF-1 axis. Having high concentrations of GH and low concentrations of IGF-1, such as during nutritional restriction brought about by the inappetence associated with weaning or activation of the acute phase immune response, suggests that GH plays a different role under these circumstances. However, it is important to note that changes in circulating concentrations of GH have been reported to be species specific (Daniel *et al.*, 2002). In humans, sheep, and pigs, GH concentrations increase following activation of the innate immune system, while in rats, cattle, and birds, GH concentrations decrease.

Uncoupling of the GH/IGF-1 axis during inappetence in the weaned pig

The purpose of this section is to highlight changes associated with the uncoupling of the GH/IGF-1 axis during periods of inappetence or voluntary feed deprivation associated with weaning and immune system activation in the young pig as these events provide a valuable insight into the endocrine status of the pigs during these periods of nutritional restriction. While it has been questioned whether the GH/IGF-1 axis is physiologically active in young pigs (Hellstern *et al.*, 1996), studies have demonstrated that the somatotrophic axis in neonatal pigs is not only functional (Wester *et al.*, 1998) but undergoes significant maturational changes (Carroll *et al.*, 1999). It has also been reported that in the neonatal pig, pituitary GH release influences IGF-1 secretion and stimulates growth (Matteri *et al.*, 1994). Developmentally, GH concentrations are elevated initially following birth and subsequently decline during the pre-weaning period (Matteri *et al.*, 1994).

Insulin-like growth factors share a similar structure with insulin and are associated with increases in DNA, RNA and protein synthesis. Insulin-like growth factor-1, considered to be the primary somatomedin, is produced in the liver, muscle and other tissues, and is responsible for several biological actions including bone and muscle growth. The primary target tissues for IGF-1 include muscle, cartilage, bone, liver, kidney, nerves, skin, and lungs. Insulin-like growth factor-1 has a potent affect on muscle growth and metabolism by suppressing protein degradation (Novakofski and McCusker, 1993). Unlike GH, IGF-1 concentrations are low during the prenatal period but increase during the perinatal period (Lee *et al.*, 1991). Concentrations of serum IGF-1 in newborn piglets increases steadily after birth, and by 3 weeks of age, serum IGF-1 concentrations exceed those found in the maternal serum (Lee *et al.*, 1991). Serum IGF-2 levels also increase following birth, but the magnitude is not the same as IGF-1. Insulin-like growth factor-2, comprised of 67 amino acids, is secreted from the liver in addition to various other tissues. The primary role of IGF-2 is not as well defined as is the role of IGF-1, but IGF-2 is more important for prenatal growth and development compared to postnatal development. The postnatal increase in both IGF-1 and IGF-2 suggests maturation of the somatotrophic axis (Lee *et al.*, 1991).

Nutritional status has a strong influence on the production and secretion of a variety of growth-related hormones, including GH (Atinmo *et al.*, 1978; Farmer *et al.*, 1992; Straus, 1994), IGF-1 (Booth *et al.*, 1994; Straus, 1994; Weller *et al.*, 1994), IGF-2 (Buonomo *et al.*, 1988), thyroid hormones (Booth *et al.*, 1994; Dauncey and Morovat, 1993), and glucocorticoids (Farmer *et al.*, 1992). Unfortunately, there have only been a relatively small number of studies that have evaluated metabolic and endocrine responses of the pig during weaning and during a dietary transition. In an early study by Carroll *et al.* (1998) the authors provided insight into the developmental and dietary related changes that occur in endocrine profiles in pigs that were weaned from their sows and placed on a commercial pre-starter ration at either 2 or 3 weeks of age. Twelve days after weaning, the pigs were transitioned from the commercial pre-starter ration to a starter ration. Specifically, these authors reported that weaning significantly reduced body weight regardless of age at weaning and regardless of gender. While average daily gain (ADG) was reduced in both groups of pigs following weaning, those weaned at 2 weeks of age experienced a greater decrease in ADG as compared to pigs weaned at 3 weeks of age. Interestingly, the authors also reported a decrease in ADG following the transition in diets at 12 days post-weaning, thus highlighting the importance of dietary consistency and the effects of abrupt changes in the ration. With regard to serum concentrations of GH, these authors reported significant developmental decreases that occurred from birth to weaning, followed by a significant increase post-weaning in both weaning groups. The authors also noted numerical increases in GH following the dietary transition. Serum concentrations of IGF-1 and IGF-2 both increased during the pre-weaning period, but fell dramatically after weaning. A significant decrease in IGF-1 was noted in the 2 week weaning group, but not the 3 week weaning group, following the dietary transition at 12 days post-weaning. As with IGF-1, serum concentrations of IGF-2 following the dietary transition revealed a weaning age interaction. Collectively, the results from this previous study clearly demonstrate that weaning and even abrupt changes in the young pig's diet post-weaning can have significant effects on the somatotrophic axis in the pig, and that the lack of feed consumption is directly related to the uncoupling of the GH/IGF-1 axis.

The results from the early study by Carroll *et al.* (1998) stimulated subsequent research by Matteri *et al.* (2000) that was aimed at distinguishing the differential effects between pig development and nutritional restriction associated with weaning as they relate to the changes in the somatotrophic axis. Additionally, these authors evaluated the potential use of spray-dried plasma as a means to stimulate feed intake and reduce the duration of voluntary feed deprivation associated with weaning. In this study, the authors weaned twenty-four pigs at exactly 14 days of age and divided them into three experimental groups: (1) pigs that were cross-fostered to another sow (not nursing any other pigs) at the same stage of lactation as the sows of the weaned pigs; (2) pigs weaned onto a commercial pre-starter ration containing no spray-dried plasma; and (3) pigs weaned onto a commercial starter ration containing 7% spray-dried plasma. For groups 2 and 3, diets were formulated to contain equal digestible essential amino acids and metabolisable energy. The results from this study revealed that during the first 3 days of the study, sow-nursed pigs gained significantly more weight than either group of pigs that were placed on the commercial pre-starter ration. However, during the last day of the study, day 4 post-weaning, weight gains in the pigs cross-fostered and nursing a sow, and those supplemented with 7% spray-dried plasma were equivalently elevated compared to the weaned pigs not supplemented with spray-dried plasma. These results are similar to previous studies that have reported increases in feed intake in weaned pigs supplemented with spray-dried plasma (Coffey and Cromwell, 1995; De Rodas *et al.*, 1995; Touchette *et al.*, 1998). However, this increase in growth for pigs supplemented with spray-dried plasma was not associated with increases in measures of somatotrophic function as the authors reported significant decreases in serum concentrations of IGF-1 and IGF-2 at 4 days post-weaning in both groups fed the commercial pre-starter ration. These results are similar to those previously reported by Carroll *et al.* (1998) for pigs weaned at either 2 or 3 weeks of age. Interestingly, Matteri *et al.* (2000) reported no differences in levels of muscle IGF-1, IGF-2, and GH receptor mRNA expression among the three weaning groups, thus suggesting that gene expression in these target tissues is related more to the developmental stage of the pig rather than the nutritional status. Consistent with a developmental aspect associated with nutritionally-linked uncoupling of the GH/IGF-1 axis are studies that demonstrate that total feed deprivation in the newborn pig actually results in increases in both IGF-1 and GH. However, the aforementioned studies do in fact demonstrate that nutritional deprivation results in an uncoupling of the GH/IGF-1 axis in pigs by at least 14 days of age.

While the uncoupling of the GH/IGF-1 axis related to immune system activation has not been as extensively researched in swine as in sheep and cattle, there is sufficient data within the literature and previous studies from our laboratory (Carroll *et al.*, unpublished data; Figure 1) that do indeed support the existence of bi-directional communication between the immune system and somatotrophic axis in swine. Initial linkages stem primarily from studies that revealed that exogenous GH treatment could alter various aspects of the activated immune system. For example, Parrott *et al.* (1995) demonstrated acute increases in circulating concentrations of GH in pigs during the first 20 minutes following an immunological challenge with endotoxin (lipopolysaccharide; LPS). Later, Hevener *et al.* (1997) reported that in finishing pigs, an acute increase in GH occurred 40 minutes after an intra-peritoneal challenge with LPS. However, the authors reported that this increase in GH concentration was short-lived and was followed by a subsequent uncoupling of the GH/IGF-1 axis that persisted for 96 hours after LPS exposure. The research by Hevener *et al.* (1997) suggested that the persistent uncoupling of the GH/IGF-1

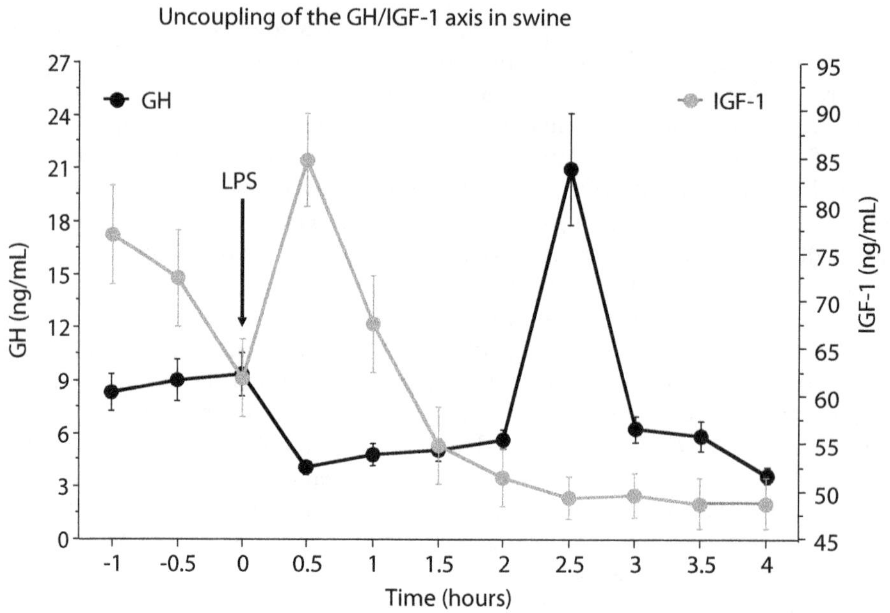

Figure 1. Serum concentrations of growth hormone (GH) and insulin-like growth factor 1 (IGF-1) in weaned pigs following an endotoxin challenge. At Time 0, pigs received an intravenous dose of lipopolysaccharide (LPS; 25 μg/kg body weight). Data from Carroll et al. *(unpublished).*

axis could be attributed to factors beyond nutritional influences as feed consumption did not differ between control and LPS-challenge pigs, a postulation consistent with that proposed for cattle (Elsasser *et al.*, 1988). A subsequent study by Wright *et al.* (2000) also supports this hypothesis as the authors reported that in pigs challenged with LPS, serum concentrations of IGF-1 remained low even after the pigs resumed normal feed consumption. In weanling pigs, we have demonstrated that during the first 30 minutes post-LPS challenge, serum concentrations of GH are dramatically reduced while serum concentrations of IGF-1 are increased. However, subsequent to this initial acute reduction in GH, GH concentrations begin to increase and spike at 3 hours post-LPS challenge while IGF-1 concentrations decline (J.A. Carroll, unpublished data). The results from these studies highlight the importance of intensive serial blood sampling when attempting to profile the dynamic aspect associated with the uncoupling of the GH/IGF-1 axis, especially uncoupling of the axis associated with acute stimulation of the immune system. Collectively, these data suggest not only a time-dependent but also age-dependent aspect associated with the uncoupling of the GH/IGF-1 axis in swine.

Stimulators of appetite

Ghrelin

Ghrelin, a ligand capable of stimulating GH secretion via binding to the GH secretagogue receptor, is also capable of *in vivo* stimulation of adrenocorticotropin-releasing hormone

(ACTH), cortisol, aldosterone, prolactin and epinephrine (Arvat *et al.*, 2001; Takaya *et al.*, 2000). While primarily produced and secreted from epithelial cells lining the fundus of the stomach, ghrelin is also produced in other organs including the pituitary, hypothalamus, small intestine, placenta, kidney, heart, thyroid gland, immune cells such as T-cells and neutrophyls, pancreas endocrine cells, and the ovary (Gaytan *et al.*, 2003; Hattori *et al.*, 2001; Korbonits *et al.*, 2001; Mori *et al.*, 2000; Volante *et al.*, 2002a,b). Even though the known biological actions of ghrelin are rather extensive, including effects on insulin secretion, glucose and lipid metabolism, gastric secretion and motility, various effects on the cardiovascular system, inhibition of reproductive processes, and potentially beneficial alterations associated with the immune system, within domestic livestock, it has received considerable attention primarily as means to stimulate GH secretion and increase food intake.

In a previous study, Salfen *et al.* (2003) characterised the pattern of ghrelin secretion during a 72 hour feed deprivation followed by a subsequent re-feeding period in a group of weaned pigs. In their initial experiment, the authors reported that circulating concentrations of ghrelin were significantly decreased by 12 hours after total feed removal, but following this initial decrease, ghrelin concentrations significantly increased from 12 to 36 hours (Figure 2). The results from this study are consistent with those that have reported that ghrelin increases in the peripheral circulation during fasting (Tschop *et al.*, 2000) in rodents, and after an overnight fast in anorectic humans (Cuntz *et al.*, 2001). In a subsequent experiment reported within the same manuscript, the authors reported that following the 36 to 48 hour increase, ghrelin concentrations once again declined by 72 hours following feed deprivation. The authors speculated that ghrelin secretion

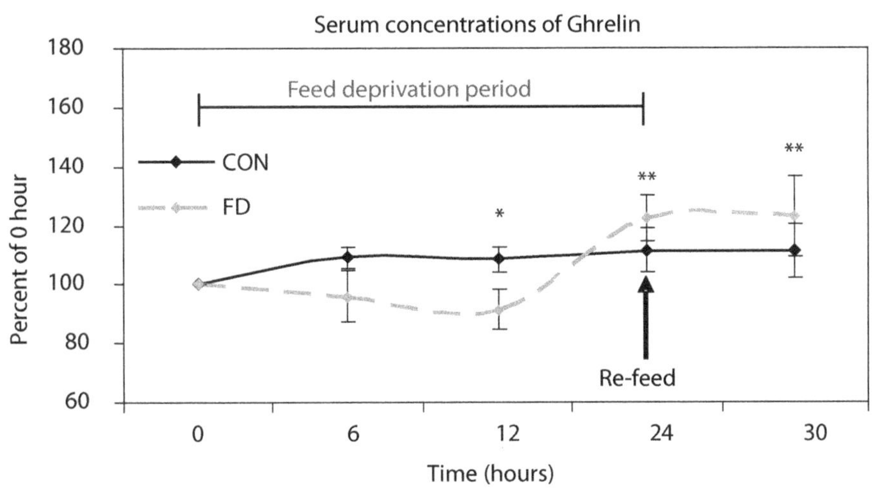

Figure 2. Recreated from Salfen et al.*, 2003. Effect of* ad libitum *feeding (CON) or feed deprivation (FD) followed by re-feeding on serum concentrations of ghrelin expressed as a percent of the 0 hour concentration in weaned pigs. Values are the mean ±S.E.M. of n=4 pigs per group. Pigs in the FD treatment were given access to* ad libitum *feed after the 24 hour sample collection.*

** Denotes difference between FD and CON treatments (P=0.08).*

*** Denotes a difference in FD pigs from the 12 hour time point to the 24 hour time point (P<0.01).*

increased from 12 to 36 hours post-feed deprivation in order to initiate feeding activity even in the absence of food, but due to the absence of food and the abatement of hunger, ghrelin concentrations again fell to concentrations lower than basal levels.

While similar responses to the increases observed by Salfen *et al.* (2003) in serum concentrations of ghrelin in response to feed deprivation in pigs have subsequently been reported by other groups (Govoni *et al.*, 2005; Inoue *et al.*, 2005), a recent study by Scrimgeour *et al.* (2008) suggests that circulating concentrations of ghrelin may be more reflective of chronic changes in energy balance as opposed to actual feeding activity. In the study by Scrimgeour *et al.* (2008), circulating concentrations of ghrelin were profiled under three different feeding protocols that included *ad libitum* feeding, twice daily feeding and once daily feeding. During all three of the feeding protocols of the experiment, the authors reported no significant changes in circulating concentrations of ghrelin related to feed consumption. The authors did however report that ghrelin concentrations increased over the 12 hour blood sampling period when pigs were fed either once or twice daily rather than *ad libitum* and that this increase in circulating concentrations of ghrelin overlapped with increases in GH and non-esterified fatty acids. Additionally, the authors reported that these increases in ghrelin concentrations were not coupled with changes in either glucose or insulin concentrations, thus supporting their conclusion that circulating concentrations of ghrelin may be more reflective of energy balance rather than feeding behaviour. Interestingly, previous studies have reported that exogenous ghrelin treatment acutely increases plasma glucose and lowers insulin (Broglio *et al.*, 2004), and alternatively, exogenous insulin as well as leptin has been reported to increase expression of ghrelin mRNA (Toshinai *et al.*, 2001).

Subsequent to their initial observations that ghrelin concentrations were reduced in feed deprived pigs, and prior reports demonstrating that ghrelin was an important variable in the initiation of feeding behavior and appetite in rats (Nakazato *et al.*, 2001; Shintani *et al.*, 2001; Wren *et al.*, 2001b), mice (Asakawa *et al.*, 2001b), and humans (Cummings *et al.*, 2001; Wren *et al.*, 2001a), Salfen *et al.* (2004) evaluated the potential use of exogenous ghrelin as a means to stimulate feed intake in 18 day old pigs that were newly weaned. In this study, the authors evaluated the relative gain, feed intake, general behaviour and endocrine profiles of newly weaned pigs for 8 days subsequent to injections with either human ghrelin (2 µg/kg body weight) or with saline three times daily via an indwelling jugular catheter for 5 days. The authors reported that while there was a temporary decrease in body weights in both the saline and ghrelin injected pigs in response to weaning, the ghrelin injected pigs regained the lost body weight more rapidly than saline injected pigs. Interestingly, the authors did not report a significant difference in feed intake between the two treatment groups, a result similar to that previously reported in mice (Tschop *et al.*, 2000). Therefore, the authors speculated that the increases in weight gain observed for ghrelin injected pigs may not only reflect an increase in energy intake, but also potential alterations in energy utilisation and partitioning. Interestingly, in human trials in which patients were continuously infused with ghrelin intravenously, voluntary energy intake was increased by 28% compared with saline infusion (Wren *et al.*, 2001a). In the study by Salfen *et al.* (2004), the authors reported that ghrelin was sufficiently metabolised from one injection to the next so there was no systemic accumulation of ghrelin throughout the experimental time period, therefore perhaps providing an explanation for the discrepancy in appetite between the human and pig trials. In addition to an increase in weight gain, Salfen *et al.* (2004) reported a significant increase in GH following the initial ghrelin injection which was followed by a period of lower serum concentrations of GH at 2 hour after injection (Figure 3). Interestingly, while serum concentrations of insulin increased within 15 minutes following the initial injection of human ghrelin, serum concentrations of glucose were not affected.

While the orexigenic action of ghrelin is a significant stimulus to motivate re-nourishment of the animal, the fact that ghrelin is also a potent GH secretagogue may be central with regard to appropriate metabolic partitioning of nutrients. Previous studies have suggested that the appetite stimulatory function(s) of ghrelin and its GH secretagogue effects are independent of each other (Nakazato *et al.*, 2001; Tschop *et al.*, 2000) and may be mediated by separate subtypes of the ghrelin receptor. Finally, another function of ghrelin worth noting is its potential to initiate anxiogenic activities, an event that has been loosely defined as 'food hunting behaviors'. Previous work has demonstrated that administration of exogenous ghrelin increases general activity in mice, and that experimentally imposed starvation stress and tail-pinch stress increased ghrelin gene expression in the stomach (Asakawa *et al.*, 2001a).

In summary, exogenous ghrelin has a variety of endocrine effects and shows potential in increasing weight gain in pigs that could prove to be beneficial with regard to the overall health and productivity of the weaned pig. Given the reported associations between ghrelin and the stress axis, one might expect greater responses in older pigs that do not have confounding stressors impinging on them like recently weaned pigs as the stress associated with weaning may in part override the potential effectiveness of ghrelin in increasing appetite. However, the appropriate dosage, route of administration and duration of exposure necessary to elicit beneficial physiological responses in the pig remain undetermined.

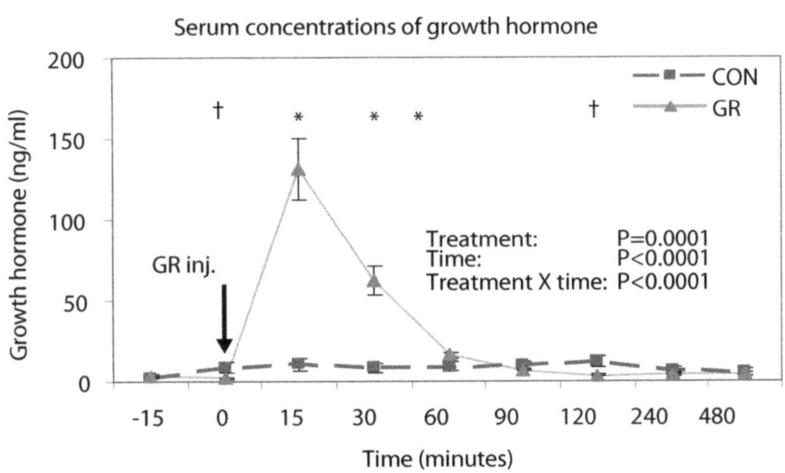

Figure 3. Recreated from Salfen et al., *2004. Serum concentrations of growth hormone (GH) before and after administration of ghrelin (GR; 2 µg/kg body weight) or saline (CON). Pigs were weaned on day 0 of study and either injected with human ghrelin or saline at 0 min. ∗ indicates that GR treatment means are higher than CON means (P<0.05). † indicates that CON treatment means are higher than GR means (P<0.05).*

Orexin

In 1998, Sakurai *et al.* (1998) described a family of hypothalamic neuropeptides, known as orexin-A and orexin-B, and their associated G protein-coupled receptors that were associated with regulation of feeding behaviour in the rat. This initial study revealed that intracerebroventricular (ICV) administration of orexin-A or orexin-B stimulated feed intake in rats and that the associated mRNA for the precursor of both of these neuropeptides, prepro-orexin, increased during food restriction. Subsequently, in a relatively short time, there have been a plethora of studies published on the topic of orexin and feed intake. In fact, in a recent review (Kirchgessner, 2002) on the subject of orexins and the brain-gut axis, the author noted that over 280 manuscripts had been published on the topic in the first four years after Sakurai and colleagues described the orexins and their receptors. Therefore, the intent of this section will be to only highlight the research that has been conducted with orexin and appetite stimulation in the pig. The reader is referred to the review by Kirchgessner (2002) for a more in depth discussion of the orexins and the orexin receptors as they relate to feeding behaviour, metabolic signalling, gut motility, as well as other important biological processes.

As early as 1999, Dyer *et al.* (1999) described the first study to successfully stimulate feed intake in weanling pigs via an intramuscular injection of synthetic porcine orexin. The study was conducted in three replicates with 8 to 10 weanling pigs per replicate. The authors reported that the pigs used had been weaned at approximately 16 days of age, and for a period of one week prior the application of treatments. To account for possible biases associated with individual feed intake in the weaned pigs, the authors monitored feed intake during a 24 hour period prior to treatment and reported that there were no differences between treatment groups. Subsequently, the pigs were injected with either synthetic porcine orexin at a dose of 3 mg/kg or sterile water. Following the treatments, the authors recorded feed intake every 2 hour for a 24 hour period. Based upon the feed intake data, the authors reported that the orexin treated pigs consumed three times as much feed as the control pigs at 12 hours after treatment, and suggested that orexin treatment resulted in an extra meal or feeding bout during this time frame. While the authors did not report any significant differences in feed intake at other time periods examined, they suggested that at a regular mealtime, the orexin-B treated pigs ingested more feed over an extended feeding period. The authors noted that this speculation was supported by the fact that cumulative feed intake was significantly greater in orexin treated pigs as compared to the control pigs from 12 to 24 hours after treatment. Overall, the authors reported that cumulative feed intake for the orexin treated pigs was approximately 18% greater than that observed in the control pigs. The authors speculated that the lag in the appetite response and feed consumption noted in their study, as compared to the initial study in rodents by Sakurai and colleagues, could be explained by the differences associated with intramuscular versus intracerebro-ventricular administered orexin. A subsequent study from our laboratory (Kojima *et al.*, 2007), demonstrated that weaning did not result in altered expression of orexin mRNA in hypothalamic tissue, however, orexin mRNA expression was highly correlated to the size of the pig at weaning. Interestingly, despite the positive results reported by Dyer *et al.* (1999) in pigs, and supporting evidence in rodents (Muroya *et al.*, 2004; Sakurai *et al.*, 1998) and sheep (Sartin *et al.*, 2001), there have been no further published studies evaluating the use of exogenous orexin as a means to enhance feed intake in the weaned pig.

Neuropeptide Y

Neuropeptide Y (NPY), originally isolated in the porcine brain (Tatemoto, 1982), is a hypothalamic peptide found in the brain and autonomic nervous system that has been associated with a number of physiological activities within the brain including memory, learning, epilepsy and appetite stimulation (Billington *et al.*, 1991; Colmers and El, 2003). It has also been associated with other physiological activities such as circadian rhythms, reproductive function, anxiety, energy expenditure and fat deposition, and vascular resistance. In rodents, sheep and pigs, NPY has even been implicated in the regulation of lutenising hormone and GH secretion (Barb *et al.*, 2006; Jain *et al.*, 1999; McShane *et al.*, 1992). Interestingly, in rodents, NPY has been reported to be an inhibitor of GH secretion (Rettori *et al.*, 1990) whereas in sheep, cattle and pigs, ICV injections of NPY stimulated GH secretion (Barb *et al.*, 2006; McMahon *et al.*, 1999; Thomas *et al.*, 1999).

With regard to appetite stimulation, NPY is a potent neuropeptide that has been previously reported to increase feeding behaviour in a number of domestic livestock including poultry (Kuenzel *et al.*, 1987), sheep (Miner *et al.*, 1989) and pigs (Parrott *et al.*, 1986). Supporting a role for NPY in appetite stimulation are data that demonstrate that injections of antibodies or antisense RNAs against NPY reduces feeding behaviour, and the fact that NPY was reported to be a more potent stimulator of feed intake than orexin in rodents (Edwards *et al.*, 1999). Subsequent studies also revealed that NPY gene expression in the hypothalamus is up-regulated in following food deprivation in rodents and sheep (Brady *et al.*, 1990; McShane *et al.*, 1993).

In pigs, Parrott *et al.* (1986) demonstrated that even in a satiated state, lateral cerebral ventricular injections of NPY significantly stimulated operant feeding behaviour during a 30 minute post-treatment period in a dose-dependent manner. In a recent study by Barb *et al.* (2006), the authors evaluated the potential interactions between ICV injections of porcine NPY and recombinant human leptin as related to feed intake in growing pigs. After a 4 hour period associated with blood collection following treatments, feeders were placed in the pens and the pigs were allowed *ad libitum* access to feed. Feed intake was subsequently monitored at 4, 20 and 44 hours after the feed was placed into the pens. At the 4 hour time point, the authors reported that cumulative feed intake was significantly reduced in those pigs injected with either leptin or the combined treatment of leptin plus NPY. Feed intake did not differ between NPY and saline injected pigs at this same time point. However, at the 20 hour time point, the authors reported that cumulative feed intake was greater in the NPY and leptin plus NPY treatment groups as compared to the saline and leptin treatment groups, and that there was no difference in cumulative feed intake between these two group. By 44 hours, cumulative feed intake for the NPY treated pig remained higher than any of the other treatment groups, but feed intake for the leptin plus NPY treatment group had returned to control levels. Leptin suppressed cumulative feed intake at all time points monitored relative to the control and NPY treatment groups. While the study by Parrott *et al.* (1986) was the first to demonstrate that NPY was capable of increasing feeding behaviour in pigs, the more recent study by Barb *et al.* (2006) was the first to demonstrate an increase in cumulative feed intake as a result of NPY administration.

Adding to the evidence that NPY is intimately involved in feed intake in swine is a recent study from our laboratory (Kojima *et al.*, 2007) that demonstrated positive relationships between preweaning body weight and post-weaning hypothalamic gene expression in young pigs. Although sufficient evidence exists demonstrating that NPY is clearly involved in control of food intake, there is apparently enough redundancy in the control mechanisms associated with appetite regulation that the presence of NPY is apparently not obligatory. For example, in studies with knockout mice targeting the disruption of the NPY gene, the mice maintained normal body weights and continued to be responsive to the effects of leptin. Additionally, mating of the NPY knockout mice to leptin deficient ob/ob mice resulted in offspring that were less obese than the ob/ob mice (Palmiter *et al.*, 1998). Results from these studies lead to the speculation that the effects of leptin on body weight is mediated, at least in part, by its effect on NPY. Collectively, these studies point to an obvious redundancy in the regulation of appetite. However, given the importance of nutrient intake to survival of the individual and the continued propagation of the species, is should not be surprising that something as essential as the regulation of food consumption would have evolved in a manner that includes overlapping mediators of feeding behaviour and energy expenditure.

Agouti-related protein (AGRP)

Agouti-related protein is an orexigenic neuropeptide that is produced exclusively in the neurons of the arcuate nucleus and has been shown to be co-localised with another potent stimulator of appetite, NPY, in rats and sheep (Adam *et al.*, 2002; Hahn *et al.*, 1998; Henry, 2003). The biological action of AGRP is mediated through its antagonistic actions on the melanocortin-4 receptor by which it blocks the appetite-suppressive effects of alpha-melanocyte stimulating hormone [alpha-MSH; (Rossi *et al.*, 1998)]. Early investigations suggested that AGRP was associated with the regulation of energy balance and body adiposity within the body as transgenic mice overexpressing AGRP became obese (Graham *et al.*, 1997). Subsequently, it was reported that the neurons that produce AGRP do indeed recognise and respond to several peripheral indicators of energy balance including insulin and leptin (Mizuno and Mobbs, 1999). In the obese line of mice, ob/ob mice, leptin deficiency has been reported to be associated with increased expression of AGRP within the arcuate nucleous (Shutter *et al.*, 1997) and in a later study Tritos *et al.* (1998) demonstrated that in brown adipose tissue-deficient mice, expression of mRNA encoding AGRP in the arcuate nucleus is decreased. Additionally, a study by Rossi *et al.* (1998) reported that ICV injection of the C-terminal fragment of AGRP increased feed intake in mice over a 24-hour period. Likewise, transferring an expression plasmid of AGRP gene into leg muscle of mice by electroporation resulted in increased body weight gain and daily food intake (Xiang *et al.*, 2004). In addition to potential mediation of AGRP via peripheral indicators of energy balance, recent work has reported that AGRP containing neruons respond to acute and chronic inflammatory cytokines (Scarlett *et al.*, 2008), perhaps suggesting a role for AGRP in disease induced inappetence.

It has been suggested that the overall biological action of AGRP may be mediated by both its ability to alleviate the appetite-suppressive effects of alpha-MSH and by allowing the endorphins to stimulate appetite. However, other studies have suggested a link between AGRP and other appetite regulators as well. In a study by Dhillo *et al.* (2002) the authors reported that in rats, administering AGRP to hypothalamic explants significantly increased the release of immunoreactive NPY and significantly decreased the release of the cocaine- and amphetamine-regulated transcript (CART). The authors' interpretation of their data was that the orexigenic neuropeptides in the arcuate nucleus stimulate the release of each other, perhaps to reinforce orexigenic behaviour via a positive-feedback loop.

In sheep, food deprivation for a period of 4 days resulted in increased hypothalamic expression of the genes encoding for NPY and AGRP (Adam *et al.*, 2002; Henry, 2003). Additionally, ICV injections of 2.0 nmol/kg body weight significantly increased food intake at 4, 6 and 12 hours post-injection in sheep that were being fed *ad libitum* prior to injection (Wagner *et al.*, 2004). In this particular study, the authors reported that co-injection of AGRP with NPY did not increase food intake beyond that achieved with AGRP or NPY alone. In addition to potential mediation of AGRP via peripheral indicators of energy balance, AGRP containing neruons have been reported to respond to acute and chronic inflammatory cytokines (Scarlett *et al.*, 2008), perhaps suggesting a role for AGRP in disease-induced inappetence. Supporting evidence for this hypothesis was recently provided by Sartin *et al.* (2008) who reported that ICV injections of 0.5 nmol/kg body weight one hour prior to endotoxin administration prevented the reduction in endotoxin-induced food intake presumably through the blockage of melanocortin-4 receptors.

While several studies have demonstrated AGRP as a potent stimulator of feed intake, relatively little is known regarding AGRP and its association with feed intake, weight gain and/or disease-induced inappetence in swine. In fact, there is only one reported study related to the effects of weaning and body weight on AGRP in swine, and that is a previous study from our laboratory that evaluated hypothalamic genes encoding several of the appetite regulating hormones including AGRP (Kojima *et al.*, 2007). In this particular study, we reported that the expression of AGRP was strongly correlated with POMC expression and moderately correlated to the long form leptin receptor in weaned pigs four days after weaning. However, we did not observe an affect of weaning on AGRP mRNA expression in the hypothalamus. Interestingly though, expression of mRNA coding for AGRP was greater in the heavier weight pigs as compared to the lighter weight pigs. To our knowledge, the positive correlations observed between body weight and expression of hypothalamic NPY, orexin, type 2 orexin receptor and agouti-related protein reported in this study represent the first known associations of body weight and neuroendocrine regulators of feed intake in weanling pigs. The authors acknowledge that a lack of a difference at a single time point following weaning in the pig does not necessarily eliminate a potential role for AGRP during this aspect of the production cycle and that further studies regarding a potential role for AGRP in the voluntary inappetence during weaning are warranted.

Appetite suppressors

Leptin

While leptin is primarily produced by adipocytes in white adipose tissue, it has been found in other tissues including brown adipose, placenta, ovaries, skeletal muscle, mammary cells, bone marrow, the pituitary gland and the liver (Margetic *et al.*, 2002). Leptin elicits its biological effect by binding to a receptor that has been classified as a member of the class 1 cytokine receptors and transmits its cellular signal via janus-activated kinases (JAK)/signal transducers and activators of transcription (STAT) and mitogen-activated protein kinase (MAPK) pathways (Houseknecht and Portocarrero, 1998; Tartaglia, 1997). In fact, leptin itself has been classified as a cytokine due to the structural homology of the leptin receptor with a member of the interleukin-6 family of receptors, specifically gp130, and common down-stream signaling pathways (Tartaglia *et al.*, 1995). Collectively, leptin has been shown to bind to at least six isoforms of the leptin receptor family that have been previously identified including a long form of the receptor, denoted as OB-rb, and four short forms of the receptor, denoted OB-ra, OB-rc, OB-rd and OB-rf (Tartaglia, 1997). Previous research characterising these isoforms revealed that the extracellular and transmembrane domains are identical between the short and long isoforms and that the differences between these isoforms arise from changes in the length of the cytoplasmic domain (Lee *et al.*, 1996). The sixth isoform of the leptin receptor, denoted OB-re, is a soluble receptor that circulates in plasma and is comprised of an extracellular loop with no intracellular motifs or transmembrane residues (Tartaglia, 1997). The long form of the leptin receptor, OB-rb, is the only receptor isoform that can signal intracellularly via the Jak-Stat and MAPK signal transduction pathways (Malendowicz *et al.*, 2006) and has been reported to be abundant in the hypothalamus, especially in the ventromedial nucleus (Tartaglia *et al.*, 1995), often referred to as the 'appetite center' of the brain. With regard to leptin's influence on appetite regulation, it has been suggested that it elicits its biological actions by inhibiting the activity of NPY and AGRP neurons and by increasing the activity of alpha-MSH neurons (Elias *et al.*, 1999). The remaining isoforms of the leptin receptor are abundant throughout the body and have been isolated in almost every tissue tested for their presence (Houseknecht and Portocarrero, 1998). In addition to its role in regulating food intake and energy balance, leptin has also been implicated in numerous other physiological processes including mediation of GH secretion, the onset of puberty, regulation of the reproductive axis, and regulation of immunological processes such as hematopoeisis, inflammation and overall immunocompetency (Barb *et al.*, 2001).

While the phenotypic manifestations associated with a leptin deficiency have been observed in laboratory mice as early as 1950 (Ingalls *et al.*, 1950) it was not until 1994 that Zhang *et al.* (1994) identified leptin as the gene product that was deficient in obese mice. In the ob/ob mice, functional leptin is not produced, resulting in an uninhibited intake of food and subsequent obesity. Similar phenotypic manifestations have also been reported in mice and rats that lack a functional leptin receptor (Chen *et al.*, 1996; Phillips *et al.*, 1996). *In vivo* studies revealed that treatment with exogenous leptin resulted in reduced feed intake and body weight in obese mice (Campfield *et al.*, 1995; Pelleymounter *et al.*, 1995). Since its discovery, leptin has received much attention in the scientific community, and due to the potential economic impact, there have been numerous studies evaluating the biological activities of leptin as they relate to appetite regulation, weight gain, body composition and reproductive performance in domestic livestock.

In swine, early studies demonstrated that leptin gene expression and plasma concentrations of leptin are higher in pigs with greater adipose mass and pigs selected for greater fat depositon (Bidwell *et al.*, 1997; Yen *et al.*, 1997) Further investigation into the relationships among weight loss and food intake restriction with leptin gene expression in pigs at various body weights revealed that the abundance of leptin mRNA in pigs correlates with fat mass and percentage of body fat under conditions of unrestricted food intake (Spurlock *et al.*, 1998a). However, the authors reported that while total food deprivation reduced the expression of leptin mRNA, feed restriction alone, even at levels associated with weight loss, did not change the abundance of leptin mRNA compared to control pigs. Specifically, the authors reported that under conditions of *ad libitum* consumption, leptin mRNA expression in castrated male pigs increased with body weight from 55 to 163 kg and that leptin mRNA expression was highly correlated to fat mass and percentage of fat. However, in 159 kg pigs that were fed *ad libitum*, maintenance or 23% of maintenance for a period of 28 days, the relative abundance of leptin mRNA did not differ despite considerable weight differences among treatment groups. However, the authors did report that in pigs weighing 60 and 136 kg, the abundance of leptin mRNA in pigs experiencing total food deprivation for a period of 3 days was reduced by more than 30% as compared to pigs fed *ad libitum*. Based upon studies relating leptin concentrations to weight loss in humans, and their data pertaining to relative abundance of leptin mRNA in pigs, the authors speculated that weight loss and food intake restrictions do not necessarily invoke parallel reductions in blood leptin or *obese* mRNA.

Evidence soon appeared demonstrating that exogenous leptin treatment was capable of stimulating GH secretion and reducing feed intake in swine (Barb *et al.*, 1998). In this study, the authors reported that ICV injection of recombinant porcine leptin significantly increase GH concentrations within 15 minutes. Interestingly, the authors reported that the GH response to 10 μg of leptin was greater than that elicited by either 50 or 100 μg. With regard to feed intake, the authors reported no differences among treatment groups at 4 hours post-injection, however, by 20 hours post-injection, feed intake had been reduced by 53%, 76% and 90% by ICV treatment with 10, 50 and 100 μg of recombinant porcine leptin, respectively. In addition to their *in vivo* data, the authors provided the first *in vitro* evidence that demonstrated a direct effect of leptin on pituitary secretion of GH, albeit, the authors did indicate that this increase only occurred at what may be considered supraphysiological concentrations.

Our laboratory has previously reported that in weaned pigs that were acclimated to solid food for a period of 10 days, subsequent feed deprivation resulted in a significant reduction in serum concentrations of leptin by 12 hours after feed removal and a decrease in leptin mRNA expression at 72 hours after feed removal when fat samples were collected (Salfen *et al.*, 2003; Figure 4). A reduction in leptin mRNA expression in adipose tissue is consistent with prior studies conducted in older pigs by Spurlock *et al.* (1998) and Barb *et al.* (2001). An interesting aspect in our study relates to the rapid recovery of serum concentrations of leptin following the re-feeding protocol after the 72 hour feed deprivation period. One might expect that in young pigs with relatively low lipid stores, the recovery of serum leptin from a 72 hour feed deprivation may take a substantial amount of time. However, our data revealed that serum concentrations of leptin had returned to control levels within 12 hours of re-feeding. The rather rapid and dramatic reduction in serum concentrations of leptin in our study may reflect differences in percent body fat and/or differences in metabolism between adult and young animals. While serum concentrations of leptin recovered rather rapidly, we did not observe a similar recovery in the expression of leptin mRNA in fat tissue. In fact, the abundance of leptin mRNA in fat tissue was not increased following the 24 re-feeding period. However, this observation is not without precedent as there is evidence that insulin may be the acute regulator of leptin in pigs (Spurlock *et al.*, 1998b) and the regulation of leptin secretion has been suggested to be mediated via neural inputs into adipose tissue (Youngstrom and Bartness, 1995). Indeed, our results actually support the speculation of

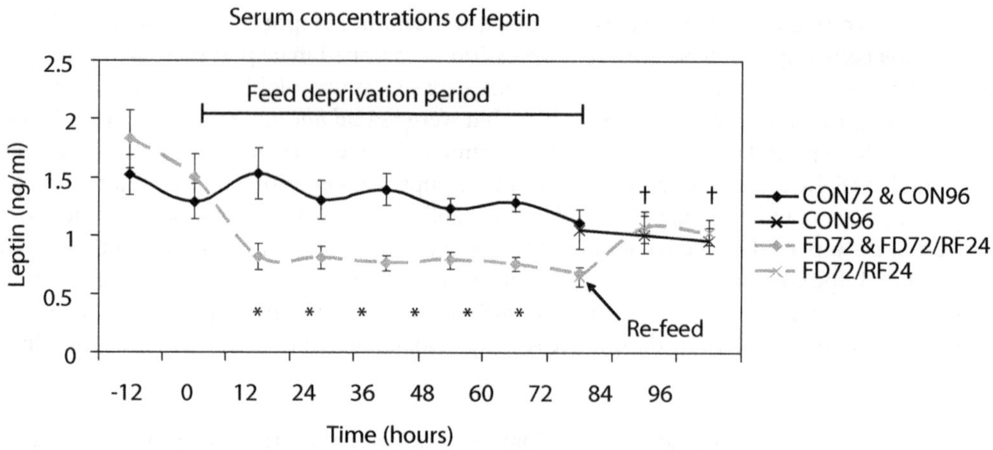

Figure 4. Recreated from Salfen et al.*, 2003. Effect of feed deprivation for 72 hours (FD72) and FD72 followed by re-feeding until 96 hours (FD72/RF24) on serum concentrations of leptin compared to* ad libitum *feeding for 72 hours (CON72) or 96 hours (CON96). Values are the mean ±S.E.M. of n=8 pigs per group.*
** Denotes difference from the 0 hour concentration within the same treatment (P<0.05).*
† Denotes change in hormone concentrations of the FD72/RF24 treatment compared to the 72 hour time point (P<0.05).

Spurlock *et al.* (1998b) that weight loss and food intake restrictions do not necessarily invoke parallel reductions in blood leptin or *obese* mRNA.

For a more detailed review of leptin and its biological role in the pig associated with appetite regulation, reproduction, mediation of GH secretion, the onset of puberty and associations with the immune system, the reader is referred to the review by Barb *et al.* (2001).

Urocortin

Urocortin is a 41 amino acid neuropeptide that is structurally related to CRH that belongs to the CRH family that includes CRH, urotensin I, sauvagine, urocortin II and urocortin III. While it binds to both the CRH type 1 and the CRH type 2 receptors, it binds to the CRH type 2 receptor with higher affinity (Spina *et al.*, 1996). Given the structural similarity to CRH and the affinity for the CRH receptors, urocortin was thought to have overlapping biological activities with CRH including suppressive effects on appetite (Parkes *et al.*, 1997; Spina *et al.*, 1996). Urocortin has been localised in regions of the brain associated with the control of fluid and electrolyte metabolism, including the supraoptic and paraventricular nuclei and the organum vasculosum of the lamina terminalis (Kozicz *et al.*, 1998; Vaughan *et al.*, 1995). A unique characteristic associated with urocortin activity is that while it elicits more of an appetite suppressive effect than CRH, it does not elicit the anxiogenic activities to the extent of CRH (Spina *et al.*, 1996).

One of the first *in vivo* evaluations of urocortin as a potential mediator of food intake was published by Spina *et al.* (1996) in which the authors demonstrated that in food deprived rats,

ICV injections of urocortin significantly suppressed food intake in a dose-dependent manner. The authors further elucidated the appetite suppressive effects of urocortin in fed animals, demonstrating that significant decreases in food and water intake were observed with doses as low as 100 ng. At the higher doses of urocortin, food and water intake were decreased for up to 12 hours. The overall effect on food consumption was attributed to both a decrease in the meal size and frequency of meals with significant effects at 10 ng for meal size and 1000 ng for meal bouts. The authors reported that at the 1000 ng dose, meal bouts were virtually abolished for 6 hours. The authors also reported that water intake paralleled that of food intake regardless of treatment. A reduction in food intake associated with ICV injections of urocortin was subsequently reported in sheep by Weisinger *et al.* (2000). However, these authors reported that urocortin administration did not elicit a decrease in water intake as seen in the previous rodent study.

Unfortunately, relatively little is known with regard to the appetite suppressive effects of urocortin in swine. In 2000, the only reported study associated with the effects of urocortin on appetite regulation in swine was published by Whitley *et al.* (2000). In this study, the authors conducted three experiments to evaluate the effects of intracerebroventicular injection of urocortion and intravenous injections of urocortion at various dosages on feed intake and serum concentrations of GH, LH and cortisol in ovariectomised gilts. In their first experiment, the authors evaluated the effects of an ICV injection of urocortin in gilts that were restrained. Their second experiment was the same as the first with the exception that the gilts were not restrained, and the third experiment evaluated intravenous injections of urocortin. In all three experiments, the gilts were fasted for a 20-hour period, and then allowed *ad libitum* access to feed at either one hour (experiment 1) or 30 minutes (experiments 2 and 3) following administration of the urocortin. The authors reported that treatment with 50 micrograms/pig decreased food consumption as compared to saline treated animals when urocortin was administered as centrally, but not when administered as a peripheral injection. With regard to serum concentrations of GH, the authors reported that urocortin treatment increased GH concentrations regardless of the route of administration. Conversely, serum concentrations of LH were decreased by urocortin administration, but only in those animals receiving the treatment via ICV injection. The effects of urocortin treatment on serum concentrations of cortisol were less consistent as the responses varied depending on the dose of urocortin and the experimental design. Given the lack of a cortisol response in experiment 1, it's tempting to speculate that any potential effects of urocortin on the hypothalamic-pituitary-adrenal axis may have been masked due to the gilts being restrained. Collectively, the data from this study provided clear evidence that urocortin administration elicited a significant suppressive effect on appetite, and significantly altered hormones associated with growth performance and reproductive capability.

A potential role for glucorticoids on neonatal pig growth and feed intake

A rather unconventional approach to stimulating feed intake and growth in neonatal swine has been explored in our laboratory through the administration of the synthetic glucocorticoid, dexamethasone (Carroll, 2001; Gaines *et al.*, 2002, 2004; Seaman-Bridges *et al.*, 2003). These studies evolved from our prior research that demonstrated that bypassing the glucocorticoid surge associated with the natural birth process, via caesarean section, reduces piglet growth (Figure 5) and alters the somatotrophic axis (Carroll *et al.*, 2000). In swine, there is an increase

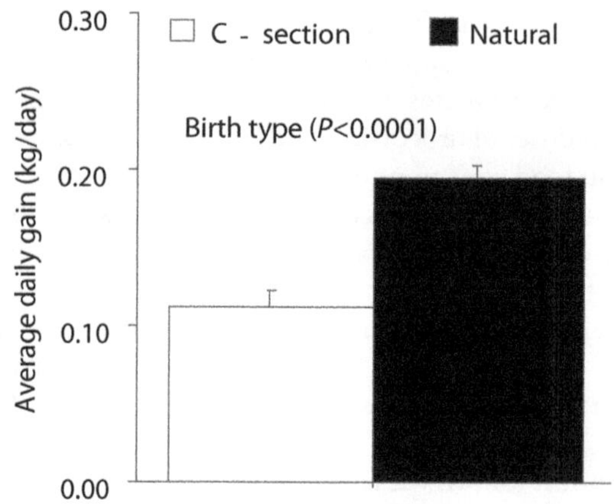

Figure 5. Recreated from Carroll et al.*, 2000. Effect of birth type on average daily gain at 2 weeks of age for C-section (n=9) and natural born pigs (n=11). Open and filled bars represent the mean ±SE for the C-section and natural born pigs, respectively.*

in maternal circulating cortisol, as well as a final fetal cortisol surge during labor (Randall, 1983). Fetal serum cortisol levels increase during the last two weeks of gestation, and levels in swine rise dramatically during the last 8 hours prior to parturition (Randall, 1983). This surge of cortisol is necessary for the initiation of the birth process (Kattesh *et al.*, 1990) and to signal the maturation of fetal tissue in preparation for life outside of the uterine environment (Seckl *et al.*, 1999). Indeed, the prepartum surge in glucocorticoids has been shown to be an important mediator of intestinal maturation and function. Sangild *et al.* (1991, 1994, 1995) have demonstrated through a series of glucocorticoid exclusion and glucocorticoid replacement studies that the prepartum surge in cortisol plays a critical role in the development of digestive capability in the pig. Additionally, they suggested that the glucocorticoid role in intestinal maturation and function is most likely limited to the immediate perinatal period (Sangild *et al.*, 1991). Glucocorticoids have also been implicated as having an important role in the fetal development of somatotrophs (Nogami *et al.*, 1997; Nogami and Tachibana, 1993) and various stimulatory effects on components of the somatotrophic axis (Miell *et al.*, 1993; Miell *et al.*, 1994; Miller and Mayo, 1997; Tamura *et al.*, 2000; Wehrenberg *et al.*, 1990). Consistent with the hypothesis that glucocorticoids mediate postnatal growth is a study by Li *et al.* (1996) which suggested that the prepartum cortisol surge associated with parturition in sheep was the maturational signal necessary for the switch from fetal to postnatal modes of somatotrophic function. Therefore, it is during this perinatal period that we hypothesised that a 'window of opportunity' may exist to permanently alter or preset physiological parameters to enhance growth and performance through exogenous glucocorticoid therapy.

In our initial study (Carroll, 2001) evaluating the use of dexamethasone as a potential mediator of postnatal growth, we injected forty crossbred pigs intramuscularly with either sterile saline or dexamethasone (1 mg/kg body weight) within 1 hour of birth. All pigs remained with their respective dams until 18 days of age and we recorded body weights weekly and on day 18. On day 17, all pigs were non-surgically fitted with an indwelling jugular catheter and placed back with the sows. On the following day, all pigs were placed in individual pens for serial blood collection to evaluate the somatotrophic axis. Body weights after the 18-day trial were 10.1% greater in pigs that were injected with dexamethasone within 1 hour of birth as compared to those injected with saline and had a corresponding increase in average daily gain of 12.2% (Figure 6). Dexamethasone treatment within 1 hour of birth also altered the basal profiles of somatotrophic hormones at 18 days of age. Serum concentrations of GH were significantly reduced in all pigs that had received the dexamethasone injection as compared to the saline injected pigs, however, the magnitude of reduction was greater in the males as compared to the females. Conversely, serum concentrations of IGF-1 were increased by dexamethasone injection. In the male pigs, dexamethasone increased serum concentration of IGF-1 by 47% as compared to saline injected male pigs, whereas in the females, dexamethasone increased serum concentration of IGF-1 by 34% as compared to saline injected female pigs (Figure 7). As with serum concentrations of GH, serum concentrations of IGF-2 were reduced in the dexamethasone treated pigs.

Interestingly, Weiler *et al.* (1997) had previously reported that chronic administration of dexamethasone for a period of 15 days to one-week old pigs actually reduced piglet growth and increased protein catabolism. The dichotomy that exists in our previous data and that reported

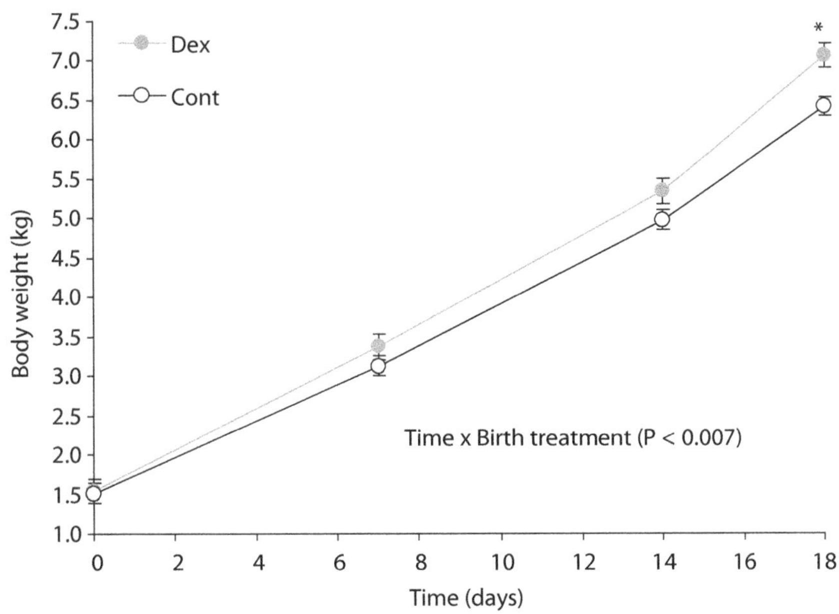

Figure 6. Recreated from Carroll (2001). Effect of dexamethasone treatment (Dex; 1 mg/kg body weight) on body weight from birth (Day 0) until 18 days of age. Body weights were not significantly different until day 18 at which time body weights were 10.1% greater for the dexamethasone treated (Dex) pigs as compared to the control (Cont) pigs (7.06±0.16 and 6.41±0.2 kg, respectively). Standard error bars represent the pooled SEM for the Dex (0.123 kg) and Cont groups (0.157 kg).
** Significant at P=0.015.*

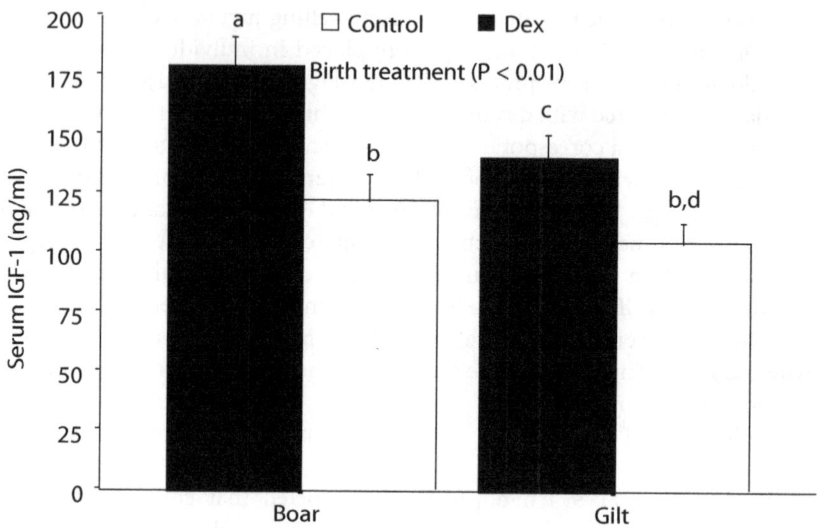

Figure 7. Recreated from Carroll (2001). Effect of birth treatment and sex on serum concentrations of insulin-like growth factor 1 (IGF-1). At 18 days of age, serum IGF-1 was higher in dexamethasone (Dex; 1 mg/kg body weight) treated boars and gilts as compared to saline (Control) treated boars and gilts (a versus b, P=0.004; c versus d, P=0.01). Overall, serum concentration of IGF-1 was lower in gilts as compared to boars (P<0.013). Bars represent the mean +SE.

by Weiler *et al.* (1997) may be attributed to the duration of glucocorticoid exposure as well as the developmental stage of the piglet. In the rat fetus, limited glucocorticoid exposure has been suggested to be an important mediator of somatotroph recruitment and development (Nogami *et al.*, 1997; Nogami and Tachibana, 1993). Previous studies have also reported that glucocorticoids alter various components of the somatotrophic axis including increased secretion of GH and IGF-1, and enhanced pituitary responsiveness to GH releasing hormone (Miell *et al.*, 1994; Miller and Mayo, 1997; Tamura *et al.*, 2000; Wehrenberg *et al.*, 1990). In our study, the dexamethasone-induced increase in pig growth was accompanied by a substantially higher concentration of circulating IGF-1, considered by many as the primary regulator of postnatal somatic growth in the pig and other domestic species. A positive relationship between circulating concentrations of IGF-1 and postnatal growth in the pig have been reported by several research groups (Breier and Gluckman, 1991; Schoknecht *et al.*, 1997; Walton *et al.*, 1995). Our data also supports a positive relationship between circulating concentrations of IGF-1 and postnatal growth in the young pig.

While we initially interpreted these data to indicate that dexmethasone treatment may have resulted in a 're-programming' of the somatotrophic axis, perhaps from an acute or short-term stress induced increase in GHRH from the hypothalamus, we could not dismiss the concept that the acute glucocorticoid treatment at birth stimulated appetite and increased milk consumption during the early postpartum period. Glucocorticoids have been reported to promote feed intake by directly stimulating neuropeptide Y (NPY) and inhibiting corticotrophin-releasing hormone (Cavagnini *et al.*, 2000). Glucocorticoids have also been shown to reverse the suppressive effects of leptin on appetite via NPY-independent pathways (Solano and Jacobson, 1999). Thus, an alternative explanation for the differences observed in the somatotrophic axis for the dexamethasone treated pigs is that these differences merely reflected the enhanced growth rate associated with greater food intake during the perinatal. This would suggest that the observed elevation in IGF-1 in the dexamethasone treated pigs may be more reflective of increased growth rather than the causal agent. If this is indeed the case in the present study, then one could speculate that circulating IGFs do not predict the future growth potential of the growing pig, and that circulating IGFs have already acted on target tissues and are in the circulation for clearance. This idea is somewhat in conformity with that proposed by Owens *et al.* (1994) who suggested that circulating IGF-1 serves a role as a reporter of growth performance, rather than a mediator of growth performance. Given that appetite and milk consumption were not monitored in our prior study, we can neither accept nor dismiss this possibility.

In our subsequent studies evaluating the use of synthetic glucocorticoids to enhance neonatal growth in pigs, we observed similar responses to that of our original study. Specifically, in the study by Gaines *et al.* (2002) that evaluated pre- and post-weaning performance of pigs injected with dexamethasone in three separate experiments at either 1 or 24 hours after birth, we observed an 8-10% increase in body weight of male pigs at weaning that had been injected with dexamethasone at both 1 and 24 hours after birth. However, we did not observe any significant difference in weaning weight for female pigs regardless of treatment. In the final experiment of this study, we observed that at the end of an 83 day period, male pigs injected with dexamethasone were on average 5.45 kg heavier than control males, but again no difference was observed among females regardless of treatment. In the following study by Seaman-Bridges *et al.* (2003), we again observed increased weaning weights and market weights in pigs that had been injected with dexamethasone within 1 hour of birth, without any negative effects on carcass or meat quality. Collectively, these studies provided convincing evidence that demonstrate that dexamethasone given within 24 hours of birth significantly improved both pre- and postweaning performance of pigs. Given previous research that potentially linked glucocorticoids to appetite stimulation, we conducted a final study to determine if providing supplemental milk to dexamethasone treated pigs while they were still nursing the sow would provide an even greater growth response during the pre-weaning period (Gaines *et al.*, 2004). Data from this study did not reveal an increase in weaning weights associated with dexamethasone treatment. However, results from this study did indicate that the management practices on some commercial farms may eliminate any positive effect associated with glucocorticoid treatment and highlights the importance of other factors that may influence the appetite and growth in neonatal pigs. Unquestionably, the collective results from these four studies warrant further investigation into the possible appetite stimulating effects of glucocorticoid therapy in the neonatal pig, and may provide further insight into the underlying mechanisms associated with appetite regulation.

Conclusion

The ability to effectively control feed intake during critical periods of development within livestock production systems will undoubtedly come to pass as scientists continue to elucidate the complex and multifaceted pathways that regulate feed intake, and develop novel approaches to manipulate these systems in a beneficial manner. The economic importance associated with achieving optimal feed intake associated with growth, health and reproduction will continue to drive research directed at understanding and conquering this complex system.

Disclaimer

Mention of trade names or commercial products in this article is solely for the purpose of providing specific information and does not imply recommendation or endorsement by the U.S. Department of Agriculture.

References

Adam, C.L., Archer, Z.A., Findlay, P.A., Thomas, L. and Marie, M. 2002. Hypothalamic gene expression in sheep for cocaine- and amphetamine-regulated transcript, pro-opiomelanocortin, neuropeptide Y, agouti-related peptide and leptin receptor and responses to negative energy balance. Neuroendocrinology 75: 250-256.

Arvat, E., Maccario, M., Di, V.L., Broglio, F., Benso, A., Gottero, C., Papotti, M., Muccioli, G., Dieguez, C., Casanueva, F.F., Deghenghi, R., Camanni, F. and Ghigo, E., 2001. Endocrine activities of ghrelin, a natural growth hormone secretagogue (GHS), in humans: comparison and interactions with hexarelin, a nonnatural peptidyl GHS, and GH-releasing hormone. Journal of Clinical Endcrinology and Metabolism 86: 1169-1174.

Asakawa, A., Inui, A., Kaga, T., Yuzuriha, H., Nagata, T., Fujimiya, M., Katsuura, G., Makino, S., Fujino, M.A. and Kasuga, M. 2001a. A role of ghrelin in neuroendocrine and behavioral responses to stress in mice. Neuroendocrinology 74: 143-147.

Asakawa, A., Inui, A., Kaga, T., Yuzuriha, H., Nagata, T., Ueno, N., Makino, S., Fujimiya, M., Niijima, A., Fujino, M.A. and Kasuga, M., 2001b. Ghrelin is an appetite-stimulatory signal from stomach with structural resemblance to motilin. Gastroenterology 120: 337-345.

Atinmo, T., Baldijao, C., Houpt, K.A., Pond, W.G. and Barnes, R.H., 1978. Plasma levels of growth hormone and insulin in protein malnourished vs normal growing pigs in response to arginine or glucose infusion. Journal of Animal Science 46: 409-416.

Barb, C.R., Hausman, G.J. and Houseknecht, K.L., 2001. Biology of leptin in the pig. Domestic Animal Endocrinology 21: 297-317.

Barb, C.R., Kraeling, R.R., Rampacek, G.B. and Hausman, G.J., 2006. The role of neuropeptide Y and interaction with leptin in regulating feed intake and luteinizing hormone and growth hormone secretion in the pig. Reproduction 131: 1127-1135.

Barb, C.R., Yan, X., Azain, M.J., Kraeling, R.R., Rampacek, G.B. and Ramsay, T.G., 1998. Recombinant porcine leptin reduces feed intake and stimulates growth hormone secretion in swine. Domestic Animal Endocrinology 15: 77-86.

Berelowitz, M., Firestone, S.L. and Frohman, L.A., 1981. Effects of growth hormone excess and deficiency on hypothalamic somatostatin content and release and on tissue somatostatin distribution. Endocrinology 109: 714-719.

Bidwell, C.A., Ji, S., Frank, G.R., Cornelius, S.G. and Willis, G.M., 1997. Cloning and expression of the porcine obese gene. Animal Biotechnology 8: 191-206.

Billington, C.J., Briggs, J.E., Grace, M. and Levine, A.S., 1991. Effects of intracerebroventricular injection of neuropeptide Y on energy metabolism. The American Journal of Physiology 260: R321-R327.

Booth, P.J., Craigon, J., Foxcroft, G.R., 1994. Nutritional manipulation of growth and metabolic and reproductive status in prepubertal gilts. Journal of Animal Science 72: 2415-2424.

Brady, L.S., Smith, M.A., Gold, P.W. and Herkenham, M., 1990. Altered expression of hypothalamic neuropeptide mRNAs in food-restricted and food-deprived rats. Neuroendocrinology 52: 441-447.

Breier, B.H. and Gluckman, P.D., 1991. The regulation of postnatal growth: nutritional influences on endocrine pathways and the function of the somatotrophic axis. Livestock Production Science 27: 77-94.

Broglio, F., Gottero, C., Prodam, F., Destefanis, S., Gauna, C., Me, E., Riganti, F., Vivenza, D., Rapa, A., Martina, V., Arvat, E., Bona, G., van der Lely, A.J. and Ghigo, E. 2004. Ghrelin secretion is inhibited by glucose load and insulin-induced hypoglycaemia but unaffected by glucagon and arginine in humans. Clinincal Endocrinology 61: 503-509.

Buonomo, F.C., Grohs, D.L. Baile, C.A. and Campion, D.R,, 1988. Determination of circulating levels of insulin-like growth factor II (IGF-II) in swine. Domestic Animal Endocrinology 5: 323-329.

Campfield, L.A., Smith, F.J., Guisez, Y., Devos, R. and Burn, P., 1995. Recombinant mouse OB protein: evidence for a peripheral signal linking adiposity and central neural networks. Science 269: 546-549.

Carroll, J.A., 2001. Dexamethasone treatment at birth enhances neonatal growth in swine. Domestic Animal Endocrinology 21: 97-109.

Carroll, J.A., Buonomo, F.C., Becker, B.A. and Matteri, R.L., 1999. Interactions between environmental temperature and porcine growth hormone (pGH) treatment in neonatal pigs. Domestic Animal Endocrinology 16: 103-113.

Carroll, J.A., Daniel, J.A., Keisler, D.H. and Matteri, R.L., 2000. Postnatal function of the somatotrophic axis in pigs born naturally or by Caesarian section. Domestic Animal Endocrinology 19: 39-52.

Carroll, J.A., Veum, T.L. and Matteri, R.L., 1998. Endocrine responses to weaning and changes in post-weaning diet in the young pig. Domestic Animal Endocrinology 15: 183-194.

Cavagnini, F., Croci, M., Putignano, P., Petroni, M.L. and Invitti, C., 2000. Glucocorticoids and neuroendocrine function. International Journal of Obesity and Related Metabolic Disorders 24 (Suppl. 2): S77-S79.

Chen, H., Charlat, O., Tartaglia, L.A., Woolf, E.A., Weng, X., Ellis, S.J., Lakey, N.D., Culpepper, J., Moore, K.J., Breitbart, R.E., Duyk, G.M., Tepper, R.I. and Morgenstern, J.P., 1996. Evidence that the diabetes gene encodes the leptin receptor: identification of a mutation in the leptin receptor gene in db/db mice. Cell 84: 491-495.

Coffey, R.D. and Cromwell, G.L., 1995. The impact of environment and antimicrobial agents on the growth response of early-weaned pigs to spray-dried porcine plasma. Journal of Animal Science 73: 2532-2539.

Colmers, W.F. and El, B.B., 2003. Neuropeptide Y and epilepsy. Epilepsy Currents 3: 53-58.

Crooker, B.A., McGuire, M.A., Cohick, W.S., Harkins, M., Bauman, D.E. and Sejrsen, K., 1990. Effect of dose of bovine somatotropin on nutrient utilization in growing dairy heifers. Journal of Nutrition 120: 1256-1263.

Cummings, D.E., Purnell, J.Q., Frayo, R.S., Schmidova, K., Wisse, B.E. and Weigle, D.S., 2001. A preprandial rise in plasma ghrelin levels suggests a role in meal initiation in humans. Diabetes 50: 1714-1719.

Cuntz, U., Leibbrand, R., Ehrig, C., Shaw, R. and Fichter, M.M., 2001. Predictors of post-treatment weight reduction after in-patient behavioral therapy. International Journal of Obesity and Related Metabolic Disorders 25 (Suppl 1): S99-S101.

Daniel, J.A., Whitlock, B.K., Wagner, C.G. and Sartin, J.L., 2002. Regulation of the growth hormone and luteinizing hormone response to endotoxin in sheep. Domestic Animal Endocrinology 23: 361-370.

Dauncey, M.J. and Morovat, A.. 1993. Investigation of mechanisms mediating the increase in plasma concentrations of thyroid hormones after a meal in young growing pigs. Journal of Endocrinology 139: 131-141.

De Rodas, B.Z., Sohn, K.S., Maxwell, C.V. and Spicer, L.J., 1995. Plasma protein for pigs weaned at 19 to 24 days of age: effect on performance and plasma insulin-like growth factor I, growth hormone, insulin, and glucose concentrations. Journal of Animal Science 73: 3657-3665.

Dhillo WS, Small CJ, Stanley SA, Jethwa PH, Seal LJ, Murphy KG, Ghatei MA, Bloom SR. 2002. Hypothalamic interactions between neuropeptide Y, agouti-related protein, cocaine- and amphetamine-regulated transcript and alpha-melanocyte-stimulating hormone in vitro in male rats. Journal of Neuroendocrinology 14: 725-730.

Dyer, C.J., Touchette, K.J., Carroll, J.A., Allee, G.L. and Matteri, R.L., 1999. Cloning of porcine prepro-orexin cDNA and effects of an intramuscular injection of synthetic porcine orexin-B on feed intake in young pigs. Domestic Animal Endocrinology 16: 145-148.

Edwards, C.M., Abusnana, S., Sunter, D., Murphy, K.G., Ghatei, M.A. and Bloom, S.R., 1999. The effect of the orexins on food intake: comparison with neuropeptide Y, melanin-concentrating hormone and galanin. Journal of Endocrinology 160: R7-12.

Eisemann, J.H., Hammond, A.C., Bauman, D.E., Reynolds, P.J., McCutcheon, S.N., Tyrrell, H.F. and Haaland, G.L., 1986. Effect of bovine growth hormone administration on metabolism of growing Hereford heifers: protein and lipid metabolism and plasma concentrations of metabolites and hormones. Journal of Nutrition 116: 2504-2515.

Eisemann, J.H., Hammond, A.C. and Rumsey, T.S., 1989. Tissue protein synthesis and nucleic acid concentrations in steers treated with somatotropin. British Journal of Nutrition 62: 657-671.

Elias, C.F., Aschkenasi, C., Lee, C., Kelly, J., Ahima, R.S., Bjorbaek, C., Flier, J.S., Saper, C.B. and Elmquist, J.K., 1999. Leptin differentially regulates NPY and POMC neurons proecting to the lateral hypothalamic area. Neuron 23: 775-786.

Elsasser, T.H., Rumsey, T.S., Hammond, A.C. and Fayer, R., 1988. Influence of parasitism on plasma concentrations of growth hormone, somatomedin-C and somatomedin-binding proteins in calves. Journal of Endocrinology 116: 191-200.

Farmer, C., Petitclerc, D., Pelletier, G., Gaudreau, P. and Brazeau, P., 1992. Carcass composition and resistance to fasting in neonatal piglets born of sows immunized against somatostatin and/or receiving growth hormone-releasing factor injections during gestation. Biology of the Neonate 61: 110-117.

Forbes, J.M., 1995. Voluntary food intake and diet selection in farm animals. CAB International, Wallingford.

Funderburke, D.W. and Seerley, R.W., 1990. The effects of postweaning stressors on pig weight change, blood, liver and digestive tract characteristics. Journal of Animal Science 68: 155-162.

Gaines, A.M., Carroll, J.A. and Allee, G.L., 2004. Evaluation of exogenous glucocorticoid injection on preweaning growth performance of neonatal pigs under commercial conditions. Journal of Animal Science 82: 1241-1245.

Gaines, A.M., Carroll, J.A. Allee, G.L. and Yi, G.F., 2002. Pre- and postweaning performance of pigs injected with dexamethasone at birth. Journal of Animal Science 80: 2255-2262.

Gaytan, F., Barreiro, M.L., Chopin, L.K., Herington, A.C., Morales, C., Pinilla, L., Casanueva, F.F., Aguilar, E., Diéguez, C. and Tena-Sempere, M., 2003. Immunolocalization of ghrelin and its functional receptor, the type 1a growth hormone secretagogue receptor, in the cyclic human ovary. Journal of Clinical Endocrinololgy and Metabolism 88: 879-887.

Georgsson, L. and Svendsen, J., 2002. Degree of competition at feeding differentially affects behavior and performance of group-housed growing-finishing pigs of different relative weights. Journal of Animal Science 80: 376-383.

Giroux, S., Martineau, G. and Robert, S., 2000. Relationships between individual behavioural traits and post-weaning growth in segregated early-weaned piglets. Applied Animal Behaviour Science 70: 41-48.

Govoni, N., De Iasio, R., Cocco, C., Parmeggiani, A., Galeati, G., Pagotto, U., Brancia, C., Spinaci, M., Tamanini, C., Pasquali, R., Ferri, G.L. and Seren, E., 2005. Gastric immunolocalization and plasma profiles of acyl-ghrelin in fasted and fasted-refed prepuberal gilts. Journal of Endocrinology 186: 505-513.

Graham, M., Shutter, J.R., Sarmiento, U., Sarosi, I. and Stark, K.L., 1997. Overexpression of Agrt leads to obesity in transgenic mice. Nature Genetics 17: 273-274.

Green, H., Morikawa, M. and Nixon, T., 1985. A dual effector theory of growth-hormone action. Differentiation 29: 195-198.

Hahn, T.M., Breininger, J.F., Baskin, D.G. and Schwartz, M.W., 1998. Coexpression of Agrp and NPY in fasting-activated hypothalamic neurons. Nature Neuroscience 1: 271-272.

Harrell, R.J., Thomas, M.J., Boyd, R.D., Czerwinski, S.M., Steele, N.C. and Bauman, D.E., 1999. Ontogenic maturation of the somatotropin/insulin-like growth factor axis. Journal of Animal Science 77: 2934-2941.

Hart, I.C. and Johnsson, I.D., 1986. Growth hormone and growth in meat producing animals. In: P.J. Buttery, D.B. Lindsay and N.B. Haynes (eds.) Control and Manipulation of Animal Growth. Buttersworths, London, pp. 135-159.

Harvey, S., 1995. Growth hormone release: Mechanisms. In: S. Harvey, C.G. Scanes and W. H. Daughaday (eds.) Growth hormone, CRC Press, Inc., Boca Raton.

Hattori, N., Saito, T., Yagyu, T., Jiang, B.H., Kitagawa, K. and Inagaki, C., 2001. GH, GH receptor, GH secretagogue receptor, and ghrelin expression in human T cells, B cells, and neutrophils. Journal of Clinical Endocrinololgy and Metabolism 86: 4284-4291.

Hattori, N., Shimatsu, A., Sugita, M., Kumagai, S. and Imura, H. 1990. Immunoreactive growth hormone (GH) secretion by human lymphocytes: augmented release by exogenous GH. Biochemimcal and Biophysical Research Communications 168: 396-401.

Hellstern, G., Reijngoud, D.J., Stellaard, F. and Okken, A., 1996. Insulin-like growth factor-I fails to reverse corticosteroid-induced protein catabolism in growing piglets. Pediatric Research 39: 421-426.

Henry, B.A., 2003. Links between the appetite regulating systems and the neuroendocrine hypothalamus: lessons from the sheep. Journal of Neuroendocrinology 15: 697-709.

Hevener, W., Almond. G.W., Armstrong. J.D. and Richards, R.G., 1997. Effects of acute endotoxemia on serum somatotropin and insulin-like growth factor I concentrations in prepubertal gilts. Americal Journal of Veterinary Research 58: 1010-1013.

Houseknecht, K.L. and Portocarrero, C.P., 1998. Leptin and its receptors: regulators of whole-body energy homeostasis. Domestic Animal Endocrinology 15: 457-475.

Ingalls, A.M., Dickie, M.M. and Snell, G.D., 1950. Obese, a new mutation in the house mouse. Journal of Heridity 41: 317-318.

Inoue, H., Watanuki, M., Myint, H.T., Ito, T., Kuwayama, H. and Hidari, H., 2005. Effects of fasting and refeeding on plasma concentrations of leptin, ghrelin, insulin, growth hormone and metabolites in swine. Animal Science Journal 76: 367-374.

Jain, M.R, Pu, S., Kalra, P.S. and Kalra, S.P., 1999. Evidence that stimulation of two modalities of pituitary luteinizing hormone release in ovarian steroid-primed ovariectomized rats may involve neuropeptide Y Y1 and Y4 receptors. Endocrinology 140: 5171-5177.

Kato, Y., Ishikawa, K., Ieiri, T., Imai, K. and Suzuki, M., 1989. Pituitary concentration of GH messenger ribonucleic acid in rats with anterolateral hypothalamic deafferentation. Endocrinologia Japonica. 36: 87-93.

Kattesh, H.G., Charles, S.F., Baumbach, G.A. and Gillespie, B.E., 1990. Plasma cortisol distribution in the pig from birth to six weeks of age. Biology of the Neonate 58: 220-226.

Kelley, K.W., 1989. Growth hormone, lymphocytes and macrophages. Biochemical Pharmacology 38: 705-713.

Kirchgessner, A.L., 2002. Orexins in the brain-gut axis. Endocrine Reviews 23: 1-15.

Kojima, C.J., Carroll, J.A., Matteri, R.L., Touchette, K.J. and Allee, G.L., 2007. Effects of weaning and weaning weight on neuroendocrine regulators of feed intake in pigs. Journal of Animal Science 85: 2133-2139.

Korbonits, M., Bustin, S.A., Kojima, M., Jordan, S., Adams, E.F., Lowe, D.G., Kangawa, K. and Grossman, A.B., 2001. The expression of the growth hormone secretagogue receptor ligand ghrelin in normal and abnormal human pituitary and other neuroendocrine tumors. Journal of Clinical Endocrinololgy and Metabolism 86: 881-887.

Kozicz, T., Yanaihara, H. and Arimura, A., 1998. Distribution of urocortin-like immunoreactivity in the central nervous system of the rat. Journal of Comparitive Neurology 391: 1-10.

Kuenzel, W.J., Douglass L.W. and Davison, B.A., 1987. Robust feeding following central administration of neuropeptide Y or peptide YY in chicks, Gallus domesticus. Peptides 8: 823-828.

Lee, C.Y., Bazer, F.W., Etherton, T.D. and Simmen, F.A., 1991. Ontogeny of insulin-like growth factors (IGF-I and IGF-II) and IGF-binding proteins in porcine serum during fetal and postnatal development. Endocrinology 128: 2336-2344.

Lee, G.H., Proenca, R., Montez, J.M., Carroll, K.M., Darvishzadeh, J.G., Lee, J.I. and Friedman, J.M., 1996. Abnormal splicing of the leptin receptor in diabetic mice. Nature 379: 632-635.

Li, J., Owens, J.A., Owens, P.C., Saunders, J.C., Fowden, A.L. and Gilmour, R.S., 1996. The ontogeny of hepatic growth hormone receptor and insulin-like growth factor I gene expression in the sheep fetus during late gestation: developmental regulation by cortisol. Endocrinology 137: 1650-1657.

Malendowicz, W., Rucinski, M., Macchi, C., Spinazzi, R., Ziolkowska, A., Nussdorfer, G.G. and Kwias, Z., 2006. Leptin and leptin receptors in the prostate and seminal vesicles of the adult rat. International Journal of Molecular Medicine 18: 615-618.

Margetic, S., Gazzola, C., Pegg, G.G. and Hill, R.A., 2002. Leptin: a review of its peripheral actions and interactions. International Journal of Obesity and Related Metabolic Disorders 26: 1407-1433.

Matteri, R.L., Dyer, C.J., Touchette, K.J., Carroll, J.A. and Allee, G.L., 2000. Effects of weaning on somatotrophic gene expression and circulating levels of insulin-like growth factor-1 (IGF-1) and IGF-2 in pigs. Domestic Animal Endocrinology 19: 247-259.

Matteri, R.L., Lamberson, W.R. and Keisler, D.H., 1994. Relationship between in vitro somatotroph function and growth in three-week-old barrows. Domestic Animal Endocrinology 11: 363-373.

McAtee, J.W. and Trenkle, A., 1971. Effect of feeding, fasting and infusion of energy substrates on plasma growth hormone levels in cattle. Journal of Animal Science 33: 612-616.

McLaughlin CL, Rogan GJ, Buonomo FC, Cole WJ, Hartnell GF, Hudson S, Kasser TR, Miller MA, Baile CA. 1991. Finishing lamb performance responses to bovine and porcine somatotropins administered by Alzet pumps. Journal of Animal Science 69: 4039-4048.

McMahon, C.D., Buxton, D.F., Elsasser, T.H., Gunter, D.R., Sanders, L.G., Steele, B.P. and Sartin, J.L., 1999. Neuropeptide Y restores appetite and alters concentrations of GH after central administration to endotoxic sheep. Journal of Endocrinology 161: 333-339.

McShane, T.M., May, T., Miner, J.L. and Keisler, D.H., 1992. Central actions of neuropeptide-Y may provide a neuromodulatory link between nutrition and reproduction. Biology of Reproduction 46: 1151-1157.

McShane, T.M., Petersen, S.L., McCrone, S. and Keisler, D.H., 1993. Influence of food restriction on neuropeptide-Y, proopiomelanocortin, and luteinizing hormone-releasing hormone gene expression in sheep hypothalami. Biology of Reproduction 49: 831-839.

Miell, J.P., Buchanan, C.R., Norman, M.R., Maheshwari, H.G, and Blum, W.F., 1994. The evolution of changes in immunoreactive serum insulin-like growth factors (IGFs), IGF-binding proteins, circulating growth hormone (GH) and GH-binding protein as a result of short-term dexamethasone treatment. Journal of Endocrinology 142: 547-554.

Miell, J.P., Taylor, A.M., Jones, J., Holly, J.M., Gaillard, R.C., Pralong, F.P., Ross, R.J. and Blum, W.F., 1993. The effects of dexamethasone treatment on immunoreactive and bioactive insulin-like growth factors (IGFs) and IGF-binding proteins in normal male volunteers. Journal of Endocrinology 136: 525-533.

Miller, T.L. and Mayo K.E., 1997. Glucocorticoids regulate pituitary growth hormone-releasing hormone receptor messenger ribonucleic acid expression. Endocrinology 138: 2458-2465.

Miner, J.L., Della-Fera, M.A., Paterson, J.A. and Baile, C.A., 1989. Lateral cerebroventricular injection of neuropeptide Y stimulates feeding in sheep. American Journal of Physiology 257: R383-R387.

Mizuno, T.M. and Mobbs, C.V.. 1999. Hypothalamic agouti-related protein messenger ribonucleic acid is inhibited by leptin and stimulated by fasting. Endocrinology 140: 814-817.

Molitch, M.E. and Hlivyak, L.E., 1980. Growth hormone short-loop feedback: anatomic specificity of growth hormone stimulation of hypothalamic somatostatin concentration. Hormone and Metabolic Research 12: 559-560.

Mori, K., Yoshimoto, A., Takaya, K., Hosoda, K., Ariyasu, H., Yahata, K., Mukoyama, M., Sugawara, A., Hosoda, H., Kojima, M., Kangawa, K. and Nakao, K., 2000. Kidney produces a novel acylated peptide, ghrelin. FEBS Letters 486: 213-216.

Moseley, W.M., Paulissen, J.B., Goodwin, M.C., Alaniz, G.R. and Claflin, W.H., 1992. Recombinant bovine somatotropin improves growth performance in finishing beef steers. Journal of Animal Science 70: 412-425.

Muller, E.E., 1987. Neural control of somatotropic function. Physiological Revievs 67: 962-1053.

Muroya, S., Funahashi, H., Yamanaka, A., Kohno, D., Uramura, K., Nambu, T., Shibahara, M., Kuramochi, M., Takigawa, M., Yanagisawa, M., Sakurai, T., Shioda, S. and Yada, T., 2004. Orexins (hypocretins) directly interact with neuropeptide Y, POMC and glucose-responsive neurons to regulate Ca 2+ signaling in a reciprocal manner to leptin: orexigenic neuronal pathways in the mediobasal hypothalamus. European Journal of Neuroscience 19: 1524-1534.

Nakazato, M., Murakami, N., Date, Y., Kojima, M., Matsuo, H., Kangawa, K. and Matsukura, S., 2001. A role for ghrelin in the central regulation of feeding. Nature 409: 194-198.

Nogami, H., Inoue, K. and Kawamura, K., 1997. Involvement of glucocorticoid-induced factor(s) in the stimulation of growth hormone expression in the fetal rat pituitary gland in vitro. Endocrinology 138: 1810-1815.

Nogami, H. and Tachibana, T., 1993. Dexamethasone induces advanced growth hormone expression in the fetal rat pituitary gland *in vivo*. Endocrinology 132: 517-523.

Novakofski, J. and McCusker, R.H., 1993. Physiology and principles of muscle growth. In: Hollis, G.R. (ed.) Growth of the pig. CAB International, Wallingford.Pages, pp 33-48.

Owens, P.C., Campbell, R.G., Francis, G.L. and Quinn, K.J., 1994. Growth hormone, gender and insulin-like growth factors: relationship to growth performance in pigs. Journal of Animal Science (Suppl. 1): 253.

Palmiter, R.D., Erickson, J.C., Hollopeter, G., Baraban, S.C. and Schwartz, M.W., 1998. Life without neuropeptide Y. Recent Progress in Hormonal Research 53: 163-199.

Parkes, D.G., Vaughan, J., Rivier, J., Vale, W. and May, C.N., 1997. Cardiac inotropic actions of urocortin in conscious sheep. American Journal of Physiology 272: H2115-H2122.

Parrott, R.F., Heavens, R.P. and Baldwin, B.A., 1986. Stimulation of feeding in the satiated pig by intracerebroventricular injection of neuropeptide Y. Physiology and Behavior 36: 523-525.

Parrott, R.F., Vellucci, S.V., Goode, J.A., Lloyd, D.M. and Forsling, M.L., 1995. Cyclo-oxygenase mediation of endotoxin-induced fever, anterior and posterior pituitary hormone release, and hypothalamic c-Fos expression in the prepubertal pig. Experimental Physiology 80: 663-674.

Pelleymounter, M.A., Cullen, M.J., Baker, M.B., Hecht, R., Winters, D., Boone, T. and Collins, F., 1995. Effects of the obese gene product on body weight regulation in ob/ob mice. Science 269: 540-543.

Phillips, L.S., 1986. Nutrition, somatomedins, and the brain. Metabolism 35: 78-87.

Phillips, M.S., Liu, Q., Hammond, H.A., Dugan, V., Hey, P.J., Caskey, C.J. and Hess, J.F., 1996. Leptin receptor missense mutation in the fatty Zucker rat. Nature Genetics 13: 18-19.

Plotsky, P.M. and Vale, W., 1985. Patterns of growth hormone-releasing factor and somatostatin secretion into the hypophysial-portal circulation of the rat. Science 230: 461-463.

Preston, R.L., Bartle, S.J., Kasser, T.R., Day, J.W., Veenhuizen, J.J. and Baile, C.A., 1995. Comparative effectiveness of somatotropin and anabolic steroids in feedlot steers. Journal of Animal Science 73: 1038-1047.

Randall, G.C., 1983. Changes in the concentrations of corticosteroids in the blood of fetal pigs and their dams during late gestation and labor. Biology of Reproduction 29: 1077-1084.

Rettori, V., Milenkovic, L., Riedel, M. and McCann, S.M., 1990. Physiological role of neuropeptide Y (NPY) in control of anterior pituitary hormone release in the rat. Endocrinology Experiments 24: 37-45.

Riley, J.E., 1989. Recent trends in pig production: the importance of intake. In: Forbes, J.M., Varley, M.A. and Lawrence, T.L.J. (eds.) The Voluntary Intake of Pigs. Occasional Publication of the British Society of Animal Production, pp 1-5.

Rossi, M., Kim, M.S., Morgan, D.G., Small, C.J., Edwards, C.M., Sunter, D., Abusnana, S., Goldstone, A.P., Russell, S.H., Stanley, S.A., Smith, D.M., Yagaloff, K., Ghatei, M.A. and Bloom, S.R., 1998. A C-terminal fragment of Agouti-related protein increases feeding and antagonizes the effect of alpha-melanocyte stimulating hormone in vivo. Endocrinology 139: 4428-4431.

Sakurai, T., Amemiya, A., Ishii, M., Matsuzaki, I., Chemelli, R.M., Tanaka, H., Williams, S.C., Richarson, J.A., Kozlowski, G.P., Wilson, S., Arch, J.R., Buckingham, R.E., Haynes, A.C., Carr, S.A., Annan, R.S., McNulty, D.E., Liu, W.S., Terrett, J.A., Elshourbagy, N.A., Bergsma, D.J. and Yanagisawa, M., 1998. Orexins and orexin receptors: a family of hypothalamic neuropeptides and G protein-coupled receptors that regulate feeding behavior. Cell 92: 1.

Salfen, B.E., Carroll, J.A. and Keisler, D.H., 2003. Endocrine responses to short-term feed deprivation in weanling pigs. Jounral of Endocrinology. 178: 541-551.

Salfen, B.E., Carroll, J.A., Keisler, D.H. and Strauch, T.A., 2004. Effects of exogenous ghrelin on feed intake, weight gain, behavior, and endocrine responses in weanling pigs. Journal of Animal Science 82: 1957-1966.

Sangild, P.T., Foltmann, B. and Cranwell, P.D., 1991. Development of gastric proteases in fetal pigs and pigs from birth to thirty six days of age. The effect of adrenocorticotropin (ACTH). Journal of Developmental Physiology 16: 229-238.

Sangild, P.T., Sjostrom, H., Noren, O., Fowden, A.L. and Silver, M., 1995. The prenatal development and glucocorticoid control of brush-border hydrolases in the pig small intestine. Pediatric Research 37: 207-212.

Sangild, P.T., Westrom, B.R., Fowden, A.L. and Silver, M., 1994. Developmental regulation of the porcine exocrine pancreas by glucocorticoids. J. Pediatr. Gastroenterol. Nutr. 19: 204-212.

Sartin, J.L., Dyer, C., Matteri, R., Buxton, D., Buonomo, F., Shores, M., Baker, J., Osborne, J.A., Braden, T. and Steele, B., 2001. Effect of intracerebroventricular orexin-B on food intake in sheep. Journal of Animal Science 79: 1573-1577.

Sartin, J.L., Marks, D.L., McMahon, C.D., Daniel, J.A., Levasseur, P., Wagner, C.G., Whitlock, B.K. and Steele, B.P., 2008. Central role of the melanocortin-4 receptors in appetite regulation after endotoxin. Journal of Animal Science 86: 2557-2567.

Scarlett, J.M., Zhu, X., Enriori, P.J., Bowe, D.D., Batra, A.K., Levasseur, P.R., Grant, W.F., Meguid, M.M., Cowley, M.A. and Marks, D.L., 2008. Regulation of agouti-related protein messenger ribonucleic acid transcription and peptide secretion by acute and chronic inflammation. Endocrinology 149: 4837-4845.

Schoknecht, P.A., Ebner, S., Skottner, A., Burrin, D.G., Davis, T.A., Ellis, K. and Pond, W.G., 1997. Exogenous insulin-like growth factor-I increases weight gain in intrauterine growth-retarded neonatal pigs. Pediatric Research 42: 201-207.

Schwarz, F.J., Schams, D., Ropke, R., Kirchgessner, M., Kogel, J. and Matzke, P., 1993. Effects of somatotropin treatment on growth performance, carcass traits, and the endocrine system in finishing beef heifers. Journal of Animal Science 71: 2721-2731.

Scrimgeour, K., Gresham, M.J., Giles, L.R., Thomson, P.C., Wynn, P.C. and Newman, R.E., 2008. Ghrelin secretion is more closely aligned to energy balance than with feeding behaviour in the grower pig. Jounral of Endocrinology. 198: 135-145.

Seaman-Bridges, J.S., Carroll, J.A., Safranski, T.J. and Berg, E.P., 2003. Short- and long-term influence of perinatal dexamethasone treatment on swine growth. Domestic Animal Endocrinology 24: 193-208.

Seckl, J.R., Nyirenda, M.J., Walker, B.R. and Chapman, K.E., 1999. Glucocorticoids and fetal programming. Biochemical Society Transactions 27: 74-78.

Shintani, M., Ogawa, Y., Ebihara, K., Aizawa-Abe, M., Miyanaga, F., Takaya, K., Hayashi, T., Inoue, G., Hosoda, K., Kojima, M., Kangawa, K. and Nakao, K., 2001. Ghrelin, an endogenous growth hormone secretagogue, is a novel orexigenic peptide that antagonizes leptin action through the activation of hypothalamic neuropeptide Y/Y1 receptor pathway. Diabetes 50: 227-232.

Shutter, J.R., Graham, M., Kinsey, A.C., Scully, S., Luthy, R. and Stark, K.L., 1997. Hypothalamic expression of ART, a novel gene related to agouti, is up-regulated in obese and diabetic mutant mice. Genes and Development 11: 593-602.

Solano, J.M. and Jacobson, L., 1999. Glucocorticoids reverse leptin effects on food intake and body fat in mice without increasing NPY mRNA. Am. J. Physiol 277: E708-E716.

Spina, M., Merlo-Pich, E., Chan, R.K., Basso, A.M., Rivier, J., Vale, W. and Koob, G.F., 1996. Appetite-suppressing effects of urocortin, a CRF-related neuropeptide. Science 273: 1561-1564.

Spurlock, M.E., Frank, G.R., Cornelius, S.G., Ji, S., Willis, G.M. and Bidwell, C.A., 1998a. Obese gene expression in porcine adipose tissue is reduced by food deprivation but not by maintenance or submaintenance intake. Journal of Nutrition 128: 677-682.

Spurlock, M.E., Ranalletta, M.A., Cornelius, S.G., Frank, G.R., Willis, G.M., Ji, S., Grant, A.L. and Bidwell, C.A., 1998b. Leptin expression in porcine adipose tissue is not increased by endotoxin but is reduced by growth hormone. J. Interferon Cytokine Res. 18: 1051-1058.

Straus, D.S., 1994. Nutritional regulation of hormones and growth factors that control mammalian growth. FASEB Journal 8: 6-12.

Takaya, K., Ariyasu, H., Kanamoto, N., Iwakura, H., Yoshimoto, A., Harada, M., Mori, K., Komatsu, Y., Usui, T., Shimatsu, A., Ogawa, Y., Hosoda, K., Akamizu, T., Kojima, M., Kangawa, K. and Nakao, K., 2000. Ghrelin strongly stimulates growth hormone release in humans. Journal of Clinical Endocrinololgy and Metabolism 85: 4908-4911.

Tamura, H., Kamegai, J., Sugihara, H., Kineman, R.D., Frohman, L.A. and Wakabayashi, I., 2000. Glucocorticoids regulate pituitary growth hormone secretagogue receptor gene expression. Journal of Neuroendocrinology. 12: 481-485.

Tartaglia, L.A., 1997. The leptin receptor. Journal of Biological Chemistry 272: 6093-6096.

Tartaglia, L.A., Dembski, M., Weng, X., Deng, N., Culpepper, J., Devos, R., Richards, G.J., Campfield, L.A., Clark, F.T., Deeds, J., Muir, C., Sanker, S. Moriarty, A., Moore, K.J., Smutko, J.S., Mays, G.G., Wool, E.A., Monroe, C.A. and Tepper, R.I., 1995. Identification and expression cloning of a leptin receptor, OB-R. Cell 83: 1263-1271.

Tatemoto, K., 1982. Neuropeptide Y: complete amino acid sequence of the brain peptide. Proceedings of the National Acadamy of Science of the USA 79: 5485-5489.

Thomas, M.G., Gazal, O.S., Williams, G.L., Stanko, R.L. and Keisler, D.H., 1999. Injection of neuropeptide Y into the third cerebroventricle differentially influences pituitary secretion of luteinizing hormone and growth hormone in ovariectomized cows. Domestic Animal Endocrinology 16: 159-169.

Toshinai, K., Mondal, M.S., Nakazato, M., Date, Y., Murakami, N., Kojima, M., Kangawa, K. and Matsukura, S., 2001. Upregulation of Ghrelin Expression in the Stomach upon Fasting, Insulin-Induced Hypoglycemia, and Leptin Administration. Biochemical and Biophysical Research Communications 281: 1220-1225.

Touchette, K.J., Allee, G.L. and Newcomb, M.D., 1998. The effect of removing spray-dried plasma from the diet of nursery pigs on subsequent feed intake and weight gain. Journal of Animal Science 76 (Suppl. 1): 180.

Tritos, N.A., Elmquist, J.K., Mastaitis, J.W., Flier, J.S. and Maratos-Flier, E., 1998. Characterization of expression of hypothalamic appetite-regulating peptides in obese hyperleptinemic brown adipose tissue-deficient (uncoupling protein-promoter-driven diphtheria toxin A) mice. Endocrinology 139: 4634-4641.

Tschop, M., Smiley, D.L. and Heiman, M.L., 2000. Ghrelin induces adiposity in rodents. Nature 407: 908-913.

Underwood, W.J., 1995. Somatotropin use in farm animals. In: Proceedings of the 13th ACVM Forum. Lake Buena Vista, FL., USA.

Vaughan, J., Donaldson, C., Bittencourt, J., Perrin, M.H., Lewis, K., Sutton, S., Chan, R., Turnbull, A.V Lovejoy, D. and Rivier, C., 1995. Urocortin, a mammalian neuropeptide related to fish urotensin I and to corticotropin-releasing factor. Nature 378: 287-292.

Volante, M., Allia, E., Gugliotta, P., Funaro, A., Broglio, F., Deghenghi, R., Muccioli, G., Ghigo, E. and Papotti, M., 2002a. Expression of ghrelin and of the GH secretagogue receptor by pancreatic islet cells and related endocrine tumors. Journal of Clinical Endocrinololgy and Metabolism 87: 1300-1308.

Volante, M., Fulcheri, E., Allia, E., Cerrato, M., Pucci, A. and Papotti, M., 2002b. Ghrelin expression in fetal, infant, and adult human lung. Journal of Histochemistry and Cytochemistry 50: 1013-1021.

Wagner, C.G., McMahon, D.D., Marks, D.L., Daniel, J.A., Steele, B. and Sartin, J.L., 2004. A role for agouti-related protein in appetite regulation in a species with continuous nutrient delivery. Neuroendocrinology 80: 210-218.

Walton, P.E., Dunshea, F.R. and Ballard, F.J., 1995. In vivo actions of IGF analogues with poor affinities for IGFBPs: metabolic and growth effects in pigs of different ages and GH responsiveness. Progress in Growth Factor Research 6: 385-395.

Wehrenberg, W.B., Janowski, B.A., Piering, A.W., Culler, F. and Jones, K.L., 1990. Glucocorticoids: potent inhibitors and stimulators of growth hormone secretion. Endocrinology 126: 3200-3203.

Weigent, D.A., Baxter, J.B., Wear, W.E., Smith, L.R., Bost, K.L. and Blalock, J.E., 1988. Production of immunoreactive growth hormone by mononuclear leukocytes. FASEB Journal 2: 2812-2818.

Weigent, D.A. and Blalock, J.E., 1990. Immunoreactive growth hormone-releasing hormone in rat leukocytes. Journal of Neuroimmunology 29: 1-13.

Weiler, H.A., Wang, Z. and Atkinson, S.A., 1997. Whole body lean mass is altered by dexamethasone treatment through reductions in protein and energy utilization in piglets. Biology of the Neonate 71: 53-59.

Weisinger, R.S., Blair-West, J.R., Burns, P., Denton, D.A., McKinley, M.J., Purcell, B., Vale, W., Rivier, J. and Sunagawa, K., 2000. The inhibitory effect of hormones associated with stress on Na appetite of sheep. Proceedings of the National Academy of Sciences of the USA 97: 2922-2927.

Weller, P.A., Dauncey, M.J., Bates, P.C., Brameld, J.M., Buttery, P.J. and Gilmour, R.S., 1994. Regulation of porcine insulin-like growth factor I and growth hormone receptor mRNA expression by energy status. American Journal of Physiology 266: E776-E785.

Wester, T.J., Davis, T.A., Fiorotto, M.L. and Burrin, D.G., 1998. Exogenous growth hormone stimulates somatotropic axis function and growth in neonatal pigs. American Journal of Physiology 274: E29-E37.

Whitley, N.C., Barb, C.R., Kraeling, R.R., Barrett, J.B., Rampacek, G.B., Carroll, J.A. and Keisler, D.H., 2000. Feed intake and serum GH, LH and cortisol in gilts after intracerebroventricular or intravenous injection of urocortin. Domestic Animal Endocrinology 19: 209-221.

Wren, A.M., Seal, L.J., Cohen, M.A., Brynes, A.E., Frost, G.S., Murphy, K.G., Dhillo, W.S., Ghatei, M.A. and Bloom, S.R., 2001a. Ghrelin enhances appetite and increases food intake in humans. Journal of Clinical Endocrinololgy and Metabolism 86: 5992.

Wren, A.M., Small, C.J., Abbott, C.R., Dhillo, W.S., Seal, L.J., Cohen, M.A., Batterham, R.L., Taheri, S., Stanley, S.A., Ghatei, M.A. and Bloom, S.R., 2001b. Ghrelin causes hyperphagia and obesity in rats. Diabetes 50: 2540-2547.

Wright, K.J., Balaji, R., Hill, C.M., Dritz, S.S., Knoppel, E.L. and Minton, J.E., 2000. Integrated adrenal, somatotropic, and immune responses of growing pigs to treatment with lipopolysaccharide. Journal of Animal Science 78: 1892-1899.

Xiang, L., Murai, A. and Muramatsu, T., 2004. The effects of agouti-related protein gene transfer in vivo by electroporation in mice. Neuroscience Letters 370: 108-113.

Yen, X., Morrison, C. and Ramsay, T.G., Porcine leptin sequence and expression in lean and obese pigs. FASEB Journal 11, A415. 1997.

Youngstrom, T.G. and Bartness, T.J., 1995. Catecholaminergic innervation of white adipose tissue in Siberian hamsters. American Journal of Physiology 268: R744-R751.

Zhang, Y., Proenca, R., Maffei, M., Barone, M., Leopold, L. and Friedman, J.M., 1994. Positional cloning of the mouse obese gene and its human homologue. Nature 372: 425-432.

Chapter 8. Metabolic regulation of feed intake in monogastric mammals

J.L. Black[1], B.A. Williams[2] and M.J. Gidley[2]
[1]John L Black Consulting, Warrimoo, NSW 2774, Australia; jblack@pnc.com.au
[2]University of Queensland, Centre for Nutrition and Food Sciences, St Lucia, Queensland 4072, Australia

Abstract

Nutrients and their metabolites control short and long term feed intake through direct or indirect endocrine secretions that interact with local and central neural processes. Carbohydrates, fats, proteins and products from both mammalian and microbial enzymic digestion directly affect the release of hormones from the gastrointestinal tract and pancreas. The quantitative and temporal release of these hormones depends on nutrient composition, site of digestion, and products released within the gastrointestinal tract. The hormones act collectively to control meal size with many suppressing intake through the jejunal, ileal and/or colonic brakes, which reduce gastric emptying, propulsive contractions along the intestine, and gastrointestinal tract secretions. However, many of these nutrient-stimulated hormones also act through the vagal nervous system or directly on specific regions of the brain to have longer-term effects on feed intake. The main metabolic control of long-term feed intake, energy metabolism, body composition and body weight is *via* two opposing energy monitoring systems, adenosine monophosphate-activated protein kinase (AMPK) and mammalian target of rapamycin (mTOR), which act both peripherally and centrally within the hypothalamus. AMPK is activated when cells are depleted of adenosine triphosphate (ATP) by monitoring the adenosine monophosphate (AMP)/ATP ratio. AMPK inhibits ATP-consuming pathways and stimulates ATP-producing pathways through the regulation of enzymes involved in lipid, carbohydrate and protein metabolism. AMPK regulates the concentration of a key intake-controlling metabolite, malonyl-CoA, within the hypothalamus. Low energy status and high AMPK activation leads to inactivation of mTOR, which contrary to AMPK, reflects high energy status of an animal. Activation of mTOR within the hypothalamus is also controlled by insulin and leptin. Basal insulin and leptin concentrations increase with animal adiposity status. Low energy or adiposity status results in low concentrations of malonyl-CoA and low activation of mTOR, stimulating the expression of melanocortin system orexigenic peptides, neuropeptide tyrosine and agouti-related peptide, and reducing the expression of anorexigenic peptides, pro-opiomelanocortin, α-melanocyte-stimulating hormone and cocaine- and amphetamine-related transcript, thereby increasing intake and reducing energy expenditure. By converse mechanisms, perceived high energy status reduces feed intake, while increasing energy expenditure. The concepts described can help explain how many dietary and non-dietary situations affect feed intake in animals.

Keywords: ileal brake, malonyl-CoA, melanocortin system, AMPK, mTOR

Introduction

The highest efficiency for use of a feed by pigs occurs when the intake of energy available for metabolism is just sufficient to allow the maximum rate of protein deposition in relation to the weight and genotype of pig (Campbell, 1988). Energy intakes below this amount result in slower growth rates and reduced efficiency of feed use, whereas energy intakes above this amount are deposited predominantly in fat with a marked decline in the efficiency of feed use. Many pigs raised commercially throughout the world are offered feed *ad libitum*. Hence, it would be a major advantage to the pig industry to have the capacity to modify diets or to use other means to either increase or decrease voluntary intake of available energy to continually maintain maximum lean tissue growth rate and maximum efficiency of feed use.

The chemical composition and physical form of a diet can have a major influence on feed intake and growth rate of pigs. For example, Cadogan *et al.* (1999) incorporated ten samples of wheat from different sources into the diets of young pigs and observed a 50% range in feed intake and growth rate, despite digestibility of the samples with extremes in intake being the same. These observations were confirmed recently in a major study in Australia for several cereal grains including wheat, barley, triticale and sorghum where more than 50% differences were measured in the intake by young pigs of diets with similar digestible energy content (MJ/kg) (Black, 2008). Although sorghum has a higher digestible energy content than wheat in pigs, the intake of digestible energy from sorghum-based diets was approximately 20% less than for wheat-based diets (Black, 2008).

Changes in the physical form of a diet for pigs or other monogastric animals by: grinding and pelleting (Wondra *et al.*, 1995); adding different types of soluble (Fleming and Lee, 1983) and insoluble fibre (Kyriazakis and Emmens, 1995); providing excess protein (L'Heureux-Bouron *et al.*, 2003) or fat (Meyer *et al.*, 1998a) have been shown to change the intake of available energy and growth rates. In addition to diet, many other factors have been reported to influence voluntary feed intake of pigs and other animals. These include: genetic propensity to deposit protein and fat (Knap, 2009); climatic conditions below or above the zone of thermal comfort; seasonal day length (Kirsch *et al.*, 1991); reproductive status and lactation (Cripps and Williams, 1975); exercise (Keeling and Martin, 1987); infectious diseases, cancer and immune responses (Marks *et al.*, 2001).

Interactions between the physical form and nutrient composition of diets, the metabolites produced and the humoral and neural systems within monogastric animals are described in this review to suggest likely mechanisms that result in the short- and long-term control of voluntary feed intake and energy expenditure. The same central mechanisms responsible for the metabolic control of feed intake may also be responsible for many of the other non-dietary factors that are known to affect intake. Voluntary feed intake in animals is controlled by two main processes: (1) rate of removal of digesta from the gastrointestinal tract (GIT) through changes in transit time, which may be associated over the longer term with changes in GIT capacity, and changes in the rate of digestion and absorption dietary constituents; and (2) brain controlled satiety and hunger responses resulting from physical consumption of feed, absorption of products from digestion and metabolites from subsequent biochemical processes.

Gastrointestinal tract transit time

There are many reports where an increase or decrease in feed intake of animals has been inversely associated with transit time of digesta through the GIT. For example, the approximately 60% increase in feed intake during pregnancy and 250% increase during lactation in rats (Cripps and Williams, 1975), has been associated with highly significant decreases in the transit time of digesta through the GIT (Chen *et al.*, 1997), as well as increases in the weight and length of GIT organs (Cripps and Williams, 1975). Similarly, increases in the intake of sheep (Westra and Christopherson, 1976), cattle (Gonyou *et al.*, 1979), pigs (Phillips *et al.*, 1982) and mice (Gue *et al.*, 1987) exposed to cold stress have been associated with marked decreases in the transit time of digesta through the GIT. Energy expenditure in humans is also associated with decreases in GIT transit time (Keeling and Martin, 1987). In another example, incorporation of pectin compared with cellulose at approximately 10% into diets for rats, which resulted in a 22% decline in feed intake and a 44% decline in growth rate, was associated with a 45% increase in transit time of digesta through the GIT (Fleming and Lee, 1983).

Effects of volume

Transit time or rate of passage of digesta through the GIT, is determined primarily by the rate of emptying of the stomach and the frequency, and effectiveness of propulsive peristaltic contractions in the intestines. The presence of feed in the stomach initially stimulates distal gastric contractions and gastric emptying. Similarly, the presence of material in the small intestines initially stimulates propulsive peristaltic contractions. These initial contractions in the distal stomach and small intestine are volume and distension dependent, and not determined by nutrient content. The activation of distal stomach and intestinal contractions is via intramural tension receptors operating through intrinsic enteric nerves and can be stimulated simply by the presence of balloons (Geliebter, 1988, Kendall *et al.*, 1987). Continuous infusion of a non-absorbable, iso-osmotic solution into the jejunum of healthy men resulted in progressive increases in propulsive intestinal peristaltic contractions and passage of material (Steadman and Kerlin, 1994). Hence, the presence of material in the stomach or small intestine will initially increase rate of passage of material along the GIT simply through its volume effect. For example, the inclusion of wheat bran into the diet of humans decreased whole gut transit time (Lewis and Heaton, 1997). Similarly, oral administration of the osmotic compound, magnesium sulphate, which increased intestinal volume, substantially decreased transit time during fasting in people (Vu *et al.*, 2000). Stanogias and Pearce (1985a) found that GIT transit times in pigs decreased with increasing neutral detergent fibre (NDF) content of diets containing a range of fibre sources. Rate of passage of material was faster for those diets containing a greater proportion of indigestible fibre. Similar findings were made by Jenkins *et al.* (1978) in humans fed diets containing a range of fibre sources, where the rate of passage of digesta was highest for diets containing wheat bran, and was negatively related to viscosity of the food.

However, the initial volume mediated decreases in GIT retention time are frequently overridden by other feedback mechanisms. Continuing gastric distension through volume increase and not nutrient composition acts via the vagal nerve to the brain to induce satiety and reduce food consumption in many animal species (Maljaars *et al.*, 2007). Thus, foods with high energy density and low volume were found to be associated with increased energy intake and body mass status in humans (Ledikwe *et al.*, 2006). Alternatively, incorporating water (Rolls *et al.*, 1998) or air (Rolls *et al.*, 2000) into milk-based drinks to increase volume, increased satiety in humans.

Nevertheless, many of the effects of gastric volume on satiety, particularly in animals offered feed *ad libitum*, may be meal related and short term. Sasaki *et al.* (1996) offered diets containing 17.4%, 30.1% or 48.5% NDF to growing hamsters for seven weeks. Feed intake increased by 44% with increasing fibre content of the diet. After seven weeks, the increase in intake was associated with a significant increase in the capacity and weight of digestive tract organs (except for small intestine weight), but no difference in GIT transit times between the diets. Similarly, increases in weight and/or length of GIT organs have been reported with increasing NDF intake in pigs (Stanogias and Pearce, 1985b), rats (Brown *et al.*, 1979) and rabbits (Gidenne, 1992).

Evidence from many integrated sources (Black, 2000) suggests that, for pigs offered feed *ad libitum*, intake increases to maintain energy intake for metabolism as the digestible energy content (MJ/kg) of a diet decreases. This increase in intake is achieved through increases in GIT capacity and/or changes in transit time of digesta. However, as the density of digestible energy in the diet declines and volume of the diet consumed increases, a point is reached where intake cannot be increased further to compensate for the decline in digestibility and the intake of energy available for metabolism falls. This negative effect of reducing digestible energy content of a diet occurs to a greater extent in young compared with older pigs (Black, 2000).

Effect of nutrients

Spiller *et al.* (1984) demonstrated that the perfusion of a partially hydrolysed triglyceride solution into the ileum of humans markedly increased transit time through a jejunal segment from 7.5 to 18.9 minutes and termed the phenomenon the 'ileal brake'. Subsequently, there have been numerous reports showing that the presence in the intestines of nutrients or their breakdown products from either mammalian or microbial digestion have negative effects on the rate of gastric emptying, propulsive peristaltic contractions of the intestines, rate of gastric acid, pancreatic and bile secretion, and on feed intake. The effectiveness of the intestinal brake, which has been divided into the jejunal, ileal and colonic brakes depending on the location of nutrients responsible for the effects, is influenced by the composition of the nutrients or degradation products present, the section of the intestines stimulated, and the volume of digesta. Although there are small differences in the activation and effectiveness of the intestinal brakes between monogastric mammalian species (Maljaars *et al.*, 2008), the phenomenon has been observed across all mammals investigated, including pigs.

Lipid based compounds or fatty acids have been used in many investigations into the function of the ileal brake. The magnitude of the ileal brake effects has been reported to increase in proportion to the amount of lipid compound administered for; intestinal transit (Lin *et al.*, 1997), pancreatic trypsin and bile acid output (Keller *et al.*, 2006); and feed intake (Meyer *et al.*, 1998a). However, the nature of the lipid compound influences the effectiveness of the ileal brake. Spiller *et al.* (1988) suggested that the hydrolysis of triglycerides was needed for stimulation of the ileal brake, when compared with the effects of free fatty acid, triglyceride and glycerol on reducing jejunal motility. Glycerol did not alter jejunal activity nor activate the ileal brake. Meyer *et al.* (1998a) confirmed that free fatty acids were needed in the intestines to induce a reduction in food intake in the rat because corn oil or a medium chain triglyceride only induced the ileal brake when not mixed with orlistat, which inhibits lipolysis of triglycerides to fatty acids. Welch *et al.* (1988), when comparing corn oil with oleic acid, suggested that 5 g fatty acid was equivalent to 12 g of triglyceride for inhibiting gastric emptying in humans. The carbon chain length of the fatty acids in triglycerides also appears to alter the effectiveness of the ileal brake, although some results are contradictory. Maljaars *et al.* (2008) concluded that medium-chain triglycerides with 12 or less carbon units induce the ileal brake more rapidly than longer-chain tryglycerides and this may be related to their absorption without the need to form chylomicrons. There is also evidence that unsaturated fats stimulate the ileal brake more strongly than saturated fats (Maljaars *et al.*, 2008).The intestinal brake has been reported to result from the presence of other major nutrient groups in the intestines in addition to fats, including proteins, soluble carbohydrates (glucose, sugars and starch) and microbial fermentation products (Maljaars *et al.*, 2008). Differences have been reported in the effectiveness of various proteins on measures of the ileal brake including, gastric emptying, intestinal transit and feed intake. Hara *et al.* (1992) found that casein was more effective at reducing gastric emptying and increasing small intestinal transit time in rats than isolated soybean protein. Other studies have shown gelatine to be more effective than casein in reducing feed intake, but the effects of whey were far greater than other protein sources (Anderson and Moore, 2004, Luhovyy *et al.*, 2007). Egg albumin has been shown to be effective in reducing feed intake in rats but not in man (Anderson *et al.*, 2004). Meyer *et al.* (1998b) found that feed intake in rats was depressed with the mid-gut infusion of phenylalanine and tryptophan, but not arginine or alanine. Several studies have shown a marked depression in feed intake in humans and rats following excessive supply of leucine (Anderson and Moore, 2004).

The presence of products of microbial fermentation in the ileum and/or colon have been reported to reduce digesta flow, increase oesophageal sphincter tension (Piche *et al.*, 2003), lower gastric tone (Ropert *et al.*, 1996), reduce gastric emptying (Cuche *et al.*, 2000), reduce gastric and pancreatic secretions (Ropert *et al.*, 1996), and reduce intestinal motility (Pimentel *et al.*, 2006) in a range of monogastric animal species. Short chain fatty acids, specifically acetate, propionate and butyrate (Ropert *et al.*, 1996), and methane (Pimentel *et al.*, 2006) have been demonstrated to activate ileal brake functions in animals. The impact of fermentation products in the distal intestines is particularly significant for slowing digesta transit through the GIT and reducing feed intake. The presence of nutrients in the distal ileum and colon has a greater effect on reducing gastric emptying, the rate of passage of digesta through the small intestine (Lin *et al.*, 1992, 1997) and feed intake (Meyer *et al.*, 1998b), than nutrients in more proximal regions of the intestines. Meyer *et al.* (1998b) infused the same amount of a lipid compound into the duodenum, ileum and colon of rats and found that feed intake was reduced by 11%, 21% and 40%, respectively. These observations confirm that the colonic brake is more effective than the ileal brake, which is more effective than the jejunal brake, in slowing the passage of digesta through the GIT and reducing feed intake.

The progressive increase in effectiveness of the intestinal brake caused by nutrients and their products of digestion as they pass from the proximal to the distal intestines means that either chemical or physical characteristics of diets that increase the flow of nutrients to the distal intestines are likely to have a greater negative effect on rate of passage of digesta through the GIT and feed intake than those that are digested in the proximal intestines. Diets containing slowly-digested starches, such as those with a high ratio of amylose to amylopectin, have been reported to have a greater suppression of food intake in humans than those with more rapid digestion (Samra and Anderson, 2007). Similarly, the introduction of insoluble fibre into human diets containing substantial proportions of starch, decreased transit time and substantially increased the formation of short-chain fatty acids in the colon (Lewis and Heaton, 1997). Although feed intake was not measured in the experiment of Lewis and Heaton, other experiments that resulted in an increase in short-chain fatty acids or other nutrients in the distal intestines have depressed food intake (Brighenti *et al.*, 2006, Koopmans, 1990, Meyer *et al.*, 1994). These observations suggest that the positive effect of indigestible fibre on rate of passage of digesta through the GIT may be overridden by the negative effects of the ileal/colonic brake when fermentable substrates flow to the distal intestines.

There is clear evidence of interactions between the positive volume and distension initiated increase in gastric emptying and propulsive peristaltic contractions in the small intestine and the negative impacts due to the intestinal brake. Lin *et al.* (2001a) investigated intestinal transit in dogs with duodenal and mid-intestine fistulae. The proximal section of the intestines was perfused with an oleate solution, while the distal section was either not perfused or perfused with a buffer to cause distension. Intestinal transit across the proximal half of the intestine was slowed substantially when the oleate solution was perfused, with marker recovery over 30 minutes decreasing from 86% to 39%. However, transit was partially restored when distension was also applied to the distal intestine, with 61% of the marker being recovered. A similar substantial increase in gastrointestinal transit time has been observed when oleate is given to patients with chronic diarrhoea (Lin *et al.*, 2001b). Transit time through a section of the small intestine measured in normal subjects was 102 mins, compared with only 29 mins for patients with diarrhoea. However, transit times in the diarrhoea patients increased to 57 minutes and 83 minutes, respectively, following a 1.6-ml or 3.2-ml dose of oleate. The observations described, confirm there is a counterbalancing action between physical components of a diet decreasing intestinal transit time and undigested nutrients increasing transit time.

Mode of action of intestinal brakes

Changes in motility

Nutrients and their digestion products in the distal intestine, reduce gastric emptying and increase the transit time of digesta in the proximal intestine, by reducing the frequency, force and propulsive distance of peristaltic contractions and increasing the proportion of stationary, non-propulsive contractions (Fone *et al.*, 1990, Huge *et al.*, 1995, Siegle *et al.*, 1990). Gastric emptying in humans after a meal was dramatically slowed by the ileal infusion of a triglyceride emulsion through a decrease in antral, duodenal and propulsive antro-pyloro-duodenal pressure waves, an increase in pyloric sphincter pressure and an increase in movement of solid material from the distal to the proximal region of the stomach (Fone *et al.*, 1990). Huge *et al.* (1995) studied how nutrients delivered to the distal intestine change the contractile motility in the proximal intestine of mini-pigs by isolating a 150 cm segment of the proximal jejunum with two fistulae. A diet containing 60% of the energy from partially hydrolysed maize starch, 18% from partially hydrolysed lactalbumin and 22% from medium- and long-chain fatty acids was infused into the distal fistula at the rate of 0 (saline control), 4.19, 6.28, 8.38, 10.47, 12.56 or 16.74 kJ/min. Increasing the rate of nutrient infusion resulted in linear changes in all measurements with the frequency of peristaltic contractions declining from 8 to 4 per minute, the force of each contraction declining from 60.5 to 44 mN, stationary contractions increasing from 47 to 75% of contractions, propulsive distance of contraction declining from 5.2 to 2.4 cm per contraction, digesta flow rate declining from 2.13 to 0.87 ml per minute and transit time increasing from 10 to 70 minutes. Although the absorption of energy from the high rate of nutrient infusion increased over the control by approximately 10%, the flow rate of digesta was depressed 2.5-fold by this rate of nutrient infusion.

Humoral and neural control

The presence of nutrients and their metabolites in the stomach and intestines of animals stimulates the release of a wide range of peptides that regulate gastric emptying, gastric acid secretion, intestinal motility and secretions from the pancreas and gall bladder. The peptides released from cells in the GIT include, among others, ghrelin, gastrin, cholecystokinin (CCK), motilin, obestatin, bombesin, gastric inhibitory protein, glucagon-like peptide-1 (GLP-1), glucagon-like peptide-2, peptide tyrosine tyrosine (PYY), oxyntomodulin (OXM) and apolipoprotein A-IV (apo A-IV). The structure, biologically active components, properties and receptors of these peptides and their actions are described in detail by Carroll and Allee (2009), Chaudhri *et al.* (2006, 2008), Maljaars *et al.* (2008), Roche *et al.* (2008) and Tso *et al.* (2004).

The most important mechanisms controlling the effects of nutrients on motility and secretions of the GIT are described briefly, although there are some differences between animal species. Feed entering the stomach stimulates the release of gastrin, which promotes the secretion of gastric acid. Nutrients flowing from the stomach into the duodenum during the early postprandial period stimulate the release of CCK from I-cells, which initially promotes gastric emptying, pancreatic and gall bladder secretions, and accelerate transit in the proximal intestine, thus enabling further meal consumption. However, the release of gastric acid and CCK in the distal GIT stimulate the release of PYY and the proglucagon peptides, GLP-1 and OXM, from L-cells in the ileum and colon long before digesta from the meal reaches this region. The release of PYY, and most likely GLP-1 and OXM, from the distal intestine appears to be mediated by both gastric acid and CCK via the CCK-1 receptor through the splanchnic and vagal neuronal systems. The release of PYY, GLP-1 and OXM from distal intestinal cells begins a feedback mechanism that starts to slow gastric emptying, gastric acid, pancreatic and bile secretions and proximal intestinal motility to initiate meal cessation.

As digesta flow continues to the jejunum during the later postprandial period, long-chain fatty acids are incorporated into chylomicrons, which are lipoprotein particles that allow absorption of these fats across intestine epithelial cells. The presence of chylomicrons stimulates the release of apo A-IV from endocyte cells of the jejunum. Further passage of digesta to the distal ileum and colon causes increased release of PYY, GLP-1 and OXM. The release of these peptides is facilitated by the nutrient classes described above that stimulate the intestinal brake and by bile acids. A recent study has shown that short-chain fatty acids resulting from the microbial fermentation of digesta constituents bind to the G-protein coupled receptor, GRP43, in the ileum and colon to stimulate release of PYY (Karaki *et al.*, 2006). Short-chain fatty acids have also been shown to increase GLP-1 secretion (Massimino *et al.*, 1998). This further release of PYY, GLP-1 and OXM in the distal intestines strengthens the features of the intestinal brake by further slowing gastric emptying and intestinal motility. PYY also stimulates release of apo A-IV. Recent studies have shown that PYY acts to slow gastric emptying via apo A-IV, which stimulates CCK release from adjacent cells in the proximal intestine and enhances CCK-1 receptors on vagal afferent neurons (Whited *et al.*, 2006). PYY actions on the GIT are via the polypeptide fold proteins, Y2 receptor. An important recent finding (Burdyga *et al.*, 2008) is that CCK regulates the expression of the Y2 receptor in the nodose ganglia of the vagal nerve via the CCK-1 receptor. Nodose ganglia Y2 receptors decreased 5-fold in fasted rats compared with rats fed *ad libitum*, but administration of CCK to fasted rats or re-feeding restored Y2 receptor concentrations.

The concentrations of two orexigenic peptides, ghrelin and motilin, rise as the postprandial period extends and nutrient absorption from the GIT declines. Ghrelin is produced in the stomach and increases gastric motility. Motilin is produced by distal cells of the duodenum and stimulates intestinal motility and secretions from stomach, pancreas and gall bladder. These actions are associated with the initiation of the next meal. The release of motilin from the previous meal along with the increase in absorbed nutrients following a new meal stimulates the release of pancreatic polypeptide above the normal circadian rhythm. One major function of pancreatic polypeptide is to delay gastric emptying following the second meal.

Increasing gastric volume during a meal and gastric distension stimulate tension-sensitive fibres of the vagal afferent neurons, which can also result in cessation of a meal. The cessation of feeding associated with distension is abolished by sub-diaphragm vagotomy. Thus, the short-term control of GIT function, meal size and frequency is influenced by the peptide and distension feedback systems in relation to specific nutrients, their absorbed products, physical structure of a diet and timing of its consumption.

Long-term control of feed intake by nutrient-stimulated GIT peptides

Many of the nutrient-stimulated peptides secreted by the GIT and pancreas, in addition to having direct effects on tract motility and secretions, act either through the vagal neuronal system or directly on specific regions of the brain to have long-term effects on feed intake. The hypothalamus and brainstem are the main areas of the brain responsible for the initial control of feed intake in animals (Chaudhri *et al.*, 2008). The hypothalamus has several nuclei involved in feed intake control, with the most important being the arcuate nucleus (ARC), the hypothalamic paraventricular nucleus (PVN) and the ventromedial nucleus (VHM). The ARC is believed to have an incomplete blood-brain barrier that allows the entry of circulating nutrients and peptides. Ablation of the PVN and VHM causes animals to be hyperphagic and to gain weight. The brain stem also has an incomplete blood-brain barrier and circulating nutrients and peptides act on the area postrema (AP) and nucleus of the tractus solitarius (NTS). The vagus nerve, which is the main link between the GIT and the brain, has cell bodies from the nodose ganglia projecting into the brainstem. The dorsal-vagal-complex of the brainstem comprising the vagal, AP and NTS neurons integrates neural and chemical signals and interfaces with the hypothalamus to control feed intake. Neural signals from the hypothalamus and brainstem interact with regions of the higher brain involved in hedonic, and other senses to have a major influence on hunger, satiety and conditioned taste responses.

The peptides released in response to nutrients, CCK, apo A-IV, PYY, GLP-1, OXM, pancreatic polypeptide and another pancreatic peptide secreted with insulin, amylin, have all been shown to reduce feed intake when administered to animals either intravenously or intraperitoneally (Chaudhri *et al.*, 2006, 2008). Although CCK stimulates gastric emptying and proximal intestinal motility at the local GIT level immediately after a meal begins, it also acts via the vagus nerve to reduce feed intake. However, its anorexigenic effects are not maintained with the appetite depression being lost after 24-hour intraperitoneal infusion and it is considered not to be an appropriate target for a therapeutic agent for reducing feed intake. In contrast, the anorexigenic effects of PYY, pancreatic polypeptide, GLP-1, OXM, apo A-IV and amylin have been shown to be of a long-term nature and their pathways of actions are being exploited as possible means for reducing feed intake in humans (Chaudhri *et al.*, 2008). Similarly, systemic administration of the orexigenic peptide, ghrelin, has been shown to have long term effects on the intake of rodents as well as healthy and cancer-suffering humans (Chaudhri *et al.*, 2008).

Monitoring of metabolic status and adiposity

The main metabolic control of long-term feed intake, energy metabolism, body composition and body weight is under the influence of systems that monitor the energy status and adiposity of the animal. There are two major systems within the body that monitor energy status, adenosine monophosphate-activated protein kinase (AMPK) and mammalian target of rapamycin (mTOR), which act both peripherally and centrally within the hypothalamus. Similarly, the fat content of animals is monitored by two hormones, leptin and insulin, which also act on peripheral tissues and via the hypothalamus. Conversely, the body lean status of animals is monitored by ghrelin.

AMPK

AMPK is an energy-sensing serine/threonine kinase, which responds to changes in the energy status of the cell and the whole body to ensure there is adequate adenosine triphosphate (ATP) for essential metabolic pathways (Hardie *et al.*, 2006). AMPK is activated by changes in its phosphorylation state in response to a low ratio of adenosine monophosphate (AMP) to ATP concentration in the cell. AMPK activation is extremely sensitive to small changes in AMP, because there are three distinct biochemical pathways through which AMP promotes its activation (Hardie *et al.*, 2006). AMPK is therefore activated whenever there is a decline in ATP and a rise in AMP status of a cell caused by either lack of nutrients or oxygen supply, or by excessive use of ATP such as during exercise or cold exposure (Van Thuijl *et al.*, 2008). Activation of AMPK inhibits anabolic ATP-consuming pathways within a cell such as gluconeogenesis and fatty acid synthesis and activates catabolic pathways that generate ATP such as glycolysis and fatty acid oxidation. AMPK stimulates glucose uptake by skeletal muscle during exercise, while it limits lipogenesis and enhances lipid oxidation.

AMPK controls the ligogenesis-lipolysis axis through inhibition of the phosphorylation of a key enzyme, acetyl-CoA carboxylase (ACC), in lipid metabolic pathways (Dowell *et al.*, 2005). There are two forms of ACC, ACC_1 which is expressed mainly in lipogenic tissues such as adipose, and ACC_2 which is expressed primarily in muscle. Both forms of ACC catalyse the formation of malonyl-CoA from acetyl-CoA. In adipose tissue, malonyl-CoA is a substrate for fatty acid synthase (FAS), the enzyme involved in carbon chain elongation to form long-chain fatty acids. Muscle tissue does not have FAS and does not form long-chain fatty acids, but it has another enzyme, malonyl-CoA decarboxylase (MCD), which converts malonyl-CoA back to acetyl-CoA. AMPK inhibits ACC_2 and stimulates MCD, which leads to a decrease in malonyl-CoA concentration.

In conditions where an animal is in high-energy status with a low AMP:ATP ratio and a low AMPK activation, glucose is converted to acetyl-CoA, which is then converted to malonyl-CoA by ACC. Malonyl-CoA along with other blood borne fatty acids in the form of fatty acyl-CoA are then converted to longer chain fatty acids through FAS and stored in adipose tissue. During this process, the concentration of malonyl-CoA increases, and inhibits the expression of carnitine palmitoyl-CoA transferase-1 (CPT1) which is required for the uptake of fatty acyl-CoA by cell mitochondria for β-oxidation and ATP production. Conversely, in conditions of low energy status where the AMP:ATP ratio and AMPK activity are both high, ACC is inactivated and MCD is activated leading to low concentrations of malonyl-CoA, which relieves the inhibition of CPT1, increases the oxidation of fatty acyl-CoA and decreases deposition in adipose tissue. Thus, AMPK has its influence over lipogenesis-lipolysis largely through altering malonyl-CoA concentrations.

AMPK regulation of feed intake and energy expenditure

AMPK is active in the hypothalamus as well as peripheral tissues (Kim *et al.*, 2004), where it has a similar role in monitoring energy status of an animal, controlling feed intake and energy expenditure largely through its influence over malonyl-CoA concentrations. When the energy status of an animal is low, hypothalamic AMPK is activated, which results in inactivation of ACC, activation of MCD and a depression in malonyl-CoA concentration. Conversely, when the energy status of an animal is high, AMPK activation is low, ACC is activated and malonyl-CoA concentrations rise. Administration of cerulenin, a fungal-produced inhibitor of FAS, or the synthetic analogue, C75, which prevents the conversion of malonyl-CoA to longer-chain fatty acyl-CoA and inactivates AMPK, also increases the concentration of malonyl-CoA within the hypothalamus (Dowell *et al.*, 2005). This high concentration of malonyl-CoA resulting from inhibition of FAS causes a marked depression of feed intake (Hu *et al.*, 2003). The high malonyl-CoA concentration also results in an increase in energy expenditure through increased expression of skeletal muscle uncoupling protein 3 (Cha *et al.*, 2004).

Malonyl-CoA has its effects on feed intake by changing the expression of neuropeptides produced by the hypothalamic melanocortin system. Low concentrations of malonyl-CoA in the hypothalamus stimulate the expression of the orexigenic peptides, neuropeptide tyrosine (NPY) and agouti-related peptide (AgRP), and decrease expression of the anorexigenic peptides, pro-opiomelanocortin (POMC), its derivative α-melanocyte-stimulating hormone (α-MSH) and cocaine- and amphetamine-related transcript (CART), thereby increasing intake and decreasing energy expenditure (Cha *et al.*, 2004; Hu *et al.*, 2003). Conversely, a high concentration of malonyl-CoA reduces the expression of the orexigenic peptides and increases the expression of the anorexigenic peptides, with a resultant reduction in feed intake and increase in energy expenditure.

mTOR

A second serine-threonine kinase, mTOR, also responds to changes in the energy status of cells. In contrast to AMPK, mTOR is activated by high nutrient and energy availability, particularly by a high ATP:AMP ratio and by amino acids, especially leucine (Wullschleger *et al.*, 2006). mTOR senses the energy status of cells through AMPK, which inhibits mTOR-dependent phosphorylation of ribosomal protein S6-kinase-1 (S6K1), S6 ribosomal protein (S6) and eukaryotic initiation factor 4E-binding protein 1 (4E-BP1). Thus, when AMPK is highly activated, mTOR activation is low and *vice versa*. mTOR is highly conserved across species and has a major role in the control of cell growth and proliferation. It regulates, in relation to nutrient supply and hormonal signals, essential aspects of cell growth including the protein synthetic processes of transcription, translation initiation and elongation, as well as cell proliferation, while inhibiting cell autophagy (Wullschleger *et al.*, 2006). Complete loss of mTOR function is lethal in mice and defects in the mTOR pathway result in smaller cells in *Drosophila*, whereas over-expression has been linked to the development of cancer, diabetes and obesity (Cota *et al.*, 2006).

In addition to peripheral tissues such as liver and muscle, mTOR is highly active in the hypothalamus. Cota *et al.* (2006) found, by using antibodies to mTOR, that it was present particularly in the PVN and ARC nuclei of the hypothalamus. In the ARC, it is localised with the melanocortin system neuropeptides controlling feed intake, being approximately 90% associated with NPY/AgRP neurons and 45% with POMC/CART neurons. Fasting causes a marked depression in mTOR expression in the ARC, reflecting low activity when the energy status of the animal is low, whereas the reverse occurs in recently fed, satiated animals (Cota *et al.*, 2006). The importance of mTOR in controlling feed intake has been demonstrated by the intracerebroventricular administration of the mTOR inhibitor, rapamycin, which significantly stimulated feed intake in pre-satiated rats (Cota *et al.*, 2006). Administration of excess leucine is known to be anorectic in animals (Anderson and Moore, 2004). When leucine, but not valine, is administered into the third ventricle of rats, feed intake was significantly reduced (Cota *et al.*, 2006). The negative effect of leucine on feed intake was overcome by co-administration of rapamycin, suggesting strongly that the effect of leucine on intake is via the mTOR pathway. The leucine- induced inhibition of feed intake was shown to be related to a significant depression in the expression of the orexigenic peptide NPY mRNA. In contrast, the negative effect on feed intake and body weight following the administration of an agonist (melanotan II) for the POMC and α-MSH melanocortin receptors 3 and 4 was not reversed by rapamycin. These observations suggest that mTOR acts by depressing orexigenic peptides rather than increasing the anorexigenic peptides of the hypothalamus.

Sensing adiposity and body weight

Insulin is released from the pancreas in response to blood glucose concentrations, which it controls within narrow limits by altering the capacity of the liver to produce glucose or the ability of muscle, liver and other tissues to take up glucose. However, basal insulin concentration is also a direct reflection of the amount of visceral fat stored by an animal (Woods *et al.*, 2006). Plasma insulin concentration provides a signal to the hypothalamus about the adiposity status of the animal and exerts control over feed intake. Insulin administered near the hypothalamus of animals has been found to reduce the amount of feed consumed and body weight in a dose dependent response (Woods *et al.*, 1979). Conversely, administration of insulin antibodies near the hypothalamus or reducing insulin receptor activity by various means increases feed intake and body weight (Woods *et al.*, 2006). Circulating insulin enters the brain via an active transport mechanism, where the concentrations convey to the brain the stored energy status of the animal and either lead to a long-term increase or decrease in feed intake.

Leptin is released primarily from white adipose tissue in proportion to the number and activity of adipocytes (Considine *et al.*, 1996, Havel, 2004). Leptin concentrations in blood directly reflect the adiposity status of an animal and particularly the amount of subcutaneous adipose tissue (Woods *et al.*, 2006). Leptin, like insulin, also acts on the ARC of the hypothalamus to reduce intake and body weight. Animals that have a genetic mutation resulting in the loss of either the leptin gene or the leptin receptor gene have elevated feed intake, reduced energy expenditure and are obese (Clement *et al.*, 1998, Montague *et al.*, 1997). Leptin replacement in animals with genetic loss of the leptin gene leads to substantial reduction in feed intake and body weight (Farooqi *et al., 1999*; Halaas *et al.*, 1995). However, long-term administration of leptin does not induce weight loss because of development of 'leptin resistance' (Roth *et al.*, 2008).

Insulin, through its responsiveness to blood glucose concentrations, reflects the ongoing energy status of an animal as well as the long-term energy status through its relationship with the amount of adipose tissue. Alternatively, leptin concentrations provide only a reflection of the long-term energy status. Short-term changes in glucose concentrations lead to immediate changes in AMPK activity through the rapid effects on hypothalamic AMP:ATP ratio (Wolfgang *et al.*, 2007). However, adiposity has an effect on intake through insulin and leptin, which are believed to increase mTOR activity and decrease AMPK activity in the hypothalamus (Minokoshi *et al.*, 2004; Myers *et al.*, 2008; Shan *et al.*, 2008; Woods *et al.*, 2008).

Although ghrelin is released primarily from the stomach and has a major role in regulating hunger during the postprandial period, it also plays an important role in monitoring the energy or weight status of animals (Chaudhri *et al.*, 2006). Plasma concentrations of ghrelin are negatively correlated with body weight in humans and increase after weight loss (Cummings, 2006). Peripheral administration of ghrelin increases feed intake in several animal species and long-term treatment induces obesity, whereas administration of ghrelin antibodies or ghrelin receptor antagonists causes a reduction in feed intake and body weight (Van Thuijl *et al.*, 2008). Thus, ghrelin action counterbalances those of leptin and insulin by increasing concentration in lean, lightweight animals. Ghrelin also acts in the hypothalamus and has been shown to increase the expression of melanocortin system orexigenic peptide, NPY (Van Thuijl *et al.*, 2008).

Integration of metabolic control of intake

A picture is now emerging about the way nutrients and metabolites, through neural and hormonal signalling pathways in the vagus nerve, brainstem and hypothalamus, combine with AMPK and mTOR to alter expression of the hypothalamic melanocortin orexigenic and anorexigenic peptides to regulate long-term feed intake, energy expenditure and body composition (Chaudhri *et al.*, 2008; Richards and Proszkowiec-Weglarz, 2007; Sandoval *et al.*, 2008). Although details of some mechanisms are still to be determined, a brief overview of current concepts follows.

A deficiency in the energy available for metabolic processes over both the short- and long-term is reflected in the hypothalamus by a high AMP:ATP ratio, which stimulates activation of AMPK. High AMPK activity results in a lowering of the concentration of malonyl-CoA which stimulates expression of the melanocortin orexigenic peptides, NPY and AgRP, and reduces expression of the anorexigenic peptides, α-MSH and CART. The high AMPK activity also lowers the activation of mTOR, which further releases inhibition on expression of NPY. Ghrelin secretion from the stomach increases as the nutrients available for metabolism decline. Basal concentrations of circulating ghrelin increase as the weight and adiposity of an individual falls. High ghrelin concentrations increase the activity of AMPK in the hypothalamus (Kola *et al.*, 2005) and either directly or indirectly, further enhance the expression of NPY (Couce *et al.*, 2006). The increase in expression of orexigenic peptides and decrease in expression of the anorexigenic peptides is transmitted via neurons to higher centres of the brain, which results in a sensation of hunger. Thus, in situations with short- or long-term shortage of available nutrients or low for age and species body mass, feed intake is stimulated and energy expenditure depressed.

Conversely, in dietary situations that impose an intestinal brake, when there are excess circulating concentrations of glucose, lipids or amino acids, or when adiposity status is high, neural and hormonal systems interact largely through the melanocortin system to reduce feed intake and increase energy expenditure. Many of the peptides released from the intestines (CCK, GLP-1, OXM) or pancreas (amylin, pancreatic polypeptide) in response to nutrients and metabolites are believed to have long-term effects on feed intake via the vagus nerve, through the AP and NTS of the brainstem to the hypothalamus (Chaudhri *et al.*, 2006; Williams *et al.*, 2006). Some of these peptides are known to be synthesised in the hypothalamus. Other peptides from the intestines (PYY, apo A-IV) and pancreas (insulin) stimulated by nutrients are believed to act directly on the hypothalamus. The anorectic action of several peptides including CCK, GLP-1, OXM and apo A-IV is strongly stimulated by the presence of leptin. There is evidence that at least some of these peptides have their anorectic effect by decreasing the expression of the melanocortin peptide NPY (PYY, CCK, pancreatic peptide) or increasing the expression of anorexigenic peptides, POMC or urocortin (PYY, GLP-1, amylin, pancreatic polypeptide). Thus, circumstances such as increased passage of undigested nutrients to the distal intestines will result in a long-term depression of feed intake due to action of peptides released by the GIT.

There is strong evidence for synergistic depression in feed intake when these peptides released in response to nutrients are administered together. For example, synergistic responses have been reported for amylin and PYY (Roth *et al.*, 2007), for PYY and GLP-1 (Neary *et al.*, 2005) and for amylin and leptin (Williams *et al.*, 2006; Roth *et al.*, 2008). The studies with amylin, either administered alone or with PYY or with leptin, show that the reduction in body mass associated with a fall in feed intake is accompanied by a marked increase in energy expenditure and loss of fat with near preservation of body protein (Roth *et al.*, 2006, 2007, 2008).

Contrary to the effects of energy deficit, a surfeit of energy available for metabolism is reflected in a low ratio of AMP:ATP and low activation of AMPK. The low AMPK activity results in increased concentrations of malonyl-CoA in the hypothalamus, with a consequent reduction in expression of the melanocortin orexigenic peptides and increase in expression of anorexigenic peptides. The low AMPK activation also stimulates activation of mTOR, which further enhances these changes in melanocortin peptide expression. In animals with high adiposity status, the increased circulating concentrations of insulin and leptin further stimulate the activation of mTOR and reduce the activation of AMPK to strengthen the expression of anorexigenic peptides. The increase in expression of anorexigenic peptides and decrease in expression of the orexigenic peptides is transmitted via neurons to higher centres of the brain, which results in the sensation of satiety. Thus, in situations with short- or long-term surfeit of available nutrients or high-adiposity status, feed intake is depressed and energy expenditure stimulated.

High concentrations of circulating nutrients can also cause a direct depression in feed intake. Intracerebroventricular administration of oleic acid, but not octanoic acid, has been shown to markedly reduce feed intake in rats (Obici *et al.*, 2002). The response was leptin dependent and the reduction in feed intake occurred in parallel with a reduced expression of NPY. Circulating fatty acids gain rapid access to the brain and equilibrate with fatty acyl-CoA in the hypothalamus to have a direct impact on the melanocortin system. Similarly, as indicated above, high concentrations of the amino acid leucine increase mTOR activation and result in depression of the expression of NPY.

The concepts presented suggest there is a fine balance between hypothalamic melanocortin orexigenic and anorexigenic peptides to control feed intake and energy expenditure in relation to the availability of energy, nutrients and adiposity status of an animal. Various mediators of the system that decrease intake can do so by either increasing the expression of anorexigenic peptides or decreasing the expression of orexigenic peptides. The reverse appears to be true for mediators that increase feed intake. The concept as presented would suggest that, when feed is freely available, animals should eat to retain a 'normal' body mass and adiposity status and 'dietary-induced obesity' should not occur. However, there is strong evidence that diets with high fat content reduce the ability of insulin, leptin and fatty acyl-CoA to depress feed intake and glucose production (Sandoval *et al.*, 2007). The down-regulation of these pathways appears to be directly related to the fat content of diets and not to obesity. The effects are observed in rats after three days of feeding a high-fat diet (Morgan *et al.*, 2004; Wang *et al.*, 2001; Woods *et al., 2004*) and occur even when rats receiving a diet with 20% fat are pair-fed with others receiving a diet containing 4% fat (Woods *et al.*, 2004). Morgan *et al.* (2004) found that the normal depression in NPY following administration of oleic acid into the third ventricle was attenuated in rats receiving a high-fat diet. In addition, Woods *et al.* (2004) observed that the negative effect on feed intake following hypothalamic apo A-IV administration was also depressed in rats consuming the high-fat compared with the low-fat ration. Diet-induced obesity has been shown to inhibit the activation of AMPK by leptin in the hypothalamus (Martin *et al.*, 2006) and prevent the rise in malonyl-CoA, which normally occurs in response to excess energy availability (Pocai *et al.*,

2006). These observations would explain the failure for expression of NPY to decrease following consumption of diets high in fat. The inability of malonyl-CoA concentration to increase in response to hypothalamic administration of fatty acids in rats with diet-induced obesity appears to be related to an increase in hypothalamic fatty acyl-CoA oxidation. Pocai *et al.* (2006) found that hypothalamic administration of a CPT1 inhibitor, which reduced the uptake of fatty acyl-CoA by mitochondria for oxidation, restored hypothalamic fatty acyl-CoA concentrations, markedly depressed feed intake and restored hepatic glucose flux. These findings indicate that obesity induced by the consumption of diets containing high amounts of fat may be due to a change in regulation of fatty acyl-CoA oxidation in the hypothalamus.

A unifying concept?

The concepts described outlining the control of feed intake, energy expenditure and body composition can help explain many of the dietary and non-dietary situations known to alter feed intake in animals. Several examples are briefly described.

Differences in intake of cereal grains

The large effect of cereal grain sample on intake of pigs observed by Cadogan *et al.* (1999) and others could be explained by differences in undigested dietary material reaching the distal ileum and colon, with a resultant long-term increase in the release of intestinal brake peptides including PYY, GLP-1 and OMX. The effect of release of these peptides would be both short term through slowing of digesta transport, and long term through the impact of the peptides on the hypothalamic melanocortin neuropeptides. However, the negative effect of the ileal brake on feed intake may be counterbalanced by a high proportion of indigestible fibre in the diet stimulating intake. These concepts are supported by studies that show cooked rice with a higher digestibility than cooked maize results in 25% greater intake of feed and 29% faster growth rate in young pigs (Vicente *et al.*, 2008). Addition of 20 or 40 g/kg oat hulls significantly decreased feed intake of pigs offered a maize-based diet, but increased intake of pigs offered a rice-based diet (Mateos *et al.*, 2007). This observation could be explained by more undigested maize starch passing to the distal intestine with increasing fibre content of the diet, increasing rate of passage of digesta, and stimulating the ileal/colonic brake. However, with the more highly digested starch in rice than in maize, little more starch would have reached the distal intestine with increasing dietary fibre content and the ileal/colonic brake would not have been activated. Conversely, intake of the rice-based diet would be increased through the positive effect of added indigestible fibre on rate of passage of digesta.

High-fat diets

The depression in energy intake observed when animals are initially offered diets high in fat could be explained by an increase in chylomicron absorption and release of apo A-IV, augmented by the other intestinal brake peptides that will slow digesta transport. The increase in absorption of long-chain fatty acids is known to increase fatty acyl-CoA in the hypothalamus and decrease feed intake, while increasing energy expenditure through activation of the melanocortin system. Apo A-IV is also a potent anorexigenic peptide in the hypothalamus. However, prolonged feeding of diets high in fat can lead to down-regulation of the processes within the hypothalamus that reduce feed intake and result in enhanced consumption and diet-induced obesity.

High- and low-protein diets

Because branched-chain amino acids are not metabolised in the liver, excess dietary protein is expressed as high concentrations of these amino acids in plasma, specifically leucine. Although a considerable amount of leucine entering the brain is used in the formation of glutamate, it also has a major function activating mTOR, which will lead to a depression in intake and increase in energy expenditure. A similar situation would occur when feeding a diet with a severely imbalanced in amino acid pattern, where excess leucine and a low AMP:ATP ratio would stimulate activity of mTOR. Morrison *et al.* (2007) has shown recently that intracerebroventricular administration of either a mixture of amino acids or leucine alone decreased the expression of AgRP via the mTOR signalling system. Alternatively, diets inadequate in protein, but with a balanced amino acid pattern, often result in an increase in feed intake. Such diets have been shown to increase in the expression of AgRP (Morrison *et al.*, 2007), which would provoke an increase in feed intake. However, excessive intake of a low protein diet would be prevented by the decrease in AMP:ATP ratio caused by an inability of the animal to use energy for protein synthesis, a decrease in activation of AMPK and an increase in hypothalamic malonyl-CoA.

Situations resulting in increased energy expenditure

Cold exposure, exercise and lactation, each result in increases in feed intake that could be explained by a high AMP:ATP ratio in cells, high AMPK activity and low hypothalamic malonyl-COA, which will stimulate expression of orexigenic peptides while depressing expression of anorexigenic peptides of the hypothalamic melanocortin system. Compensatory gain that is often observed in animals following refeeding after a prolonged period of nutrient restriction, is accompanied by a marked increase in feed intake (Heyer and Lebret, 2007). This marked increase in feed intake could also be explained partially by low energy status of the animal reflected in high AMP:ATP, but may also be enhanced by high circulating ghrelin concentrations resulting from the low body mass for age. Both changes would be expected to increase feed intake under the mechanisms described above.

Genetic propensity for fat and protein deposition, and basal energy expenditure

Feed intake differs for animals with different genetic propensity to deposit energy as fat and protein, and to use energy for essential metabolic functions (Knap, 2009). Similarly, an increase in the propensity of an animal to deposit protein and decrease lipogenesis through administration of growth hormone leads to a decrease in total energy deposition and a concomitant decrease in feed intake (Campbell, 1998). Changes in feed intake in relation to energy expenditure for different 'genetically prescribed' functions can be readily explained by hypothalamic monitoring of energy status via the AMPK and mTOR signalling systems and of body mass and adiposity via ghrelin, insulin and leptin.

Seasonal day length and torpor

Many small mammals survive severe cold and inadequate feed supply during short winter days by entering a daily torpor for two to eight hours, which results in a fall in body temperature by up to 20 °C (Kirsch *et al.*, 1991). Torpor occurs in the Siberian hamster only when body fat reserves fall to an amount that permits circulating leptin concentrations to fall below 2.6 ng/ml compared with values around 9 ng/ml during periods of long daylight (Geiser *et al.*, 1998). Leptin receptors are co-localised in the ARC of the hypothalamus with NPY neurons. Stimulation of NPY, Y1 receptor, by intracerebroventricular administration of NPY completely suppresses neural activation of brown adipose tissue and inhibits non-shivering thermogenesis, decreases metabolic rate and markedly reduces core body temperature (Pelz and Dark, 2007). However, administration of NPY when leptin concentrations are high during long-day periods reduces body temperature by only 1-3 °C.

Infectious diseases, immune response and cancer

Infectious diseases including acquired immunodeficiency syndrome (AIDS), immune response stimulation by lipopolysaccharides (LPS) from the cell walls of gram-negative bacteria, and many forms of cancer, result in a syndrome called cachexia. The syndrome is characterised by a marked depression in feed intake and activity, an increase in metabolic energy expenditure often associated with fever and preferential loss of body protein. Cachexia contrasts with starvation where metabolic energy expenditure is reduced and muscle mass is preserved at the expense of body fat (Marks *et al.*, 2001). The syndrome is associated with increases in the release of pro-inflammatory cytokines and can be created experimentally in animals by peripheral administration of LPS or cytokines such as interleukin-1 (IL-1) (Fong *et al.*, 1989). The melanocortin peptide, α-MSH, shows a marked rise in the plasma of patients suffering from AIDS (Catania *et al.*, 1991). Intracerebroventricular administration of α-MSH has been found to significantly increase the reduction in feed intake and activity associated with LPS administration to rats (Huang *et al.*, 1999). However, administration of antagonists, either natural (AgRP) or synthetic (SHU-9119), to the α-MSH receptors, melanocortin receptor 3 and 4 (MC3-R/MC4-R), reversed the symptoms caused by LPS (Huang *et al.*, 1999; Marks *et al.*, 2001). Similarly, the symptoms did not appear in MC4-R knockout mice following administration of LPS. These results show a central role for the hypothalamic melanocortin system in fever and cachexia and that a blockade of MC4-R signalling normalises feed intake and activity without an increase in morbidity or mortality (Marks *et al.*, 2001; Catania, 2007).

Conclusions

A great deal of progress has been made over recent years in understanding how characteristics of a diet in association with products from digestion and metabolism control meal eating and long-term intake in animals. There is certainly more to learn. However, it has become clear that the physical structure of a diet, the nutrients it contains and the metabolites produced interact with the vagus nerve, peptides secreted from the stomach, intestines and pancreas, and the brainstem, hypothalamus and higher centres of the brain to influence hunger, satiety, feed intake, energy expenditure and the fat and lean tissues of animals. Control of feed intake is complex with many counterbalancing features.

One critical factor determining feed intake is control of the rate of passage of digesta through the GIT and rate of digestion. Rate of passage is stimulated by peptides released from the stomach and duodenum in the postprandial period, and by mild distension of the GIT resulting from meal consumption. However, passage rate through the GIT is inhibited by excessive distension or by the ileal and colonic brakes activated by undigested material entering sections of the intestines. The extent of digestion generally increases as the rate of passage of digesta declines. The second critical control of intake is through two serine/threonine kinases, AMPK and mTOR, which monitor the current energy status of an animal. These two energy-monitoring systems interact with insulin, leptin and ghrelin, which monitor either long-term adiposity or lean body mass of an animal. The third control is through the hypothalamic melanocortin system in the hypothalamus. This system is acted on by AMPK, mTOR, insulin, leptin, ghrelin and many of the GIT-pancreas peptides to express both orexigenic and anorexigenic peptides, which pass a sensation of hunger or satiety to higher sections of the brain. These peptides are also responsible for altering rates of energy expenditure within animal tissues and stimulating catabolic or anabolic processes to modulate growth and body composition.

The concepts described provide possible explanations for some of the fundamental mechanisms that may underlie a wide range of dietary and non-dietary situations known to affect feed intake in animals. An understanding of these mechanisms also provides an opportunity to develop strategies for either increasing or decreasing feed intake by animals, through manipulating the physical and/or chemical composition of diets, or by use of a range of pharmaceutical products currently being investigated to control intake in humans.

Acknowledgements

The review was supported by funds from the Australian Co-operative Research Centre for an Internationally Competitive Pork Industry.

References

Anderson, G.H. and Moore, S.E., 2004. Dietary proteins in the regulation of food intake and body weight in humans. The American Society for Nutritional Sciences 134: 974S- 979S.

Anderson, G.H., Tecimer, S.N., Shah, D. and Zafar, T.A., 2004. Protein source, quantity, and time of consumption determine the effect of proteins on short-term food intake in young men. The American Society for Nutritional Sciences 134: 3011-3015.

Black, J.L., 2000. Amino acid and energy requirements. In: Moughan, P.J., Verstegen, M.W.A. and Visser-Reyneveld, M.I. (eds.) Feed evaluation: principles and practice. Wageningen Pers, the Netherlands, pp. 189-207.

Black, J.L., 2008. Premium Grains for Livestock Program, Component 1: Co-ordination. Final Report. Grains Research and Development Corporation, Canberra, Australia, 237 pp.

Brighenti, F., Benini, L., Del Rio, D., Casiraghi, C., Pellegrini, N., Scazzina, F., Jenkins, D.J.A. and Vantini, I., 2006. Colonic fermentation of indigestible carbohydrates contributes to the second-meal effect. American Journal of Clinical Nutrition 83: 817-822.

Brown, R.C., Kelleher, J. and Losowsky, M.S., 1979. The effect of pectin on the structure and function of the rat small intestine. British Journal of Nutrition 42: 357-365.

Burdyga, G., de Lartigue, G., Raybould, H.E., Morris, R, Dimaline, R., Varro, A., Thompson, D.G. and Dockray, G.J., 2008. Cholecystokinin regulates expression of Y2 receptors in vagal afferent neurons serving the stomach. Journal of Neuroscience 28: 11583-11592.

Cadogan, D.J., Choct, M., Campbell, R.G. and Kershaw, S., 1999. Effects of new season's wheat on the growth performance of young pigs. In: Cranwell, P.D. (ed.) Manipulating pig production VII. Australian Pig Sciences Association, Werribee Victoria, Australia. p. 40.

Campbell, R.G., 1988. Nutritional constraints to lean tissue accretion in farm animals. Nutrition Research Reviews 1: 233-253.

Carroll, J.A. and Allee, G.L., 2009. Hormonal control of feed intake in swine. In: Torrallardona, D. and Roura, E. (eds.) Voluntary feed intake in pigs. Wageningen Academic Publishers, the Netherlands, pp. 155-187.

Catania, A., 2007. The melanocortin system in leukocyte biology. Journal of Leukocyte Biology 81: 383-392.

Catania, A., Lipton, J.M., Airaghi, L., Manfredi, M.G., Vivirito, M.C., Milazzo, F. and Zanussi, C., 1991. The neuropeptide alpha-MSH and other POMC-derivatives in patients with acquired immunodeficiency syndrome. International Conference on AIDS 7: 286 (Abstract).

Cha, S.H., Hu, Z. and Lane, M.D., 2004. Long-term effects of a fatty acid synthase inhibitor on obese mice: food intake, hypothalamic neuropeptides, and UCP3. Biochemical and Biophysical Research Communications 317: 301-308.

Chaudhri, O.B., Salem, V., Murphy, K.G. and Bloom, S.R., 2008. Gastrointestinal satiety signals. Annual Review of Physiology 70: 239-255.

Chaudhri, O., Small, C. and Bloom, S., 2006. Gastrointestinal hormones regulating appetite. Philosophical Transactions of the Royal Society B: Biological Sciences 361(1471): 1187-1209.

Chen, T.S., Doong, M.L., Wang, S.W., Tsai, S.C., Lu, C.C., Shih, H.C., Chen, Y.H., Chang, F.Y., Lee, S.D. and Wang, P.S., 1997. Gastric emptying and gastrointestinal transit during lactation in rats. American Journal of Physiology Gastrointestinal and Liver Physiology 272: G626-G631.

Clement, K., Vaisse, C., Lahlou, N., Cabrol, S., Pelloux, V., Cassuto, D., Gourmelen, M., Dina, C., Chambaz, J., Lacorte, J.M., Basdevant, A., Bougneres, P., Lebouc, Y., Froguel, P. and Guy-Crand, B., 1998. A mutation in the human leptin receptor gene causes obesity and pituitary dysfunction. Nature 392: 330-331.

Considine, R.V., Sinha, M.K., Heiman, M.L., Kriauciunas, A., Stephens, T.W., Nyce, M.R., Ohannesian, J.P., Marco, C.C., McKee, L.J, Bauer, T.L. and Caro, J.F., 1996. Serum immunoreactive-leptin concentrations in normal-weight and obsess humans. New England Journal of Medicine 334: 292-295.

Cota, D., Proulx, K., Blake Smith, K.A., Kozma, S.C., Thomas, G., Woods, S.C. and Seeley, R.J., 2006. Hypothalamic mTOR signalling regulates food intake. Science 312: 927-930.

Couce, M.E., Cottam, D., Esplen, J., Teijeiro, R, Schauer, P. and Burguera, B., 2006. Potential role of hypothalamic ghrelin in the pathogenesis of human obesity. Journal of Endocrinological Investigations 29: 599-605.

Cripps, A.W. and Williams, V.J., 1975. The effect of pregnancy and lactation on food intake, gastrointestinal anatomy and the absorptive capacity of the small intestine in the albino rat. British Journal of Nutrition 33: 17-32.

Cuche, G., Cuber, J.C. and Malbert, C.H., 2000. Ileal short-chain fatty acids inhibit motility by a humoral pathway. American Journal of Physiology Gastrointestinal and Liver Physiology 279: G925- G930.

Cummings, D.E., 2006. Ghrelin and the short- and long-term regulation of appetite and body weight. Physiology and Behavior 89: 71-84.

Dowell, P., Hu, Z. and Lane, M.D., 2005. Monitoring energy balance: metabolites of fatty acid synthesis as hypothalamic sensors. Annual Review of Biochemistry 74: 515-534.

Farooqi, I.S., Jebb, S.A., Langmack, G., Lawrence, E., Cheetham, C.H., Prentice, A.M., Hughes, I.A., McCamish, M.A. and O'Rahilly, S., 1999. Effects of recombinant leptin therapy in a child with congenital leptin deficiency. New England Journal of Medicine 341: 879-884.

Fleming, S.E. and Lee, B., 1983. Growth performance and intestinal transit time of rats fed purified and natural fibers. Journal of Nutrition 113: 592-601.

Fone, D.R., Horowitz, M., Read, N.W., Dent, J. and Maddox, A., 1990. The effects of terminal ileal triglyceride infusion on gastroduodenal motility and the intragastric distribution of a solid meal. Gastroenterology 98: 568-575.

Fong, Y., Moldawer, L.L., Marano, M., Wei, H., Barber, A., Manogue, K., Tracey, K.J., Kuo, G., Fischman, D.A., Cerami, A. and Lowry, S.F., 1989. Cachectin/TNF or IL-1 alpha induces cachexia with redistribution of body proteins. American Journal of Physiology Regulatory Integrative Comparative Physiology 256: R659-R665.

Geiser, F., Kortner, G. and Schmidt, I., 1998. Leptin increases energy expenditure of a marsupial by inhibition of daily torpor. American Journal of Physiology Regulatory Integrative Comparative Physiology 275: R1627-R1632.

Geliebter, A., 1988. Gastric distension and gastric capacity in relation to food intake in humans. Physiology and Behavior 44: 665-668.

Gidenne, T., 1992. Effect of fibre level, particle size and adaptation period on the digestibility and rate of passage as measured at the ileum and in the faeces in the adult rabbit. British Journal of Nutrition 7: 133-146.

Gonyou, H.W., Christopherson, R.J. and Young B.A., 1979. Effects of cold temperatures and winter conditions on some aspects of behaviour of feedlot cattle. Applied Animal Ethology 5: 113-124.

Gue, M., Fioramonti, J. and Bueno, L., 1987. Comparative influences of acoustic and cold stress on gastrointestinal transit in mice. Journal of Physiology Gastrointestinal and Liver Physiology 253: G124-G128.

Halaas, J.L., Gajiwala, K.S., Maffei, M., Cohen, S.L., Chait, B.T., Rabinowitz, D., Lallone, R.L., Burley, S.K. and Friedman, J.M., 1995. Weight-inducing effects of the plasma protein encoded by the obsess gene. Science 269: 543-546.

Hara, H., Nishikawa, H. and Kiriyama, S., 1992. Different effects of casein and soyabean protein on gastric emptying of protein and small intestinal transit after spontaneous feeding of diets in rats. British Journal of Nutrition 68: 59-66.

Hardie, D.G., Hawley, S.A. and Scott, J.W., 2006. AMP-activated protein kinase-development of the energy sensor concept. Journal of Physiology 574: 7-15.

Havel, P.J., 2004. Update on adipocyte hormones – regulation of energy balance and carbohydrate/lipid metabolism. Diabetes 53: 143-151.

Heyer A. and Lebret B., 2007. Compensatory growth response in pigs: Effects on growth performance, composition of weight gain at carcass and muscle levels, and meat quality. Journal of Animal Science 85: 769-778.

Hu, Z., Cha, S.H., Chohnan,S. and Lane, M.D., 2003. Hypothalamic malonyl-CoA as a mediator of feeding behaviour. Proceedings of the National Academy of Sciences of the USA 100: 12624-12629.

Huang, Q-H., Hruby, V.J.. and Tatro, J.B., 1999. Role of central melanocortins in endotoxin-induced anorexia. American Journal of Physiology Regulatory Integrative Comparative Physiology 276: R864-R871.

Huge, A., Weber, E. and Ehrlein, H-J., 1995. Effects of enteral feedback inhibition on motility, luminal flow, and absorption of nutrients in proximal gut of minipigs. Digestive Diseases and Sciences 40: 1024-1034.

Jenkins, D.J.A., Wolever, T.M.S., Leeds, A.R., Gassull,M.A., Haisman, P., Dilawari, J., Goff, D.V., Metz, G.L. and Alberti, K.G.M.M., 1978. Dietary fibres, fibre analogues and glucose tolerance: importance of viscosity. British Medical Journal 1: 1392-1394.

Karaki, S., Mitsui, R., Hayashi, H., Kato, I., Sugiya, H., Iwanaga, T., Furness, J.B. and Kuwahara, A., 2006. Short-chain fatty acid receptor, GPR43, is expressed by enteroendocrine cells and mucosal mast cells in rat intestine. Cell and Tissue Research 324: 353-360.

Keeling, W.F. and Martin, B.J., 1987. Gastrointestinal transit during mild exercise. American Journal of Applied Physiology 63: 978-981.

Keller, J., Holst, J.J. and Layer, P., 2006. Inhibition of human pancreatic and biliary output but not intestinal motility by physiological intraileal lipid loads. Journal of Physiology Gastrointestinal and Liver Physiology 290: G704-G709.

Kendall, G.P.N., Tompson, D.G. and Day, S.J., 1987. Motor responses of the small intestines to intraluminal distension in normal volunteers and a patient with visceral neuropathy. Gut 28: 714-720.

Kim, E.H., Miller, I., Aja, S., Landree, L.E., Pinn, M., McFadden, J., Kuhajda, F.P., Moram, T.H. and Ronnett, G.V., 2004. C75, a fatty acid synthase inhibitor, reduces food intake via hypothalamic AMP-activated protein kinase. Journal of Biological Chemistry 279: 19970-19976.

Kirsch, R., Ouarour, A. and Pevet, P., 1991. Daily torpor in the Djungarian hamster *(Phodopus sungorus)*: photoperiod regulation, characteristics and circadian organization. Journal of Comparative Physiology A 168: 121-128.

Knap, P.W., 2009. Voluntary feed intake and pig breeding. In: Torrallardona, D. and Roura, E. (eds.) Voluntary feed intake in pigs. Wageningen Academic Publishers, the Netherlands, pp. 13-35.

Kola, B., Hubina, E., Tucci, S.A., Kirkham, T.C., Garcia, E.A., Mitchell, S.E., Williams, L.M., Hawley, S.A., Hardie, D.G., Grossman, A.B. and Korbonits, M., 2005. Cannabinoids and ghrelin have central and peripheral metabolic and cardiac effects via AMP-activated protein kinase. Journal of Biological Chemistry 280: 25196-25201.

Koopmans, H.S., 1990. The role of the ileum in the control of food intake and intestinal adaptation. Canadian Journal of Physiology and Pharmacology 68: 650-655.

Kyriazakis, I. and Emmans, G.C., 1995. The voluntary feed intake of pigs given feeds based on wheat bran, dried citrus pulp and grass meal in relation to measurements of feed bulk. British Journal of Nutrition 73: 191-207.

Ledikwe, J.H., Blanck, H.M., Khan, K., Serdula, M.K., Seymour, J.D, Tohill, B.C. and Rolls, B.J., 2006. Dietary energy density is associated with energy intake and weight status in US adults. American Journal of Clinical Nutrition 83: 1362-1368.

Lewis, S.J. and Heaton, K.W., 1997. Increasing butyrate concentration in the distal colon by accelerating intestinal transit. Gut 41: 245-251.

L'Heureux-Bouron, D., Tome, D., Rampin, O., Even, P.C., Larue-Achagiotis, C. and Fromentin, G., 2003. Total subdiaphramatic vagotomy does not suppress high protein diet-induced food intake depression in rats. Journal of Nutrition 133: 2639-2642.

Lin, H.C., Kim, B.H., Elashoff, J.D., Doty, J.E., Gu, Y.G. and Meyer, J.H., 1992. Gastric emptying of solid food is most potently inhibited by carbohydrate in the canine distal ileum. Gastroenterology 102: 793-801.

Lin, H.C., Perdomo, O.L. and Zhao, X.T., 2001a. Intestinal transit in dogs is accelerated by volume distension during fat-induced jejuna brake. Digestive Diseases and Sciences 46, No.1: 19-23.

Lin,H.C., Van Citters, G.W., Heimer, F. and Bonorris, G., 2001b. Slowing of gastrointestinal transit by oleic acid. A preliminary report of a novel, nutrient-based treatment in humans. Digestive Diseases and Sciences 46: 223-229.

Lin, H.C., Zhao, X-T. and Wang, L., 1997. Intestinal transit is more potently inhibited by fat in the distal (ileal brake) than in the proximal (jejuna brake) gut. Digestive Diseases and Sciences 42: 19-25.

Luhovyy, B.L., Akhavan, T. and Anderson, G.H., 2007. Whey proteins in the regulation of food intake and satiety. Journal of the American College of Nutrition 26, No. 6 704S-712S.

Maljaars, J., Peters, H.P.F. and Masclee, A.M. 2007. Review article: the gastrointestinal tract: neuroendocrine regulation of satiety and food intake. Alimentary Pharmacology and Therapeutics 26: 241-250.

Maljaars, P.W.J., Peters, H.P.F., Mela, D.J. and Masclee, A.A.M., 2008. Ileal brake: A sensible food target for appetite control. A review. Physiology and Behaviour 95: 271-281.

Marks, D.L., Ling, N. and Cone, R.D., 2001. Role of the central melanocortin system in cachexia. Cancer Research 61: 1432-1438.

Martin, T.L., Alquier, T., Asakura, K., Furukawa, N., Preitner, F. and Kahn, B.B., 2006. Diet-induced obesity alters amp kinase activity in hypothalamus and skeletal muscle. Journal of Biological Chemistry 281: 18933-18941.

Massimino, S.P., McBurney, M.I. and Field, C.J., 1998. Fermetable dietary fibre increases GLP-1 secretion and improves glucose homeostasis despite increased intestinal glucose transport capacity in healthy dogs. Journal of Nutrition 128: 1786-1793.

Mateos, G.G., Martin, F., Latorre, M.A., Vicente, B. and Lazaro, R., 2007. Inclusion of oat hulls in diets of young pigs based on cooked maize or cooked rice. Animal Science 82: 57-63.

Meyer, J.H., Elashoff, J.D., Doty, J.E. and Gu, Y.G., 1994. Disproportionate ileal digestion on canine food consumption. A possible model for satiety in pancreatic insufficiency. Digestive Diseases and Sciences 39: 1014-1024.

Meyer, J.H., Hlinka, M., Khatibi, A., Raybould, H.E. and Tso, P., 1998a. Role of small intestine caloric ompensations to oil premeals in rats. American Journal of Physiology Regulatory, Integrative and Comparative Physiology 275: R1320-R1333.

Meyer, J.H., Hlinka, M., Tabrizi, Y., DiMaso, N. and Raybould, H.E., 1998b. Chemical specificities and intestinal distributions of nutrient-driven satiety. American Journal of Physiology Regulatory, Integrative and Comparative Physiology 275: R1293-R1307.

Minokoshi, Y., Alquier, T., Furukawa, N., Kim, Y-B., Lee, Q., Xue, B., Mu, J., Foufelle, F., Ferre, P., Birnbaum, M.J., Stuck, B.J. and Kahn, B.B., 2004. AMP-kinase regulates food intake by responding to hormonal and nutrient sgnals in the hypothalamus. Nature 428: 569-574.

Montague, C.T., Farooqi, I.S., Whitehead, J.P., Soos, M.S., Rau, H., Wareham, N.J., Sewter, C.P., Digby, J.E., Mohammed, S.N., Hurst, J.A., Cheetham, C.H., Earley, A.R., Barnett, A.H., Prins J.B. and O'Rahilly, S., 1997. Congenital leptin deficiency is associated with severe onset of obesity in humans. Nature 387: 903-908.

Morgan, K., Obici, S. and Rossetti, L., 2004. Hypothalmic responses to long-chain fatty acids are nutritionally regulated. Journal of Biological Chemistry 279: 31139-31148.

Morrison, C.D., Xi, X., White, C.L., Ye, J. and Martin, R.J., 2007. Amino acids inhibit *Agrp* gene expression via an mTOR-dependent mechanism. American Journal of Physiology Endocrinology and Metabolism 293: E165-E171.

Myers, M.G., Cowley, M.A., and Munzburg, H., 2008. Mechanisms of leptin action and leptin resistance. Annual Review of Physiology 70: 537-556.

Neary, N.M., Small, C.J., Druce, M.R., Park, A.J., Ellis, S.M., Semjonous, N.M., Dakin, C.L., Filipsson, K., Wang, F., Kent, A.S., Frost, G.S., Ghatei, M.A., and Bloom, S.R., 2005. Peptide YY_{3-36} and glucagon-like peptide-1_{7-36} inhibit food intake additively. Endocrinology 146: 5120-5127.

Obici,S., Feng, Z., Morgan, K., Stein, D., Karkanias, G. and Rossetti, L., 2002. Central administration of oleic inhibits glucose production and food intake. Diabetes. 51: 271-275.

Pelz, K.M. and Dark, J., 2007. IVC NPY Y1 receptor agonist but not Y5 agonist induces torpor-likr hypothermia in cold-acclimatised Siberian hamsters. American Journal of Physiology Regualtory Integrative Comparative Physiology 292: R2299-R2311.

Phillips, P.A., Young, B.A. and McQuitty, J.B., 1982. Liveweight, protein deposition and digestibility responses in growing pigs exposed to low temperature. Canadian Journal of Animal Science 62: 95-108.

Piche, T., Varannes des, B. S., Sacher-Huvelin,S., Holst, J.J., Cuber, J.C. and Galmiche, J.P., 2003. Colonic fermentation influences lower esophageal sphincter function in gastroesophageal reflux disease. Gastroenterology 124: 894-902.

Pimentel, M., Lin, H.C., Enayati, P., van den Burg, B., Lee, H-R., Chen, J.H., Park, S., Kong, Y. and Conklin, J., 2006.Methane, a gas produced by enteric bacteria, slows intestinal transit and augments small intestinal contractile activity. American Journal of Physiology Gastrointestinal and Liver Physiology 290: G1089-G1095.

Pocai, A., Lam, T.K.T., Obici, S., Gutierrez-Juarez, R., Muse, E.D., Arduini, A. and Rossetti, L., 2006. Restoration of hypothalamic lipid sensing normalises energy and glucose homeostasis in overfed rats. Journal of Clinical Investigation 116: 1081-1091.

Richards, M.P. and Proszkowiec-Weglarz, M. 2007. Mechanisms regulating feed intake, energy expenditure, and body weight in poultry. Poultry Science 86: 1478-1490.

Roche, J.R., Blanche, D., Kay, J.K., Miller, D.R., Sheahan, A.J. and Miller, D.W., 2008. Neuroendocrine and physiological regulation of intake with particular reference to domesticated ruminant animals. Nutrition Research Reviews 21: 207-234.

Rolls, B.J., Bell, E.A. and Waugh, B.A., 2000. Increasing the volume of a food by incorporating air affects satiety in men. American Journal of Clinical Nutrition 72: 361-368.

Rolls, B.J., Castellanos, V.H., Halford, J.C., Kilara, A., Panyam, D., Plekman, C.L., Smith, G.P. and Thorwart, M.L., 1998. Volume of food consumed effects satiety in men. American Journal of Clinical Nutrition 67: 1170-1177.

Ropert, A., Cherbut, C., Roze, C., Le Quellec, A., Holst, J.J., Fu-Cheng, X., Varannes, des, B. S. and Galmiche, J.P., 1996. Colonic fermentation and proximal gastric tone in humans. Gastroenterology 111: 289-296.

Roth, J.D., Coffey, T., Jodka, C.M., Maier, H., Athanacio, J.R., Mack, C.M., Weyer, C. and Parkes, D.G., 2007. Combination therapy with amylin and peptide YY[3-36] in obese rodents: anorexigenic synergy and weight loss additivity. Endocrinology 148: 6054-6061.

Roth, J.D., Hughes, H., Kendall, E., Baron, A.D. and Anderson, C.M., 2006. Anti-obesity effects of the β-cell hormone amylin in diet-induced obese rats: effects on food intake, body weight, composition, energy expenditure and gene expression. Endocrinology 147: 5855-5864.

Roth, J.D., Roland, B.L., Cole, R.L., Trevaskis, J.L., Weyer, C., Koda, J.E., Anderson, C.M., Parkes, D.G. and Baron, A.D., 2008. Leptin responsiveness restored by amylin agonism in diet-induced obesity: evidence from nonclinical and clinical studies. Proceedings of the National Academy of Sciences of the USA 105: 7257-7262.

Samra, R.A. and Anderson, G.H., 2007. Insoluble cereal fiber reduces appetite and short-term food intake and glycemic response to food consumed 75 min later by healthy men. American Journal of Clinical Nutrition 86: 972-979.

Sandoval, D., Cota, D. and Seeley, R.J., 2007. The integrative role of CNS fuel-sensing mechanisms in energy balance and glucose regulation. Annual Review of Physiology 70: 513-535.

Sasaki, O., Shinohara, H., Yamagishi, T., Nishida, A. and Mizuma, Y., 1996. Application of agar-fill method to estimate compartment capacity of gastrointestinal tract in Syrian hamsters. British Journal of Nutrition 76: 833-840.

Shan, J., Nguyen, T.B., Totary-Jain, H., Dansky, H., Marx, S.O. and Marks, A.R., 2008. Leptin-enhanced neointimal hyperplasia is reduced by mTOR and PI3K inhibitors. Proceedings of the National Academy of Sciences of the USA 105: 19006-19011.

Siegle, M.L., Schmid, H.R. and Ehrlein, H.J., 1990. Effects of ileal infusions of nutrients on motor patterns of canine small intestine. American Journal of Physiology Gastrointestinal and Liver Physiology 259: G78-G85.

Spiller,R.C., Trotman, I.F., Adrian, T.E., Bloom, S.R., Misiewicz, J.J. and Silk, D.B.A., 1988. Further characterisation of the 'ileal brake' reflex in man – effect of ileal infusion of partial digests of fat, protein, and starch on jejunal motility and release of neurotensin, enteroglucagon, and peptide YY. Gut 29: 1042-1051.

Spiller,R.C., Trotman, I.F., Higgins, B.E., Ghatei, M.A., Lee, Y.C., Bloom, S.R., Misiewicz, J.J. and Silk, D.B., 1984. The ileal brake-inhibition of jejunal motility after ileal fat perfusion in man. Gut 25: 365-374.

Stanogias, G. amd Pearce, G.R., 1985a. The digestion of fibre by pigs. 1. The effects f amount and type of fibre on apparent digestibility, nitrogen balance and rate of passage. British Journal of Nutrition 53: 513-530.

Stanogias, G. amd Pearce, G.R., 1985b. The digestion of fibre by pigs. 3. Effects of the amount and type of fibre on physical characteristics of segments of the gastrointestinal tract. British Journal of Nutrition 53: 537-548.

Steadman, C. and Kerlin, P., 1994. Response of the human intestine to high volume infusion. Gut 35: 641-645.

Tso, P., Sun, W. and Liu, M., 2004. Gastrointestinal Satiety Signals IV. Apolipoprotein A-IV. American Journal of Physiology Gastrointestinal and Liver Physiology 286: G885-G890.

Van Thuijl, H., Kola, B. and Korbonits, M., 2008. Appetite and metabolic effects of ghrelin and cannabinoids: involvement of AMP-activated protein kinase. In: Litwack, G. (ed.) Ghrelin, Vitamins and Hormones, volume 77. Elsevier Inc. Amsterdam, the Netherlands, pp.121-148.

Vicente, B., Valencia, D.G., Perez-Serrano, M., Lazaro, R. and mateos, G.G., 2008. The effects of feeding rice in substitution of corn and the degree of starch gelatinisation of rice on the digestibility of dietary components and productive performance of young pigs. Journal of Animal Science 86: 119-126.

Vu, M.K., Nouwens, A.G., Biemond, I., Lamers, C.B.H.W. and Masclee, A.A.M., 2000. The osmotic laxative magnesium sulphate activates the ileal brake. Alimentary Pharmocology and Therapeutics 14: 587-595.

Wang, J., Obici, S., Morgan, K., Barzilai, N., Feng, Z. and Rossetti, L. 2001. Overfeeding rapidly induces leptin and insulin resistance. Diabetes 50: 2786-2791.

Welch, I.M., Cunningham, K.M. and Read N.M., 1988. Regulation of gastric emptying by ileal nutrients in humans. Gastroenterology 94: 401-404.

Westra, R. and Christopherson, R.J., 1976. Effects of cold on digestibility, tetention time of digesta, reticulum motility and thyroid hormones in sheep. Canadian Journal of Animal Science 56: 699-708.

Whited, K.L., Tso, P and Raybould. H.E., 2006. Involvement of apolipoprotein A-IV and cholecystokinin$_1$ receptors in exogenous peptide YY$_{3\text{-}36}$-induced stimulation of intestinal feedback. Endocrinology 148: 4695-4703.

Williams, D.L.., Baskin, D.G. and Schwartz, M.W., 2006. Leptin regulation of the anorexic response to glucagon-like peptide-1 receptor stimulation. Diabetes 55: 3387-3393.

Wolfgang, M.J., Cha, S.H., Sidhaye, A., Chohnan, S., Cline, G., Shulman, G.I. and Lane, M.D., 2007. Regulation of hypothalamic malonyl-CoA by central glucose and leptin. Proceedings of the National Academy of Sciences of the USA 104: 19285-19290.

Wondra, K.J., Hancock, J.D., Behnke, K.C. and Stark, C.R., 1995. Effects of mill type and particle size uniformity on growth performance, nutrient digestibility, and stomach morphology in finishing pigs. Journal of Animal Science 73: 3564-2573.

Woods, S.C., Benoit, S.C. and Clegg, D.J., 2006. The brain-gut-islet connection. Diabetes 55: S114-S121.

Woods, S.C., D'Alessio, D.A., Tso, P., Rushing, P.A., Clegg, D.J., Benoit, S.C., Gotoh, K., Liu, M. and Seeley, R.J., 2004. Consumption of a high-fat diet alters the homeostatic regulation of energy balance. Physiology and Behavior 83: 573-578.

Woods, S.C., Lotter, E.C., McKay, D. and Porte D. Jr., 1979. Chronic intracerebroventricular infusion of insulin reduces food intake and body weight of baboons. Nature 282: 503-505.

Woods, S.C., Seeley, R.J. and Cota, D., 2008. Regulation of food intake through hypothalamic signaling networks involving mTOR. Annual Review of Nutrition 28: 295-311.

Wullschleger, S., Loewith, R. and Hall, M.N., 2006. TOR signalling in growth and metabolism. Cell 124: 471-484.

Chapter 9. Evaluation of free-choice feedstuffs preference by pigs

D. Torrallardona and D. Solà-Oriol
IRTA, Centre Mas de Bover, Ctra. Reus-El Morell, km. 3.8, 43120 Constantí, Spain;
david.torrallardona@irta.es

Abstract

In newly weaned piglets, ensuring a fast initiation to feeding is essential to maintain gut function and growth. Formulating diets with ingredients for which piglets have an innate preference might facilitate feed intake initiation around weaning. Free-choice feeding protocols have been widely used to assess pig's preferences for different feeds. Feed preference values are strongly influenced by the methodology used (i.e. reference feed used, experience of the pigs, length of test, etc.) and this should be considered when interpreting the results. In this paper data from pigs' preferences trials is reviewed. Data is grouped according to the nature of the feed ingredients studied (cereals, protein sources, free amino acids, fats and oils, fibre sources and other additives). The relationship between preference and the physical properties of the feeds (i.e. particle size and texture parameters) and their nutritive value is discussed, and practicalities for the use of free-choice preference data under practical (no-choice) conditions are also considered.

Keywords: palatability, feed intake, choice feeding

Introduction

Initiating dry feed consumption by piglets at weaning can be difficult. The removal of piglets from the sow combined with being offered separately dry feed and water and other stress factors, can result in long periods of low or inadequate feed intake (Dong and Pluske, 2007). It has been reported that it takes up to 3 weeks for piglets to re-establish pre-weaning levels of energy intake (Le Dividich and Sève, 2000) and meanwhile their gut integrity can be compromised (Lallès *et al.*, 2007; Pluske *et al.*, 1997). Ensuring a fast initiation of feeding immediately after weaning is essential to maintain gut function and piglet growth (Kelly *et al.*, 1991; Pluske *et al.*, 1996a,b). As the weanling pig mostly depends on its sense organs to evaluate the novel feed offered, using ingredients for which it has an innate preference might facilitate feed intake initiation at weaning.

Feed preference in pigs can be influenced by all five senses, although not all of them are equally important. For instance it appears that sight in pigs is rather poor. Their colour perception is not well developed in comparison to other species like poultry (Kennedy, 1980), and therefore significant feed preferences in response to colour stimuli should not be expected. Edge *et al.* (2004) tested the effect of three different colour troughs on feed intake. They tested yellow, red and black colours as they provided high, intermediate and low levels of contrast with the pelleted feed offered. No differences in feed intake between the three colour troughs were observed. Hearing can also be related to feeding behaviour in pigs. It is well known that pigs can associate noises with time of feeding or that piglets recognise the sounds that lactating sows make at the time of milk let down and suckling. Petrie and Gonyou (1988) found that auditory stimulation with recorded sow nursing and piglet vocalisations increased feeding and drinking behaviour in newly weaned pigs. The sense of touch appears to be developed (Roura and Tedó, 2009) and it is accepted that pigs are able to differentiate feed temperature and texture. Newly weaned piglets offered feed mixed with water heated at 36 °C had a higher feed intake and growth than pigs offered the same feed mixed with non-heated water at 14 °C (Reiners *et al.*, 2008). Also changes in feed preference due to differences in presentation form may be associated with texture sensing. Maize, barley or oats feeds presented in pelleted form have been reported to be clearly preferred over their corresponding mash form (Salmon Legagneur and Février, 1955; Skoch *et al.*, 1983; Solà-Oriol *et al.*, 2009a). Also a higher preference has been observed for feed oferred in wet than in dry form (Aumaitre and Salmon Legagneur, 1961).

However, the two most important senses in pigs are taste and olfaction (Klasing and Humphrey, 2009; Roura and Tedó, 2009). The pig's sense of taste is well developed; taste buds are found throughout the oral cavity although they are found predominantly on the top of the tongue. It appears that taste is more developed in pigs than in humans as suggested by the number of taste buds in their mouth (Leibetseder, 1980). Both species, however, have a well developed taste sense compared to other animals like chicken. Finally, olfaction in the pig is clearly more developed than in fowl and humans (Leibetseder, 1980). Because of their sensitive sense of smell pigs are used for locating truffles, which grow at 15-30 cm underground. Smell plays an important role in food seeking, but also for other very important functions from the evolutionary point of view such as reproduction and identification of possible predators. It also plays a role in identifying volatile compounds released from the feed during mastication, which together with the taste characteristics contribute to the overall flavour of the feed.

Taste and smell have evolved to associate beneficial (or nutritious) compounds and detrimental (or toxic) compounds with pleasant and unpleasant sensations, respectively (Goff and Klee, 2006), resulting in innate preferences or aversions for different feeds regardless of previous experiences or post-ingestive mechanisms. Nutritive characteristics are usually associated with receptors of pleasant tastes: sweet receptors with carbohydrate energy sources and some amino acids, umami with amino acids, and salty with mineral sources. In contrast, the stimulation of sour and bitter receptors is unpleasant (Nelson and Sanregret, 1997) and warns the animal against deteriorated feed or toxic compounds, respectively. Although with less intensity, the other senses may also provide information on the nutritional value of feeds. For example, texture may give an indication of the fibre or fat content and feed colour may be associated with their vitamin contents.

Figure 1 summarises the possible preferential responses of an animal to a novel feed. As already mentioned, naïve animals may prefer one feed over another (innate preference) or may not like it (innate aversion). It is also possible that the two are equally preferred (indifference) or that no differences between them can be perceived, in which case indifference would also be manifested. It is also possible that the feed is simply recognised as 'novel' and refused for this reason (neophobia), in which case an innate aversion would also be shown. In a situation of 'no-choice', however, pigs may be forced to nibble and experiment with feeds for which they had an innate aversion. If beneficial (or no adverse) effects result from the consumption of this feed, an acquired preference (or indifference) may develop. Similarly, consuming an innately preferred feed may turn to an acquired aversion if it is associated with negative or harmful experiences.

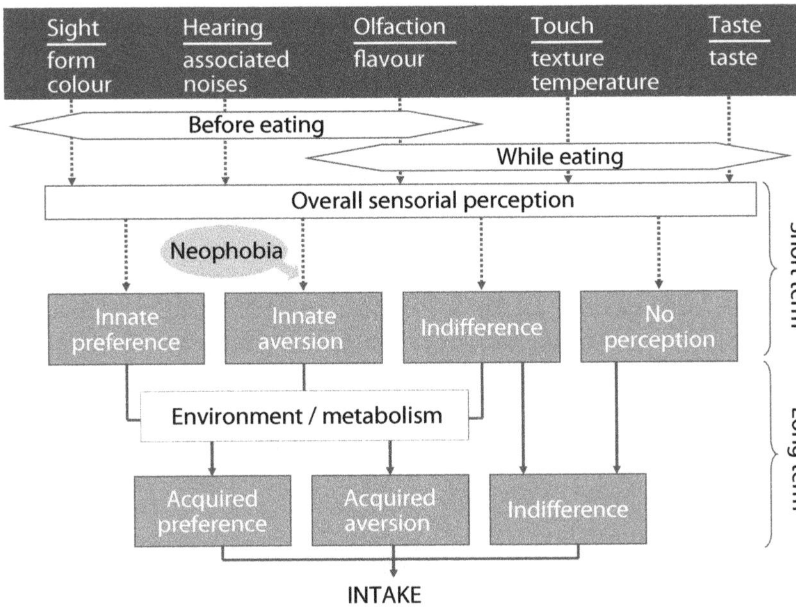

Figure 1. Sensorial perception and feed preference (see text for explanation).

While innate preference or aversions are important in the short term (i.e. at first contact with a new feed), in the long term, acquired preferences may prevail.

This paper reviews free-choice trials in which the preferences of pigs for different feed ingredients have been measured, focusing on aspects that may be relevant for their application in the formulation of piglet diets that facilitate the initiation to feeding before and after weaning.

Measurement of feed preference

Pigs, unlike humans, cannot explain the sensations experienced when eating different feeds. Therefore, we rely on behavioural studies, such as choice-feeding tests, to study their preferences. The preference or aversion of pigs for a particular feed over another (reference feed) can be estimated from the proportional feed intake of that feed when offered as a free choice against the reference feed (Baldwin, 1980). The estimation of feed preference is relative and varies depending on the chosen reference feed.

Free-choice tests

The preference for one feed to another can be measured with a free-choice test, in which the pigs are offered both feeds simultaneously and the proportional intake of each feed is measured. To establish a palatability ranking for a set of feeds, ideally all the feeds should be tested against each other. This is possible for a limited number of feeds but as the number of feeds to be tested increases, the number of possible double choice combinations becomes unaffordably large.

To overcome this limitation, some authors have performed multiple-diet free-choice tests in which several diets are offered simultaneously. This may be a valid approach to identify the most preferred feed from a given group, but the interpretation of the relative preference between the less preferred feeds may be difficult. Another approach is to evaluate each feed in a double choice test against a common reference feed. This way the relative preference between any pair of the feeds could be estimated indirectly without the need to perform an unaffordably large number of double choice tests.

Preference data from trials in which four different diets were tested against each other (Bruneau and Chavez, 1995; Leibbrandt *et al.*, 1975; Partanen *et al.*, 2002; Reis de Souza *et al.*, 2005) were used to plot preferences taking each one of them as the reference diet (Figure 2). In most

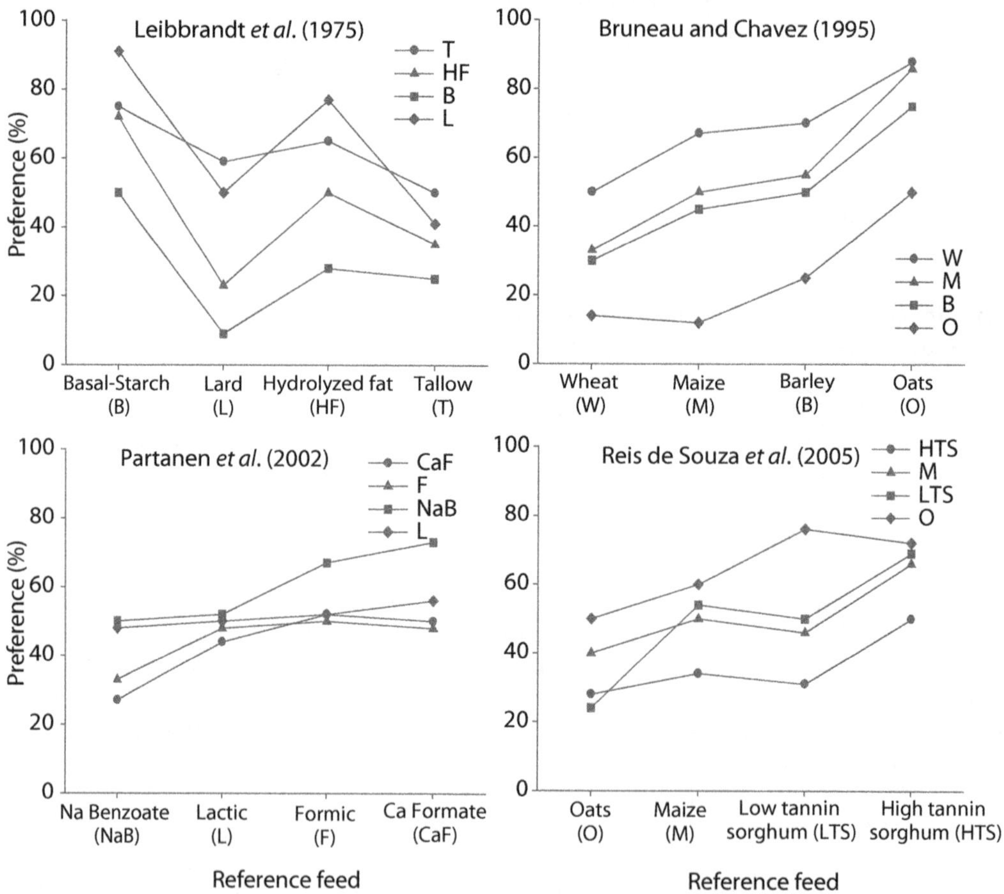

Figure 2. Free double choice preferences (% of total feed intake) of different feeds over different reference feeds. A value of 50% was used for the comparison of each ingredient against itself.

situations, the ranking of preference between the different diets is independent of which diet has been chosen as reference. Therefore, measuring feed preference indirectly with the use of a common reference feed appears to be a valid approach.

Free-choice tests to measure feed preference can be conducted in many different ways and this must be taken into consideration when interpreting the results. A few points to consider are:

1. *Previous experience of the animals with one (or more) of the feeds under test.* Previous experience should be avoided if we are interested in studying the pigs' 'innate' preference for a given feed. However, if we want to evaluate acquired preferences that take into account possible metabolic or toxic effects of the feeds, previous experience with all the feeds tested would be desirable.
2. *Length of the trial.* Possible adverse effects to the ingestion of a particular feed may require a few days to manifest. If a given feed is innately preferred but an acquired aversion is developed in the medium-long term, the observed preference will depend on the length of the test. In our opinion the length of the test should be in line with the conditions under investigation. For example if preference values are to be used to select ingredients that facilitate feed intake during the first three or four critical days after weaning, a trial length of up to four days may be optimal, whereas if data is to be used on grower-finisher pigs, long-term tests may be preferred. Short term choice may be driven mainly by the sensorial characteristics of the feeds, whereas nutritional, metabolic and physiological characteristics may become more relevant in mid-long term choice.
3. *Individual or group housing.* An individually housed pig may choose to eat from a particular trough, because he is attracted by the feed that it contains or just because he has got into the habit of doing so (even if no preference for one feed over the others exists). In the latter situation, even though the pig does not have a preference for one feed over the other, extreme preference values may be obtained for each individual animal. Although when several pigs (observations) are considered the average preference value may be close to 50% (i.e. two feeds with the same preference), this mean value will have a large variability. When pigs are kept in groups, however, some competition (depending on the size of the group and the feeder space available) will appear and this may balance the feed intake from different troughs, provided that feed preferences are similar. Care should be taken however to control this competition, so that less dominant pigs do not have excessive difficulty in consuming a particular feed they may prefer. The degree of difficulty a pig is willing to go through to be able to eat a particular feed may be an interesting way of measuring feed preference, but this is not within the scope of this paper.
4. *Alternating feeder position.* Pigs may have acquired the habit of eating from a particular location before the start of the test, and this may affect the outcome of the test. To avoid this some authors swap the position of the feeders at regular time intervals. This may be advantageous when feeds with very similar preferences are tested, as it would reduce possible variability due to the pig's habits (for example in the previously described situation with individually kept pigs). Although this practice may also facilitate nibbling the less preferred feed and enhance the development of an acquired preference, it could also confuse the piglets and lead to unreliable preference values in the first few hours (days) after each change. McLaughlin *et al.* (1983) observed that pigs showing preference for a particular feed, ate more of the less preferred feed after the position of the feeders was changed, and that up to four days were

required to get back to the same preference values as before. Thus, an adaptation period after each feeder swap is recommended when conducting this practice. Nevertheless, if pigs are kept in small groups and a small degree of competition for feeder space exists, they may be encouraged to eat from all the feeders and therefore, this practice may only be necessary for individually kept pigs.

5. *Free-choice* vs. *single stimulus protocols.* Feed preference has also been measured with the use of single stimulus protocols, in which the feeds under test are not offered simultaneously (free-choice) but alternatively at regular time intervals. Wahlstrom *et al.* (1974) compared both approaches and concluded that this approach was preferable as it resulted in less variable values. However, the magnitude of the difference in preference between feeds was also reduced, which suggests that, with this approach, pigs had probably been forced to consume the less preferred feed (as in the previously described situation of excessive competition for feeder space), and preference values may have been affected by this. With the single stimulus protocol, time interval length may be critical and may also influence the recorded preference values. Although the use of a single stimulus protocol may be interesting under a given situation, in our opinion the free-choice is preferable.

However, in practice, pigs are not offered a choice of feeds but only one. Preference test conditions are far-removed from reality. Thus preference values may not agree with the feed intakes observed in no-choice situations where pigs are forced to eat the feed provided.

'No-choice' tests

Another way of comparing the effect that different feeds have on feed intake is to conduct trials in which different groups of animals are offered different feeds. This is a direct approach that resembles real farm situations and should be the ultimate test for feed acceptability. However, it involves demanding trials in which only a few feeds can be tested at a time, and using it to test a large number of feeds may not be possible.

Figure 3 shows the relationship between the measurements of feed intake for different diets (expressed as percentage of the control diet intake) under free-choice or no-choice conditions. Overall, a positive Pearson's correlation was observed between both sets of data (r=0.40; $P<0.05$). Taken individually, however, for some of the diets values are in reasonably good agreement, but for others the relationship is very poor. Of particular interest is the good agreement for the low intake of oats and the high intake of dehulled rolled oats. On the contrary, whey, barley and sorghum feeds showed a low feed intake under free-choice, but a higher intake than the control under the no-choice situation. Similarly, glucose and maize diets also showed a low feed intake under free-choice, but equivalent intakes to the control under the no-choice situation. The effect of supplementing 4% excess of different amino acids is also shown. Under free-choice the diets containing aminoacids in excess were avoided, and their intake did not exceed 8% of that of the basal diet for any of them. Under the no-choice situation however, a reduction in feed intake was observed for Met, Trp and Arg, whereas Lys and Thr did not affect feed intake.

It must be considered that these were long-term (4-5 weeks) trials, in which pigs in the no-choice situation may have been forced to eat the only feed provided. Under these conditions, animals

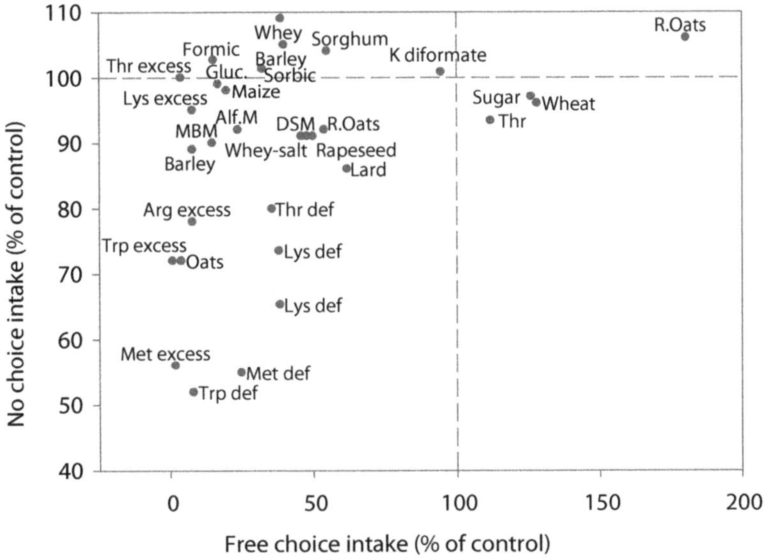

Figure 3. Comparison of relative feed intakes (% of control) of different feeds offered to pigs in a free-choice (vs. control) or in a no-choice situation. Data taken from Aumaitre (1975, 1980); Edmonds et al. *(1987); Ettle and Roth (2004, 2005, 2009); Ettle* et al. *(2004); Kirchgessner* et al. *(1999); Orr (1980); Roth* et al. *(2006); Schöne* et al. *(1994) and Wahlstrom* et al. *(1974).*

would be expected to consume even low-preference feeds to meet their nutritional requirements, unless doing so were to cause them discomfort. In the free-choice situation the pig just selects the most preferred feed and this may explain the differences observed between the two situations for some of the comparisons. It appears, however, that in the case of some diets that may cause discomfort to the pig, such as in the case with amino acid deficient diets, diets with excess Arg, Trp or Met and the oat based diets (due to their bulky effect), no adaptation to the less preferred diet occurs and there is a good agreement between both situations.

At weaning, however, the pig needs to identify the new feed and learn to consume it. Here an innate attraction of the pigs for a particular feed may be very important. Short-term studies would shed light on whether the most preferred feed in a free-choice situation facilitates the initiation of piglets to feeding around weaning.

Preferences for different feed ingredients

Cereals

Free-choice preference studies to assess the palatability of cereals in pigs have been used for many years. Aubel (1935) conducted two tests, and in each of them nine different varieties of sorghum were offered to individually housed pigs as a free choice. The order of preference for the different sorghum varieties was consistent between the different pigs and a reliable preference ranking for the different varieties of sorghum was established. Although not reported, the differences in preference were probably due to differences in the tannin contents of the different varieties.

Table 1 compiles studies in which the palatability of cereals by pigs has been evaluated with double-choice preference tests. Wheat has been directly compared to barley, maize, dehulled oats and oats (Aumaitre, 1980; Orr, 1980; Reis de Souza *et al.*, 2005). Apart from dehulled oats, it can be concluded that the preference for wheat is higher than for the other 3 cereals. This agrees with Salmon Legagneur and Février (1959) that observed a higher preference for dehulled oats (68%) than for barley (8%), maize (7%) or the basal feed (17%) in a 4-diet free-choice test and Solà-Oriol *et al.*, 2009c that reported rice preference over sorghum, maize or rye. Preference for barley has also been shown to be higher than for oats, but lower than maize (Bruneau and Chavez, 1995; Orr, 1980). Maize is also preferred to oats, dehulled oats or sorghum (Bruneau and Chavez, 1995; Orr, 1980; Reis de Souza *et al.*, 2005). Finally, a higher preference for oats than for sorghum has also been reported (Reis de Souza *et al.*, 2005).

We have conducted a series of double free-choice tests to evaluate pig preference for 24 different cereal sources (Solà-Oriol *et al.*, 2009a). A common reference diet based on white broken rice was used and the relative preferences between all the cereals were estimated indirectly. Cereals were tested at low (L; 15%), medium (M; 30%) and high (H; 60%) inclusion rates. Both the nature of the cereal and its inclusion rate affected preference as shown in Figure 4. The diets containing L inclusion of extruded rice, and L and M inclusion rates of extruded naked oats, or naked oats were preferred by pigs to the reference rice diet. Pigs also preferred this diet to the diets containing

Table 1. Double choice feed preference studies testing palatability of cereal sources for pigs.

Reference	Pig's age	Days in test	Products compared (dose)[1]	Preference (% total intake)
Aumaitre (1980)	weaners	28	wheat (50%) – barley (50%)	71–29[a]
			wheat (50%) – maize (50%)	83–17[a]
			wheat (50%) – rolled oats (50%)	36–64[a]
Bruneau and Chavez (1995)	suckling	12	maize (70%) – wheat (70%)	49–51
			maize (70%) – barley (70%)	53–47
			wheat (70%) – oats (70%)	73–27***
			barley (70%) – wheat (70%)	43–57
			maize (70%) – oats (70%)	66–34
			barley (70%) – oats (70%)	51–49
	weaners	13	maize (70%) – wheat (70%)	33–67**
			maize (70%) – barley (70%)	55–45
			wheat (70%) – oats (70%)	86–14***
			maize (70%) – oats (70%)	88–12***
			barley (70%) – wheat (70%)	30–70**
			barley (70%) – oats (70%)	75–25***

Table 1. Continued.

Reference	Pig's age	Days in test	Products compared (dose)[1]	Preference (% total intake)
Orr (1980)	weaners	14	maize (63%) – sorghum (63%)	65–35[b]
			maize (63%) – wheat (63%)	44–56
			maize (63%) – oats (30%)	79–21*
			maize (63%) – barley (63%)	92–8**
			maize (63%) – oatmeal (63%)	96–4**
			maize (63%) – dehulled rolled oats (63%)	65–35*
Reis de Souza *et al.* (2005)	weaners	14	sorghum low TN (50%) – sorghum high TN (50%)	69–31[b]
			oats (50%) – sorghum low TN (50%)	76–24*
			sorghum low TN (50%) – maize (50%)	54–46
			maize (50%) – sorghum high TN (50%)	66–34*
			oats (50%) – maize (50%)	60–40
			oats (50%) – sorghum high TN (50%)	72–28**
Salmon Legagneur and Février (1959)	suckling	39	basal feed – wheat grain	31–69[a]
			basal feed – rye grain	39–61[a]
			basal feed – barley grain	53–47[a]
			basal feed – oats grain	70–30[a]
			basal feed – faba beans	76–24[a]
			basal feed – maize grain	80–20[a]
	suckling	39	basal feed – dehulled oats	21–79[a]
			basal feed – mashed maize	91–9[a]
			basal feed – mashed faba beans	81–19[a]
Solà-Oriol *et al.* (2009c)	weaners	8	rice (50%) – sorghum (25%)	71–29*
			rice (50%) – sorghum (50%)	79–21*
			rice (100%) – sorghum (100%)	84–16*
	weaners	10	rice (50%) – maize (25%)	65–35*
			rice (50%) – maize (50%)	72–28*
			rice (100%) – maize (100%)	84–16*
	weaners	10	rice (50%) – rye (25%)	73–27*
			rice (50%) – rye (50%)	51–49
			rice (100%) – rye (100%)	81–19*

[1] TN tannins.

[a] Statistics not reported.

[b] Proportions differ significantly ($P<0.1$).

* Proportions differ significantly ($P<0.05$).

** Proportions differ significantly ($P<0.01$).

*** Proportions differ significantly ($P<0.001$).

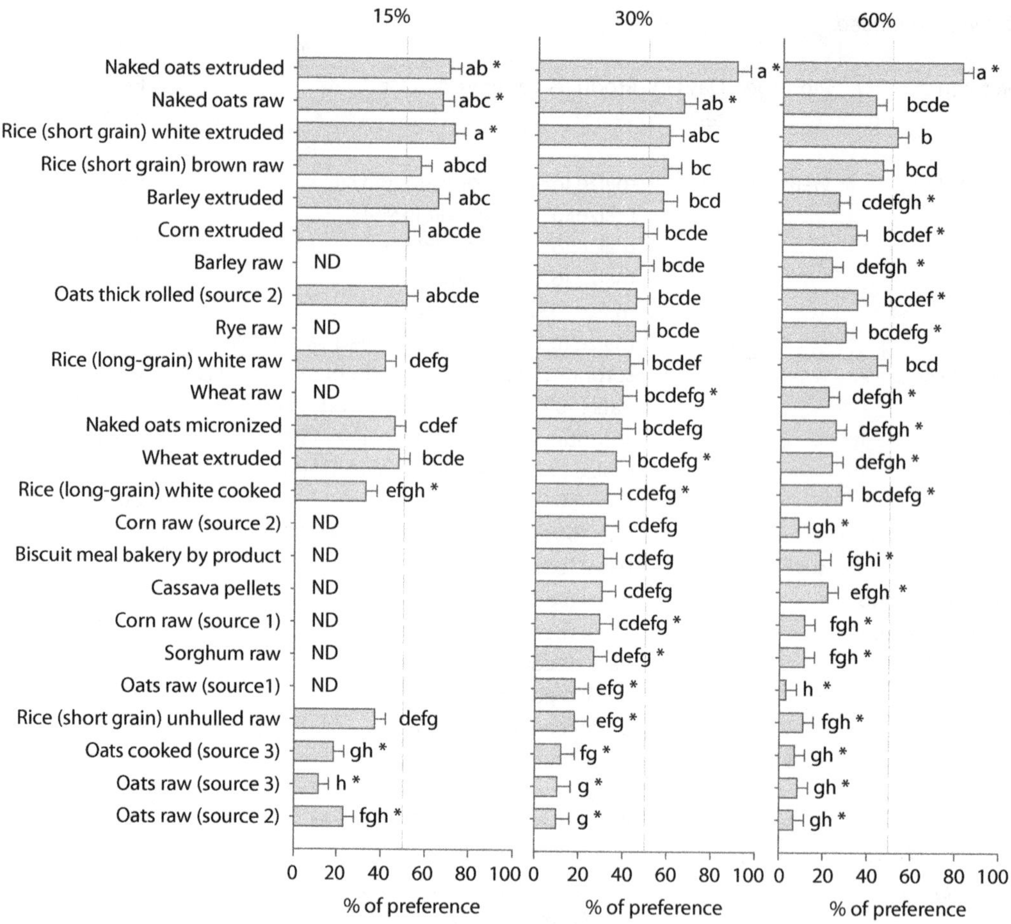

Figure 4. Pig's preference (% of total feed intake) for diets containing 15, 30 or 60% of different cereals offered in a two-way choice with a common reference diet based on white broken rice (Solà-Oriol et al., *2009a). Preferences were measured over a period of four days. Within each inclusion rate, columns with different letters indicate significantly different prefence values (P<0.05). Columns with the symbol * indicate a preference value significantly different from 50% (P<0.05).*

cooked long-grain rice, oats or cooked oats at all three inclusion rates; extruded wheat, wheat, corn, sorghum, or unhulled short-grain rice at L and H inclusion rates; thick rolled oats, extruded corn, rye, extruded barley, micronized naked oats, barley, cassava, or biscuit meal at H inclusion rates. Cereal extrusion resulted in improved preference for corn and naked oats, but it had no effect on barley, rice, or wheat.

This supports the good productive results observed in weaning pigs offered diets containing highly preferred feed ingredients such as naked oats and rice (Brand and Van der Merwe,, 1996; Maslanek *et al.* 2001; Mateos *et al.*, 2001; Pluske *et al.*, 2002, 2007).

We observed that extrusion significantly improved corn and naked oats preferences, but had no effect on barley, rice and wheat. This could be due to an improvement in the availability of their starch and a faster glucose release (Medel *et al.*, 1999; Piao *et al.*, 1999; Solà-Oriol *et al.*, 2008). Improved feed intake and performance of piglets due to the extrusion of corn was also reported by Hongtrakul *et al.* (1998), but Medel *et al.* (1999) observed that corn extrusion improved weight gain in piglets without affecting feed intake. The latter authors suggested that extrusion affected nutrient digestibility rather than palatability. For naked oats, we observed reduced feed preference due to micronisation, despite the improved nutrient digestibility that might be expected with this treatment (Huang *et al.*, 1998).

Protein sources

Table 2 summarises some double-choice preference tests that evaluated the palatability of different protein sources in pigs. Soybean meal was preferred over corn gluten feed, dried skim milk, soy protein concentrate or meat and bone meal (Devilat *et al.*, 1970; Lenehan *et al.*, 2007; Orr, 1980). Although feeds containing 10% of cottonseed meal were preferred to soybean meal, an aversion was seen for inclusion levels of 15-22% (Cooper *et al.*, 1979; Ferguson *et al.*, 2002). No significant preferences were observed between soybean meal and fishmeal or lupine (Bradford and Gous, 1992; Solà-Oriol *et al.*, 2009c). Fishmeal was preferred over rapeseed meal (Kyriazakis and Emmans, 1993), but no difference was observed when compared to an enzymatically digested animal protein source (Lindemann *et al.*, 2000). Finally, spray dried porcine plasma was preferred to dried skim milk (Ermer *et al.*, 1994). Aspects such as presence of antinutritional factors (glucosinolates or tannins), damage by heat treatment, or nutrient imbalance were also shown to have a negative effect on feed preference (Ferguson *et al.*, 2002; Hansen *et al.*, 1987; Kyriazakis and Emmans, 1993; Schöne *et al.*, 1994). However, the addition of 1% tannins onto a soybean based diet did not manage to reverse its preference over a cottonseed based diet (Ferguson *et al.*, 2002).

We also evaluated the pig's preferences for 15 different protein sources against the same reference diet used in the cereal studies (Solà-Oriol, 2008). The different protein sources replaced different amounts of a soybean meal product with 56% CP (SBM-56), the reference ingredient for these studies. Proteins were tested at low (L; 5%, medium (M; 10%) and high (H; 20%) inclusion rates. As for cereals, type of protein and its inclusion rate influenced feed preference (Figure 5). Dried porcine solubles (at L inclusion rate), fishmeal (at L and M rates) and lupine, soybean meal-44 and dried skimmed milk (at M inclusion rate) were preferred to the reference diet with SBM-56. However, an aversion compared to SBM-56 was observed for potato protein at all the inclusion rates, for rapeseed meal and acid milk whey at the L and M inclusion rates, and for dried porcine solubles, soybean protein concentrate, wheat gluten and sunflower meal at the H inclusion rate. We have also shown that storage of dry skimmed milk, soy protein concentrate and potato protein for 10 months resulted in a significant reduction of preference from 55 to 23%, 42 to 22% and 16 to 7%, respectively (Solà-Oriol *et al.*, 2007a). This is probably due to unpalatable metabolites resulting from the degradation of some nutrients.

The use of highly preferred protein sources like fishmeal or dried porcine solubles at adequate doses have also been shown to encourage feed intake at weaning (Kim and Easter, 2001; Lindemann *et al.*, 1998; Zimmerman *et al.*, 1997). However despite the very low feed palatability of potato

Table 2. Double choice feed preference studies testing palatability of protein sources for pigs.

Reference	Pig's age	Days in test	Products compared (dose)[1]	Preference (%)
Aumaitre and Salmon Legagneur (1964)	suckling	50	basal – SBM (10%)	33–67
			basal – PNM (10%)	51–49
			basal – SFM (10%)	58–42
			basal – CCM (10%)	73–27[d]
Bradford and Gous (1992)	weaners	28	FM (21%) - SBM (35%)	42–58
			FM (21%) – PNM (21%)+SFM (20%)+CSM (25%)	74–26***
			SBM (35%) – PNM (21%)+SFM (20%)+CSM (25%)	87–13***
Cooper *et al.* (1979)	weaners	7	basal (20% SBM) – CSM (10%)	30–70*
			basal – CSM (15%)	88–12*
			basal – CSM (18%)	90–10*
	weaners	7	basal pelleted – CSM (10%) pelleted	38–62
			basal pelleted – CSM (14%) pelleted	69–31
			basal pelleted – CSM (18%) pelleted	48–52
Devilat *et al.* (1970)	growers	35	basal (20% SBM) – basal	55–45[a]
		34	basal – CGF (unbalanced 45%)	94–6[b]
		36	basal – protein free diet	69–31[a]
	finishers	38	basal – basal	47–53[a]
		41	basal – CGF (unbalanced 45%)	89–11[b]
		42	basal – protein free diet	63–37[a]
	finishers	38	basal – basal	43–57[a]
		41	basal –CGF (unbalanced 45%)	97–3[b]
		42	basal – protein free diet	72–28[a]
	growers	30	CGF (unbalanced 45%) – protein free diet	30–70[a]
			protein free (unbalanced) – protein free diet	46–54[b]
Kyriazakis and Emmans (1993)	growers	18-26	basal (low protein) – FM (22%)	54–46[c]
			basal – SBM (22%)	47–53[c]
			basal – RSM low G (22%)	59–41[c]
			basal – RSM high G (22%)	71–29[c]
			FM (22%) – RSM low G (22%)	68–32[c]
			FM (22%) – RSM high G (22%)	98–2[c]
			RSM low G (22%) – RSM high G (22%)	74–26[c]
Ermer *et al.* (1994)	weaners	21	SDPP (8.5%) – DSM (20%)	71–29**
Ferguson *et al.* (2002)	growers	28	CSM (balanced 22%) – CSM (unbalanced 22%)	86–14*
			SBM (balanced 26%) – SBM (unbalanced 26%)	52–48
			CSM (balanced 22%) – SBM (balanced 26%)	13–87*
			CSM (unbalanced 22%) – SBM (unbalanced 26%)	5–95*
	growers	28	CSM (unbalanced 22%) – CSM (balanced 22%)	45–55
			CSM (unbalanced 22%) – CSM+TN (unbalan. 22%)	71–29*
			CSM (balanced 22%) – CSM+TN (balanced 22%)	59–41*
			CSM (balanced 22%) – SBM+TN (balanced 26%)	38–62
			CSM (balanced 22%) – SBM (balanced 26%)	31–69*

Table2. Continued.

Reference	Pig's age	Days in test	Products compared (dose)[1]	Preference (%)
Hansen *et al.* (1987)	weaners	28	SBM (28%) – overheated SBM (27%)	63–37
			SBM (30%) – overheated SBM (30%)	62–38
Lenehan *et al.* (2007)	weaners	7	SBM (40%) – SPC (29%)	97–3***
Lindemann *et al.* (2000)	weaners	21	FM (6%) – EDAP (8%)	63–37
			FM (6%) – EDAP (4%)	47–53
Orr (1980)	weaners	14	basal (maize soy) – MBM (7%)	87–13**
			basal – DSM (7%)	68–32**
Salmon Legagneur and Aumaitre (1961)	suckling	46	basal (wheat) – FM (shark 8%)	17–83**
			basal – FM (tuna 8%)	51–49
			basal – FM (8%)	21–79**
			basal – FM (trout 8%)	10–90**
			basal – FM (herring 8%)	63–37[d]
			basal – FM (sardine 8%)	35–65*
			basal – FM (anchovy 8%)	53–47
			basal – MM (fresh 10%)	49–51
			basal – MM (10%)	41–59*
Schöne *et al.* (1994)	growers	5	RSM low G (1%) – RSM high G (1%)	50–50
			RSM low G (2%) – RSM high G (2%)	48–52
			RSM low G (4%) – RSM high G (4%)	55–45
			RSM low G (8%) – RSM high G (8%)	59–41[d]
			RSM low G (16%) – RSM high G (16%)	66–34*
Solà-Oriol *et al.* (2009c)	weaners	11	basal (25% SBM 56) – lupine (7.5%)	55–45
			basal – lupine (15%)	44–56
			SBM 56 (100%) – lupine (100%)	61–39

[a,b] Values within a trial with different letters are significantly different ($P<0.05$).

[c] Statistics not reported;

[d] Proportions differ significantly ($P<0.1$).

* Proportions differ significantly ($P<0.05$).

** Proportions differ significantly ($P<0.01$).

*** Proportions differ significantly ($P<0.001$).

[1] SBM soybean meal; CGF corn gluten feed; FM fishmeal; EDAP enzymatically digested animal protein; SDPP spray dried porcine plasma; DSM dried skim milk; CSM cottonseed meal; SPC soy protein concentrate; MBM meat and bone meal; PNM peanut meal; SFM sunflower meal; CCM coconut meal; MM meat meal; RSM rapeseed meal; G glucosinolate; TN tannins.

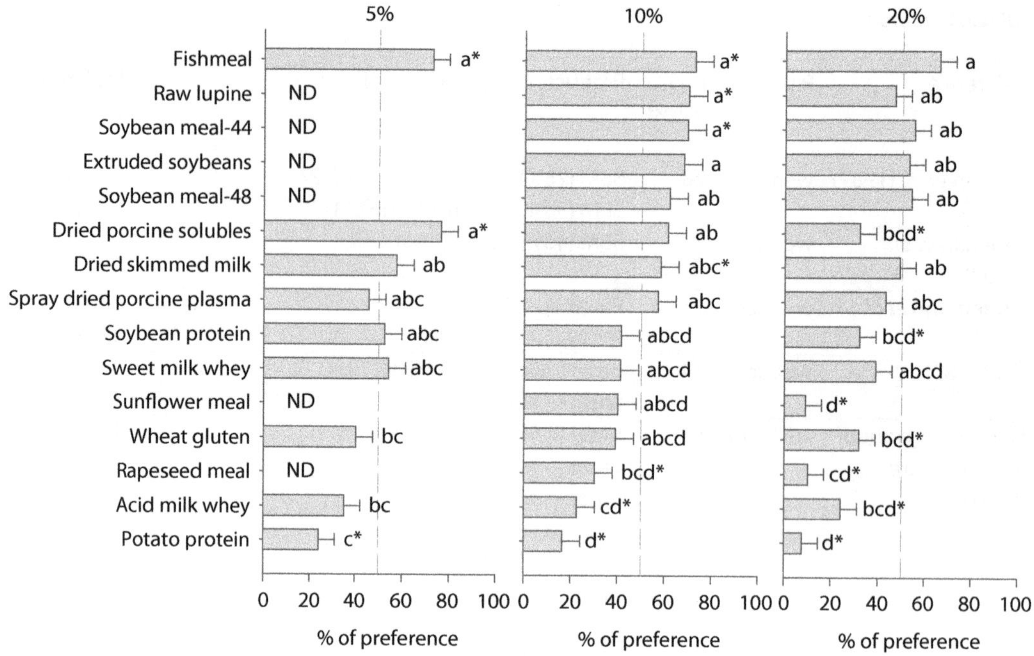

*Figure 5. Pig's preference (% of total feed intake) for diets containing 5, 10 or 20% of different protein sources offered in a two-way choice with a common reference diet based on a soybean meal product with 56% CP (Solà-Oriol, 2008). Preferences were measured over a period of four days. Within each inclusion rate, columns with different letters indicate significantly different prefence values (P<0.05). Columns with the symbol * indicate a preference value significantly different from 50% (P<0.05).*

protein it has been shown to provide excellent productive results in feeding trials (Pedersen and Lindberg, 2004). The low preference for potato protein may arise from an innate aversion to residual amounts of the toxic compound solanin. However, as no adverse effects appear under a no-choice situation (feeding trial) an acquired preference may have developed. Further studies should be conducted to confirm this possibility. In addition, the relevance of the extreme impact of potato on some texture parameters cannot be ignored (Solà-Oriol *et al.*, 2009b).

Amino acids

Edmonds *et al.* (1987) conducted a series of free choice trials with pigs to test the effect of excessive amounts of free amino acids on feed preference. In a six-way free choice setup, a balanced diet was clearly preferred (81%) to diets containing a 4% excess of tryptophan (1%), methionine (2%), arginine (6%), lysine (7%) or threonine (3%). When the same choice was offered but without the balanced diet, diets with 4% excess of lysine (38%), threonine (37%) and arginine (22%) were preferred to diets with the same amount of methionine (3%) or tryptophan (0%). The free choice between the three better accepted diets did not reveal significant differences in preference between them (43, 34 and 23% for threonine, lysine and arginine, respectively). However, when compared as a double choice, excess methionine was significantly preferred to tryptophan (67 vs. 33%).

When the same free choice between the balanced diet and the diets with 4% excess of amino acids was offered to chicken Edmonds and Baker (1987) no aversion to the amino acids in excess was observed (Figure 6). These differences between the two species may be associated with the poor development of the chemical senses (taste and smell) in the chicken compared to the pig (Leibetseder, 1980). If this aversion had been induced by the amino acid's toxicity, similar responses could have been expected for pigs and chicken.

Several double free-choice trials studying the effect of adding free amino acids on pig's feed preference are reviewed in Table 3. Preference studies between lysine deficient diets and the same diets with added free lysine show preferential intakes for the higher levels of this amino acid (Ettle and Roth, 2009; Henry, 1987, 1993; Kirchgessner *et al.*, 1999). With equivalent experimental conditions, the same was observed for tryptophan (Ettle and Roth, 2004), threonine (Ettle and Roth, 2005) and methionine (Roth *et al.*, 2006).

Piglets offered the free choice between diets supplemented with DL-methionine and the same diet supplemented with liquid methionine hydroxy-analogue showed a preference for the DL-methionine form (Ettle *et al.*, 2007). However, differences in feed intake between the two sources of methionine when offered under no-choice conditions were not observed.

Owen *et al.* (1994) also conducted experiments in which growing pigs were offered the free choice between low- and high-lysine diets that with different combinations of sorghum and soybean

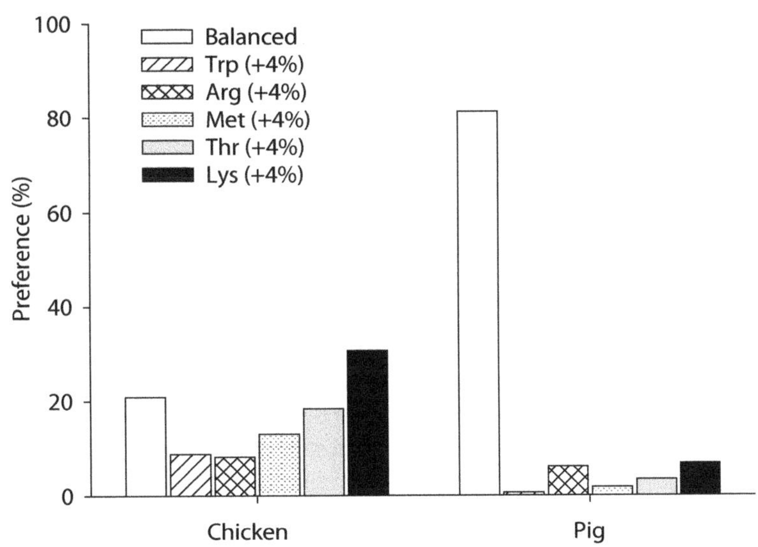

Figure 6. Pig and chicken free choice preferences between a balanced basal diet and diets supplemented with 4% excess of different amino acids (drawn after Edmonds and Baker, 1987 and Edmonds et al., *1987).*

Table 3. Pig's double choice feed preference studies of the effect of supplementing amino acid deficient diets with free amino acids.

Reference	Pig's age	Days in test	AA levels compared (content in diet)	Preference (% total intake)
Ettle and Roth (2009)	weaners	35	Lys (0.70%) – Lys (1.0%)	28–72*
			Lys (0.70%) – Lys (1.0%)	18–82*
Ettle and Roth (2004)	weaners	42	Trp (0.11%) – Trp (0.16%)	12–88*
			Trp (0.11%) – Trp (0.20%)	8–92*
Ettle and Roth (2005)	weaners	42	Thr (0.57%) – Thr (0.62%)	43–57*
			Thr (0.57%) – Thr (0.67%)	53–47
	weaners	42	Thr (0.50%) – Thr (0.56%)	25–75*
			Thr (0.50%) – Thr (0.62%)	26–74*
Henry (1987)	growers	42	Lys (0.60%) – Lys (0.83%)	39–61[a]
			Lys (0.60%) – Lys (1.16%)	48–52[b]
			Lys (0.60%) – Lys (1.40%)	48–52[b]
	growers	42	Lys (0.86%) – Lys (1.21%)	50–50[a]
			Lys (0.72%) – Lys (1.21%)	42–58[b]
			Lys (0.60%) – Lys (0.86%)	39–61[b]
Henry (1993)	growers females	42	Lys (0.61%) – Lys (0.85%)	54–46
			Lys (0.74%) – Lys (0.85%)	46–54
			Lys (0.61%) – Lys (1.21%)	41–59*
			Lys (0.74%) – Lys (1.21%)	46–54
	growers castrates	42	Lys (0.61%) – Lys (0.85%)	45–55*
			Lys (0.74%) – Lys (0.85%)	45–55
			Lys (0.61%) – Lys (1.21%)	48–52
			Lys (0.74%) – Lys (1.21%)	48–52
Kirchgessner *et al.* (1999)	weaners	42	Lys (0.70%) – Lys (1.15%)	40–60*
			Lys (0.70%) – Lys (1.25%)	28–72*
Roth *et al.* (2006)	weaners	42	Met (0.19%) – Met (0.23%)	24–76*
			Met (0.19%) – Met (0.26%)	20–80*

[a,b] Values within a trial with different letters are significantly different ($P<0.05$).

* Proportions differ significantly ($P<0.05$).

meal. It was observed that pigs selected a combination of feeds providing a lysine content that was in excess of their requirements. However, it must be considered that low and high lysine diets were associated with higher contents of sorghum and soybean meal, respectively and this by itself may have explained to some degree the preferences observed.

Fats

Fats have also been shown to affect feed preference in pigs. Pigs offered a choice between a basal diet and the same diet in which 10% of starch was replaced by lard, preferred this diet to the basal diet (91 vs. 9% of intake; $P<0.05$). When starch was replaced by hydrolysed fat feed, preferences were 72 vs. 28% ($P<0.1$) and when it was replaced by tallow they were 75 vs. 25% ($P>0.1$), respectively (Leibbrandt *et al.*, 1975). In the same study, lard was preferred over hydrolysed fat (77 vs. 23%; $P<0.05$) when compared directly against each other, whereas the comparison between lard and tallow or tallow and hydrolysed fat showed no significant differences (41 vs. 59% and 65 vs. 35%, respectively). In contrast, a decrease in feed preference when a maize-soybean diet was supplemented with 3% of lard (62 vs. 28%; $P<0.01$) has also been reported (Orr, 1980). Cats offered the free-choice between purified diets containing 25% of different fats and a diet with 25% bleached tallow, preferred the diets containing bleached tallow to diets containing butter or chicken fat ($P<0.001$), but no significant differences in preference were observed with lard, unbleached tallow, yellow grease or partially hydrogenated vegetable fat (Kane *et al.*, 1981). It has to be considered that fats are susceptible to rancidification and this may have an important effect on feed acceptability (De Rouchey *et al.*, 2004).

Our group has also evaluated the pig's preferences for 6 different fat sources using the same reference diet as for the cereal and protein sources (Solà-Oriol *et al.*, 2006). The different fat sources replaced different amounts of sunflower oil, the reference ingredient. Palm oil (at 3% of inclusion) was significantly preferred to sunflower oil in the reference diet, whereas 3 and 10% linseed oil and 10% soybean oil presented significantly lower preferences (Figure 7).

Fibre sources

Gut capacity may be a limiting factor for energy intake in young pigs offered bulky fibrous feeds (Kyriazakis and Emmans, 1995). Fibre sources may also affect feed preference according to their

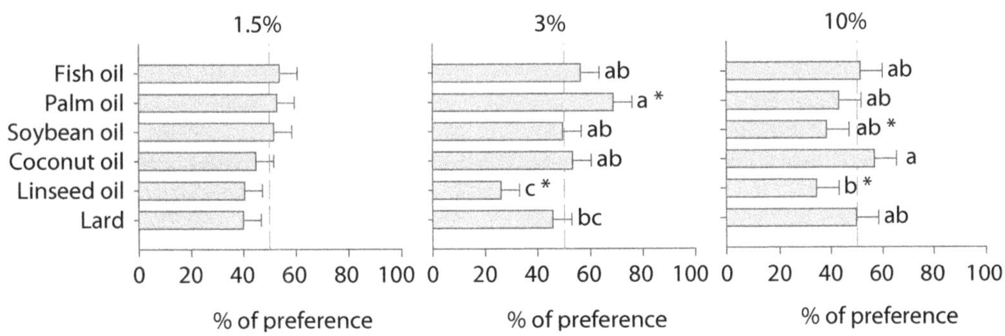

Figure 7. Pig's preference (% of total feed intake) for diets containing 1.5, 3 or 10% of different fat sources offered in a two-way choice with a common reference diet based on sunflower oil (Solà-Oriol et al., *2006). Preferences were measured over a period of four days. Within each inclusion rate, columns with different letters indicate significantly different prefence values (P<0.05). Columns with the symbol * indicate a preference value significantly different from 50% (P<0.05).*

physical characteristics (e.g. water holding capacity, effect on viscosity etc.). The addition of 3% dehydrated alfalfa meal to a maize-soybean diet for piglets has been shown to reduce feed preference (81 vs. 19%; $P<0.01$; Orr, 1980).

Using the same reference diet as for the cereal, proteins and fats, we measured the pig's preferences for three different fibre sources using wheat bran in the reference diet as the replaceable fibre (Solà-Oriol, 2008). At inclusion levels of 13%, dehydrated alfalfa and sugar beet pulp were less preferred than wheat bran or carob bean pulp (Figure 8).

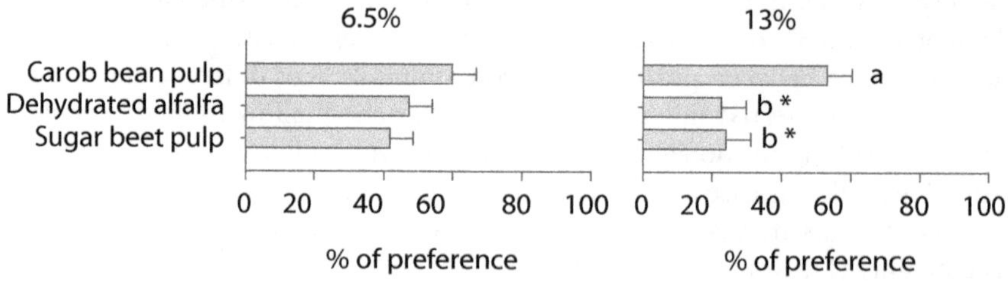

*Figure 8. Pig's preference (% of total feed intake) for diets containing 6.5 or 13% of different fibre sources offered in a two-way choice with a common reference diet based on wheat bran (Solà-Oriol, 2008). Preferences were measured over a period of four days. Within each inclusion rate, columns with different letters indicate significantly different prefence values (P<0.05). Columns with the symbol * indicate a preference value significantly different from 50% (P<0.05).*

Palatability enhancers and other feed additives

The addition of sweeteners such as sucrose has been known for many years to improve feed preference in pigs. The addition of as little as 0.62% sucrose resulted in a preferential choice of 89% over 11% for the non supplemented diet (Grinstead *et al.*, 1961). Increasing amounts of sucrose improve feed preference in a dose response manner. Pigs offered the free choice among feeds containing 0, 5, 10, 15 or 20% sucrose (Lewis *et al.*, 1955) and 3, 5 or 10% sucrose (Salmon Legagneur and Février, 1956) always preferred the diet with the highest rate of added sugar (0-1-2-4-93% and 18-24-58%, respectively).

Pigs have been shown to have a higher preference for a pelleted feed coated with 10% sucrose than for the same feed with the sugar added into the mixer prior to pelleting (83-17%, Salmon Legagneur and Février, 1956). This is probably due to a more direct contact of the sugar with the taste buds in the mouth and a sweeter taste sensation. Weaning pigs offered the choice between diets containing 5% sucrose, 5% glucose or 0.22% saccharin, showed a clear preference for the sugar diet (66-11-23%, respectively; Aumaitre, 1975). The higher preference for sucrose than for glucose agrees with their reported sweetening potences (Glaser *et al.*, 2000) supporting a taste mediated preference. It has also been reported that under a no-choice situation suckling pigs offered creep feed supplemented with 7.5% sucrose initiated feeding earlier and achieved a higher weaning weight than piglets offered the same feed without sucrose (Salmon Legagneur and Février, 1956).

Non-carbohydrate high intensity sweeteners have also been used to enhance the sweet taste of feeds (Roura and Tedó, 2009). However their sweetening potencial for pigs is usually much lower than that described in humans (Glaser *et al.*, 2000). The addition of 150 to 2000 ppm saccharin did not result in any changes of feed preference by pigs (Roura *et al.*, 2007).

Similarly, umami and salty tastes have also been targeted as possible ways of improving feed preference. Henson *et al.* (1962) showed that additions of monosodium glutamate (umami taste enhancer) to piglet's feed ranging between 0.1-1% resulted in preferences over the unsupplemented feed ranging from 85 to 97%. Wahlstrom *et al.* (1974) studied the effect of increasing the salt content of low lactose whey on diet palatability, but did not obtain consistent preference responses.

Weanling pigs have been shown to have a preference for some flavors and aversion for others (McLaughlin *et al.,* 1983). The use of flavours in piglets' feed has been shown to improve feed intake and performance after weaning (Torrallardona *et al.*, 2001). However, it could be expected that only preferred flavours may be effective in facilitating iniciation of feed intake at weaning. Trials in which lactating sows have been fed a flavouring agent (aiming to flavour their milk), and the same flavour has been added to the piglet's feed have shown increased feed intakes of the piglets offered flavoured over non-flavoured feeds (Campbell, 1976; King, 1979; Langendijk *et al.*, 2007). Other authors have tried this same approach but without success (Kennedy, 1980). The nature of the feed flavour used may be important in the success or failure of this strategy, but this is an area that requires further investigation. For more information on the use of flavours to condition feed acceptance by piglets and on the effects of palatability enhancers on feed appetence refer to Bolhuis *et al.* (2009) and Roura and Tedó (2009).

In-feed antibiotics are also known to affect feed palatability. Tomlin *et al.* (1958) offered four feeds containg 40 ppm erythromycin, oxytetracycline, chlortetracycline or penicillin as a free choice to finisher pigs and observed a clear aversion to erythromycin (0.6% of intake), intermediate acceptance for oxytretracycline and penicillin (20.4 and 29.4%) and a preference for oxytetracycline (49.6%). These observations were confirmed in a second trial in which a basal antibiotic free diet was also included in the free-choice. Erythromycin was avoided whereas oxytetracycline was preferred over the basal diet.

Organic acids and essential oils, products that have been postulated as possible aternatives to the banned antibiotic growth promoters can also have an important effect on feed palatability. Doses of 1.2% of formic acid and sorbic acid resulted in significant reductions in feed preference compared to the unsupplemented feed (13-87% and 25-75%, respectively), whereas the salt form potassium diformate (at 1.2 and 2.4% of inclusion) did not affect feed preference (Ettle *et al.*, 2004). Feeds containing 8% lactic acid, formic acid or calcium formate did not show differences in preference when compared among each other, nor did the comparison between lactic acid and sodium benzoate (Partanen *et al.*, 2002). The same authors, however, showed that sodium benzoate was preferred over formic acid (67-33%) and calcium formate (73-27%). Among the essential oils, 100 ppm additions of fennel or caraway oils have been shown to reduce feed preference in growing pigs. However, when the same essential oils were added at a dose of 25 ppm, feed preference was not affected (Schöne *et al.*, 2006).

Combination of ingredients and feed preference

The common reference diet used in our free-choice preference tests allows the contrast of the effects of the cereal and protein sources by themselves, but also that of different cereal-protein combinations. We measured the preference for different combinations between barley and oats as cereal sources and lupine and rapeseed as protein sources (Figure 9). Generally, the preferences for the different cereal-protein combinations agreed well with those expected from the preferences of cereals and proteins tested on their own. However, including 10% lupine improved preference on its own, but reduced it when combined with barley or oats (Solà-Oriol, 2007c). Further research is needed to judge the real potential of using single feedstuff preference values to predict the preference of different feedstuff combinations.

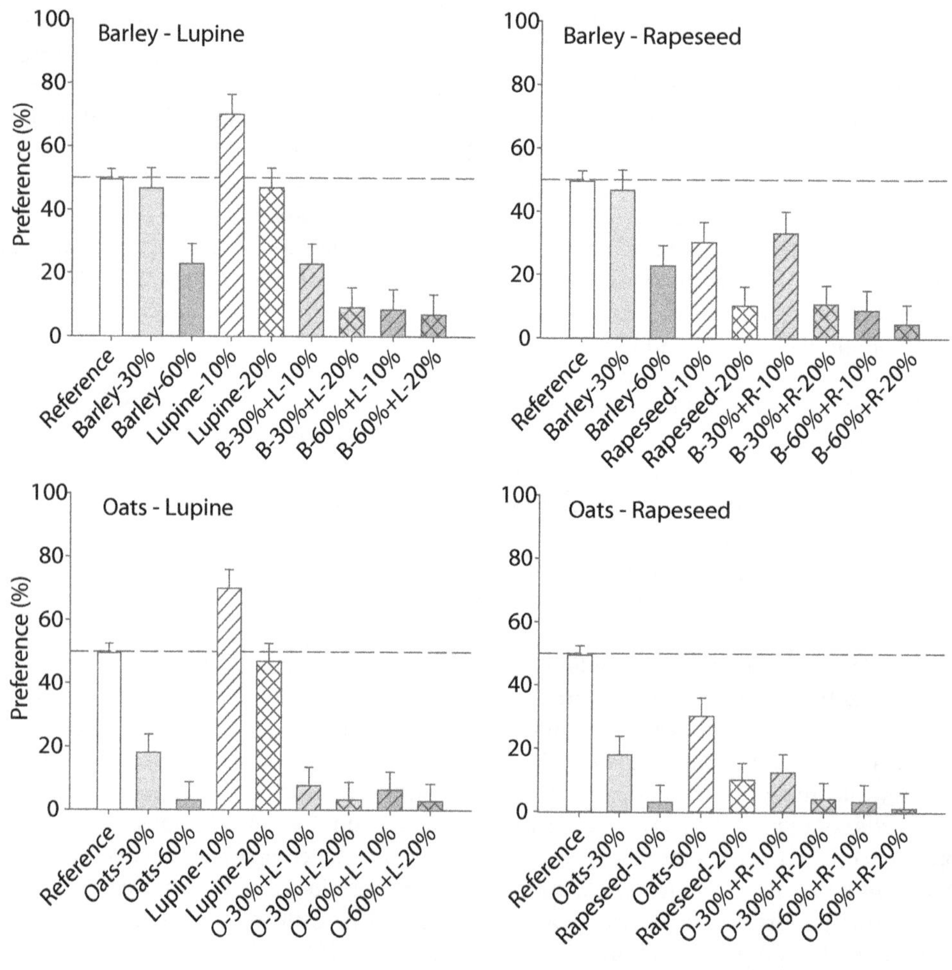

Figure 9. Preferences (% of total feed intake) for diets containing different cereals and protein sources alone or in combination (Solà-Oriol et al.*, 2007c).*

In our studies (Solà-Oriol, 2008; Solà-Oriol *et al.*, 2009a), feed preferences for cereals ranged between 3-90%, for protein sources between 8-76%, for fat sources between 26-29% and those for fiber sources between 23-60% (Figures 4, 5, 7 and 8). Relative to the neutral value of 50%, the changes in feed preference per percentual unit of ingredient inclusion ranged between: -2.6 and +1.5 for cereals, -5.2 and +5.3 for proteins, -0.8 and +0.6 for fats and between -2.1 and +1.5 for fibres. This suggests that protein sources have a higher relative impact on feed preference per unit of feed ingredient than cereals, fats or fibre sources.

Preference and feed's particle size, texture and nutritive value

We studied the relationship between the preferences of all the feeds tested and their particle size and texture profiles. Poor correlations were observed between preference and particle size, uniformity of particle size, the total number of particles, surface area and the percentages of coarse or fine particles. Piglet studies in which reducing particle size did not improve the intake of sorghum, corn or wheat based diets support these observations (Healy *et al.* 1994; Mavromichalis *et al.*, 2000). Furthermore, in grower-finisher pigs, lowering particle size of corn and wheat based diets has been reported to reduce feed intake (Mavromichalis *et al.*, 2000; Wondra *et al.*, 1995). For the texture parameters, however, a significant negative correlation was found between preference and the feed's hardness (r=-0.21), fragility (r=-0.20) and chewing work (r=-0.33) that was measured with a texturometer (Sola-Oriol *et al.*, 2009b).

The preferences observed for the 25 different cereal feeds (60% inclusion) correlated significantly with the cereals content of crude fibre (r=-0.61), digestible energy (r=0.54) (Solà-Oriol *et al.*, 2007b) and with their rate of glucose release estimated in an *in vitro* digestion system (r=0.48). No significant correlations were observed for dry matter, ash, organic matter, fat, protein or gross energy (Solà-Oriol *et al.*, 2008).

The ileal and faecal digestibility of twelve of the cereal feeds were also measured *in vivo*. It was observed that preference correlated significantly with the content of digestible dry matter, organic matter and starch at the ileal (r=0.64, 0.67 and 0.82) and at the faecal level (r=0.82, 0.80 and 0.81, respectively). The flow of digesta at the ileal level was estimated as the total recovery of indigestible marker during the first 12 hours after administration, and it was also found to be significantly correlated with feed preference (r=0.62). Finally, the non-digestible (faecal) NSP content of eight cereal feeds was found to be negatively correlated with feed preference (r=-0.84; Solà-Oriol, 2008).

Although we have shown that the 'mouth feel' sensations associated with the texture characteristics of the feed explain in part some of the preferences observed, for cereals we have observed that their nutritive value plays a more important role. In particular, we have observed a positive effect of digestible energy and starch content and of glycemic index on cereal preference, whereas crude fiber and indigestible NSP content had a negative effect. As mentioned previously, it is possible that there is an oro-sensorial component to identify these nutritives characteristics (Goff and Klee, 2006).

Despite their major relative impact on feed preference, the nutritive characteristics and digestibility of the feeds with the different protein sources were not studied. It is possible that other factors than those considered for cereals may be important; particularly, their amino acid composition and digestibility. Although antinutritional compounds are usually inactivated with technological treatments (Schöne *et al.*, 1990), their bitter-tasting properties may still persist and affect palatability. It is well documented that raw soybean meal has to be heat treated to inactivate trypsin inhibitor and hemaglutinins. Similarly, it has been shown that tannins and sinapine in rapeseed decrease the palatability of the diets for pigs (Baidoo *et al.* 1986). Legume seeds also contain high levels of NSP, especially if their hull is not fully removed (Van Barneveld, 1999; Hollung *et al.*, 2005), and as in cereals they may have a big impact on feed preference. Refer to Clasadonte and Van der Poel (2009) for more details on the effects of ANF on feed intake.

Finally, feed palatability is not a constant feature and may change under different situations. We have shown that storage of some proteins resulted in a loss of palatability (Solà-Oriol *et al.*, 2007a). Fats may also be especially susceptible to changes in palatability as a result of rancidification. Offering piglets feeds with increasing levels of fat oxidation resulted in a linear decrease in feed intake (De Rouchey *et al.*, 2004).

Conclusions

Feed palatability for pigs is clearly driven by the nature of the feedstuffs that compose it. Therefore, the accurate selection of feed ingredients with a high preference and at the most preferred inclusion rate could be a strategy to improve feed palatability. Although under practical conditions piglets are not offered a choice between diets, it should be expected that better preferred feeds reduce the time to first feed ingestion at weaning. If so, this should also alleviate the negative impact of post weaning anorexia on the maintenance of a healthy gut mucosa by the piglets. There may be situations, however, in which preference may not be a good indicator for the practical situation. Conventional feeding trials simulating commercial conditions should always be the ultimate test to validate the efficacy of a given ingredient to improve feed intake.

Acknowledgements

Some of the results presented were obtained with support from Lucta, S.A (CDTI project 050369). Support from the Spanish government (grant AGL2005-07438-C02-02/GAN) is also acknowledged.

References

Aubel, C.E., 1935. Testing the comparative palatability of different grains. Journal of Animal Science 1935: 107-109.

Aumaitre, A., 1975. Sur l'appetibilite des aliments destines aux porcelets: importance du sel et du sucre. Journees de la Recherche Porcine en France 7: 169-176.

Aumaitre, A., 1980. Palatability of piglet feeds: trial methods and practical results. Advances in Animal Physiology and Animal Nutrition 11: 86-95.

Aumaitre, A. and Salmon Legagneur, E., 1961. Les préférences alimentaires du porcelet V. - Comparaison de divers modes de distribution de l'aliment may be associated with texture sensations. Annales de Zootechnie 10: 1497-203.

Aumaitre, A. and Salmon Legagneur, E., 1964. Les préférences alimentaires du porcelet. VII. Appétibilité des tourteaux. Annales de Zootechnie 13: 237-240.

Baidoo, S.K., McIntosh, M.K. and Aherne, F.X., 1986. Selection preference of starter pigs fed canola-meal and soybean-meal supplemented diets. Canadian Journal of Animal Science 66: 1039-1049.

Baldwin, B.A., 1980. Operant studies and preference tests on the role of olfaction and taste in the ingestive behaviour of pigs and ruminants. Advances in Animal Physiology and Animal Nutrition 11: 36-42.

Bolhuis, J.E., Oostindjer, M., Van den Brand, H., Gerrits, W.J.J. and Kemp, B., 2009. Voluntary feed intake in piglets: potential impact of early experience with flavours derived from the maternal diet. In: Torrallardona, D. and Roura, E. (eds.) Voluntary feed intake in pigs. Wageningen Academic Publishers, the Netherlands, pp. 37-60.

Bradford, M.M.V. and Gous, R.M., 1992. The response of weaner pigs to a choice of foods differing in protein content. Animal Production 55: 227-232.

Brand, T.S. and van der Merwe, J.P., 1996. Naked oats (avena nuda) as a substitute for maize in diets for weanling and grower-finisher pigs. Animal Feed Science and Technology 57: 139-147.

Bruneau, C.D. and Chavez, E.R., 1995. Dietary preferences for cereals of nursing and weaned piglets. Livestock Production Science 41: 225-231.

Campbell, R.G., 1976. A note on the use of a feed flavour to stimulate the feed intake of weaner pigs. Animal Production 23: 417-419.

Clasadonte, L. and Van der Poel, A.F.B., 2009. Antinutritional factors: impact on pigs' voluntary feed intake. In: Torrallardona, D. and Roura, E. (eds.) Voluntary feed intake in pigs. Wageningen Academic Publishers, the Netherlands, pp. 243-275.

Cooper, R.G., Lennon, A.M., Tribble, L.F. and Buddingh, F., 1979. LCP Cottonseed Flour as a Protein Source for Swine. Journal of Animal Science 48: 1216-1222.

DeRouchey, J.M., Hancock, J.D., Hines, R.H., Maloney, C.A., Lee, D.J., Cao, H., Dean, D.W. and Park, J.S., 2004. Effects of rancidity and free fatty acids in choice white grease on growth performance and nutrient digestibility in weanling pigs. Journal of Animal Science 82: 2937-2944.

Devilat, J., Pond, W.G. and Miller, P.D., 1970. Dietary amino acid balance in growing-finishing pigs: effect on diet preference and performance. Journal of Animal Science 30: 536-543.

Dong, G.Z. and Pluske, J.R., 2007. The low feed intake in newly-weaned pigs: problems and possible solutions. Asian Australasian Journal of Animal Science 20: 440-452.

Edge, H.L., Varley, M.A. and Rowlinson, P., 2004. The effect of trough colour on the voluntary food intake of postweaned pigs – a short communication. Livestock Production Science 88: 193-197.

Edmonds, M.S. and Baker, D.H., 1987. Comparative effects of individual amino acid excesses when added to a corn-soybean meal diet: effects on growth and dietary choice in the chick. Journal of Animal Science 65: 699-705.

Edmonds, M.S., Gonyou, H.W. and Baker, D.H., 1987. Effect of excess levels of methionine, tryptophan, arginine, lysine or threonine on growth and dietary choice in the pig. Journal of Animal Science 65: 179-185.

Ermer, P.M., Miller, P.S. and Lewis, A.J., 1994. Diet preference and meal patterns of weanling pigs offered diets containing either spray-dried porcine plasma or dried skim milk. Journal of Animal Science 72: 1548-1554.

Ettle, T. and Roth, F.X., 2004. Specific dietary selection for tryptophan by the piglet. Journal of Animal Science 82: 1115-1121.

Ettle, T. and Roth, F.X., 2005. Dietary preferences for feeds varying in threonine concentration by the piglet. Physiology and Behavior 85: 289-295.

Ettle, T. and Roth, F.X., 2009. Dietary selection for lysine by piglets at different feeding regimen. Livestock Science (in press).

Ettle, T., Mentschel, K. and Roth, F.X., 2004. Dietary self-selection for organic acids by the piglet. Archives of Animal Nutrition 58: 379-388.

Ettle, T., Rademacher, M., Roth, F.X. and Payne, R.L., 2007. Dietary preference for methionine sources in 8 to 25-kg nursery pigs. Journal of Animal Science 85 (Suppl. 1): 81 (Abstr.).

Ferguson, N.S., Bradford, M.M.V. and Gous, R.M., 2002. Diet selection priorities in growing pigs offered a choice of feeds. South African Journal of Animal Science 32: 136-143.

Glaser, D., Wanner, M., Tinti, J.M. and Nofre, C., 2000. Gustatory responses of pigs to various natural and artificial compounds known to be sweet in man. Food Chemistry 68: 375-385.

Goff, S.A. and Klee, H.J., 2006. Plant volatile compounds: Sensory cues for health and nutritional value? Science 311: 815-819.

Grinstead, L.E., Speer, V.C. and Hays, V.M., 1961. Baby pigs preference for diets containing varying levels of sucrose. Journal of Animal Science 20: 934 (Abstr.).

Hansen, B.C., Flores, E.R., Tanksley Jr., T.D. and Knabe, D.A., 1987. Effects of different heat treatments during processing of soybean meal on nursery and growing pig performance. Journal of Animal Science 65: 1283-1291.

Healy, B.J., Hancock, J.D., Kennedy, G.A., Bramel-Cox, P.J., Behnke, K.C. and Hines, R.H., 1994. Optimum particle-size of corn and hard and soft sorghum for nursery pigs. Journal of Animal Science 72: 2227-2236.

Henry, Y., 1987. Self-selection by growing pigs of diets differing in lysine content Journal of Animal Science 65: 1257-1265.

Henry, Y., 1993. Self-selection of lysine by growing pigs: choice combinations between deficient or suboptimal and adequate or superoptimal dietary levels according to sex. Reproduction Nutrition Development 33: 489–502.

Henson, J.N., Bogdonoff, P.D. and Thrasher, G.W., 1962. Levels of monosodium glutamate in pig starter preference. Journal of Animal Science 21: 999(Abstr.).

Hollung, K., Øverland, M., Hrustić, M., Sekulić, P., Miladinović, L., Martens, H., Narum, B., Sahlstrøm, S., Sørensen, M. and Storebakken, T., 2005. Evaluation of nonstarch polysaccharides and oligosaccharide content of different soybean varieties (glycine max) by near-infrared spectroscopy and proteomics. Journal of Agricultural and Food Chemistry 53: 9112-9121.

Hongtrakul, K., Goodband, R.D., Behnke, K.C., Nelssen, J.L., Tokach, M.D., Bergström, J.R., Nessmith Jr., W.B. and Kim, I.H., 1998. The effects of extrusion processing of carbohydrate sources on weanling pig performance. Journal of Animal Science 76: 3034-3042.

Huang, S.X., Sauer, W.C., Pickard, M., Li, M. and Hardin, R.T., 1998. Effect of micronization on energy, starch and amino acid digestibility in hulless barley for young pigs. Canadian Journal of Animal Science 78: 81-87.

Kane, E., Morris, J.G. and Rogers, Q.R., 1981. Acceptability and digestibility by adult cats of diets made with various sources and levels of fat. Joal of Animal Science 53: 1516-1523.

Kelly, D., Smyth, J.A. and McCracken, K.J., 1991. Digestive development of the early-weaned pig. 1. Effect of continuous nutrient supply on the development of the digestive-tract and on changes in digestive enzyme-activity during the 1st week post-weaning. British Journal of Nutrition 65: 169-180.

Kennedy, J.M., 1980. The development of dietary preferences in pigs and poultry. Advances in Animal Physiology and Animal Nutrition 11: 141-147.

Kim, S.W. and Easter, R.A., 2001. Nutritional value of fish meals in the diet for young pigs. Journal of Animal Science 79: 1829-1839.

King, R.H., 1979. The effect of adding a feed flavour to the diets of young pigs before and after weaning. Australian Journal of Experimental Agriculture and Animal Husbandry 19: 695-697.

Kirchgessner, M., Stangl, G.I. and Roth, F.X., 1999. Evidence of a specific dietary selection for lysine by the piglet. Journal of Animal Physiology and Animal Nutrition 81: 124-131.

Klasing, K.C. and Humphrey, B.D, 2009. Taste receptor biology. In: Torrallardona, D. and Roura, E. (eds.) Voluntary feed intake in pigs. Wageningen Academic Publishers, the Netherlands, pp. 87-103.

Kyriazakis, I. and Emmans, G.C., 1993. The effect of protein source on the diets selected by pigs given a choice between a low and high protein food. Physiology and Behavior 53: 683-688

Kyriazakis, I. and Emmans, G.C., 1995. The voluntary feed intake of pigs given feeds based on wheat bran, dried citrus pulp and grass meal, in relation to mesurements of feed bulk. British Journal of Nutrition 73: 191-207.

Lallès, J.P., Bosi, P. Smidt, H. and Stokes, C.R., 2007. Nutritional management of gut health in pigs around weaning. Proceedings of the Nutrition Society 66: 240-268.

Langendijk, P, Bolhuis, J.E. and Laurenssen, B.F.A., 2007. Effects of pre- and postnatal exposure to garlic and aniseed flavour on pre- and postweaning feed intake in pigs. Livestock Science 108: 284-287.

Le Dividich, J. and Sève, B., 2000. Effects of underfeeding during the weaning period on growth, metabolism, and hormonal adjustments in the piglet. Domestic Animal Endocrinology 19: 63-74.

Leibbrandt, V.D., Hays, V.W., Ewan, R.C. and Speer, V.C., 1975. Effect of fat on performance of baby and growing pigs. Journal of Animal Science 40: 1081-1085.

Leibetseder, J., 1980. Introduction to animal olfactory and taste physiology. Advances in Animal Physiology and Animal Nutrition. 11: 11-20.

Lenehan, N.A., DeRouchey, J.M., Goodband, R.D., Tokach, M.D., Dritz, S.S., Nelssen, J.L., Groesbeck, C.N. and Lawrence, K.R., 2007. Evaluation of soy protein concentrates in nursery pig diets. Journal of Animal Science 85: 3013-3021.

Lewis, C.J., Catron, D.V., Combs Jr., G.E., Ashton, G.C. and Culbertson, C.C., 1955. Sugar in Pig Starters. Journal of Animal Science 14: 1103-1115.

Lindemann, M.D., Van de Ligt, J.L.G., Monegue, H.J., Keller, G. and Cromwell, G.L., 1998. Evaluation of dried porcine solubles (DPS) as a feed ingredient for weanling pigs. Journal of Animal Science 76(Suppl.1): 181(Abstr.).

Lindemann, M.D., Cromwell, G.L., Monegue, H.J., Cook, H., Soltwedel, K.T., Thomas, S. and Easter, R.A., 2000. Feeding value of an enzymatically digested protein for early-weaned pigs. Journal of Animal Science 78: 318-327.

Maslanek, A., Flis, M. and Falkowski, J., 2001. A note on naked oats as a substitute for wheat and barley for young pigs. Journal of Animal and Feed Sciences 10: 215-218.

Mateos, G.G., Alcantarilla, A., Latorre, M.A., Lazaro, R., Gomez, E. and Laso, N., 2001. Influence of type of cereal and level of fiber on performance of early-weaned piglets. Journal of Animal Science 79(Suppl. 1): 106-107(Abstr.).

Mavromichalis, I., Hancock, J.D., Senne, B.W., Gugle, T.L., Kennedy, G.A., Hines, R.H. and Wyatt, C.L., 2000. Enzyme supplementation and particle size of wheat in diets for nursery and finishing pigs. Journal of Animal Science 78: 3086-3095.

McLaughlin, C.L., Baile, C.A., Buckholtz, L.L. and Freeman, S.K., 1983. Preferred flavors and performance of weaned pigs. Journal of Animal Science 56: 1287-1293.

Medel, P., Salado, S., de Blas, J.C. and Mateos, G.G., 1999. Processed cereals in diets for early-weaned piglets. Animal Feed Science and Technology 82: 145-156.

Nelson, S.L. and Sanregret, J.D., 1997. Response of pigs to bitter-tasting compounds. Chemical Senses 22: 129-132.

Orr Jr., D.E., 1980. Determination of individual feed ingredient and total ration palatability. Advances in Animal Physiology and Animal Nutrition 11: 53-63.

Owen, K.Q., Knabe, D.A., Burgoon, K.G. and Gregg, E.J., 1994. Self-selection of diets and lysine requirements of growing-finishing swine. Journal of Animal Science 72: 554-564.

Partanen K., Siliander-Rasi, H. and Suomi, K., 2002. Dietary preferences of weaned piglets offered diets containing organic acids. Agricultural and food science in Finland 11: 107-119.

Pedersen, C. and Lindberg, J.E., 2004. Comparison of low-glycoalkaloid potato protein and fish meal as protein sources for weaner piglets. Acta Agriculturae Scandinavica Section A-Animal Science 54: 75-80.

Petrie, C.L. and Gonyou, H.W., 1988. Effects of auditory, visual and chemical stimuly on the ingestive behavior of newly weaned piglets. Journal of Animal Science 66: 661-668.

Piao, X.S., Chae, B.J., Kim, J.H., Jin, J., Cho, W.T. and Han, I.K., 1999. Effects of extrusion condition of barley on the growth and nutrient utilization in growing pigs. Asian-Australasian Journal of Animal Sciences 12: 783-787.

Pluske, J.R., Williams, I.H. and Aherne, F.X., 1996a. Maintenance of villous height and crypt depth in piglets by providing continuous nutrition after weaning. Animal Science 62: 131-144.

Pluske, J.R., Williams, I.H. and Aherne, F.X., 1996b. Villous height and crypt depth in piglets in response to increases in the intake of cows' milk after weaning. Animal Science 62: 145-158.

Pluske, J.R., Hampson, D.J. and Williams, I.H., 1997. Factors influencing the structure and function of the small intestine in the weaned pig: a review. Livestock Production Science 51: 215-236.

Pluske, J.R., Pethick, D.W., Hopwood, D.E. and Hampson, D.J., 2002. Nutritional influences on some major enteric bacterial diseases of pigs. Nutrition Research Reviews 15: 333-371.

Pluske, J.R., Montagne, L., Cavaney, F.S., Mullan, B.P., Pethick, D.W. and Hampson, D.J., 2007. Feeding different types of cooked white rice to piglets after weaning influences starch digestion, digesta and fermentation characteristics and the faecal shedding of beta-haemolytic escherichia coli. British Journal of Nutrition 97: 298-306.

Reiners, K., Hessel, E.F. and Van den Weghe, H.F.A., 2008. The effect of heatedmas on performance and feeding behavior of newly weaned piglets. Journal of Animal Science 86: 3600-3607.

Reis de Souza, T.C., Mariscal Landín, G. and Aguilera Barreyro, A., 2005. Effect of different cereals on nutrient digestibility and dietary preference in starter diets for piglets. Veterinaria México 36: 11-24.

Roth, F.X., Meindl, C. and Ettle, T., 2006. Evidence of a dietary selection for methionine by the piglet. Journal of Animal Science 84: 379-386.

Roura, E. and Tedó, G., 2009. Feed appetence in pigs: an oronasal sensing perspective. In: Torrallardona, D. and Roura, E. (eds.) Voluntary feed intake in pigs. Wageningen Academic Publishers, the Netherlands, pp. 105-140.

Roura, E., Risley, C., Solà-Oriol, D. and Torrallardona, D., 2007. Responses to increasing doses of saccharin in piglet feeds on preference values and performance. Journal of Animal Science 85 (Suppl.2): 108 (Abstr.).
Salmon Legagneur, E. and Aumaitre, A., 1961. Les préférences alimentaires du porcelet. VI. Appétibilité des farines animales. Annales de Zootechnie 10: 313-319.
Salmon Legagneur, E. and Février, R., 1955. Les preferences alimentaires du porcelet. I Influence du mode de présentation des aliments: granulés ou farine. Annales de Zootechnie 4: 215-218.
Salmon Legagneur, E. and Février, R., 1956. Les preferences alimentaires du porcelet. II Le sucre dans les aliments pour porcelets. Annales de Zootechnie 5: 73-79.
Salmon Legagneur, E. and Février, R., 1959. Les preferences alimentaires du porcelet. III Appétence de quelques céréales. Annales de Zootechnie 8: 139-146.
Schöne, F., Hennig, A. and Lange, R., 1990. Selection of feed containing rapeseed meals with different glucosinolate content by growing pigs. Agribiological Research 43: 260-270.
Schöne, F., Kirchheim, U. and Schumann, W., 1994. Glucosinolate degradation by rapeseed myrosinase and effect on rapeseed acceptability by growing pigs. Animal Feed Science and Technology 48: 229-235.
Schöne F., Vetter, A., Hartung, H., Bergmann, H., Biertümpfel, A., Richter, G., Müller, S. and Breitschuh, R., 2006. Effects of essential oils from fennel (*Foeniculi aetheroleum*) and caraway (*Carvi aetheroleum*) in pigs. Journal of Animal Physiology and Animal Nutrition 90: 500-510.
Skoch, E.R., Biner, S.F., Deyoe, C.W., Allee, G.L. and Behnke, K.C., 1983. Effects of pelleting conditions on performance of pigs fed a corn-soybean meal diet. Journal of Animal Science 57: 922-928.
Solà-Oriol, D., Roura, E. and Torrallardona, D., 2006. Palatability of diets with different oil and fat sources in piglets. Journal of Animal Science 84 (Suppl. 1): 242 (Abstr.).
Solà-Oriol, D., Roura, E. and Torrallardona, D., 2007a. Storage affects the palatability of protein sources in piglet diets. Journal of Animal Science 85 (Suppl. 1): 77 (Abstr.).
Solà-Oriol, D., Roura, E. and Torrallardona, D., 2007b. Cereal nutrient composition correlates with feed oro-sensorial perception in piglets. Journal of Animal Science 85 (suppl. 1): 77 (Abstr.).
Solà-Oriol, D., Roura, E. and Torrallardona, D., 2007c. Interaction between the palatabilities of cereal and protein source in diets for pigs. Journal of Animal Science 85 (Suppl. 2): 108 (Abstr.).
Solà-Oriol, D., 2008. Quantitative evaluation of the palatability of feed ingredients in swine. PhD thesis. Universitat Autònoma de Barcelona, Bellaterra, Spain.
Solà-Oriol, D., Pujol, S., Van Kempen, T., Roura, E. and Torrallardona, D., 2008. Glucose release correlates with piglet's preference for cereal based feeds. Journal of Animal Science 86 (suppl. 3): 97 (Abstr.).
Solà-Oriol, D., Roura, E. and Torrallardona, D., 2009a. Feed preference in pigs: Effect of cereal sources at different inclusion rates. Journal of Animal Science 87: 562-570.
Solà-Oriol, D., Roura, E. and Torrallardona, D., 2009b. Feed preference in pigs: Relationship with feed particle size and texture. Journal of Animal Science 87: 571-582.
Solà-Oriol, D., Roura, E. and Torrallardona, D., 2009c. Use of double-choice feeding to quantify feed ingredient preferences in pigs. Livestock Science (in press).
Tomlin, D.C., Wallace, H.D. and Combs, Jr., G.E., 958. The influence of certain antibiotics on the palatability of swine rations. Journal of Animal Science 17: 42-46.
Torrallardona, D., Llauradó, L., Matas, J., Fort, F. and Roura, E., 2001. The use of flavours in feed improves performance of piglets weaned at 21 days of age. Cahiers OPTIONS Méditerranéennes 54: 213-215.
Van Barneveld, R.J., 1999. Understanding the nutritional chemistry of lupin (*Lupinus* spp.) seed to improve livestock production efficiency. Nutrition Research Reviews 12: 203-230.

Wahlstrom R.C., Hauser, L.A. and Libal, G.W., 1974. Effects of low lactose whey, skim milk and sugar on diet palatability and performance of early weaned pigs. Journal of Animal Science 38: 1267-1271.
Wondra, K.J., Hancock, J.D., Behnke, K.C., Hines, R.H. and Stark, C.R., 1995. Effects of particle-size and pelleting on growth-performance, nutrient digestibility, and stomach morphology in finishing pigs. Journal of Animal Science 73: 757-763.
Zimmerman, D.R., Sparks, J.C. and Cain, C.M., 1997. Carry-over responses to an intestinal hydrolysate in weanling pig diets. Journal of Animal Science 75(Suppl.1): 71(Abstr.).

Chapter 10. Antinutritional factors: impact on pigs' voluntary feed intake

L. Clasadonte and A.F.B. van der Poel
Wageningen University, Wageningen Institute of Animal Sciences, Animal Nutrition Group, Marijkeweg 40, 6709 PG Wageningen, the Netherlands; thomas.vanderpoel@wur.nl

Abstract

In this paper, an overview is given into the effects of antinutritional factors on the voluntary feed intake in pigs. Tabled values of research are presented including the inclusion level of the ingredient, diet information, weight/age of the pigs and voluntary feed intake. If available, the ingredient level of an antinutritional factor and its daily intake are also provided. From the data presented in this paper, no clear linear relationships or consistent effects were noticed between the level of antinutritional factors or its intake and voluntary feed intake in pigs. This might be due to the simultaneous presence of other undesired factors in a seed, to constituents (e.g. amino acids), or to factors originating from processing (e.g. Maillard products) in a way that the pig's sensorial perception cannot be believed to be discriminative to only a single antinutritional factor. If this is generally true, factors hampering or stimulating feed intake in swine can be better considered from the point of view that all diet ingredients – raw or processed – have their specific problems. We identified two groups of ingredients that have a rather direct influence on feed intake when fed raw due to gut troubles or that influence palatability and thus are less consumed. Moreover, we identified two groups of antinutritional factors that either have short-term or long-term effects. The consequences for a number of diet ingredients and the validity of studying only feed intake are shortly discussed.

Keywords: antinutritional factors, feed intake, diet consumption, swine

Introduction

Plant proteins are a major source of dietary proteins in farm animals' feeding patterns. After several food crisis (e.g. BSE) animal proteins have been prohibited in pig feed and plant proteins have become more important. Most cereal grains (e.g. wheat, corn) have a low level of proteins (6-14%) and are deficient in certain amino acids such as lysine. On the other hand oilseeds (e.g. soybean, rapeseed) and legume seeds (e.g. pea, beans) contain 20-50% proteins and can be mixed with cereal proteins to produce a high nutritive mixture of proteins/amino acids.

Nutritionists, however, have to face a problem since plants synthesise substances that may exert deleterious effects when ingested by man and animals. These constituents are naturally occurring and referred to as antinutritional factors (ANFs). The physico-chemical nature of most of these ANFs is known: some of them are protein in nature (enzyme inhibitors; lectins), some of them are relatively low molecular weight constituents (tannins; oligosaccharides; saponins). The mechanism of their physiological action - their *modus operanti* - is extremely diverse. They are best classified on the basis of the type of nutrients they affect, either directly or indirectly, and the biological response produced in the animal. One rather general classification is as follows (Chubb, 1982):

- Factors depressing digestion or depressing metabolic utilisation of proteins (protease inhibitors, lectins, saponins and polyphenolic compounds).
- Factors depressing the digestion of carbohydrates (amylase inhibitors, polyphenolic compounds, flatulence factors).
- Factors inactivating vitamins or increasing the requirements of certain vitamins (anti vitamins).
- Factors that stimulate the immune system (antigenic proteins).

Huisman (1990a) has defined ANFs as non-fibrous natural substances causing negative effects on growth or health in man and animals. ANFs have a natural function in plants since they contribute to crop's resistance to attacks of predators. An example of the plant defence function of ANFs is the presence of tannins in sorghum. Jansman (1993) described high-tannins sorghum varieties that were less attractive for birds caused by an astringent taste of tannins and were less susceptible to fungi and pests in comparison with low-tannin sorghum varieties.

In 2004 during the 4th International Workshop on Antinutritional Factors in Legume Seeds and Oilseeds (Muzquiz *et al.*, 2004), researchers submitted evidence of the beneficial effects of low dietary levels of some ANFs. Lajolo and Genovese (2002), in their review about the antiproliferative effects of lectins, trypsin inhibitors, phytic acid and saponins, reported tumour weight reduction when PHA (phytohaemagglutinin i.e. a type of lectin present in kidney beans) were included in the diet of mice (up to 7 mg/g diet). The authors explained these results by the competition induced by lectins between hyperplasic growth of the small intestine and the tumour growth. Consequently, a new concept emerged named nutritional active factors (NAFs contrary to ANFs) the same naturally occurring constituents that may have positive effects on e.g. health. In this article we still use the term antinutritional factors in pig nutrition for a discussion on their impact on feed intake.

Feed intake is – among others – regulated by physiological mechanisms through feedbacks. This information directly affects the level of feeding motivation (Day *et al.*, 1998). Palatable feed (the characteristic of an ingredient easy to eat for the animal) will elevate feed intake whereas a toxic ingredient may inhibit food ingestion. Moreover, feed intake is influenced by the learning abilities of the animal. If the animal experienced post-ingestive effects after first ingredient's consumption he will link his digestive trouble to the feed and will decrease feed intake (Meunier-Salaün and Picard, 1996; Day *et al.*, 1998; Laitat *et al.*, 2004).

The impact of pure ANFs on feed intake have been studied in small animals such as rats but only very few experiences on pigs have been reported. Huisman (1990b) found that pigs are more sensitive to ANFs than rats, so it is difficult to predict ANF activities on pigs using rats as a model. In addition, the costs of this type of investigation are high in pigs due to large amounts of pure ANF to be isolated. Therefore most of studies on pigs have reported on the effects of ANFs in a matrix (i.e. diet composed by several ingredients) where one of the ingredients is known to contain a high level of a certain ANF. It is known also, that specific diet ingredients do have one or more specifically related ANF for which these foodstuffs are recognised. The simultaneous presence of ANFs in one diet ingredient imposes a further reason that it is difficult to analyse a certain antinutrient as the causative factor to affect feed intake.

The analysis of the influence of ANF on feed intake, therefore, is rather complex and certain decisions had to be taken with respect to the manner in which the material should be presented in this paper. A first classification was on a number of major antinutritional factors (protease inhibitors, lectins, alkaloids, glucosinolates, etc.), whereas within these factors we structured the effects on voluntary feed intake as found in literature reported values and according to the relevant pig diet ingredient.

Protease inhibitors

Characteristics of protease inhibitors, their mechanism and general effects on pigs

Hundreds of protease inhibitors (PIs) have been characterised in plants. Important nutritional plants contain a high level of protease inhibitors e.g. peas, common beans and soybean. Protease inhibitors are proteins able to inhibit other proteins i.e. enzymes, by forming stabilised complexes. Thereby they reduce the growth and the utilisation of nutrients, enhance the production of endogenous proteins and cause hypertrophy of the pancreas especially in rats and chicken; the latter effect is smaller in pigs (Huisman, 1990a). The main inhibitors in cereals and legume seeds are trypsin and chymotrypsin inhibitors. Two families of trypsin inhibitors (TIs) have been identified: the Kunitz trypsin inhibitor, proteins with a molecular weight of about 20,000 Da and a strong specificity for trypsin, and the Bowman-Birk trypsin inhibitors with a molecular weight between 6,000 and 12,000 Da able to inhibit chymotrypsin as well as trypsin at independent binding sites (Palacios *et al.*, 2004). Some products (common bean, pea, linseed) contain more of the Bowman-Birk (BB) inhibitor while soybeans contain about the same amounts of Kunitz and BB inhibitors. Further information can be derived from the review of Lajolo and Genovese (2002).

Effects of trypsin inhibitor on piglets' feed intake

In Table 1 the effect on feed intake of 30% pea inclusion in a diet fed to pigs from 28 up to 42 kg has been given (Huisman, 1990b).

In this experiment, no significant differences in feed intake were observed between a group of piglets fed a control diet and a group of piglets fed peas at 30% inclusion level. Weight gain, however, was greatly impaired: it was reduced by 21% when feeding level was 2.2 times maintenance and 7% when feeding level was 3.2 times maintenance. Compared to results from experiments with rats and with chicken, Huisman (1990b) concluded that piglets are more sensitive to dietary antinutritional factors causing an impaired performance compared to rats and chicken; the latter animal species are not considered adequate models to forecast the pigs' response to ANFs. Data referring to the inclusion of full fat soybeans and their effect on feed intake are shown in Tables 2 and 3.

White *et al.* (1989) performed a 35-day feeding trial using piglets with an average initial body weight (BW) of 5 kg (Table 2). Four dietary treatments were used in this study: diets with

Table 1. Effects of diet pea inclusion on piglets' feed intake (Huisman, 1990b).

Source of protease inhibitors	Inclusion level (g/kg)	TIA (g/kg pea)	Diet	Pigs' age (days)	Feed intake (g/day)	TI intake (g/day)
Pea				28 up to 42		
			feeding level 2.2 maintenance			
	0	-	control		202.9[a]	-
	300	-	test diet		198.2[a]	-
			feeding level 3.2 maintenance			
	0	-	control		274.5[a]	-
	300	-	test diet		282.3[a]	-

[a] Mean values within a column with different superscripts are significantly different (P<0.05).

Table 2. Effects of soybean thermal processing on piglets' feed intake (White et al., *1989).*

Source of protease inhibitors	Inclusion level (g/kg)	TIA (g/kg soybean)	Diet	Pigs' weight (kg)	Feed intake (g/day)	TI intake (g/day)
Soybeans	399			5 kg + 35 days		
		40.0	soybeans at 10.3% moisture and roasted at 110 °C		320[a]	5.11
		24.0	soybeans at 20.5% moisture and roasted at 110 °C		410[a]	3.93
		22.9	soybeans at 10.3% moisture and roasted at 125 °C		390[a]	3.56
		9.2	soybeans at 20.5% moisture and roasted at 125 °C		550[b]	2.01

[a, b] Mean values within a column with different superscripts are significantly different (P<0.05).

soybeans processed at two temperatures or moisture levels were fed. Feed intake was observed to be significantly higher when soybean moisture content was increased prior to the heat treatment. The factorial analysis showed that increased temperature from 110 °C to 125 °C or increased moisture content from 10.3% to 20.5% significantly improved daily feed intake at the highest temperature when a concomitant decrease was observed in the level of trypsin inhibiting activity per kg soybean. (Table 2). Moreover, an increased temperature improved feed conversion ratio whereas increased moisture content did not.

Zarkadas and Wiseman (2005) performed two trials in which they fed heat processed soybeans to young piglets for 27 days (Table 3). In trial 1 full fat soybean (FFSB) were micronised with low steam (LS) or full steam (FS) at high or low cooking (HC vs. LC) conditions. Piglets fed the low steam treated FFSB had similar daily feed intakes. The higher temperatures associated with full steam showed a higher but non-significant voluntary feed intake. A factorial analysis, done by the authors, shows that the differences between LS and FS are significant for all measures ($P<0.001$). However, temperature associated with cooking time (HC vs. LC) did not have significant effects on the voluntary feed intake.

In trial 2 extruded FFSB with varying extrusion exit product temperatures ranging from 90-150 °C were analysed for their TIA content and fed to piglets. The analysis of variance revealed that there is a non-linear effect of treatment on voluntary feed intake indicating that there was an initial reduction in feed intake followed by a subsequent increase as temperature increased (Zarkadas and Wiseman, 2005).

Table 4 is an overview of the experiment performed by Valencia *et al.* (2008). They used four different dietary treatments in which pea or soybean protein products were examined in pigs between 26 and 48 days of age.

Table 3. Effects of processing full fat soybean (FFSB) on piglets' feed intake (Zarkadas and Wiseman, 2005).

Source of protease inhibitors	Inclusion level (g/kg)	TIA (g/kg soybeans)	Diet	Pigs' weight (kg)	Feed intake (g/day)	TI intake (g/day)
Soybeans	300			10 up to 27		
		2.0	full steam high cook FFSB		812	0.49
		4.1	full steam low cook FFSB		771	0.95
		5.2	low steam high cook FFSB		678	1.06
		16.0	low steam low cook FFSB		680	3.26
	300	24.1	extruded FFSB (70 °C)		776	5.61
		16.8	extruded FFSB (90 °C)		642	3.24
		6.4	extruded FFSB (110 °C)		639	1.23
		2.9	extruded FFSB (150 °C)		844	0.73

Table 4. Effects of pea protein and soybean protein sources on piglets' feed intake (Valencia et al.*, 2008).*

Source of protease inhibitors	Inclusion level (g/kg)	TIA (g/kg product)	Diet	Pigs' age (days)	Feed intake (g/day)	TI intake (g/day)
Pea protein concentrate	105	4.9	basal + PPC	26 up to 36	207	0.11
Soya protein concentrate	100.8	1.6	basal + SPC		234	0.04
Soybean meal	121.7	2.7	basal + SBM		232	0.08
Full fat soya bean	157.6	4.7	basal + FFSB		230	0.17
Pea protein concentrate	105	4.9	basal + PPC	37 up to 48	580[ab]	0.30
Soya protein concentrate	100.8	1.6	basal + SPC		613[a]	0.10
Soybean meal	121.7	2.7	basal + SBM		614[a]	0.20
Full fat soya bean	157.6	4.7	basal + FFSB		551[b]	0.41
Pea protein concentrate	105	4.9	basal + PPC	26 up to 48	411	0.21
Soya protein concentrate	100.8	1.6	basal + SPC		441	0.07
Soybean meal	121.7	2.7	basal + SBM		441	0.14
Full fat soya bean	157.6	4.7	basal + FFSB		405	0.30

[a, b] Mean values within a column with different superscripts are significantly different ($P<0.05$).

The trypsin inhibitor content was highest for raw pea protein concentrate (PPC) (4.9 g/kg) and full fat soybean (FFSB) (4.7 g/kg) and lowest for soybean protein concentrate (SPC) (1.6 g/kg) with soybean meal (SBM) being intermediate (2.7 g/kg). Results show that from 26 up to 36 days of age, the feed intake of pigs was similar for each diet. At the age ranging between 37 to 48 days, piglets fed the FFSB diet showed the lowest feed intake ($P<0.05$). At this age, pigs fed SPC and SBM diets ate significantly more ($P<0.05$) than pigs fed the FFSB diet. For the overall experiment pigs fed SPC and SBM tended to have higher average daily feed intake ($P<0.10$). Valencia *et al.* (2008) noticed a strong bitter aftertaste of the PPC indicating the presence of saponins. Hence the bitter taste of the PPC probably caused an initial rejection of the feed but normal intakes were observed within a few days after feeding. At a higher age (37 to 48 days) and over the total period under investigation, the FFSB diet showed the lowest voluntary feed intake. Pigs fed this diet had the highest trypsin inhibitor intake.

Effects of trypsin inhibitor on growing finishing pigs' feed intake

In Table 5, the results of the study of Mathé *et al.* (2003) are given on the impacts of pea inclusion of pigs' feed intake.

In a cross-over design, during eleven weeks, Mathé *et al.* (2003) studied the impact of pea inclusion rate (0, 13, 26, and 39%, respectively) on the eating behaviour and growth of fattening pigs. Four transitions were studied: T0/T13, T13/T26, T26/T39 and T39/0. From their study they concluded that feed intake was significantly different ($P<0.001$) between the different pea inclusion levels. The inclusion of pea at a level of 26% revealed the highest voluntary intake whereas 13% inclusion level the lowest. With this investigation Mathé *et al.* (2003) showed that the presence of pea in the

Table 5. Effects of pea inclusion level on feed intake of fattening pigs during dietary transitions (Mathé et al.*, 2003).*

Source of protease inhibitors	Inclusion level (g/kg)	TIA (g/kg pea)	Diet	Pigs' weight (kg)	Feed intake (g/day)	TI intake (g/day)
Pea				33 up to 114		
	0	-	Control		2,510[b]	-
	130	-	Experimental		2,461[a]	-
	260	-	Experimental		2,636[d]	-
	390	-	Experimental		2,563[c]	-

[a, b, c, d] Mean values within a column with different superscripts are significantly different ($P<0.001$).

feed has not only an effect on feed intake but also on eating behaviour i.e. the overall eating time is increased but the number of visits per day at the feeder is reduced. Pigs fed a pea diet eat more feed at each visit at the feeder in comparison with pigs in the control group.

Cook *et al.* (1988) compared the consumption and utilisation by piglets of diets containing supplemental protein sources being either heated, solvent-extracted soybean meal (SBM) or raw low-Kunitz trypsin inhibitor soybean (LT) or raw high-Kunitz trypsin inhibitor full fat soybean (HT) varieties. The intake of trypsin inhibitors was calculated from the data given by the authors assuming a soybean dry matter content of 90%. As shown in Table 6, feed intake was not significantly different ($P>0.05$) among dietary treatments. On the other hand daily growth was shown to be highly distinct ($P<0.01$) between treatments and represented 436, 286, 211 g/day for piglets fed the SBM, LT and HT diet, respectively. This represents a loss of daily gain of 34% when LT diet is fed and 52% when HT diet is fed. Feed intake is reduced by 14% between the SBM diet and the LT diet whereas feed intake is enhanced when the HT diet is fed. So, the adverse effects of Kunitz trypsin inhibitor could not be related to differences in feed intake between diets. However,

Table 6. Feed intake of piglets fed diets with raw soybeans either low or high in Kunitz trypsin inhibitor or with heated and processed soybean meal (Cook et al.*, 1988).*

Source of protease inhibitors	Inclusion level (g/kg)	TIA (mg/kg SB[1])	Diet	Pigs' weight (kg)	Feed intake (g/day)	TI intake (mg/day)
Soybean products				7 kg + 28 days		
	260	0.0	soybean meal		864	0.0
	320	0.1	raw soybeans LT[2]		745	0.0
	320	5.0	raw soybeans HT[2]		895	1.3

[1] Fat-free, dry matter basis.

[2] LT = soybeans with low Kunitz TI level; HT = soybeans with high Kunitz TI level.

other growth inhibitors may still be present in the LT soybeans. Kunitz trypsin inhibitors proved to have no negative effects on feed intake but reduced growth by a far lower protein utilisation. The effect of the Kunitz trypsin inhibitor has been examined in finishing pigs as shown in Table 7.

In similar experimental settings as explained previously (Table 6), Cook *et al.* (1988) examined the effect of soybean products in studies with finishing pigs (Table 7). Once more, feed intake was shown not to be significantly affected by the dietary treatments. Moreover it was observed that the growing pigs' intake of the LT diet was reduced by 9% compared to the SBM diet whereas the HT diet was reduced by 6%. The effects of both trypsin and chymotrypsin inhibitor have been examined by Batterham *et al.* (1993). Results are summarised in Table 8. Since it is costly to add

Table 7. Feed intake of finishing pigs fed diets with raw soybeans either low or high in Kunitz trypsin inhibitor or with heated and processed soybean meal (Cook et al.*, 1988).*

Source of protease inhibitors	Inclusion level (g/kg)	TIA (mg/kg SB[1])	Diet	Pigs' weight (kg)	Feed intake (g/day)	TI intake (mg/day)
Soybean products				67 up to 105		
	140	0.0	Soybean meal		3270	0.0
	180	0.1	Raw soybeans LT[2]		2970	0.0
	180	5.0	Raw soybeans HT[2]		3070	2.5

[1] Fat-free, dry matter basis.
[2] LT = soybeans with low Kunitz trypsin inhibitor level; HT = soybeans with high Kunitz trypsin inhibitor level.

Table 8. Feed intake of pigs given diets varying in trypsin inhibitor (TI) and chymotrypsin inhibitor (CTI) levels (Batterham et al.*, 1993).*

Source of protease inhibitors	Inclusion level (g/kg)	Diet	Pigs' weight (kg)	Feed intake (g/day)	TI intake (g/day)	CTI intake (g/day)
Chick pea			20 up to 50			
		control		1,697	0.3	0.3
	250	chick pea 1		1,764	2.1	1.6
	500	chick pea 1		1,720	3.8	2.6
	750	chick pea 1		1,744	5.6	3.8
	250	chick pea 2		1,755	3.0	2.8
	500	chick pea 2		1,737	5.6	5.4
	750	chick pea 2		1,755	8.2	7.9
Pigeon pea	250	pigeon pea		1,846	2.6	1.5
	500	pigeon pea		1,533	3.8	2.3
	750	pigeon pea		1,579	5.7	3.3

pure inhibitors in diets for pigs 'one alternative approach is to add graded levels of a feed source rich in the protease inhibitors but free of other antinutritional factors and determine threshold levels'(Batterham *et al.*, 1993). In their study the authors examined the addition of chick pea and pigeon pea to the pigs' diet.

Chick pea 1 was a low fibre 'Opal' cultivar while chick pea 2 was a high fibre 'Tyson' cultivar. The pigeon pea used in this experiment was the 'Hunt' cultivar. The authors selected chick peas and pigeon peas with high levels of protease inhibitors but free of other antinutritional factors. Both types of peas were fed as meals and finally ground through a hammer mill. Pigs from the control group were fed soybean meal instead of peas.

From Table 8 we can deduce that feed intake of growing pigs is not significantly (P>0.05) impaired by the inclusion of chick pea or pigeon pea in the diet. However, Batterham *et al.* (1993) found a linear depressing effect of the addition of pigeon pea meal on feed intake (P<0.05). The authors suppose other antinutritional factors in pigeon pea meal do depress pigs' performances.

Lectins

Characteristics of lectins, their mechanism and general effects on pigs

Many plant species contain carbohydrate-binding proteins called lectins or agglutinins. These biologically active substances are resistant to gut proteolysis (Pusztai *et al.*, 1990). Lectins are of special interest in pig nutrition since they are present in important dietary ingredients such as common beans (*Phaseolus vulgaris* beans), peas (*Pisum sativum*) and soybeans (*Glycine max*). Lectins' discovery goes back to 1888 with Stillmark, who discovered a haemagglutinating factor in castor bean extracts. Nowadays, several hundreds of lectins have been identified and form a group of proteins characterised by a noncatalytic domain that binds reversibly to a specific carbohydrate (Peumans and Van Damme, 1995). Thus lectins bind to specific mono- or oligosaccharides of glycoproteins, glycolipids or polysaccharides and interfere with the normal functioning of the organism. These lectins are secondary plant metabolites and assure the plant's defence against predators such as bacteria, fungi, insects and mammalians. Several arguments support this plant defence theory. Firstly, lectins have a high affinity for oligosaccharides, which are not common or totally absent in plants (e.g. sialic acid-binding lectins from elderberry bind to a sugar that is absent in plants but a major carbohydrate of animal glycoproteins). Most lectins are resistant to proteolytic breakdown thus they remain active in the intestine and are able to bind to surface receptors of the intestinal epithelium (Huisman, 1990a). Consequently, physiological processes such as digestion and absorption are disturbed. Lectins are mainly found in storage organs and seeds, particularly vulnerable to parasites and predators attacks (e.g. soybean agglutinin and pea lectin represent up to 8% of total protein in the cotyledons) (Chrispeels and Raikhel, 1991; Peumans and Van Damme, 1995).

Lectins vary considerably in molecular weight and chemical structure and are characterised by the ability to bind to specific sugars. The interaction of lectins with the glycosylated receptors of the gut epithelial cells results in gut damage and consequently, body metabolism, immune and endocrine systems. The gut wall damage may e.g. disturb the absorption of nutrients and may affect the activity of brush border membrane enzymes. It should be emphasised, however, that the main negative effects vary with the origin of lectins: especially the PHA lectins from some species of the common bean (*Phaseolus vulgaris* L.) have been demonstrated to be very toxic (Sgarbieri and Whitaker, 1982).

Effects of lectins on piglets' feed intake

In the Tables 9 and 10 the effects of *Phaseolus* bean inclusion in pig diets on feed intake are shown.

In his experiment, Huisman (1990b) tested the voluntary feed consumption of beans by piglets and by rats. As seen in Table 9 feed intake was dramatically depressed when unheated *Phaseolus*

*Table 9. Effect of beans (*Phaseolus vulgaris *L.) on piglets' feed intake (Huisman, 1990b).*

Source of lectin	Inclusion level (g/kg)	Lectins (g/kg beans)	Diet	Pigs' age (days)	Feed intake (g/day)	Lectin intake (g/day)
Phaseolus vulgaris beans				28 up to 49		
	0	0.0	control diet		240[a]	0.0
	200	30.0	unheated beans		97[b]	0.6
	200	1.9	heated beans		232[a]	0.1
				56 up to 61		
	0	0.0	control diet		320[a]	0.0
	200	30.0	unheated beans		174[b]	1.0
	200	1.9	heated beans		299[a]	0.1

[a, b] Mean values within a column with different superscripts are significantly different ($P<0.05$).

*Table 10. Effects of variable protein contents in diets containing beans (*Phaseolus vulgaris *L.) on feed intake (Huisman, 1990b).*

Source of lectin	Inclusion level (g/kg)	Lectins (g/kg beans)	Diet	Pigs' age (days)	Feed intake (g/day)	Lectin intake (g/day)
Phaseolus vulgaris beans				28 up to 49		
	0	0.0	control diet		234[a]	0.0
	200	40.0	untoasted beans + casein		146[b]	1.2
	200	40.0	untoasted beans		135[b]	1.1
	200	0.8	toasted beans		225[a]	0.0

[a, b] Mean values within a column with different superscripts are significantly different ($P<0.05$).

beans were fed. The observed reduction in feed intake reaches some 60% when raw beans with a high level of lectins were given to piglets between 28 and 49 days old. The observed reduction in feed intake in rats was only 9%. Piglets seem to be more sensitive to the bean antinutritional factors compared to rats. Moreover, growth was negative in piglets fed raw beans and was reduced by 19% compared to the pigs fed toasted beans (Huisman, 1990b). For pigs between 56 and 61 days, similar effects were found (Table 9).

From the overall study it could be deduced that a 'toxic' factor in the bean played a role in addition to the effect of the reduced feed intake. This toxic factor might be lectins because a reduced feed intake was observed in rats when purified lectins from the *Phaseolus* bean were included in their diet (Pusztai *et al.*, 1981).

Inclusion of 20% raw beans caused a markedly reduced feed intake and weight gain in piglets of 28 up to 49 days of age (Table 10). These results are in adequacy with those from Table 9. Huisman (1990b) tested the following hypothesis: the negative effects of inclusion of raw beans on live weight gain is due to an insufficient protein and amino acid supply. From the results of the investigation above (Table 10) it can be concluded that another factor must be responsible for the negative effects on feed intake and protein deposition than an insufficient amino acid supply, since casein added to the untoasted beans, showed no improvement in feed intake compared to the diet with only untreated beans. Since heating the beans showed a feed intake comparable to the level of the control diet, the causative factor for reducing feed intake is a heat labile one.

In Table 11 the effects of PHA lectins are shown in connection to *Phaseolus* bean inclusion level in the diet and feed intake.

Kluess *et al.* (2007) observed a significant negative impact on feed intake as well as growth performances of growing pigs (33 up to 48 days of age) by the inclusion of PHA bean lectins and high levels of fermentable carbohydrates (CHO) in the diet. Compared to a control diet without lectins feed intake was reduced severely which was even more depressed with the inclusion of high levels of rapidly or slowly fermentable carbohydrates in the last week of growth.

Effects of lectins on growing finishing pigs' feed intake

In Table 12 the effects of bean lectins in finisher pigs are demonstrated. In this experiment, the response of pigs of different ages to a 20 percent raw bean inclusion level in the diet was investigated.

Results of Table 12 show that feed intake was reduced in pigs at all three ages when raw beans were included in the diet. It was reported, that in all three cases pigs lost weight. So, Huisman (1990b) concludes that there is no age dependency in pigs up to 4 months of age with respect to the effects of the inclusion of raw beans on performances.

In Table 13, the feed intake of pigs fed isogenic lines of soybeans and the effect of extrusion processing on feed consumption is shown. Unfortunately, levels of lectins were not indicated in the work of Palacios *et al.* (2004).

Table 11. Effects of PHA-lectins and diet composition on feed intake in piglets (Kluess et al., *2007).*

Source of lectin	Inclusion level (g/kg)	PHA (g/ kg beans)	Diet	Pigs' weight (days)	Feed intake (g/day)	PHA intake (g/day)
Phaseolus vulgaries beans				33 up to 40		
	0	0.0	low fermentable CHO without PHA		329[a]	0.00
	50	75.0	low fermentable CHO + PHA		251[b]	0.94
	50	75.0	high rapidly fermentable CHO + PHA		253[b]	0.95
	50	75.0	high slowly fermentable CHO + PHA		235[b]	0.88
Phaseolus vulgaries beans				41 up to 48		
	0	0.0	low fermentable CHO without PHA		581[a]	0.00
	50	75.0	low fermentable CHO + PHA		393[b]	1.47
	50	75.0	high rapidly fermentable CHO + PHA		323[c]	1.21
	50	75.0	high slowly fermentable CHO + PHA		331[c]	1.24

[a, b, c] Mean values within a column with different superscripts are significantly different (*P*<0.001).

*Table 12. Effect of beans (*Phaseolus vulgaris *L.) on pigs' feed intake and age dependency (Huisman, 1990b).*

Source of lectin	Inclusion level (g/kg)	Lectins (g/kg beans)	Diet	Pigs' age (days)	Feed intake (g/day)	Lectin intake (g/day)
Phaseolus vulgaris beans				56		
	0	0.0	control diet		836[a]	0.0
	200	42.0	raw beans		325[b]	2.7
				84		
	0	0.0	control diet		1,437[a]	0.0
	200	42.0	raw beans		621[b]	5.2
				112		
	0	0.0	control diet		2,074[a]	0.0
	200	42.0	raw beans		907[b]	7.6

[a, b] Mean values within a column with different superscripts are significantly different (*P*<0.05).

Table 13. Effect of soybean variety and processing on piglets' feed intake (Palacios et al.*, 2004).*

Source of lectins	Inclusion level (g/kg)	Lectins (g/kg)	Diet	Pigs' age (days)	Feed intake (g/day)	Lectin intake (g/day)
Soybeans				41 up to 69		
	274	-	soybean meal		833[a]	-
	354	-	conventional soybeans, raw		513[b]	-
	354	-	conventional soybeans, extruded		810[a]	-
	379	-	lectin-free soybeans, raw		596[c]	-
	379	-	lectin-free soybeans, extruded		757[a]	-
	391	-	lectin & Kunitz trypsin inhibitor-free soybeans, raw		708[d]	-
	391	-	lectin & Kunitz trypsin inhibitor-free soybeans, extruded		776[a]	-

[a, b, c, d] Mean values within a column with different superscripts are significantly different ($P<0.01$).

Soybeans – conventional or a special variety – were included at 35-40 percent inclusion levels in diets for pigs of 41 up to 69 days of age. In two treatments the soybeans were thermally processed by extrusion. From this study it is obvious, that both the absence of lectins or the absence of Kunitz trypsin inhibitor in raw soybeans is a guarantee for a higher feed intake compared to that of conventional soybean meal. Only extrusion reveals a feed intake by piglets comparable to soybean meal indicating that other heat-labile factors are present in (modified) raw soybeans that diminish feed intake and growth performance of pigs.

Tannins

Characteristics of tannins, their mechanism and general effects on pigs

The tannins or tannic acid are water-soluble polyphenolic compounds. They have a wide range of molecular weight and variable chemical complexicity. Tannins are capable of precipitating alkaloids, gelatin and other proteins from aqueous solutions (Marquardt, 1989). Although tannins are chemically not well defined, they are usually divided into two groups. Hydrolysable tannins are tannins that have a central carbohydrate core (hydroxyl groups) which are esterified to various phenolic carboxylic acids. This group of tannins is easily hydrolysed to give e.g. glucose and phenolic acids. Condensed tannins are tannins that are mainly polymerised products of catechin and flavan 3,4-diol or a mixture. Bioavailability, absorption or metabolism of dietary tannins are primarily determined by its physico-chemical properties such as structure, molecular size, solubility and potential to conjugate with other molecules. Important plants in animal nutrition contain high rates of tannins, e.g. sorghum, barley, rapeseed or legume seeds like common beans and faba beans.

In general, the dietary effects of tannins are not exactly known but it has been recognised that primarily the negative effects of condensed tannins are caused by the complexation with dietary proteins, and sometimes with digestive enzymes (Jansman, 1993; Chung *et al.*, 1998).

As a result, nutrient digestibility is decreased. Some other effects that have been attributed to tannins are reduced feed intake, increased damage to the gut wall, toxicity of absorbed tannins and reduced absorption of some minerals. Animals (and humans) that consume high tannin diets may develop a physiological means to counteract the adverse effects of tannins. This occurs through the production of a family of proline-rich proteins in the salivary gland (Jansman *et al.*, 1994). These proteins bind tannins in a highly specific manner, thereby reducing their effects. However, pigs for example are unable to completely eliminate these negative effects of dietary tannins in this way.

Effects of tannins on piglets' feed intake

In Table 14 the results of Lizardo *et al.* (1995) are presented. They studied the impact of pig diets containing different levels of tannins originating from sorghum on feed intake and performance.

No significant differences in pig feed intake were detected due to the different dietary treatments in two age classes. It was observed that castrated males eat significantly more than females ($P<0.05$) during the second phase of the experiment (42 up to 63 days). This observation might be explained by hunger and by differences in the muscular tissue development between both sexes. From the presented data, no obvious relation between tannin level (g/kg catechine) or tannin intake (g/day) with voluntary feed intake can be observed.

Table 14. Effects of sorghum tannins on feed intake in growing pigs (Lizardo et al., *1995).*

Source of tannins	Inclusion level (g/kg)	Tannins (g/kg catechine)	Diet	Pigs' age (days)	Feed intake (g/day)	Tannins intake (g/day)
Sorghum				21 up to 42		
	560	0.0	maize		302	0.00
	560	0.9	sorghum Bianco		320	0.16
	560	2.4	sorghum Rosso I		315	0.42
				42 up to 63		
	650	0.0	maize		923	0.00
	650	1.0	sorghum Bianco		929	0.60
	650	2.8	sorghum Rosso I		937	1.71
				21 up to 63		
	-	-	maize		665	-
	-	-	sorghum Bianco		655	-
	-	-	sorghum Rosso I		682	-

In Tables 15 and 16 the effect of pea tannins and pea variety are shown for growing and for finishing pigs from studies of Grosjean *et al.* (1991).

When tested on piglets, peas with different levels of tannins (up to 3.4 g/kg pea) were observed to be consumed similarly. No tannins at all could be detected in the French pea variety, while Australian peas caused pigs to have an intake of about 0.90 g tannins per day.

Table 15. Effects of pea variety on feed intake of growing pigs (Grosjean et al., *1991).*

Source of tannins	Inclusion level (g/kg)	Tannins (g/kg peas)	Diet	Pigs' age (days)	Feed intake (g/day)	Tannins intake (g/day)
Pea				28 up to 63		
	0	0.0	control		900	0.00
	300	0.0	French pea		900	0.00
	300	3.4	Australian pea 1988		910	0.93
	300	3.1	Australian pea 1989		930	0.87

Table 16. Effects of pea variety on feed intake of growing and of finishing pigs (Grosjean et al., *1991).*

Source of tannins	Inclusion level (g/kg)	Tannins (g/kg peas)	Diet	Pigs' weight (kg)	Feed intake (g/day)	Tannins intake (g/day)
Pea				25 up to 60		
	0	0.0	control		1,630^{b}	0.00
	300	0.0	pea variety 'Solara'		1,640^{b}	0.00
	300	3.1	pea variety 'Gali'		1,650^{a}	1.55
	300	3.4	Australian pea 1988		1,630^{b}	1.66
				60 up to 103		
	0	0.0	control		2,320^{b}	0.00
	300	0.0	pea variety 'Solara'		2,320^{b}	0.00
	300	3.1	pea variety 'Gali'		2,350^{a}	2.21
	300	3.4	Australian pea 1988		2,300^{b}	2.34
				25 up to 103		
	0	0.0	control		2,010^{ab}	0.00
	300	0.0	pea variety 'Solara'		2,010^{ab}	0.00
	300	3.1	pea variety 'Gali'		2,030^{b}	1.91
	300	3.4	Australian pea 1988		2,000^{a}	2.04

$^{a, b}$ Mean values within a column with different superscripts are significantly different (P<0.01).

Effects of tannins on growing finishing pigs' feed intake

In peas, *var.* Solara, no tannins could be detected whereas in the *var.* Gali, 3.1 g tannins per kg of pea were analysed. When tested in studies with pigs, peas with different levels of tannins (up to 3.4 g/kg) were shown to be consumed similarly but diets with the Gali pea variety, having a relatively high tannin content, had the highest feed intake. Australian peas with the highest tannin level were intermediate.

Glucosinolates

Characteristics of glucosinolates, their mechanism and general effects on pigs

Full fat seeds and seed meals of the genus *Brassica* are the main source of glucosinolates (GSLs) in animal diets. Also in crambe meal (*Crambe abyssinica*) GSLs are present (Liu, 1994). More than 120 different GSLs have been identified and all share a common structure namely β-D-thioglucose, a sulphonated oxime moiety and a variable side-chain derived from methionine, tryptophan or phenylalanine (Tripathi and Mishra, 2007).

Rapeseed and rapeseed meal (RSM) contain three major glucosinolates: progoitrin or epi-progoitrin, gluconapin and glucobrassicanapin. The content of these glucosinolates varies according to climate (e.g. in India RSM contains more gluconapin than in Europe), to harvest time (e.g. autumn harvested foliage contains higher glucosinolates than winter harvest), to plant varieties. In the 80's selection made it possible to decrease average (erucic acid levels and) glucosinolate levels in RSM; GSL levels were decreased from 166 μmol/g to 38 and less μmol/g dry oil-free meal (Tripathi and Mishra, 2007). Glucosinolates are biologically inactive molecules but their degradation products are biologically active. Glucosinolates are hydrolysed by the enzyme myrosinase present in cruciferous plants and in some bacteria from the gastro-intestinal tract. Hydrolysis products are then formed such as isothiocyanates which are responsible for bitterness. Thiocyanate ions inhibit iodine uptake by the thyroid gland for the production of triodothyronine (T3) and thyroxine (T4) leading to lower plasma levels of these compounds. Nitriles may cause liver and kidney lesions. Some glucosinolates can lower palatability since they have a bitter aftertaste (e.g. sinigrin) however some glucosinolates are not bitter but give bitter compounds after hydrolysis or heat treatment (e.g. progoitrin). In general, glucosinolates and derived products do negatively affect thyroid and growth in pigs (Schöne *et al.*, 1997).

Effects of glucosinolates on piglets' feed intake

Liu (1994) indicates that piglets of 20 kg refuse raw crambe meal (RCM) from 3% dietary inclusion level corresponding to 5 μmol epi-progoitrin per gram diet. When the RCM is withdrawn and the animals are fed with the control diet, feed intake and consequently daily live weight gain are recovered. In the contrast, animals receiving treated CM consumed as much feed as those of the control diet. Consequently RCM is unpalatable to piglets even at minimum dietary inclusion level due to its high level of GSLs but this can be overcome by toasting.

Table 17 shows that feed intake is greatly impaired when 15.5% untreated rapeseed meal (RSM) is included to piglet's diet. The pigs' feed intake is reduced by half compared to the control diet. Levels of OZT+ICT intakes have been calculated assuming a dry mater content in RSM of 90%.

Roura *et al.* (2007) performed a palatability test to compare the preferences of piglets for soybean meal (SBM), low glucosinolates rapeseed meal (RSM) or sunflower meal (SFM) following the animal model described by Solà-Oriol *et al.* (2005). Piglets of 20 kg live weight were used in this study. The results are given in Table 18.

In this study a reference diet (diet not shown) was used with 20% inclusion of a soybean protein concentrate (SBM-56), rice (60%), wheat bran (13%), sunflower oil (3%) and other ingedients (4%). Three test diets have been formulated identical to the reference diet but with partial (10%; trial 1) or complete (20%; trial 2) substitution of the SBM-56 by either SBM-48, sunflower meal or rapeseed meal. When rapeseed meal is included at a level of 10%, preference value is 30% which was reported significantly lower than 50% (P<0.01). When SBM-56 was completely

Table 17. Effects of rapeseed meal processing or supplementation on feed intake of growing pigs (Liu, 1994).

Source of glucosinolates	Inclusion level (g/kg)	OZT+ITC (g/ kg RSM DM)	Diet	Piglets' weight (kg)	Feed intake (g/day)	OZT+ITC intake (g/day)
Rapeseed meal				16 up to 55		
	0	0.00	control		1,510[b]	0.00
	32	0.20	washed RSM		1,480[b]	0.01
	63	0.20	washed RSM		1,470[b]	0.02
	63	0.20	washed RSM + Lysine		1,600[a]	0.02
	155	0.20	washed RSM		1,230[c]	0.03
	155	5.70	untreated RSM		800[d]	0.64

[a, b, c, d] Mean values within a column with different superscripts are significantly different (P<0.05).

Table 18. Effects of preference values in double choice feed tests in piglets as effected by different protein sources (Roura et al., *2007).*

Protein source	Trial 1 (10% substitution)		Trial 2 (20% substitution)	
	Preference value	Significance	Preference value	Significance
Soybean meal (48% CP)	62	NS	54	NS
Sunflower meal	40	NS	10	P<0.01
Rapeseed meal	30	P<0.01	11	P<0.01

replaced by rapeseed meal in the diet, this resulted in a significant lower preference value of 11%. Consequently, the authors conclude that even low-glucosinolate rapeseed meal still results in a decreased feed intake because of its poor palatability. In addition, a good correlation between preference values and voluntary feed intake was observed.

No data on the effects of glucosinolates on growing finishing pigs were encountered.

Effects of glucosinolates on sows' feed intake

Table 19 refers to the study performed by Etienne *et al.* (1993) on sows in order to determine the effects of low-glucosinolates RSM.

The effects of glucosinolates were studied using sows in the weight range of 30-60 kg, 60-100 kg and 100-130 kg. For all weight classes, no differences in feed intake were observed. Moreover no effects on sows' reproductive functions and on piglets' health were found. The authors conclude that, together with previous experiments, 2 μM GSL per gram of feed can be considered as the maximal level of glucosinolates that can be fed daily to pregnant sows without detrimental consequences.

Table 19. Effects of rapeseed meal inclusion and sow weight on feed intake (Etienne et al.*, 1993).*

Source of glucosinolates	Inclusion level (g/kg)	Glucosinolates (μM/g RSM)	Diet	Sows' weight (kg)	Feed intake (g/day)	Glucosinolates intake (μM/day)
Rapeseed meal				30 up to 60		
	0	0.0	control		1,524	0.0
	200	1.8	rapeseed meal		1,517	549.2
				60 up to 100		
	0	0.0	control		2,446	0.0
	200	1.8	rapeseed meal		2,429	879.3
				100 up to 130		
	0	0.0	control		2,561	0.0
	200	1.8	rapeseed meal		2,558	926.0

Alkaloids

Characteristics of alkaloids, their mechanism and general effects on pigs

Alkaloids are compounds that contain nitrogen in a heterocyclic ring, are generally basic and often have a bitter taste. The word alkaloid simply means 'alkali-like'. Alkaloids are present in many plants wherein they are thought to serve as a chemical defence against herbivory. Alkaloids are oxidised in the liver resulting in metabolites, such as dehydrosparteine, which are responsible for the observed toxicity. The level of toxicity is influenced by the structure of the alkaloids. There is a high degree of variation in the ability of different animal species to deal with these compounds. The toxic effects of alkaloids and their metabolites are primarily mediated through the nervous system, although they are also reported to stimulate copper uptake by liver cells thus leading to copper toxicity. The negative effect of lupins on feed intake appears to be mediated through the neural system as well. Pigs appear more sensitive to alkaloids compared to poultry. New varieties of lupins have become available that contain low alkaloid levels as well as low levels of other constituents such as trypsin inhibitors and lectins (De Lange *et al.*, 2000).

Earlier reports have indicated that inclusion levels of lupins imply adverse effects on pig daily gain and feed intake (Fernandez and Batterham, 1995; Bugnacka and Falkowski, 2001). For example, quinolizidine alkaloids negatively affect palatability and affect amino acid metabolism with a concomitant reduction in feed intake (Cheeke and Kelly, 1989). Ferguson *et al.* (2003) however, stated that growing pigs (15 up to 46 kg live weight) do have a stronger dislike for antinutritional factors compared to imbalanced rations of amino acids in a diet. In their study with new improved low-alkaloid lupin varieties, pigs showed an overwhelming preference for a diet without dehulled lupin seeds.

Effects of alkaloids on piglets' feed intake

In Table 20, the results of feeding lupin to pigs are shown.

In the experience of Bourdon *et al.* (1980) two different types of lupins have been used. The Lublanc variety does contain alkaloids whereas the Kalina variety does not. Both lupin types contain alpha-galactosides. As we can gather from Table 20, feed intake in young pigs is depressed by 56% in the case of lupin (*var.* Lublanc) inclusion and by 36% in the case of lupin (*var.* Kalina) inclusion. The low feed intake observed when the animals are fed Lublanc lupin variety may be explained by its alkaloid content. These compounds are known to have a bitter taste. However the level of alpha-galactosides in both varieties also contributes to decreased feed intake. According to Bourdon *et al.* (1980) alpha-galactosides promote flatulence and a quick satiety feeling after ingestion of low quantities of feed.

Table 20. Effects of white lupin variety on the feed intake of young pigs (Bourdon et al.*, 1980).*

Source of alkaloïds	Inclusion level (g/kg)	Diet	Pig' weight (kg)	Feed intake (g/day DM)	α-Galactosides intake	Alkaloïds intake
Lupin			25 up to 30			
	0	control		1,160	no	no
	400	variety Lublanc		511	yes	yes
	400	variety Kalina		743	yes	no

According to the results of Cherrière *et al.* (2003; Table 21) 10% lupin inclusion in the diet has no impact on piglet's feed intake. At 15% inclusion level results are not significantly different compared to the control diet but feed intake decreases by 3.9% in the case of white lupin whereas feed intake is only lowered by 1.8% in the case of blue lupin *var.* Boltensia.

When blue lupin *var.* Bora is included at a level of 20% and 30%, feed intake decreases (compared to the control diet fed pigs) by 10.6% and 17.1%, respectively. Feed conversion ratio (data not shown) was not significantly different. Thus Cherrière *et al.* (2003) state that the palatability of the diet is reduced by the presence of alkaloids in the third trial. On the other hand feed conversion ration in the first trial is significantly impaired at 15% lupin inclusion. This indicates metabolic troubles which are linked to the presence of alpha-galactosides causing metabolic interference (Froidmont *et al.*, 2003). Moreover white lupin contains the highest level of stachyose (66 mg/g DM) compared to blue lupin (*var.* Boltensia 45.7 mg/g DM and *var.* Bora 39.7 mg/g DM). According to Cherrière *et al.* (2003) the reduced performances of piglets at the level of 15% white lupin inclusion may be due to its stachyose level where its alkaloid level is not the causative constituent.

Table 21. Feed intake of piglets as affected by diet inclusion of different levels and variety of lupins (Cherrière et al.*, 2003).*

Source of alkaloïds	Inclusion level (g/kg)	Diet	Pig' weight (kg)	Feed intake (g/day)	α-Galactosides intake (g/day)	Alkaloïds intake (g/day)
Lupin			12 up to 25			
Trial 1	0	white lupin 'Ares'		753	0.0	0.0
	50			765	2.9	4.7
	100			749	5.6	9.2
	150			724	8.1	13.3
Trial 2	0	blue lupin 'Boltensia'		705	0.0	0.0
	50			705	2.3	19.3
	100			704	4.5	38.6
	150			692	6.7	56.9
Trial 3	0	blue lupin 'Bora'		750[a]	0.0	0.0
	100			742[a]	4.6	68.9
	200			670[b]	8.3	125.7
	300			622[b]	11.6	175.1

[a, b] Mean values within a column with different superscripts are significantly different ($P<0.01$).

Effects of alkaloids on growing- finishing pigs' feed intake

Ferguson *et al.* (2003) determined the source of antinutritional factors in two lupin varieties (*L. albus* and *L. angustifolius*) in pigs from 15 up to 47 kg live weight (Table 22). Three basal diets were formulated in this study: soya-based no lupins; full-fat dehulled lupin meal and one diet replacing sunflower husk with lupin husk. A series of different diets were then obtained by mixing the basal diets with dehulled lupin and lupin hull contents.

Feeding treatments show that replacement of soybean meal (SBM) by dehulled lupin meal reduce the feed intake linearly ($P<0.05$) as concentrations of *L. albus* increased, but not when dehulled *L. angustifolius* concentrations increased (Tables 22 and 23). These results are suggesting that the antinutritional factors in the *L. angustifolius* and *L. albus* varieties are located in the seed cotyledon rather than in the seed coat.

There was no effect on feed intake for pigs in the growing period when low-alkaloid (sweet) lupins were fed. However, in the finishing period up to 123 kg body weight, a higher feed intake was observed for pigs fed the lupin diets with no differences between the two lupin varieties, which had a low-alkaloid level of 0.05 and 0.23% for *L. angustifolius* and *L. luteus*, respectively.

Overall results in the literature of feeding lupins at levels of inclusion <200 g/kg diet (Van Nevel *et al.*, 2000) and feeding diets containing 300 g of sweet lupin seeds (*L. Albus*) per kg diet to pigs (Table 25) showed that feed intake was lowered from 2.32 to 2.05 kg in the total growing period of 24-102 kg live weight. The presence of a low level of quinolizidine alkaloids (0.0123%) could not be responsible for the lowered feed intake (Van Nevel *et al.*, 2000) nor were levels of manganese (Pearson and Carr, 1977).

*Table 22. Effects of diet type based on lupins (*L. angustifolius*) on feed intake in growing pigs between 15 and 46 kg (Ferguson* et al.*, 2003).*

Source of alkaloïds	Inclusion level (g/kg)	α-Galactosides (g/kg lupin DM)	Alkaloïds (mg/kg lupin DM)	Diet	Pig' weight (kg)	Feed intake (g/day)
Lupin					15 up to 46	
	0	44.5	0	dehulled lupin '*L. angustifolius*'		1,530
	75	44.5	<100	dehulled lupin '*L. angustifolius*'		1,469
	150	44.5	<100	dehulled lupin '*L. angustifolius*'		1,503
	225	44.5	<100	dehulled lupin '*L. angustifolius*'		1,391
	300	44.5	<100	dehulled lupin '*L. angustifolius*'		1,446
	0	19.5	<100	lupin hulls '*L. angustifolius*'		1,485
	12.5	19.5	<100	lupin hulls '*L. angustifolius*'		1,520
	25	19.5	<100	lupin hulls '*L. angustifolius*'		1,519
	37.5	19.5	<100	lupin hulls '*L. angustifolius*'		1,538

*Table 23. Effects of diet type based on lupins (*L. albus*) on feed intake in growing pigs between 20 and 47 kg (Ferguson* et al.*, 2003).*

Source of alkaloïds	Inclusion level (g/kg)	α-Galactosides (g/kg lupin DM)	Alkaloïds (mg/kg lupin DM)	Diet	Pig' weight (kg)	Feed intake (g/day)
Lupin					20 up to 47	
	0	49.1	0	dehulled lupin '*L. albus*'		1,367
	75	49.1	<100	dehulled lupin '*L. albus*'		1,402
	150	49.1	<100	dehulled lupin '*L. albus*'		1,251
	225	49.1	<100	dehulled lupin '*L. albus*'		1,153
	300	49.1	<100	dehulled lupin '*L. albus*'		1,167
	0	15.5	<100	lupin hulls '*L. albus*'		1,263
	12.5	15.5	<100	lupin hulls '*L. albus*'		1,304
	25	15.5	<100	lupin hulls '*L. albus*'		1,267
	37.5	15.5	<100	lupin hulls '*L. albus*'		1,122

Table 24. Effects of replacement of soybean meal by low-alkaloid lupins on feed intake of grower/finisher pigs (Roth-Maier et al.*, 2004).*

Source of alkaloïds	Inclusion level (g/kg)	Diet	Pig' weight (kg)	Feed intake (g/day)	α-Galactosides intake (g/day)	Alkaloïds intake (g/day)
Lupin			30 up to 60			
	0	control		1,910	-	0.00
	200	lupin 'Angustifolius'		1,940	-	0.20
	200	lupin 'Luteus'		1,960	-	0.90
			60 up to 123			
	0	control		$2,720^{b}$	-	0.00
	200	lupin 'Angustifolius'		$2,740^{a}$	-	0.27
	200	lupin 'Luteus'		$2,750^{a}$	-	1.27
			30 up to 123			
	0	control		$2,400^{c}$	-	0.00
	200	lupin 'Angustifolius'		$2,430^{a}$	-	0.24
	200	lupin 'Luteus'		$2,440^{a}$	-	1.12

$^{a, b}$ Mean values within a column with different superscripts are significantly different ($P<0.05$).

Table 25. Feed intake of growing pigs as affected by diets containing different levels of L. albus *seeds (Van Nevel* et al.*, 2000).*

Source of alkaloïds	Inclusion level (g/kg)	Diet	Pig' weight (kg)	Feed intake (g/day)	α-Galactosides intake (g/day)	Alkaloïds intake (g/day)
Lupins			24 up to 43			
	0	control		1,470	-	0.00
	150	15% *L. albus*		1,350	-	0.02
	300	30% *L. albus*		1,210	-	0.04
			43 up to 102			
	0	control		2,710	-	0.00
	150	15% *L. albus*		2,780	-	0.05
	300	30% L. *albus*		2,440	-	0.09
			24 up to 102			
	0	control		2,320	-	0.00
	150	15% *L. albus*		2,300	-	0.04
	300	30% *L. albus*		2,050	-	0.08

(Poly)saccharides

As pointed out earlier (Brenes *et al.*, 2004) low-alkaloid lupin seeds do contain non-starch polysaccharides and oligosaccharides from the raffinose family, that limit its inclusion in the diet for pigs.

Characteristics of (poly)saccharides, their mechanism and general effects on pigs

Flatulence is a digestive disorder characterised by an abnormal level of gas formation in the intestinal tract following the intake of oligosaccharides that are not enzymatically digested by mammals and birds. The major factors concern alpha-galactosides such as raffinose, stachyose, ajugose and verbascose, all of which belong to the raffinose family of oligo-saccharides.

Fermentation of oligosaccharides in the hindgut results in the production of carbon dioxide, hydrogen and methane as well as volatile fatty acids. In high concentrations, these products may cause flatulence, diarrhoea, cramp and general discomfort (Saini, 1989).

Effects of (poly)saccharides on piglets' feed intake

Performance responses to the inclusion of lupin seeds/seed meals in diet for pigs can be impaired in the case that the overall content of non-starch polysaccharides is high: reduction in voluntary feed intake was observed, an increase in the feed-to-gain ratio and a decrease in nutrient digestibility which could be – at least in part – attributed to the indigestible fibre content (Brenes *et al.*, 2004).

Effects of (poly)saccharides on growing finishing pigs' feed intake

Bourdon *et al.* (1980) realised an experience on growing-finishing pigs fed a lupin variety Kalina free of alkaloids (Table 26).

The aim of the study performed by Bourdon *et al.* (1980) was to evaluate the possibilities of substitution of soybean meal with sweet lupin in association or not with alfalfa protein concentrate. Remarkable is to observe that two pigs fed the diet 'Lupin + Lysine' and one pig fed the diet 'Lysine + antibiotics+ Lysine' died only 24 hours after ingestion of the experimental diet. The authors observed flatulence symptoms and gut transit stop. They linked pigs' death to a high level of alfa-galactosides present in lupins. Consequently, pigs fed those two diets received in addition linseed 7 up to 10.5 g/dag) to avoid further flatulence problems. As can be seen in

Table 26. Effect of lupin inclusion and supplementation on feed intake of growing-finishing pigs (Bourdon et al., *1980).*

Source of α-Galactosides	Inclusion level (g/kg)	Diet	Pig' weight (kg)	Feed intake (g/day)	α-Galactosides intake (g/day)	Alkaloïds intake (g/day)
Lupin			25 up to 60			
		control		1,820 [a]	no	no
	150	lupin		1,830 [a]	yes	no
	300	lupin + Lysine		1,690 [b]	yes	no
	300	lupin + antibiotics + Lysine		1,730 [b]	yes	no
	100	lupin + 10% alfalfa PC[1] + Lysine		1,810 [a]	yes	no
			60 up to 100			
		control		2,610 [a]	no	no
	150	lupin		2,550 [a]	yes	no
	300	lupin + Lysine		2,390 [b]	yes	no
	300	lupin + antibiotics + Lysine		2,500 [b]	yes	no
	100	lupin + 10% alfalfa PC[1] + Lysine		2,620 [a]	yes	no
			25 up to 100			
		control		2,230 [ab]	no	no
	150	lupin		2,290 [a]	yes	no
	300	lupin + Lysine		2,060 [c]	yes	no
	300	lupin + antibiotics + Lysine		2,120 [bc]	yes	no
	100	lupin + 10% alfalfa PC[1] + Lysine		2,220 [ab]	yes	no

[a, b, c] Mean values within a column with different superscripts are significantly different (*P*<0.01).

[1] PC = protein concentrate.

Table 26, feed intake is impaired when the inclusion level of lupin reaches 30%. The addition of antibiotics does neither significantly enhance feed intake nor growth. Antibiotics do not seem to reduce flatulence. Bourdon *et al.* (1980) concluded that sweet lupin may partially substitute soybean meal (i.e. 20% SBM in the control diet) at an inclusion level of 10-15% but the association of 10% lupin and 10% alfalfa protein concentrate may constitute a valuable alternative to the total substitution of soybean meal without any decrease in the pig's performance.

Saponins

Characteristics of saponins, their mechanism and general effects on pigs

Saponins are naturally occurring surface-active glycosides and their name comes from their ability to form stable, soap-like foams in aqueous solutions (Francis *et al.*, 2002). Saponins are non-volatile compounds consisting of a sugar moiety linked to a hydrophobic aglycone (sapogenin) which may be triterpenoid or steroid. Triterpenoid saponins are generally found in cultivated crops such as soybeans, beans, peas, alfalfa, sunflower. Steroid saponins can be found in oats or yam. In different pea cultivars saponin content may range between 0.8 and 2.5 g/kg (Heng *et al.*, 2006). Heng *et al.* (2006) found out that saponin A and saponin B have both a bitter taste which increases with concentration. However saponin A was observed to be significantly more bitter than saponin B. This experiment was conducted with humans but pigs generally perceive and refuse diets that humans find bitter (Laitat *et al.*, 2004). Francis *et al.* (2002) in their review reported negative effects linked to saponins such as the reduction of intestinal motility, reduction in protein digestibility, damage to the intestinal membrane and inhibition of nutrient transport. Their antinutritional properties seem related to their ability to form complexes with sterols, in particular those in membranes of animal cells. This appears to result in increased permeability of the intestinal mucosa.

Effects of saponins on piglets' feed intake

The mechanism by which growth depression is caused by saponins was measured by Cheeke *et al.* (1977) when studying the role of palatability of saponins using different lines of alfalfa. The results of high and low saponin lines on feed intake in young piglets are shown in Table 27.

Table 27. Effects of high and low saponin alfalfa on piglets' feed intake (Cheeke et al.*, 1977).*

Source of saponins	Inclusion level (g/kg)	Saponin (g/kg alfalfa)	Diet	Pigs' age (days)	Feed intake (g/day)	Saponin intake (g/day)
Alfalfa				22 up to 28		
	0	0	control diet		240	0.00
	150	4	low saponin alfalfa		240	0.14
	150	17	high saponin alfalfa		220	0.56

The pigs showed similar intakes for the alfalfa-free diet (i.e. control diet) and the 15% alfalfa diet with a low or a high saponin level. Daily gain, however, was decreased ($P<0.05$) with a 15% high saponin diet. From this study it was concluded that the low saponin alfalfa supports a higher growth rate in swine than high saponin alfalfa. Feed intake by the piglets was lower in the high saponin alfalfa ($P>0.05$). It was suggested that increased palatability was a major causative factor.

No data on the effect of saponins on growing-finishing pigs or sows were encountered.

Discussion

Naturally occurring antinutritional factors are considered as a mechanism developed by plants to insure protection against predators. These factors are secondary metabolites (i.e. they are not directly essential for basic photosynthetic or respiratory metabolism (Theis and Lerdau, 2003)) and might be an evolutionary adaptation of plants to resist predators (Chrispeels and Raikhel, 1991; Peumans and Van Damme, 1995). Proteins in the phytohemagglutinin family for instance arose through duplication and divergence of an ancestral gene (Chrispeels and Raikhel, 1991). Plant constituents such as lectins are non-toxic to very toxic proteins; non-toxic forms may not completely protect plants from consumption. The reaction of avoidance by the animal, however, may be beneficial for the survival of the species (Peumans and Van Damme, 1995). We may then question ourselves why plant materials do not manufacture a large amount of constituents that are very toxic, so they can exert severe effects on animals, when eaten. This question might be answered by the physiological trade-off between growth and differentiation processes, which determines the level of primary and secondary metabolites. Herms and Mattson (1992) explained that plants face a dilemma namely 'grow fast enough to compete, and yet maintain the physiological adaptations (defences) necessary for survival in the presence of herbivores and pathogens'. Plants have limited resources and trade-offs occur among growth, storage, maintenance, reproduction and defence. Moreover, primary and secondary metabolites share common precursors. Alkaloid precursors e.g. use the same amino acids than protein precursors, thus competition between growth and defences takes place. Chrispeels and Raikhel (1991) observed that both storage proteins and lectins are broken down during the periods of germination and seedling growth to provide amino acids for the growing seedling. Thus secondary metabolites have other functions in plants than only the defensive role but the functions are related to the germination and growing period (Savelkoul, 1992).

Antinutritional factors do affect the voluntary feed intake in pigs, although acting by different mechanisms. Antinutritional factors may deteriorate the palatability of an ingredient and lead to the refusal of the diet when it tastes bitter for example. The animal's sensorial perception is a way to discriminate between feeds with desired or undesired constituents. A further reason for a decreased feed intake may be learning: experiencing bad consequences after ingestion or linking digestive trouble to feed may also cause feed refusal (Day *et al.*, 1998; Laitat *et al.*, 2004).

Growing pigs do have a stronger dislike for the presence of antinutritional factors in the diet compared to an imbalanced amino acid ratio as previously discussed by Ferguson *et al.* (2003) and Kyriazakis and Emmans (1992). It is therefore of importance to study the effects of specific antinutritional factors on feed preference. Only limited research, however, has been done to examine the effects on antinutritional factors using pure (isolated) factors. Since it has been shown that pigs are more sensitive than rats to digestive troubles (Huisman, 1990b) or technological treatments (Van der Poel, 1990) it is necessary to use pigs as the target animal in studies. For that reason it is relatively expensive to produce sufficient quantities of isolated factors to conduct a pig and even a piglet trial. Consequently a large number of trials has tested antinutritional factors in the matrix of the diet ingredients.

Two issues are important in this matter. Firstly, the simultaneous presence of different antinutritional factors is a complexing factor when researchers want to investigate the effects of a specific antinutritional factor on pigs' performances and measuring the impacts of a reduction of the latter. It is noted that possible interactions between different antinutritional factors in a matrix of a specific diet ingredient may affect palatability for example. Moreover, when thermal processing eliminates certain proteinaceous antinutritional factors, other, more heat-resistant constituents (e.g. alkaloids; condensed tannins) may still be present after processing and still exert their negative effects on palatability.

For breeding purposes, the availability of near-isogenic lines in which the partner lines are distinct in the presence or absence of an antinutritional factor has allowed research to study the natural resistance to micro-organisms as well as the effects of antinutritional factors *in vivo* (Helsper, 1998). Nowadays, there is a shift from classical plant breeding to the application of molecular genetic technology to improve crops.

Ingredient palatability and preference for a certain diet ingredient is affected by a large number of factors such as age, previous dietary exposure, ingredient inclusion levels (Solà-Oriol *et al.*, 2007). Moreover, nutrients such as free amino acids or Maillard products formed after thermal processing are further factors to affect palatability. It is therefore difficult to attribute effects on the pig's voluntary feed intake to one or more specific antinutritional factors.

Soybeans

Both protease inhibitors and lectins in unprocessed soybeans negatively affect the voluntary feed intake in pigs. The Kunitz inhibitor from unheated soybeans was shown not to significantly affect voluntary feed intake but to severely impaire growth in young piglets (Cook *et al.*, 1988). Similar non-significant effects were observed for finishing pigs although feed intake was lowered by 6% when raw soybeans with high Kunitz trypsin inhibitor level were fed and 9% when raw soybeans with low Kunitz trypsin inhibitor level were fed. The extrusion of raw soybeans enhances feed intake when conventional soybeans or lectin-free soybeans or lectin & Kunitz inhibitor-free soybeans were fed to pigs of 41 up to 69 days of age (Palacios *et al.*, 2004). Compared to the raw conventional soybeans, voluntary feed intake was significantly increased when no lectins were present and even more in the case both lectins and Kunitz inhibitor were absent from the raw soybeans. In only some trials, a direct negative relationship is observed between the level of trypsin inhibitors in heat-processed soybeans and feed intake in young piglets (White *et al.*, 1989; Zarkadas and Wiseman, 2005).

Phaseolus *beans*

The presence of lectins in these types of beans severely impair the voluntary feed intake in pigs (Huisman, 1990b; Kluess *et al.*, 2007), reason why all these beans have to be heat processed prior to their inclusion into pig diets. Feed intake in unprocessed beans (200 g/kg diet) was observed to be depressed in pigs of 56, 84 and 112 days of age but was restored by thermal processing to the feed intake of piglets fed a control diet (Huisman, 1990b).

Pea

The inclusion of 300 g/kg peas in the diet did not change feed intake, although growth in young piglets was impaired (Huisman, 1990b). In fattening pigs, 260 g/kg inclusion of peas showed the highest voluntary feed intake but the presence of peas changed eating behaviour in a way that eating time was increased with less visits per day at the feeder (Mathé *et al.*, 2003). Grosjean *et al.* (1991) showed that pea variety may have a large impact on voluntary feed intake that is not entirely related to the level of tannins.

Lupins

The inclusion of seeds from older varieties of *Lupinus* spp. in diets for pigs has always been shown to be detrimental because of its bitter-tasting properties (alkaloids) and because of the accumulation of the element manganese (Cheeke and Kelly, 1989; Roth-Maier *et al.*, 2004). As for the quinolizidine alkaloids, they affect metabolism of amino acids and additionally, there is a reduction observed for feed intake up to the point of feed refusal (Van Nevel *et al.*, 2000). Both the inclusion of lupins into the diet as certain lupin varieties ('Bora'; 'Albus') have a decisive role in the observed lower voluntary feed intake in (Bourdon *et al.*, 1980; Cherrière *et al.*, 2003; Van Nevel *et al.*, 2000). The fact that lupin ANF are located in the seed cotyledon rather than in the seed coat (Ferguson *et al.*, 2003) indicates the change of the target ANF to be studied from alkaloids (in alkaloid-rich, bitter lupins) to oligosaccharides (in sweet lupins).

Rapeseed/crambe

Untreated rapeseed or crambe meal (Liu, 1994), even from low-glucosinolates varieties, exerts negative effects on the pigs' preference for diets containing 10 and 20% rapeseed (Roura *et al.*, 2007). Processing improves feed intake but no clear relationship is observed between the level of intact glucosinolates or its splitting products with the voluntary feed intake. A good relation between preference value and voluntary feed intake with pigs was observed.

Sorghum

Lizardo *et al.* (1995) could not find a direct relationship between the sorghum tannin level in the diet (560 or 650 g/kg diet) and the voluntary feed intake in pigs from 21 up to 63 days of age. In all cases feed intake of the sorghum diets was (non significant) higher compared to the control maize diets.

Alfalfa

There is only a small evidence that 15% inclusion of *Alfalfa* in diets for young piglets reduces voluntary feed intake when a high-saponin variety is used (Cheeke *et al.*, 1977).

In this review, no clear linear relationships or consistent effects were noticed between the level of antinutritional factors or its intake with the voluntary feed intake in pigs. This might be due to the simultaneous presence of other undesired factors in a seed or to constituents (e.g. amino acids) or to factors originating from processing (e.g. Maillard products) in a way that the pig's sensorial perception cannot be believed to be attributed to a sole antinutritional factor. If this is generally true, factors hampering or stimulating feed intake in swine can be better considered from the point of view that all diet ingredients – raw or processed – all have their specific problems. Thanks to this review we might, however, identify two groups of ingredients. Firstly ingredients that do not have a consistent effect on feed intake such as peas, rapeseed and lupins and secondly ingredients that impair feed intake systematically when fed in the raw state such as soybeans and *Phaseolus* beans. The latter ingredient group shows a significant negative effect on feed intake when ingredients are not heat treated. Normally, soybeans and *Phaseolus* beans are only consumed after removal of heat-labile antinutritional factors by processing (Palacios *et al.*, 2004). In the case of *Phaseolus* beans, the reduced feed intake is directly related to the presence of toxic lectins (Kluess *et al.*, 2007) but when examining soybeans, no significant results were found to attribute reduced feed intake to lectins, Kunitz inhibitor or other antinutritional factors. As examining the first group of ingredients we cannot conclude on a common pattern of their effects. Peas e.g. do not show a significant negative effect on feed intake (Huisman, 1990b; Grosjean *et al.*, 1991) and even enhance feed intake at an inclusion level of 26% (Mathé *et al.*, 2003). These observations cannot be linked directly to the presence of a specific antinutritional factor since peas were not tested for the simultaneous presence of trypsin inhibitor and tannins: only one of these factors was examined and analysed. Similarly, lupins were studied by analysing the effect of either alkaloids or alpha-galactosides. The study of rapeseed meal has shown that glucosinolates are not the only cause to explain a depressed feed intake since even low-glucosinolates rapeseed meal is less preferred in pigs (Roura *et al.*, 2007).

We might also define two groups of antinutritional factors. The first group of antinutritional factors, e.g. trypsin inhibitors, lectins and polysaccharides, has a direct impact on gut functions whereas the second group of antinutritional factors, e.g. tannins, alkaloids, glucosinolates, saponins, reduce the palatability of the ingredient and pigs instinctively lower their feed intake. These antinutritional factors will impair other functions such as reproduction later on (e.g. glucosinolates).

Finally, examining only feed intake provides not sufficient information when studying the effects of antinutritional factors on animal performances. In most of the studies, feed intake is not significantly impaired by the presence of antinutritional factors whereas growth and feed conversion ratio are clearly affected by the presence of these elements. Peas are, in most studies, not lowering feed intake but growth is greatly impaired (Huisman, 1990b). Sows for example do not show a significantly lower preference for rapeseed meal but their reproduction functions will be affected by the ingestion of glucosinolates. These compounds decrease reproduction capabilities and are detrimental for the development of the piglets. As a consequence, antinutritional factors do have long-term effects on animals that cannot always be detected by an observed significant decrease of diet consumption.

References

Batterham, E.S., Saini, H.S, Andersen, L.M. and Baigent, R.D., 1993. Tolerance of growing pigs to trypsin and chymotrypsin inhibitors in chickpeas (*Cider arietinum*) and pigeonpeas (*Cajanus* cajan). Journal of the Science of Food and Agriculture 61: 211-216.

Bourdon, D., Perez, J.M. and Calmes, R., 1980. Le lupin (*Lupinus albus L.*) dans l'alimentation du porc. Valeur énergétique et azotée et conditions d'utilisation. Journées Recherche Porcine en France 12: 245-264.

Brenes, A., Jansman, A.J.M., and Marquardt, R.R., 2004. Recent progress on research on the effects of antinutritional factors in legume and oil seeds in monogastric animals. In: M. Muzquiz, G.D. Hill, C. Burbano, C. Cuadrado and M.M. Pedrosa (eds.) Recent advances of research in antinutritional factors in legume seeds and oilseeds. EAAP No. 110, Wageningen Academic Publishers, Wageningen, the Netherlands, pp. 195-217.

Bugnacka, D. and Falkowski, J., 2001. The effect of dietary levels of yellow lupin seeds (*Lupinus luteus* L.) on feed preferences and growth performance of young pigs. Journal of Animal and Feed Sciences 10: 133-142.

Cheeke, P.R., Kinzell, J.H. and Pedersen, M.W., 1977. Influence of saponins on alfalfa utilization by rats, rabbits and swine. Journal of Animal Science 45: 476-481.

Cheeke, P.R. and Kelly, J.D., 1989. Metabolism, toxicity and nutritional implications of quinolizidine (lupin) alkaloids. In: J. Huisman, T.F.B. van der Poel and I.E. Liener (eds.) Recent advances in antinutritional factors in legume seeds. Pudoc Wageningen, the Netherlands, 189-201.

Cherrière, K., Albar, J., Noblet, J., Skiba, F., Granier, R. and Peyronnet, C., 2003. Utilisation du lupin bleu (*Lupinus angustifolius*) et du lupin blanc (*Lupinus albus*) par les porcelets en post sevrage. Journées Recherche Porcine en France 35: 97-104.

Chrispeels, M. and Raikhel, N.V., 1991. Lectins, lectin genes, and their role in plant defense. The Plant Cell 3: 1-9.

Chubb, L.G., 1982. Anti-nutritive factors in animal feedstuffs. In: W. Haresigns (ed.), Recent Advances in Animal Nutrition, University of Nottingham, Butterworths, London, pp. 21-37.

Chung, K.T., Wong, T.Y., Wei, C.I., Huang, Y.W. and Lin, Y., 1998. Tannins and human health: a review. Critical Reviews in Food Science and Nutrition 38: 421-464.

Cook, D.A., Jensen, A.H., Fraley, J.R. and Hymowitz, T., 1988. Utilization by growing and finishing pigs of raw soybeans of low kunitz trypsin inhibitor content. Journal of Animal Science 66: 1686-1691.

Day, J.E.L., Kyriazakis, I. and Rogers, J., 1998. Food choice and intake: towards a unifying framework of learning and feeding motivation. Nutrition Research Reviews 11: 25-43.

De Lange, C.F.M., Nyachoti, C.M. and Verstegen, M.W.A., 2000. Significance of antinutritional factors in feedstuffs for monogastric animals. In: P.J. Moughan, M.W.A. Verstegen and M.I. Visser-Reyneveld, (eds.) Feed evalution, principles and practice. Wageningen Pers, Wageningen, the Netherlands, pp. 169-188.

Etienne, M., Dourmad J.Y. and Evrard, J., 1993. Effets de la consommation de tourteau de colza à très basse teneur en glucosinolates pendant la croissance et la gestation chez la truie. Journées Recherche Porcine en France 25: 193-202.

Ferguson, N.S., Gous, R.M. and Iji, P.A., 2003. Determining the source of anti-nutritional factor(s) found two species of lupin (*L. Albus* and *L. Angustifolius*) fed to growing pigs. Livestock Production Science 84: 83-91.

Fernandez, J.A. and Batterham, E.S., 1995. The nutritive value of lupin-seed and dehulled lupin-seeds meals as protein sources for growing pigs as evaluated by different techniques. Animal Feed Science and Technology 53: 279-296.

Francis, G., Kerem, Z., Makkar, H.P.S and Becher, K., 2002. The biological action of saponins in animal systems: a review. British Journal of Nutrition 88: 587-605.

Froidmont, E., Schoeling, O., Deliège, F., Wathelet, B., Wavreille, J. and Barthiaux-Thill, N., 2003. Influence de la substitution du tourteau de soja par des graines de lupin, avec ou sans complément d'alpha-galactosidase, sur la digestibilité des régimes et la rétention azotée du porc en croissance. Journées Recherche Porcine en France 35: 105-112.

Grosjean, F., Bourdon, D., Kiener, T., Castaing, J. and Gatel, F., 1991. Valeur alimentaire pour les porcs des pois français et importés. Journées Recherche Porcine en France 23: 53-60.

Helsper, J.P.F.G., 1998. Recent developments on the improvement of ANF levels in legume seeds by breeding. In: A.J.M. Jansman, G.D. Hill, J. Huisman and A.F.B. van der Poel (eds), Recent advances of research in antinutritional factors in legume seeds and rapeseed. Wageningen Pers, Wageningen, The netherlands, pp. 351-361.

Heng, L., Vincken, J.P., Van Koningsveld, G., Legger, A., Gruppen, H., Van Boekel, T., Roozen, J. and Voragen, F., 2006. Bitterness of saponins and their content in dry peas. Journal of the Science of Food and Agriculture 86: 1225-1231.

Herms, D.A. and Mattson, W.J., 1992. The dilemma of plants: to grow or defend. The Quarterly Review of Biology 67: 283-335.

Huisman, J., 1990a. Dietary effects and some analytical aspects of antinutritional factors in peas (*Pisum sativum*), common beans (*Phaseolus vulgaris*) and soybeans (*Glycine max*) in monogastric farm animals. A review. In: Antinutritional effects of legume seeds in piglets, rats and chickens. PhD Thesis, Wageningen University, Ponsen & Looijen Publishers, the Netherlands, 147 pp.

Huisman, J., 1990b. Animal species difference research with piglets, rats and chickens fed peas and common beans. In: Antinutritional effects of legume seeds in piglets, rats and chickens. PhD Thesis, Wageningen University, Ponsen & Looijen Publishers, the Netherlands, 147 pp.

Jansman, A.J.M, 1993. Tannins in feedstuffs for simple-stomached animals. In: Tannins in faba beans (*Vicia faba* L.). PhD Thesis, Wageningen University, Ponsen & Looijen Publishers, the Netherlands, 207 pp.

Jansman, A.J.M., Verstegen, M.W.A. and Huisman, J., 1994. Effects of dietary inclusion of hulls of faba beans (*Vicia faba* L.) with a low and high content of condensed tannins on digestion and some physiological parameters in piglets. Animal Feed Science and Technology 43: 239-257.

Kluess, J.W., Koninkx, J., Bikker, P., Fledderus, J., Dirkzwager, A. and Veldman, B., 2007. The effect of PHA-lectins and diet composition on performance and intestinal development in weaning piglets. Livestock Science 108: 288-291.

Kyriazakis, I. and Emmans, G.C., 1992. Selection of a diet by growing pigs given choices between foods differing in contents of protein and rapeseed meal. Appetite 19: 121-132.

Laitat, M., De Jaeger, F., Vandenheede, M. and Nicks, B., 2004. Facteurs influençant la consommation alimentaire et les performances zootechniques du porc sevré: perception et caractéristiques de l'aliment. Annales de Médecine Vétérinaire 148: 15-29.

Lajolo, F.M. and Genovese, M.I., 2002. Nutritional significance of lectins and enzyme inhibitors from legumes. Journal of Agriculture and Food Chemistry 50: 6592-6598.

Liu, Y.G., 1994. Crambe meal: digestibility in pigs and rats in comparison with rapeseed meal. In: Crambe meal: evaluation, improvement and comparison with rapeseed meal. PhD Thesis, Wageningen University, Ponsen & Looijen Publishers, the Netherlands, 135 pp.

Lizardo, R., Peiniau, J., Lebreton, Y. and Aumaitre, A., 1995. Influence des tannins du sorgho sur la croissance, la digestibilité des constituants de la ration et l'activité des enzymes pancréatiques et intestinales chez le porcelet. Journées Recherche Porcine en France 27: 209-216.

Marquardt, R.R., 1989. Dietary effects of tannins, vicine and convicine. In: J. Huisman, A.F.B. van der Poel and I.E. Liener (eds), Recent advances of research in antinutritional factors in legume seeds. Pudoc Wageningen, the Netherlands, pp. 141-155.

Mathé, D., Monéger, R. and Guillou, D., 2003. Effet du pois protéagineux sur les performances et le comportement du porc lors des transitions alimentaires. Journées Recherche Porcine en France 35: 127-132.

Meunier-Salaün, M.C. and Picard, M., 1996. Les facteurs de choix alimentaires chez le porc et les volailles. INRA Productions Animales 9: 339-348.

Muzquiz, M., Hill, G.D., Cuadrado, C., Pedrosa, M.M. and Burbano, C., 2004. Recent advances of research in antinutritional factors in legume seeds and oilseeds. EAAP No. 110. Wageningen Academic Publishers, Wageningen, the Netherlands, 370 pp.

Palacios, M.F., Easter, R.A., Soltwedel, K.T., Parsons, C.M., Douglas, M.W., Hymowitz, T. and Pettigrew, J.E., 2004. Effect of soybean variety and processing on growth performance of young chicks and pigs. Journal of Animal Science 82: 1108-1114.

Pearson, G. and Carr, J.R., 1977. A comparison between meals prepared from the seeds of different varieties of lupin as protein supplements to barley-based diets for growing pigs. Animal Feed Science and Technology 2: 49-58.

Peumans, W.J. and Van Damme, J.M., 1995. Lectins as plant defense proteins. Plant Physiology 109: 347-352.

Pusztai, A., Clarke, E.M.W., Grant, G. and King, T.P., 1981. The toxicity of *Phaseolus vulgaris* lectins. Nitrogen balance and immunochemical studies. Journal of the Science of Food and Agriculture 32: 1037-1046.

Pusztai, A., Ewen, S.W.B., Grant, G., Peumans, W.J., Damme, E.J.M., Rubio, L. and Bardocz, S., 1990. Relationship between survival and binding of plant lectins during small intestinal passage and their effectiveness as growth factors. Digestion 46: 308-316.

Roura, E., Solà-Oriol, D., Mallo, J., Van Hees, H., Tedo, G. and Torrallardona, D., 2007. Utilisation de tourteau de colza et de tourteau de tournesol à doses élevées dans les aliments porc. Evaluation des préférences alimentaires et de la consommation volontaire d'aliment. Journées Recherche Porcine en France 39: 163-166.

Roth-Maier, D.A., Böhmer, B.M. and Roth, F.X., 2004. Effects of feeding canola meal and sweet lupin in amino acid balanced diets on growth performance and carcass characteristics of growing-finishing pigs. Animal Research 53: 21-34.

Saini, H.S., 1989. Legume seed oligosaccharides. In: J. Huisman, A.F.B. van der Poel and I.E. Liener (eds), Recent advances of research in antinutritional factors in legume seeds. Pudoc Wageningen, the Netherlands. pp. 329-341.

Savelkoul, F.H.M.G., 1992. Effects of bioprocessed antinutritional factors on bean protein quality. PhD Thesis, Wageningen University, Wageningen, the Netherlands, 111 pp.

Sgarbieri, V.C. and Whitaker, J.R., 1982. Physical, chemical and nutritional properties of common bean proteins. Advances in Food Research 28: 94-160.

Schöne, F., Groppel, B., Hennig, A. and Jahreis, G., 1997. Rapeseed meals, methimazole, thiocyanate and iodine affect growth and tyroid. Investigations into glucosinolate tolerance in the pig. Journal of the Science of Food and Agriculture 74: 69-80.

Solà-Oriol, D., Roura, E. and Torrallardona, D., 2005. An in vivo model development to know feed ingredient preferences in weanling pigs. EEAP-56th Annual Meeting, pp. 235.

Solà-Oriol, D., Roura, E. and Torrallardona, D., 2007. Pig preference for cereal based diets, relationship with their digestibility and physical properties. Livestock Science 108: 190-193.

Theis, N. and Lerdau, M., 2003. The evolution of function in plant secondary metabolites. International Journal of Plant Sciences 164: 93-102.

Tripathi, M.K. and Mishra, A.S., 2007. Glucosinolates in animal nutrition: a review. Animal Feed Science and Technology 132: 1-27.

Valencia, D.G., Serrano, M.P., Centeno, C., Lazaro, R. and Mateos, G.G., 2008. Pea protein as a substitute of soya bean protein in diets for young pigs: Effects on productivity and digestive traits. Livestock Science 118: 1-10.

Van der Poel. A.F.B., 1990. Effects of processing on bean (*Phaseolus vulgaris* L.) protein quality. PhD Thesis, Animal Nutrition group, Wageningen University, Ponsen & Looijen Publishers, 160 pp.

Van Nevel, C., Seynaeve, M., van De Voorde, G., De Smet, S., Van Driessche, E. and De Wilde, R., 2000. Effects of increasing amounts of *lupinus albus* seeds without or with whole egg powder in the diet of growing pigs on performance. Animal Feed Science and Technology 83: 89-101.

White, C.E., Campbell, D.R. and Combs, G.R., 1989. Effects of moisture and processing temperature on activities of trypsin inhibitor and urease in soybeans fed to swine. Recent Advances in antinutritional factors in legume seeds. Pudoc, Wageningen, the Netherlands, pp. 230-234.

Zarkadas, L.N. and Wiseman., J., 2005. Influence of processing of full fat soya beans included in diets for piglets. I. Performance. Animal Feed Science and Technology 118: 109-119.

Chapter 11. Feed manufacturing technology and feed intake in young pigs

R.T. Zijlstra[1], S. Tibble[2] and T.A.T.G. van Kempen[2]
[1]Department of Agricultural, Food and Nutritional Science, University of Alberta, Edmonton, Canada; ruurd.zijlstra@ualberta.ca
[2]Provimi Research and Innovation Centre, Brussels, Belgium

Abstract

The immediate challenge in pigs post-weaning remains the stimulation of voluntary feed intake. Through time, the feed manufacturing industry has implemented a substantial array of technologies to stimulate feed intake and continues to search for further innovation. The present review will focus on two main aspects of feed manufacturing: (1) selection of feedstuffs, with a focus on raw materials (>1% of the diet) or macro ingredients and (2) feed processing technology. Selection of raw materials plays a critical role in the stimulation of voluntary feed intake after weaning. A key example is spray-dried plasma protein, a co-product from slaughter plants, which has become a dominant palatability stimulant in diets for nursery pigs. More recently, crude glycerol, a co-product from the biodiesel industry, has gathered attention, while research into the glycemic index of feedstuffs is also demonstrating a strong effect on feed intake. The heat treatment of feedstuffs or complete feeds including pelleting and extrusion also plays an important role. In summary, feed manufacturing and feedstuff selection allow for opportunities to stimulate feed intake of young pigs, but must coincide with proper feed management to ensure that target feed intake levels are achieved.

Keywords: weaning, feed manufacturing, pig

Introduction

Feed intake of pigs is affected by an array of factors: environmental, social, animal, and feed (Nyachoti *et al.*, 2004). Especially in newly weaned pigs, the impact of feed composition and feed form is important. In the compound feed industry, feed manufacturing technology is used to combine multiple feedstuffs into a final feed for presentation to the pig. The topic on the relationship between technology and feed intake will be covered with information that is mostly publically available with occasional examples of information from our private domain. We understand that numerous variables are changed in feed when selection of raw materials or feed manufacturing is changed, making it impossible to make universal recommendations for best practices.

For the pork producer, a critical time for achieving a high feed intake of pigs is in the period immediately after weaning, because intake drives subsequent health. As an example, artificial infection of piglets with E. coli only compromised growth performance if feed consumption was less than 20 g/d of feed at the time of infection (T.A.T.G. Van Kempen, personal communications). Feed intake during this post-weaning period can be managed effectively by switching piglets from the sow onto a milk-replacer liquid feeding program (Zijlstra *et al.*, 1996). However, liquid feeding of weaned pigs is expensive and may result in a second weaning onto a dry feed later and thus delays and may not solve the problems related to post-weaning lag, although, with careful management, beneficial effects of liquid feeding of weaned pigs might be carried on until market weight (Kim *et al.*, 2001). Most swine producers thus select a direct switch from the sow onto dry feed for a variety of reasons such as convenience or a direct link to crops grown on the farm. Knowledge on diets for weaned pigs has improved drastically compared to a decade ago (Lallès *et al.*, 2007) while increased weaning age in Europe may also have reduced the post-weaning lag. These positive factors, however, may well be offset by increases in disease pressure, e.g. circovirus, and concurrent removal of antimicrobial and specialty feedstuffs such as plasma protein from diets in specific. The net effect is that nutrition of the newly weaned pig remains a major challenge.

Achieving a high level of feed intake quickly after weaning is essential. Obviously, a low level of feed intake means that pigs are not filling their genetic capacity to grow, and may also severely comprise the integrity of the gastro-intestinal tract (Pluske *et al.*, 1997). Although specific feed additives such as flavours, organic acids, pre- and pro-biotics may play an important role attracting piglets to feed, raw materials and feed form are also important, especially in critical phases such as immediately post-weaning. In the energy-dependent phase of growth of grower-finisher pigs, voluntary feed intake is the main driver for growth, but feed manufacturing technology plays a different role, being foremost a component of efforts for reducing cost per unit of lean gain, or for allowing incorporation of co-products of less nutritional value and reducing feed wastage. Generally, once pigs eat dry feed well, feed manufacturing technology becomes less of a concern for achieving high feed intake.

Selection of raw materials

In feed formulation, raw materials are typically matched cost-effectively to the required nutrient profile in the diet. For some raw materials, minimum and maximum inclusion rates are used, especially in diets for weaned pigs: minimum inclusion levels may force a feedstuff with a value-added [not related to nutrient content] attribute into the diet, and maximum inclusion levels may be required to reduce effects that are negative. A stimulus of voluntary feed intake is a value-added attribute, and spray-dried plasma protein, crude glycerol, and oat groats [dehulled oats] are examples of such feedstuffs. Raw materials that contain antinutritional factors may reduce voluntary feed intake, and their inclusion may be limited to manage the risk of reduced growth, and protein feedstuffs of legume origin are examples. Compounded feeds may also contain feed additives to enhance nutrient digestibility or voluntary feed intake of swine, but these are not part of the present overview. Finally, the use of by-products of the agro industry may have specific effects on feed intake, due to dietary factors such as viscosity and water-holding capacity (Kyriazakis and Emmans, 1995), a relationship that has not been properly characterised.

Protein feedstuffs

Protein feedstuffs contain more than 20% crude protein. Detailed overviews of the nutritional attributes of such feedstuffs have been provided previously (e.g. Chiba *et al.*, 2003; Sauvant *et al.*,

2004; Thacker and Kirkwood, 1990). Protein feedstuffs can be divided into animal- and plant-based protein, and furthermore into entire crop that can be simply ground for incorporation into swine diets or that are fractionated. Field pea or faba bean are example of an entire crop, whereas field pea or faba bean protein concentrate or even corn dried distiller's grain plus solubles are co-products following a fractionation process. The fractionation processes are gaining more attention recently, because they provide additional markets for crop producers and offer opportunities to reduce feed costs (Zijlstra *et al.*, 2004) or reduce environmental impact of swine production (Muley *et al.*, 2005). Fractionation of plant-based protein feedstuffs will typically generate protein feedstuffs with a protein content that is higher than the feedstock of origin, partly due to removal of components such as starch (Zijlstra and Beltranena, 2009). Some examples including their mechanism of action are described below.

The prime example of a protein feedstuff that enhances voluntary feed intake of weaned pigs is spray-dried plasma protein. Spray-dried plasma protein is commonly included in the first one or two diets after weaning, because it stimulates feed intake during the first one or two weeks after weaning (Kats *et al.*, 1994; Maxwell and Carter, 2001; Van Dijk *et al.*, 2001), in either pelleted or mash feed (Steidinger *et al.*, 2000). Originally, a direct effect of spray-dried plasma protein to stimulate feed intake was suggested (Ermer *et al.*, 1994); however, the importance of immunoglobulins for animal health contained in spray-dried plasma has become clear (Pérez-Bosque *et al.*, 2008; Pierce *et al.*, 2005). As a result, spray-dried plasma protein is regarded as a potent feedstuff to partially ameliorate potential detrimental effects of the removal of antibiotics as a growth promoter from swine diets (Pettigrew *et al.*, 2003). The positive effect of spray-dried plasma protein on voluntary feed intake is especially noticeable in barn environments that are not clean (Coffey and Cromwell, 1995). In other words, spray-dried plasma protein may mask poor management of the barn environment, and especially a resulting high pathogen load. The immunoglobulins in spray-dried plasma protein inactive intestinal pathogenic bacteria in commercial practice or following a specific challenge with endotoxins (Pérez-Bosque *et al.*, 2008) or pathogenic bacteria such as E. coli K 88 or K 99 (Bhandari *et al.*, 2008; Torrallardona *et al.*, 2003).

Plant-based feedstuffs contain an array of antinutritional factors (ANF) that can reduce voluntary feed intake of pigs (for overviews, see Huisman *et al.*, 1990; Chiba, 2001; Van Heugten, 2001; Clasadonte and Van der Poel, 2009). Some ANF are heat-labile, such as trypsin inhibitors in legume seeds, indicating that feed processing technologies that use heat may reduce the content of heat-labile ANF in feedstuffs. For example, heat processing using atmospheric steam prevented entirely the 42% reduced feed intake observed when 20% raw *Phaseolus vulgaris* beans are included in diets fed to weaned pigs (van der Poel *et al.*, 1990) or increase feed intake of pig fed diets containing 30% full fat soy bean (Zarkadas and Wiseman, 2005). Glucosinolates contained in seeds of Brassica origin may reduce feed intake of pigs, especially young pigs (Tripathi and Mishra, 2007). Moderate levels of glucosinolates contained in canola meal do not appear to reduce feed intake of grower pigs (Roth-Mailer *et al.*, 2004), although original data indicates that canola meal inclusion will reduce feed intake of young pigs (Thacker, 1990). Heat treatment may also reduce glucosinolates in feedstuffs (Tripathi and Mishra, 2007).

Apart from inactivating ANF, heat processing may also cause Maillard reactions between amino acid and reducing sugars that will reduce bioavailability of amino acids, particularly for lysine (Pahm *et al.*, 2009). The Maillard reactions also cause the formation of a large array of flavour compounds (Van Boekel, 2006), critical components for the taste of food for human consumption. However, the importance of these Maillard flavour compounds for feed intake of swine is not clear, because a paucity of meaningful information on this topic exists.

Limitations posed by antinutritional factors can be reduced long-term via genetic selection of crops or short-term via feed processing technology, specifically crop fractionation. For example, 'zero-tannin' faba bean that contain less than 1% tannin were developed instead of more than 1% tannin normally contained in faba bean (Jansman, 1993). The newly developed faba bean or field pea can be ground in a pin mill and subsequently air-classified. Faba bean protein concentrate may contain 60 to 65% crude protein that can compete in protein quality against regular specialty protein feedstuffs, such as a blend of soy protein concentrate, corn gluten meal, and fish meal (Figure 1). Feed intake of weaned pigs fed these novel plant-based feedstuffs can be equal to pigs fed current specialty protein sources.

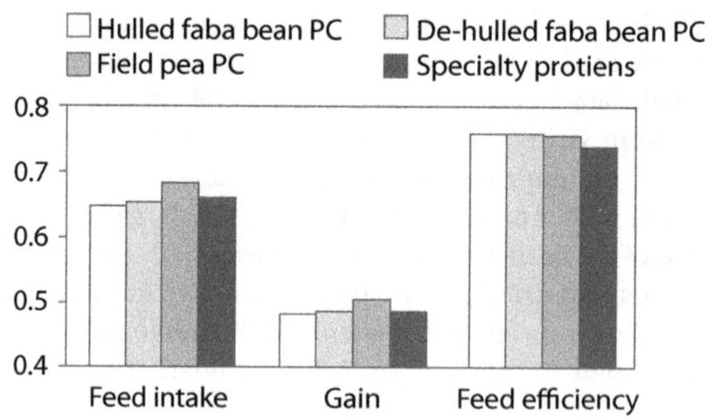

Figure 1. The lack of effect of protein source (PC = protein concentrate) included at 16 to 17.5% in the diet on feed intake (kg/d), gain (kg/d), and feed efficiency of weaned pigs; adapted from Gunawardena et al. *(2008).*

Fat sources

Adding up to 10% fat to diets low in fat may increase feed intake of pigs 2 to 5 weeks post-weaning (Li *et al.*, 1990); however, a sustained high dietary fat content may reduce feed intake in swine (Azain, 2001). Moderate changes in fatty acid profile by adding conjugated linoleic acid and monounsaturated fatty acids do not affect feed intake (Martin *et al.*, 2008). Still, fat digestibility differs among fat sources: long-chain unsaturated fatty acids have a higher digestibility than long-chain saturated fatty acids. Within the saturated fatty acids, digestibility is higher for medium and short-chain fatty acids than for long-chain fatty acids (Duran-Montgé *et al.*, 2007). Plant-based fats are thus preferential for diets for young pigs due to increased fat digestibility, although effects on feed intake might be neutral (Lawrence and Maxwell, 1983). Interestingly, feeding of a purified medium-chain triglyceride product may reduce feed intake of weaned pigs compared to corn, coconut, and soybean oil (Cera *et al.*, 1990). Oxidative damage to fat (rancidity) and elevated free fatty acid content are other factors with a negative effect on palatability (DeRouchey *et al.*, 2004). Finally, fat addition may reduce dustiness of feed, which is important because chronic exposure to dust may reduce feed intake (Wathes *et al.*, 2004).

Recently, crude glycerol as a co-product from the biodiesel industry has gathered increased interest from the feed industry as a potential raw material. Interestingly, crude glycerol will increase feed intake of pigs fed pelleted wheat-based diets quadratically, with a maximum response at around 4% dietary inclusion (Figure 2). In corn-based diets, the effects of crude glycerol on feed intake were not detected for weaned pigs (Groesbeck *et al.*, 2008) and grower-finisher pigs (Lammers *et al.*, 2008), perhaps due to the higher intrinsic oil content in corn than wheat, and thus lower fat content of wheat-based control diets. The mechanism for crude glycerol stimulating feed intake is not entirely clear. Effects of crude glycerol on hardness of pellets might play a role, because adding 4% glycerol made pellets softer from 3.5 to 1.1 kg (Kahl) at Akey Research, and glycerol may potentially increase pellet durability (Groesbeck *et al.*, 2008). However, effects of glycerol on pellet quality will differ depending on purity of the glycerol (Südekum *et al.*, 2008). Glycerol also directly improves taste due to its sweetness (Kijora *et al.*, 1995.), 0.6 × the sweetness of sugar cane, and potentially indirectly due to residual salts.

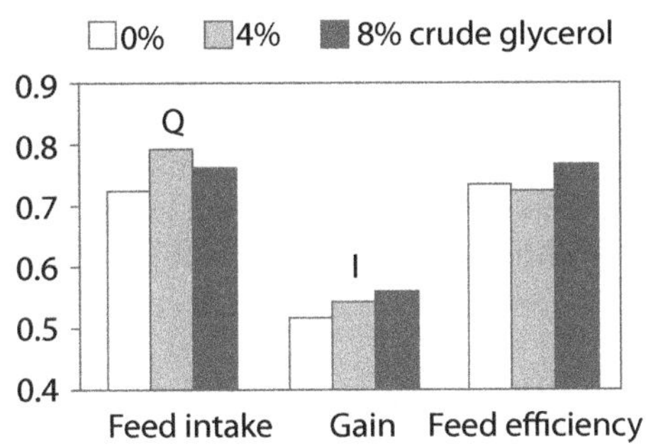

Figure 2. The addition of crude glycerol to pelleted wheat-based diets fed to weaned pigs will increase feed intake (kg/d) quadratically (Q; P<0.05) and may enhance gain (kg/d) linearly (l; P<0.10); adapted from Zijlstra et al. *(2009).*

Cereal grains

Cereal grains or their co-products are the largest component of diets for pigs, including for weaned pigs. Field pea and similar legume seeds may also serve as a source of starch in diets. Recently, attention is provided to detect taste preferences of weaned pigs for specific cereal grains included in diets without raw materials with a strong overriding taste such as spray-dried plasma or fish meal in a 2-way choice preference test against a common reference diet. For example, weaned pigs preferred brown rice and barley over oats (Figure 3), not surprising due to the indigestible oat hull. Naked oats (oats without the hull) was the feedstuff that was most preferred by young pigs. Naked oats or oat groats are thus considered a superior raw material for inclusion

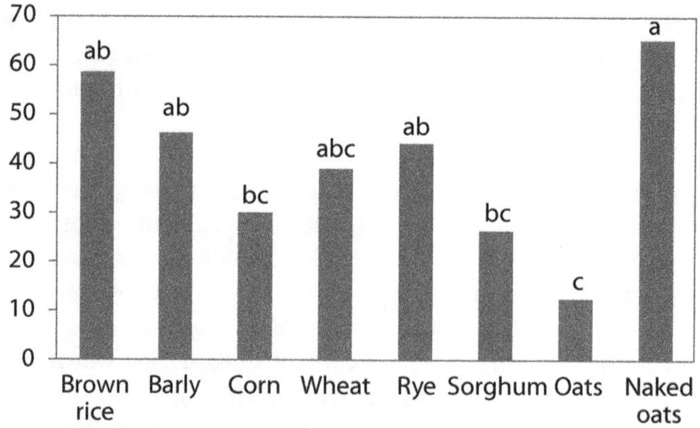

Figure 3. Preference (% of total feed intake) of diets based on 30% inclusion rate of cereal offered in a 2-way choice with a common reference diet (50% is no preference); adapted from Solà-Oriol et al. *(2009a).*

into diets fed to young pigs, due to their high nutrient digestibility and palatability (Zijlstra *et al.*, 2002). Surprisingly, corn has a relatively low preference.

Weaned pigs fed cereal grains other than corn and wheat, such as triticale, can achieve a similar feed intake and growth performance (Beltranena *et al.*, 2008), indicating that the current narrow scope of cereal grains used in the feeding of swine can be expanded. Finally, the capacity of pigs to consume diets high in barley and thus low in energy content is sometimes underestimated. In a practical setting of pig production, pigs fed a diet containing 25 to 30% barley were able to increase their feed intake by 13%, although intake of digestible energy could not be maintained (Beaulieu *et al.*, 2006).

One important factor within cereals that affects feed intake is the nature of the starch. This effect is likely mediated through the rate of starch digestion which in turn affects the rate of glucose absorption. Blood glucose and the resulting insulin response subsequently affect satiety. Hence, cereals with slowly digestible starch sources provide less stimulation of insulin and thus a higher feed intake and less lipogenesis (Van Kempen *et al.*, 2007, P.R. Regmi, personal communication, 2009). Using an in vitro assay, commonly used cereals and cereal byproducts have been characterised for starch digestion kinetics and NIRS calibrations have been developed for routine use. This work shows that raw barley provides the lowest insulin stimulus while extruded rice provides the highest.

Tests with diets based on purified starch with varying rates of starch digestion using piglets have yielded a 9% improvement of feed intake, and a 27% improvement in average daily gain by switching from rapidly to slowly digesting starch sources (Figure 4). The maximal rate of glucose release explained an impressive 61% of the variation in feed intake and 56% of the variation in weight gain (Van Kempen *et al.*, 2007). These effects have been repeated by feeding pigs diets formulated based on maximal rate of glucose release but using conventional raw materials (T.A.T.G. Van Kempen, personal communications).

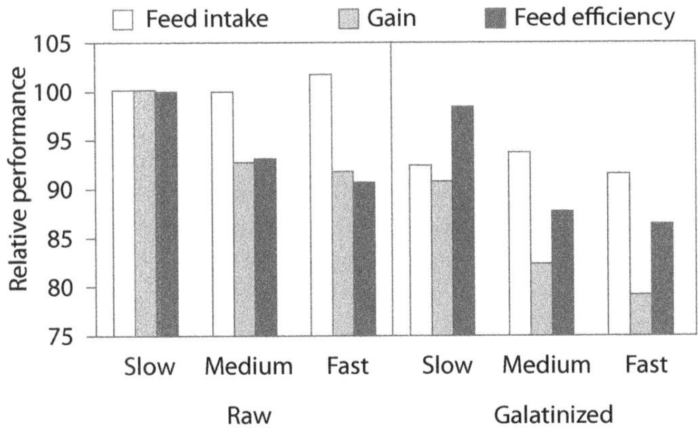

Figure 4. Relative growth performance in piglets fed feeds based on 6 different starch sources; adapted from van Kempen et al. *(2007).*

Feed manufacturing

The general assumption is that young pigs prefer a pelleted diet over a mash diet, although a large body of published data does not exist for this topic (Maxwell and Carter, 2001). Particle size reduction and extrusion may also impact feed intake. Generally, effects of feed processing on nutrient digestibility or characteristics of starch have been described in detail (e.g. Svihus *et al.*, 2005), whereas effects of feed processing on feed intake have gathered less attention in peer-reviewed research.

Particle size reduction

In feed mills, cereals and legume seeds are ground prior to mixing the feed, generally using a hammer mill or roller mill. Some feed mills grind the entire diet after mixing. Grinding can be accomplished using different screen sizes or roller mill settings, respectively, and thus create different particles sizes of the ground material. Mill type (roller or hammer) does not seem to impact digestibility variables much (Laurinen *et al.*, 2000; Hancock and Behnke, 2001), and effects of mill type on feed intake are not expected. Among a wide range of feedstuffs including cereals and oilseed meals, particle size was poorly correlated to feed preference in pigs (Solà-Oriol *et al.*, 2009b), indicating chemical factors and texture might be more important factors for feed intake. In contrast, reducing particle size within a cereal grain will improve nutrient digestibility, and hence feed efficiency (Hancock and Behnke, 2001), whereas consistent effects of reduced particle size on feed intake are less clear.

Grinding more finely or reducing mean particle size from 1,300 to 400 μm linearly reduced feed intake from weaning to 3 wk after weaning in pigs fed wheat-based diets (Mavromichalis *et al.*, 2000). Likewise, reducing mean particle size from 900 to 300 μm linearly reduced feed intake from 2 wk after weaning in pigs fed corn and sorghum-based diets, but did not affect feed intake immediately after weaning (Healy *et al.* 1994). Moreover, in piglets fed creep feed from day 6 to 27 prior to weaning at Provimi Polska, grinding coarser significantly increased feed intake by 77% from 65 to 115 g/d, indicating that piglets prefer more coarsely ground feed as well (R. Burek, personal communications). Finally, in grower-finisher pigs, reducing mean particle size of corn from 1,000 to 400 μm in corn-based diets (Wondra *et al.*, 1995a), of wheat from 600 to 400 μm in wheat-based diets (Mavromichalis *et al.*, 2000), of barley (Laurinen *et al.*, 2000) or grinding an entire barley and wheat-based diet more fine by using a 3- instead of 6-mm screen in the hammer mill (Callan *et al.*, 2007) consistently reduced feed intake. These results might be partly explained by the paradigm that pigs will control digestible energy intake by altering feed intake (Nyachoti *et al.*, 2004) or by the described effects of maximal rate of glucose release on feed intake (Van Kempen *et al.*, 2007). However, this paradigm contrasts observations in lactating sows that linearly increased feed intake when mean particle size was reduced from 1,200 to 400 μm in corn-based diets (Wondra *et al.*, 1995b). The reason for the discrepancy is not clear. In contrast to the importance of particle size for the large cereal portion of the diet, further grinding of a small dietary proportion of an already ground protein meal such as soybean does not affect feed intake at all (Lawrence *et al.*, 2003).

Apart from affecting feed intake via energy intake, reducing particle size may cause changes in feed intake indirectly via gastro-intestinal responses. For example, reducing particle size may potentially cause ulceration and keratinisation in the stomach (Mavromichalis, 2000; Wondra *et al.*, 2003a,b) that may, indirectly, reduce feed intake. Furthermore, pigs fed a coarse diet size may have a relatively heavier stomach, small intestine, and caecum and a longer small intestine, indicating that coarse diets may require more muscular and enzymatic action for processing by the stomach and small intestine than fine diets (Hedemann *et al.*, 2005), thereby increasing the energy requirement of pigs. Finally, changing particle size may alter the interaction of feed with intestinal microbial populations. For example, coarse grinding will increase fermentation in stomach (Mikkelsen *et al.*, 2004) and hind gut (Callan *et al.*, 2007). The changes in the physicochemical and microbial properties of digesta in the porcine stomach by feeding coarsely ground meal may decrease survival of Salmonella during passage through the stomach (Mikkelsen *et al.*, 2004). Increased organic matter reaching the hind gut by feeding pigs coarsely ground meal may increase Lactobacilli and reduce Enterobacteria populations observed in faeces (Callan *et al.*, 2007). The impact of these indirect effects on feed intake is less clear.

Pelleting

Steam pelleting is a technology that heats the mash feed using steam, compresses the feed into pellets, and then cools the pellets to remove the added moisture and heat. Steam conditioning before pelleting will result in more durable pellets and less starch damage than dry pelleting. Steam will also act as a lubricant between the mash and surface of the die hole in the pellet mill (Skoch *et al.*, 1983). Pelleting thus changes the physical form of the feed, but the added heat and resistance may also alter physico-chemical properties of the feed. Feeding pelleted feed to pigs immediately after weaning is the standard in the swine industry, a reasonable approach, because pigs fed pellets may achieve a higher feed intake in the first one week after weaning than pigs fed mash (Steidinger *et al.*, 2000). However, other research indicates that feeding in the form of pellets versus mash may reduce feed intake of pigs immediately after weaning, while resulting in equal growth (Medel *et al.*, 2003).

Both mash and pelleted feeds are fed globally to grower-finisher pigs and sows, and decisions to pellet feed are made for financial or logistical reasons, such as reduced feed wastage or increased ease of transportation. Generally, pelleting of feed will reduce feed intake of grower-finisher pigs and improves feed efficiency, partly due to improved nutrient digestibility (Hancock and Behnke, 2001). However, part of this difference in feed intake is simply due to feed wastage being higher in pigs fed mash diet.

Pellet quality is measured using indicators such as pellet durability index and pellet hardness (Thomas and Van der Poel, 1996). In North America, pellet durability index is used most, whereas pellet hardness is also used in Europe. Whereas pellet quality is important for appeal for the customer and can be measured accurately using sophisticated equipment, the pig likely perceives quality attributes differently. For example, young pigs may have a preference for softer pellets (Jensen and Becker, 1965), because extreme hardness of the pellet reduced voluntary feed intake particularly for the first few days. Hardness of pellets can be reduced by, for example, changing starch source from gelatinised starch to raw starch (Mavromichalis *et al.*, 2002) or by adding crude glycerol. In fact, selection of raw material, in particular the functional properties of the starch and protein fractions plays a larger role than steam pelleting on hardness and durability quality of pellet (Wood, 1987). Obviously, changing starch characteristics of raw materials may cause confounding effects of pellet hardness (Mavromichalis *et al.*, 2002) and starch digestion rates on feed intake (Van Kempen *et al.*, 2007).

Conventional wisdom may dictate that young, small pigs prefer small pellets or crumbled pellets and older, large pigs prefer larger pellets; however, published supporting evidence is weak (Hancock and Behnke, 2001). There may be a preference of piglets for smaller pellets (Aumaitre and Salmon-Legagneur, 1961 as referenced by Van der Poel *et al.*, 1998); however, careful screening of non peer-reviewed publications indicates that young pigs may not have a preference for pellet size ranging from 2 to 12 mm (Traylor *et al.*, 1996), from 1.8 to 9 mm Edge *et al.*, 2001) or differing between 3 and 5 mm (O'Connell and Lynch, 2006). Sometimes, feed intake does not differ between a pellet size of 1.8 and 2.0 mm, yet weight gain increased by 13% at Provimi BV (W. Markert, personal communications). Recently, reducing pellet size from 2.4 to 2.0 mm increased feed intake from 198 to 210 g/d in pigs immediately after weaning at SCA (S. Tibble, personal communications). The development of the 1.8-mm micropellet may provide further opportunities to enhance feed intake after weaning. During the first 25-d post-weaning in three subsequent trials, reducing pellet size from 2.4, 5, or 9-mm to 1.8 m increased feed intake from 407 to 422 g/d, from 2.0 to 1.8 mm increased feed intake from 448 to 479 g/d, and from 2.0 to 1.8 mm increased feed intake from 468 to 489 g/d (S. Tibble, personal communications). The combined body of research indicates that although controversy remains on the ideal pellet size, the technical challenge of producing small and durable pellet that is not too hard can be met.

Steam conditioning prior to the pellet mill can be replaced with shear conditioning, also known as expander processing. Expander processing may also entirely replace steam pelleting. Shear conditioning may alter the physico-chemical properties of mash feeds and improve the physical and hygienic characteristics of the compounded feed (Van der Poel *et al.*, 1998). Feed intake of grower pigs offered a choice between steam pelleted, expanded, or expanded plus steam pelleted feed strongly preferred steam pelleted or expanded feed. The pellets of expanded plus steam pelleted feed were harder than steam pelleted feed (Van der Poel *et al.*, 1998), illustrating the importance of physical quality of feed for feed intake of pigs. In grower-finisher pigs, expansion of feed based on wheat and barley without subsequent pelleting either did not alter feed intake or nutrient digestibility (O'Doherty and Keady, 2001; Callan *et al.*, 2007), indicating that in these cereals expansion in itself may not be effective to change physical properties of feed.

Similar to grinding, pelleting may cause changes in feed intake indirectly via gastro-intestinal responses, including via changes in intestinal microbial populations. For example, pelleting can reduce bacteria in feed and bacterial activity in the stomach (Mikkelsen *et al.*, 2004). The changes in the physicochemical and microbial properties of digesta in the porcine stomach by feeding pelleted feed may increase survival of Salmonella during passage through the stomach (Mikkelsen *et al.*, 2004). The impact of these indirect effects on feed intake is less clear.

Extrusion

Extrusion provides a prolonged period of heat and pressure compared to pelleting. Changes in physico-chemical properties are thus more severe; however, the process is also more expensive, resulting in less application for general swine feeding. In contrast to pet food and aquaculture application, use of extrusion might be limited to the young pig. Extrusion has been studied mostly for specific feedstuffs, like cereal grains to gelatinise starch and improve energy digestibility and protein feedstuffs to remove ANF and improve amino acid digestibility.

Effects of extruding on feed intake of pigs have been inconsistent among studies. Extrusion of maize included at 60% into diets for weaned pigs did not improve feed intake of weaned pigs consistently (Amornthewaphat and Attamangkune, 2008), and neither did gradual increasing the proportion of corn that was extruded (Lv *et al.*, 2006). Extrusion of carbohydrate sources included at 42 to 47% in diets for weaned pigs did not alter feed intake (Hongtrakul *et al.*, 1998). Extrusion of maize will improve digestibility of N and energy, and will increase butyrate production in the hindgut, likely because more retrograded starch will reach the hindgut (Lv *et al.*, 2006). Extrusion of wheat or sorghum did not affect feed intake of pigs, even though extrusion enhanced starch digestibility in the wheat-based diets (Durmic *et al.*, 2002), and falls in line with the theory that not nutrient digestibility but glycemic index is a prime driver of feed intake (Van Kempen *et al.*, 2007). Extrusion improved the preference value for corn and naked oats, but did not affect barley, rice, or wheat (Solà-Oriol *et al.*, 2009a). Finally, extrusion cooking of barley fed to weaned piglets significantly increased feed intake by 17% from day 0 to 14 after weaning at Provimi Research (A. van Wesel, personal communications), indicating that extrusion may play a critical role in improving feed intake of diets high in fibre immediately after weaning, although these effects of cooking cereal grains have not been observed by others (Medel *et al.*, 2002, 2003).

Extrusion of protein sources and whole diets has been studied as well. Extrusion of soy flakes or soy protein concentrate but not soybean meal enhanced feed intake of weaned pigs for the first 2 weeks after weaning (Friesen *et al.*, 1993; Woodworth *et al.*, 2001), partly due to reduced ANF content. Extrusion of a complex diet improved feed intake of weaned pigs compared to a pelleted diet, but extrusion did not improve feed intake of the simple diet (Hongtrakul *et al.*, 1998).

Non-traditional feed processing

This review is mostly focused on the role of traditional feed manufacturing on feed intake of pigs. However, we recognise the possibilities that components of non-traditional manufacturing and delivery may offer. For example, liquid feeding may drastically improve feed efficiency, and as a result, pigs may achieve the same gain with a reduced feed intake (Choct *et al.*, 2004). Feeding feed as a paste as opposed to a dry feed (51 instead of 90% dry matter, respectively) statistically increased the intake of creep feed during the suckling period from 114 to 197 g/d (R. Burek, personal communications), further highlighting the importance of feed delivery. Prolonged steeping, i.e. soaking, of the cereal-based feed prior to final delivery to pigs appears to be an interesting tool for enhancing feed intake of pigs fed a liquid diet (Choct *et al.*, 2004). Enzymes intrinsic to the raw materials may be activated during the steeping process (Brooks *et al.*, 1996), and thereby stimulate nutrient digestibility. Finally, fermentation of feed has been suggested as an approach to control intestinal microbial populations; however, even though non-fermented liquid feed may increase feed intake of grower pigs, fermentation of liquid feed may actually reduce feed intake and thereby reduce weight gain (Canibe and Jensen, 2003), indicating that fermentation of liquid feed does not consistently influence feed intake (Scholten *et al.*, 1999).

Conclusions and implications

Feed manufacturing technology is an important component for stimulating feed intake in young pigs. The selection of raw materials and feed manufacturing are two items that can be carefully controlled by the feed manufacturing industry. Raw materials will influence taste and functional properties of feed, thereby altering feed intake of young pigs. Using manufacturing, physical shape, chemical structure and functional properties of feed can be altered, providing further opportunities to enhance feed intake of young pigs. However, manufacturing can also be too extreme, indicating that there is a delicate balance between ingredients and manufacturing technology that needs to be managed in concert for stimulating feed intake of pigs.

Acknowledgements

The authors thank Dr. R. Burek (Provimi Polska), Ing. Ad van Wesel (Provimi Research), Dr. Wolfgang Markert (Provimi BV), and Dr. Doug Cook (Akey) for providing materials for inclusion in this review.

References

Amornthewaphat, N. and Attamangkune, S., 2008. Extrusion and animal performance effects of extruded maize quality on digestibility and growth performance in rats and nursery pigs. Animal Feed Science and Technology 144: 292–305.

Aumaitre, A. and Salmon-Legagneur, E., 1961. Les preferences alimentaires du porcelet. Annales de Zootechnique 10: 197-203.

Azain, M.J., 2001. Fat in swine nutrition. In: Lewis, A.J., and Southern, L.L. (eds.) Swine Nutrition. CRC Press, Boca Raton, FL, USA, pp. 95-105.

Beaulieu, A.D., Levesque, C.L. and Patience, J.F., 2006. The effects of dietary energy concentration and weaning site on weanling pig performance. Journal of Animal Science 2006. 84: 1159-1168.

Beltranena, E., Salmon, D.F., Goonewardene, L.A. and Zijlstra, R.T., 2008. Triticale as a replacement for wheat in diets for weaned pigs. Canadian Journal of Animal Science 88: 631-635.

Bhandari, S.K., Xu, B., Nyachoti, C.M., Giesting, D.W. and Krause, D.O., 2008. Evaluation of alternatives to antibiotics using an Escherichia coli K88+ model of piglet diarrhea: Effects on gut microbial ecology. Journal of Animal Science 86: 836-847.

Brooks, P.H., Geary, T.M., Morgan, D.T. and Campbell, A., 1996. New developments in liquid feeding. Pig Veterinary Journal 36: 43-64.

Callan, J.J., Garry, B.P. and O'Doherty, J.V., 2007. The effect of expander processing and screen size on nutrient digestibility, growth performance, selected faecal microbial populations and faecal volatile fatty acid concentrations in grower–finisher pigs. Animal Feed Science and Technology 134: 223-234.

Canibe, N. and Jensen, B.B., 2003. Fermented and nonfermented liquid feed to growing pigs: Effect on aspects of gastrointestinal ecology and growth performance. Journal of Animal Science 81: 2019-2031.

Cera, K.R., Mahan, D.C. and Reinhart, G.A., 1990. Evaluation of various extracted vegetable oils, roasted soybeans, medium-chain triglyceride and an animal-vegetable fat blend for postweaning swine. Journal of Animal Science 68: 2756-2765.

Choct, M., Selby, E.A.D., Cadogan, D.J. and Campbell, R.G., 2004. Effect of liquid to feed ratio, steeping time, and enzyme supplementation on the performance of weaner pigs. Australian Journal of Agricultural Research 55: 247-252.

Chiba, L.I., 2001. Protein supplements. In: Lewis, A.J. and Southern, L.L. (eds.) Swine nutrition, 2nd Edition. CRC Press, Boca Raton, FL, USA, pp. 803-837.

Clasadonte, L. and Van der Poel, A.F.B., 2009. Antinutritional factors: impact on pigs' voluntary feed intake. In: Torrallardona, D. and Roura, E. (eds.) Voluntary feed intake in pigs. Wageningen Academic Publishers, the Netherlands, pp. 243-275.

Coffey, R.D. and Cromwell, G.L., 1995. The impact of environment and antimicrobial agents on the growth response of early-weaned pigs to spray-dried porcine plasma. Journal of Animal Science 73: 2532-2539.

DeRouchey, J.M., Hancock, J.D., Hines, R.H., Maloney, C.A., Lee, D.J., Cao, H., Dean, D.W. and Park, J.S., 2004. Effects of rancidity and free fatty acids in choice white grease on growth performance and nutrient digestibility in weanling pigs. Journal of Animal Science 82: 2937-2944.

Duran-Montgé, P., Lizardo, R., Torrallardona, D. and Esteve-Garcia, E., 2007. Fat and fatty acid digestibility of different fat sources in growing pigs. Livestock Science 109: 66-69.

Durmic, Z., Pethick, D.W., Mullan, B.P., Accioly, J.M., Schulze, H. and Hampson, D.J. 2002. Evaluation of large-intestinal parameters associated with dietary treatments designed to reduce the occurrence of swine dysentery. British Journal of Nutrition 88: 159-169.

Edge, H.L., Dalby, J.A., Rowlinson, P., Varley, M.A., 2001. The effect of pellet size on the voluntary food intake and performance of young pigs. Proceedings of BSAS Mtg., pp. 167 [Abstract].

Ermer, P.M., Miller, P.S. and Lewis, A.J., 1994. Diet preference and meal patterns of weanling pigs offered diets containing either spray-dried porcine plasma or dried skim milk. Journal of Animal Science 72: 1548-1554.

Friesen, K.G., Nelssen, J.L., Goodband, R.D., Behnke, K.C., Kats, L.J., 1993. The effect of moist extrusion of soy products on growth performance and nutrient utilization in the early-weaned pig. Journal of Animal Science 71: 2099-2109.

Groesbeck, C.N., McKinney, L.J., DeRouchey, J.M., Tokach, M.D., Goodband, R.D., Dritz, S.S., Nelssen, J.L., Duttlinger, A.W., Fahrenholz, A.C. and Behnke, K.C., 2008. Effect of crude glycerol on pellet mill production and nursery pig growth performance. Journal of Animal Science 86: 2228-2236.

Gunawardena, C.K., Zijlstra R.T., and Beltranena, E., 2008. Faba bean and field pea protein concentrates as replacements for specialty proteins in nursery diets for weaned pigs. Journal of Animal Science 86 (e-Suppl. 3): 45.

Hancock, J.D. and Behnke, K.C. 2001. Use of ingredient and diet processing technologies (grinding, mixing, pelleting, and extruding) to produce quality feeds for pigs. In: Lewis, A.J., and Southern, L.L. (eds.) Swine Nutrition, 2nd Edition. CRC Press, Boca Raton, FL, USA, pp. 469-497.

Healy, B.J., Hancock, J.D., Kennedy, G.A., Bramel-Cox, P.J., Behnke, K.C. and Hines R.H., 1994. Optimum particle size of corn and hard and soft sorghum for nursery pigs. Journal of Animal Science 72: 2227-2236.

Hedemann, M.S., Mikkelsen, L.L., Naughton, P.J. and Jensen B.B., 2005. Effect of feed particle size and feed processing on morphological characteristics in the small and large intestine of pigs and on adhesion of *Salmonella enterica* serovar Typhimurium DT12 in the ileum in vitro. Journal of Animal Science 83: 1554-1562.

Hongtrakul, K., Goodband, R.D., Behnke, K.C., Nelssen, J.L., Tokach, M.D., Bergstrom, J.R., Nessmith, Jr.,W.B. and Kim, I.H., 1998. The effects of extrusion processing of carbohydrate sources on weanling pig performance. Journal of Animal Science 76: 3034-3042.

Huisman, J., Van der Poel, A.F.B., Verstegen, M.W.A. and Van Weerden, E.J., 1990. Antinutritional factors (ANF) in pig nutrition. World Review of Animal Production 25: 77-82.

Jansman, A.J., 1993. Tannins in feedstuffs for simple-stomached animals. Nutrition Research Reviews 6: 209-236.

Jensen, A.H. and Becker, D.E., 1965. Effect of pelleting diets and dietary components on the performance of young pigs. Journal of Animal Science 24: 392-397.

Kats, L.J., Nelssen, J.L., Tokach, M.D., Goodband, R.D., Hansen, J.A. and Laurin, J.L., 1994. The effect of spray-dried porcine plasma on growth performance in the early weaned pig. Journal of Animal Science 72: 2075-2081.

Kijora, C., Bergner, H., Kupsch, R.D. and Hagemann, L., 1995. Glycerol as a feed component in diets of fattening pigs. Archives of Animal Nutrition. 47: 345-360.

Kim, J.H., Heo, K.N., Odle, J., Han, K. and Harrell, R.J., 2001. Liquid diets accelerate the growth of early-weaned pigs and the effects are maintained to market weight. Journal of Animal Science 79: 427-434.

Kyriazakis, I. and Emmans, G.C. 1995. The voluntary feed intake of pigs given feeds based on wheat bran, dried citrus pulp and grass meal, in relation to measurements of feed bulk. British Journal of Nutrition 73: 191–207.

Lallès, J.P., Bosi, P., Smidt, H. and Stokes, C.R., 2007. Nutritional management of gut health in pigs around weaning. Proceedings of the Nutrition Society 66: 240-268.

Lammers, P.J., Kerr, B.J., Weber, T.E., Bregendahl, K., Lonergan, S.M., Prusa, K.J., Ahn, D.U., Stoffregen, W.C., Dozier, W.A., III and Honeyman, M.S., 2008. Growth performance, carcass characteristics, meat quality, and tissue histology of growing pigs fed crude glycerin-supplemented diets. Journal of Animal Science.86: 2962-2970.

Laurinen, P., Siljander-Rasi, H., Karhunen, J., Alaviuhkola, T., Näsi, M. and Tuppi, K., 2000. Effects of different grinding methods and particle size of barley and wheat on pig performance and digestibility. Animal Feed Science and Technology 83: 1-16.

Lawrence, N.J. and Maxwell, C.V., 1983. Effect of dietary fat source and level on the performance of neonatal and early weaned pigs. Journal of Animal Science 57: 936-942.

Lawrence, K.R., Hastad, C.W., Goodband, R.D., Tokach, M.D., Dritz, S.S., Nelssen, J.L., DeRouchey, J.M. and Webster, M.J., 2003. Effects of soybean meal particle size on growth performance of nursery pigs. Journal of Animal Science 81: 2118-2122.

Li, D.F., Thaler, R.C., Nelssen, J.L., Harmon, D.L., Allee, G.L. and Weeden, T.L., 1990. Effect of fat sources and combinations on starter pig performance, nutrient digestibility and intestinal morphology. Journal of Animal Science 68: 3694-3704.

Lv, S., Li, D., Xing, J., Yongxi, M., Huang, D. and Li, J., 2006. Effects of extrusion of corn on growth performance, nutrient digestibility and short-chain fatty acid profiles in the hindgut of weaned piglets. Archives of Animal Nutrition 60: 170-179.

Martin, D., Muriel, E., Gonzalez, E., Viguera, J. and Ruiz, J., 2008. Effect of dietary conjugated linoleic acid and monounsaturated fatty acids on productive, carcass and meat quality traits of pigs. Livestock Science 117: 155-164.

Mavromichalis, I., Hancock, J.D., Senne, B.W., Gugle, T.L., Kennedy, G.A., Hines, R.H. and Wyatt, C.L., 2000. Enzyme supplementation and particle size of wheat in diets for nursery and finishing pigs. Journal of Animal Science 78: 3086-3095.

Mavromichalis, I., Cook, D.R., Ward, M.M. and Paton, N.D. 2002. Effect of pellet hardness on growth performance of weaned pigs. Journal of Animal Science 80 (Suppl. 2): 29.

Maxwell, C.V. and Carter, S.D., 2001. Feeding the weaned pig. In: Lewis, A.J., and Southern, L.L. (eds.) Swine nutrition, 2nd Edition. CRC Press, Boca Raton, FL, USA, pp. 691-715.

Medel, P., Baucells, F., Gracia, M. I., De Blas, C. and Mateos, G.G., 2002. Processing of barley and enzyme supplementation in diets for young pigs. Animal Feed Science and Technology 95: 113–122.

Medel, P., Latorre, M.A., de Blas, C., Lázaro, R. and Mateos, G.G., 2004. Heat processing of cereals in mash or pellet diets for young pigs. Animal Feed Science and Technology 113: 127-140.

Mikkelsen, L.L., Naughton, P.J., Hedemann, M.S. and Jensen, B.B., 2004. Effects of physical properties of feed on microbial ecology and survival of salmonella enterica serovar typhimurium in the pig gastrointestinal tract. Applied Environmental Microbiology 70: 3485-3492.

Muley, N.S., Van Heugten, E., Moeser, A.J., Rausch, K.D. and Van Kempen, T.A.T.G., 2007. Nutritional value for swine of extruded corn and corn fractions obtained after dry milling. Journal of Animal Science 85: 1695-1701.

Nyachoti, C.M., Zijlstra, R.T., De Lange, C.F.M and Patience, J.F., 2004. Voluntary feed intake in growing-finishing pigs: a review of the main determining factors and potential approaches for accurate predictions. Canadian Journal of Animal Science 84: 549-566.

O'Connell, M.K. and Lynch, P.B., 2006. Effect of pellet size on pellet quality and pig performance. Agricultural Research Forum, Tullamore, Ireland.

O'Doherty, J.V., Keady, U., 2001. The effects of expander processing and extrusion on the nutritive value of peas for pigs. Animal Science 72: 43-53.

Pahm, A.A., Pedersen, C. and Stein, H.H., 2009. Standardized ileal digestibility of reactive lysine in distillers dried grains with solubles fed to growing pigs. Journal of Agricultural and Food Chemistry 57: 535-539.

Pérez-Bosque, A., Miro, L., Polo, J., Russell, L., Campbell, J., Weaver, E., Crenshaw, J. and Moreto, M., 2008. Dietary plasma protein modulate the immune response of diffuse gut-associated lymphoid tissue in rats challenged with *Staphylococcus aureus* enterotoxin. British Journal of Nutrition 138: 533-537.

Pettigrew, J. E., 2006. Reduced use of antibiotic growth promoters in diets fed to weanling pigs: dietary tools, part 1. Animal Biotechnology 17: 207-215.

Pierce, J. L., Cromwell, G.L., Lindemann, M.D., Russell, L.E. and Weaver, E.M., 2005. Effects of spray-dried animal plasma and immunoglobulins on performance of early weaned pigs. Journal of Animal Science 83: 2876-2885.

Pluske, J.R., Hampson, D.J. and Williams, I.H., 1997. Factors influencing the structure and function of the small intestine in the weaned pig: A review. Livestock Production Science 51: 215-236.

Roth-Maier, D.A., Böhmer, B.M., Roth, F.X., 2004. Effects of feeding canola meal and sweet lupin (*L. leteus L. angustifolus*) in amino acid balanced diets on growth performance and carcass characteristics of growing and finishing pigs. Animal Research 53: 21-34.

Sauvant, D., Perez, J.-M. and Tran, G. 2004. Tables of composition and nutritional value of feed materials. Wageningen Academic Publishers, Wageningen, the Netherlands and INRA, Paris, France. 304 pp.

Scholten, R.H.J., Van der Peet-Schwering, C.M.C., Verstegen, M.W.A., Den Hartog, L.A., Schrama, J.W. and Vesseur, P.C., 1999. Fermented co-products and fermented compound diets for pigs: a review. Animal Feed Science and Technology 82: 1-19.

Skoch, E.R., Binder, S.F., Deyoe, C.W., Allee, G.L. and Behnke, K.C., 1983. Effects of pelleting conditions on performance of pigs fed a corn-soybean meal diet. Journal of Animal Science 57: 922-928.

Solà-Oriol, D., Roura, E. and Torrallardona D., 2009a. Feed preference in pigs: Effect of cereal sources at different inclusion rates. Journal of Animal Science 87: 562-570.

Solà-Oriol, D., Roura, E. and Torrallardona, D., 2009b. Feed preference in pigs: Relationship with feed particle size and texture. Journal of Animal Science 87: 571-582.

Steidinger, M.U., Goodband, R.D., Tokach, M.D., Dritz, S.S., Nelssen, J.L., McKinney, L.J., Borg, B.S. and Campbell, J.M., 2000. Effects of pelleting and pellet conditioning temperatures on weanling pig performance. Journal of Animal Science 78: 3014-3018.

Südekum, K.-H., Schröder, A., Fiebelkorn, S., Schwer, R. and Thalmann, A., 2008. Quality characteristics of pelleted compound feeds under varying storage conditions as influenced by purity and concentration of glycerol from biodiesel production. Journal of Animal and Feed Sciences 17: 120-136.

Svihus, B., Uhlen, A.K., Harstad, O.M., 2005. Effect of starch granule structure, associated components and processing on nutritive value of cereal starch: A review. Animal Feed Science and Technology 122: 303-320.

Thacker, P.A. and Kirkwood, R.N., 1990. Non-traditional feed sources for use in swine production. Butterworth, Stoneham, MA, USA. 515 pp.

Thacker, P.A., 1990. Canola meal. In: Thacker, P.A. and Kirkwood, R.N. (eds.) Non-traditional feed sources for use in swine production. Stoneham, MA, USA, pp. 69-78.

Thomas, M. and Van der Poel, A.F.B., 1996. Physical quality of pelleted animal feed 1. Criteria for pellet quality. Animal Feed Science and Technology 61: 89-112.

Torrallardona, D., Conde, M.R., Badiola, I., Polo, J. and Brufau, J. 2003. Effect of fishmeal replacement with spray-dried animal plasma and colistin on intestinal structure, intestinal microbiology, and performance of weanling pigs challenged with Escherichia coli K99. Journal of Animal Science 81: 1220-1226.

Traylor, S.L., Behnke, K.C., Hancock, J.D., Sorrell, P. and Hines, R.H., 1996. Effects of pellet size on growth performance in nursery and finishing pigs. Journal of Animal Science 74 (Suppl. 1): 67.

Tripathi, M.K. and Mishra, A.S., 2007. Glucosinolates in animal nutrition: a review. Animal Feed Science and Technology: 132: 1-27.

Van Boekel, M.A., 2006. Formation of flavour compounds in the Maillard reaction. Research review paper. Biotechnology Advances 24: 230-233.

Van der Poel, A.F.B., Mollee, P.W., Huisman, J. and Liener, I.E., 1990. Variations among species of animals in response to the feeding of heat-processed beans (*Phaseolus vulgaris L.*). 1. Bean processing and effects on growth, digestibility and organ weights in piglets. Livestock Production Science 25: 121-135.

Van der Poel, A.F.B., Schoterman, A. and Bosch, M.W., 1998. Effect of expander conditioning and/or pelleting of a diet on the ileal digestibility of nutrients and on feed intake after choice feeding of pigs. Journal of the Science of Food and Agriculture 76: 87-90.

Van Dijk, A.J., Everts, P.H., Nabuurs, M.J.A., Margry, R.J.C.F. and Beynen, A.C., 2001. Growth performance of weanling pigs fed spray-dried animal plasma: A review. Livestock Production Science 68: 263-274.

Van Heugten, E., 2001. Mycotoxins and other antinutritional factors in swine feeds. In: Lewis, A.J. and Southern, L.L. (eds.) Swine nutrition, 2nd Edition. CRC Press, Boca Raton, FL, USA, pp. 563-583.

Van Kempen, T., Pujol, S., Tibble, S. and Balfagon, A. 2007. In vitro characterization of starch digestion and its implications for pigs. In: Wiseman, J., Varley, M.A., McOrist, S. and Kemp, B. (eds.) Paradigms in pig science. Nottingham University Press, Nottingham, UK. pp. 515-526.

Wathes, C.M., Demmers, T.G.M., Teer, N., White, R.P., Taylor, L.L., Bland, V., Jones, P., Armstrong, D., Gresham, A.C.J., Hartung, J., Chennells, D.J. and Done, S.H., 2004. Production responses of weaned pigs after chronic exposure to airborne dust and ammonia. Animal Science 78: 87-97.

Wondra, K.J., Hancock, J.D., Hines, R.H. and Stark, C.R. 1995a. Effects of particle size and pelleting on growth performance, nutrient digestibility, and stomach morphology in finishing pigs. Journal of Animal Science 73: 757-763.

Wondra, K.J., Hancock, J.D., Kennedy, G.A., Hines, R.H. and Behnke, K.C., 1995b. Reducing particle size of corn in lactation diets from 1,200 to 400 micrometers improves sow and litter performance. Journal of Animal Science 73: 421-426.

Wood, J.F., 1987. The functional properties of feed raw materials and their effect on the production and quality of feed pellets. Animal Feed Science and Technology 18: 1-17.

Woodworth, J.C., Tokach, M.D., Goodband, R.D., Nelssen, J.L., O'Quinn, P.R., Knabe, D.A. and Said, N.W., 2001. Apparent ileal digestibility of amino acids and the digestible and metabolizable energy content of dry extruded-expelled soybean meal and its effects on growth performance of pigs. Journal of Animal Science 79: 1280-1287.

Zarkadas, L.N. and Wiseman, J., 2005. Influence of processing of full fat soya beans included in diets for piglets. I. Performance. Animal Feed Science and Technology 118: 109-119.

Zijlstra, R.T., Whang, K.Y., Odle, J. and Easter, R.A., 1996. Effect of milk replacer feeding to early-weaned pigs on growth, body composition and small-intestinal morphology, compared to suckling littermates. Journal of Animal Science 74: 2948–2959.

Zijlstra, R.T., B.G. Rossnagel, D.D. Maenz, J.F. Patience, V.J. Racz, 2002. Nutritional value of high-oil oat groats. 2001 Annual Research Report, Prairie Swine Centre Inc., Saskatoon, SK. pp 26.

Zijlstra, R.T., Van Kessel, A.G. and Drew, M.D., 2004. Ingredient fractionation: The value of value-added processing for animal nutrition. 'The worth of the sum of parts versus the whole'. In: Proceedings of the 25th Western Nutrition Conference, Saskatoon, SK, Canada, pp. 41-53.

Zijlstra, R.T. and Beltranena, E., 2008. Variability of quality in biofuel co-products. In: Garnsworthy, P.C. and Wiseman, J (eds.) Recent Advances in Animal Nutrition 2008. Nottingham Academic Press, Nottingham, UK, pp. 313-326.

Zijlstra, R.T., Menjivar, K., Lawrence, E. and Beltranena, E., 2009. The effect of crude glycerol on growth performance and nutrient digestibility in weaned pigs. Canadian Journal of Animal Science 89: 85–89.

Chapter 12. Feeding behaviour and social influences on feed intake

X. Manteca[1] and S. Edwards[2]
[1]School of Veterinary Science, Universitat Autònoma de Barcelona, 08193 Bellaterra (Barcelona), Spain; xavier.manteca@uab.cat
[2]School of Agriculture, Food and Rural Development, University of Newcastle, Newcastle Upon Tyne, NE1 7RU, United Kingdom; sandra.edwards@ncl.ac.uk

Abstract

The food intake and growth of pigs are affected by both physical and social constraints influencing feeding behaviour. Feeder design has received some empirical study, but only limited systematic scientific investigation despite its importance for both intake and feed wastage. Feeding space requirements are subject to interacting influences of pig age, diet form, feeder design and social constraints. The flexibility of feeding behaviour in the pig makes it difficult to synthesise firm predictions of the relationship between feeding space and performance in different contexts from current knowledge. Other social stressors including group size, mixing of unfamiliar animals and stocking density can adversely affect performance by inhibiting or disrupting access to the feeder, by indirect effects on feed intake through a physiological stress response and by exacerbating the effects of heat stress when pigs are crowded. The interactions between these factors require further research to develop quantitative predictive models to inform practical decisions.

Keywords: feeder design, feeding space, feed intake, growth, social stressors

Introduction

Pigs evolved as an omnivorous, foraging species. Wild boar or domestic pigs kept under semi-natural conditions, spend more than 50% of their active time during the day in foraging behaviour as family groups, spacing themselves widely during this activity (Graves, 1984; Stolba and Wood-Gush, 1989). This contrasts with the farmed situation in which groups of pigs, usually of similar age and size, are fed nutrient dense diets under conditions of restricted space. These diets may be provided in discrete meals, when pigs must compete spatially for a restricted feed supply, or fed ad libitum from hoppers where pigs must share access over time. In both circumstances the social interactions between animals may have an important effect on food intake and growth rate. The objectives of this chapter are (1) to review spatial and temporal aspects of feeding behaviour, (2) to review the effects of competition for access to the feeder on food intake and growth rate, and discuss the possible effects of group size (i.e. number of pigs per pen) (3) to review the effects of other social influences such as social mixing and floor space allowance on food intake and growth rate, and (4) to suggest areas deserving further research with an special emphasis on the positive effects of social interactions on welfare, food intake and growth rate.

Feeder design

The equipment used for feed provision varies widely in commercial practice. At its simplest, feed is dispensed in discrete meals onto the floor or into open troughs, leaving individuals within the group to compete for access. Under these conditions, ensuring that all animals obtain a fair share of the allocation is the commercial priority, to prevent uneven growth and high group variability in liveweight. Feed may also be provided ad libitum from linear hoppers, circular pans or single space feeders. In these circumstances, it is important to ensure maximum intake by all individuals, but to simultaneously minimise feed wastage. When feed supply is not limited, wastage can in many situations be extensive with levels in excess of 10% being not uncommon (Baxter, 1989).

A great diversity of feeder designs is found in practice, but there is very little scientific information on their critical design features. Whilst a number of articles in industry publications suggest the merits of different types of feeders (e.g. Taylor and Curtis, 1989), few critical evaluations appear in refereed scientific publications. In response to the suggestion of Vargus *et al.*, (1987) that circular feeding troughs might offer advantages in managing agonistic interactions during feeding, Blackshaw and Blackshaw (1990) compared two feeder designs for growing pigs: a round feeder and a linear feeder. The round feeder provided 9 feeding spaces, separated only by simple rods, at a 94 cm diameter trough, whilst the 50 cm linear feeder provided 2 rectangular feeding places with partitions providing full head and neck protection. With groups of 12 pigs fed ad libitum, both feeders had similar occupation time, low levels of aggression and did not differ in pig growth rate or feed utilisation efficiency. Subsequent work has focussed mainly on feeding space and water provision. Comparisons of single-space and multi-space feeders for growing and finishing pigs have generally shown no difference in pig performance (Walker and Overton, 1989; Walker, 1990; Gonyou and Lou, 2000). However, feeders which incorporated a water supply gave higher intakes for growing and finishing pigs than those supplying only dry feed (Walker, 1990; Gonyou and Lou, 2000). This result was supported by a review of 48 separate experiments in US institutions (Wastell and Pettigrew, 2000).

A review by Laitat *et al.* (2005) highlighted the diversity of feeder designs for weaned piglets, identifying 18 different categories based on size, shape and presence or absence of water provision. Again, few of these have been rigorously evaluated in comparative experiments. In contrast to results in finishing pigs, Pluske and Williams (1996) found no benefit of a wet-and-dry single space feeder when compared to dry feed for weaned piglets. Laitat *et al.* (1999) compared a wet-and-dry multi-space tube feeder with a number of separate single-space dry feeder units separated by water nipples, and again found no consistent difference in pig performance. O'Connell *et al.* (2002) compared five feeder types: two multi-space designs, with dry or wet-and-dry feed, two feeders with communal troughs, rectangular and circular in shape, and a single-space feeder. All feeders supplied water except the multi-space dry design. This dry feed design gave less feeder competition, least aggression and lowest feeder occupancy. Feed conversion ratio was best with the single-space feeder, and poorest with the wet-and-dry multi-space design. In a subsequent evaluation of a wet-and-dry single space feeder against a multi-space dry feeder, the dry feeder again gave better feed efficiency during the weaner phase (Magowan *et al.*, 2008). In this factorially designed experiment, a change in feeder type after the weaner stage, at 10 weeks of age, gave a transient reduction in feed intake, but pigs changing from the dry to the wet-and-dry feeder at this time showed better subsequent growth rate. However, feeder type had no significant effect on feed intake or feed efficiency in the finishing stage, or overall from wean to finish.

Comparisons of this nature are often simple product tests, since the feeders used in a given experiment often differ in many respects including space allowance, trough shape and feed form. This makes it difficult to draw any generalised conclusions. The optimal design for feeding troughs has received relatively little systematic scientific research. Baxter (1989) was the first to study feeding behaviour in detail with the objective of optimising trough design. By photographing pigs of different ages during natural feeding, he developed a mathematical description of the posture of the animals in terms of the dimensions and angles of the head and neck region. This information was then used to describe a space envelope which the interior of a feeder should provide to allow pigs of different liveweights to feed in their preferred position. Designing a feeder profile using these criteria, he demonstrated that feed wastage could be reduced to <2%.

Feeding space requirement

In addition to the cross sectional profile of a feeder, modelled in the work of Baxter (1989), the lateral feeding space required by a pig can also be calculated mathematically. When groups of pigs are restrict-fed in discrete meals, they must be allowed at least a body width of trough space for all to be able to eat simultaneously. This body width can be related to liveweight using allometry. The width of a pig at the shoulders can be defined as:

$W = 0.064 \times \text{liveweight}^{0.33}$ (Petherick, 1983)

and this defines the minimum necessary feeding space. When pigs are fed from circular feeders, the trough circumference for one pig feeding place can be less than the linear shoulder width, since the radial effect means that only a head width is required. This makes circular feeders theoretically more space efficient for a given number of feeding places, although this has not been rigorously tested.

However, the provision of the minimum trough length for restrict-fed pigs does not ensure that all animals will access adequate feed. Baxter (1983) recorded the behaviour of pigs when feeding at a trough space allowance 9% greater than shoulder width. He found that all pigs fed simultaneously for only 24% of the feeding time, as a result of aggression at the trough. A pig withdrew from feeding because of aggression on average every 2 minutes throughout the feeding period. This aggression was initiated by feeding pigs, often socially subordinate animals, whilst defending their position at the trough rather than by pigs trying to gain access. Such aggression could be markedly reduced, and the simultaneous utilisation of trough space improved, by the inclusion of trough dividers. Head barriers reduced aggression by 60%, and head and shoulder barriers by almost 100%, with associated reduction in feed wastage (Baxter, 1989).

Where pigs are fed ad libitum, the calculation of necessary feeding space for a group is a more complex issue, since the possibility of feed space sharing over time exists. In the wild, pigs need to forage for much of the day to obtain sufficient food. However, when given more concentrated feed under farm conditions, the time required to locate food is minimal, and the time required to ingest the daily feed requirement is relatively small. Thus pigs spend only 60-80 minutes per day in ingestive behaviour (De Haer and De Vries, 1983; Hyun *et al.*, 1997; Bruininx *et al.*, 2001). These data would suggest that 18-24 pigs could share a feeding place if trough space was used with perfect efficiency throughout the 24 hour period. Such efficiency would seem unlikely, since pigs show both diurnal patterns of feeding behaviour and also social facilitation of feed intake. In unconstrained situations, pigs generally show a bimodal distribution of feeding bouts, analogous to behaviour in the wild, with peaks at the beginning and end of the light period (De Haer and Merks, 1992; Nielsen *et al.*, 1995; Hyun *et al.*, 1997; O'Connell *et al.*, 2002), and feed relatively little over the night period. They also show social facilitation, such that feeding activity by one pig promotes initiation of the behaviour by others and results in group synchrony of behaviour and increased competition a peak times (Hsia and Wood-Gush, 1983, 1984).

Despite these apparent contrary influences, it has been found in practice that high feeder utilisation rates are possible in some situations, with no detrimental effects on performance being seen when 18, 20 or even 30 finishing pigs have been required to share a feeding space (Walker and Overton, 1989; Walker, 1991; Nielsen *et al.*, 1995; Spoolder *et al.*, 1999), despite increased levels of queuing and aggression at the feeder. This is because pigs show great flexibility in feeding behaviour. Feeding rate depends on the age of the pig and the form of the food, being fastest with liquid food, intermediate for pellets, and slowest for meal (Gonyou and Lou, 2000). It is therefore rather surprising that a study of fattening pigs fed via sensor-controlled liquid feeding showed detrimental effects on growth rate as the number of pigs per feeding place increased from 4 to 7 to 13 (Rasmussen *et al.*, 2006). This might be explained by the fact that feed availability was not truly ad libitum, since feed was available on demand only in 5 pre-defined 80-minute feeding periods spaced throughout the day.

Pigs are able to show adaptive changes in feeding behaviour and feeding rate when social constraints on feeding time are increased. In comparison with individually-housed pigs, group-housed pigs eat fewer meals per day of longer duration (De Haer and Merks, 1992; De Haer and De Vries, 1993). When social pressure is increased, either by increasing group size (Nielsen *et al.*, 1995), reducing the number of feeding places (Nielsen *et al.*, 1996; Gonyou and Lou, 2000), or increasing stocking density (Hyun *et al.*, 1998), pigs show a further reduction in meals per day, increased feeding at night and an increase in eating rate in order to maintain feed intake. This flexibility makes it difficult to predict the feeding pressure at which performance will start to be adversely affected. Research suggests that productivity is maintained at feeder occupation rates of up to 80-90% of the 24 hour period (Walker, 1991; Gonyou and Lou, 2000), but the point at which such rates are reached will depend on feeder design, feed form, age of pig and possibly other social factors within the group. A safe recommendation would seem to be up to 12 growing/finishing pigs per feeding place with dry feed, and up to 20 with wet-and-dry feeders. Data for weaned piglets is less comprehensive, but lower feeding pressure would seem appropriate. However, in a series of trials with different numbers of individual feeding places for different group sizes, Lindemann *et al.* (1987) found no effect on feed intake or growth of weaned piglets over the range from 1 to 6 pigs per feeding space.

Where linear troughs, rather than feeders providing discrete feeding places with head protection, were used for *ad libitum* feeding, Turner *et al.* (2002) showed that a feed space allowance of 32.5 mm per pig reduced intake and growth in pigs of 40-56 kg liveweight when compared with an allowance of 42.5 mm per pig. This suggests that a pressure of more then 5 pigs per feeding place reduced performance. The contrast with the studies on single-space feeders summarised previously may reflect the earlier observation that two pigs would only eat simultaneously at a feeder space allowance that was almost double that required to accommodate their physical dimensions (McGlone *et al.*, 1983). It was of interest to note that Turner *et al.* (2002) found no interaction between feeding space allowance and group size over the range from 20 to 80 pigs, suggesting that social facilitation effects at larger group sizes did not lead to a requirement for different feed space recommendations.

Wellock *et al.* (2003) attempted to model the effects of changing feed space allowance on pig performance, taking account of changes with age in feed requirement and maximum feeding rate, and the effect of feed composition as described by water holding capacity. The model assumed that pigs do not avoid feeding immediately adjacent to one another, which may not in fact be the case with all designs of feeder. The model also sought to take account of the influence of other social stressors, as reviewed in the following sections, on food intake and performance by incorporating these relationships into a general growth simulation model. The conceptual equations developed to describe these relationships gave quantitative predictions which were in reasonable agreement with previous empirical studies on social stressors such as space allowance and group size, but it was not possible to compare the predictions for changing feed space allowance since no comprehensive previous published studies existed.

Effects of group size

Conventional group size in fattening pigs is 10-30 animals, but large groups are increasingly common due to their economic advantages (Turner and Edwards, 2004). The effect of large group sizes on welfare has been addressed in several papers, although results are somewhat contradictory.

Some studies have found a negative effect of large groups on several parameters that are relevant to welfare, including production and health traits. Turner *et al.* (2003) used regression analyses of data from 20 earlier studies and almost 22,000 animals to study the implications of group size for growth performance. A significant, negative, approximately linear relationship in average daily gain with increases in group size was recorded during the weaner (weaning to 30 kg) and grower (31-68 kg) stages. Weaners, but not growers, showed a reduction in food intake with increasing group size; consequently the efficiency of growth was compromised during the grower stage. No influence of group size on performance was found during the finisher stage (>68 kg). None of the other traits measured were consistently affected by changes in group size. The authors concluded that a large group size may compromise the growth performance of young pigs, although the long-term consequences for other traits are likely to be slight.

Wolter *et al.* (2000) used a 2×2 factorial arrangement of treatments to determine the effects of group size (20 or 100 pigs per group) and floor-space allowance in weanling piglets. Piglets in large groups were lighter at the end of weeks 1, 4 and 9 of treatment and had lower feed intake from week 1 to 4 than piglets in small groups. Also, pigs in large groups had a greater within-pen coefficient of variation in body weight at the end of the week 9 than pigs in small groups. Finally, McGlone and Newby (1994) found that injury and morbidity rates were greater for pigs held in groups of 40 animals than for pigs kept in groups of 10 or 20.

Other studies, however, have failed to find any significant effect of large groups. O´Connell *et al.* (2004) compared groups of 10, 20, 30, 40 and 60 pigs from weaning at 4 weeks of age until 10 weeks of age. Group size did not affect overall levels of aggressive behaviour, growth rate or feed intake. The authors did find that food conversion tended to be poorer in groups of 40 and 60 pigs than in smaller group sizes, although the effect did not reach statistical significance. The only significant adverse effect of large groups were found when groups of 40 and 60 animals were split in groups of 20, as food conversion ratio was poorer in groups split from groups of 60 pigs than in groups split from groups of 40 pigs, or in groups that were kept in a group of 20 pigs throughout the growing and finishing periods. The authors concluded that large groups of up to 60 pigs do not have any adverse effect on welfare when compared with smaller groups, although if pigs are to be housed in groups of 20 during the finishing period, it may be better to house pigs in groups of 40 rather than in groups of 60 during the post-weaning period.

Schmolke *et al.* (2003) compared group sizes of 10, 20, 40 and 80 pigs from 23.2 kg BW to 95.5 kg BW and did not find any consistent effect of group size on daily gain, feed intake, feed efficiency or variability in final BW. Similar proportions of pigs were removed from the trail for health reasons in all treatments. Turner *et al.* (2000) studied the effect of group size (20 vs. 60) and drinker: pig ratios (1:10 vs. 1:20) on skin lesions and found that the number of lesions was not affected by the four treatment combinations. The treatments did not affect performance. In another study, Turner *et al.* (2001) compared groups of 20 and 80 pigs and found that aggression between pen mates was unaffected by group size.

Finally, some studies have found positive effects of large group sizes. In the study cited above, O'Connell *et al.* (2004) found that the coefficient of variation for growth was greater in groups of 10 than in groups of 20, 40 or 60 pigs. Wolter *et al.* (2001) compared groups of 25, 50 and 100 pigs. The effect of group size on production parameters was not consistent, but the proportion of pigs removed due to poor health or injury was higher in groups of 25 pigs compared to the other group sizes.

Group size may have an effect on how pigs react to being mixed with unacquainted individuals. Turner *et al.* (2001) found that pigs from large groups (80 pigs per group) displayed a marked reduction in aggression towards pigs from other groups compared with pigs from small groups (20 pigs per group). Turner and Edwards (2004) suggest that in large groups less reliance is placed on aggression during the immediate post-mixing phase than in small groups.

Social mixing

Mixing of unfamiliar animals (often with a change of physical environment) is a common practice in pig husbandry and happens commonly at weaning and at the beginning of the growing-finishing period. Mixing of unacquainted pigs has adverse effects on welfare and production, mainly because pigs fight with each other in order to establish dominance relationships, most aggressive interactions being typically shown during the first few hours after grouping (Meese and Ewank, 1972).

The frequency, duration and intensity of aggression interactions after mixing varies depending on several variables, such as enrichment of the environment, whether food is provided as libitum or restricted and time of day when pigs are mixed. Weaned pigs offered tyres and chain devices within the pen displayed a reduction in the frequency of aggressions (Simonsen, 1990). Barnett *et al.* (1994) found that offering food ad libitum and regrouping after sunset reduced the number of aggressive interactions among the group.

In order to reduce the amount of aggression at mixing the use of tranquillising drugs has been widely advocated for many years. Although they effectively reduce aggression among grouped pigs, their effect is limited over time and they cannot avoid a rise of the frequency of agonistic interactions to develop the social hierarchy at the end of the drug effect (Gonyou *et al.*, 1988). Pheromones from pigs can modulate aggressive behaviour after regrouping (McGlone *et al.*, 1987; Guy *et al.*, 2009).

It has also been shown that socialised piglets (piglets that have been mixed with piglets from another litter before weaning) learn social skills that allow them to more rapidly form stable hierarchies when regrouped after weaning (D'Eath, 2005). Group size may also have an effect on how pigs react to being mixed with unacquainted individuals, as described previously.

Although some studies have found a negative effect of social mixing on production parameters (e.g. Stookey and Gonyou, 1994), others have failed to do so (e.g. Coutellier *et al.*, 2007). As stressors have an additive effect (Hyun *et al.*, 1998), it is likely that social mixing will have a greater effect when combined with other stressors. This is probably the case at weaning, when piglets are subjected to nutritional, environmental and psychological stressors (including regrouping) that are imposed simultaneously (Pluske *et al.*, 1995). The combined effect of these stressors is that piglets usually show a period of reduced feed intake that may have long-lasting effects on performance (Pollmann, 1993).

Space allowance

Space allowance is important from both an economic and animal welfare standpoint. Several studies have reported a decrease in feed intake caused by reduced floor space allowance (e.g. Randolph *et al.*, 1981; Kornegay and Notter, 1984; Kornegay *et al.*, 1993). Other studies have failed to find such an effect, although decreasing space allowance caused a reduction in growth rate (Hyun *et al.*, 1998). In general, individual pig productivity decreases as space allowance is reduced, signifying a welfare concern, but production per unit area may increase (Edwards *et al.*, 1988).

The most common means to express space allowance is as space per animal, but this has the limitation that space requirements increase with body weight. A second option is to express space allowance as weight density (i.e. kg/m^2), but space requirements are not directly proportional to body weight. A third means is to use an allometric approach in which $A = k\ BW^{0.667}$ where A is floor space allowance and k is a space allowance coefficient (Petherick, 1983, Baxter, 1984). The allometric expression can be applied over a wide range of weights (Gonyou *et al.*, 2006) and has been supported by several studies (e.g. Hurnik and Lewis, 1991).

Petherick and Baxter (1981) estimated k values for sternally recumbent pigs (k=0.019) and for lateral recumbent lying pigs (k=0.047). By applying these k values in the equation above it can be calculated how much physical space a pig needs for each posture at a given body weight. The actual space needed is dependent on how many pigs want to lie down at any given time and on the posture the pigs adopt when lying. Lying posture is to a large extent determined by air temperature (Petherick, 1983): at high temperatures, pigs will try to lose body weight by increasing the contact area between the body and the floor and therefore lateral lying will be preferred (Ducreaux *et al.*, 2002).

Ekkel *et al.* (2003) addressed the issue of huddling or space sharing and number of pigs lying at any given time. Their results showed that pigs of all weight categories lie down for a great part of the day, but spend little time in contact with conspecifics. This seems to confirm Petherick's suggestion that at thermoneutral conditions the floor area occupied by lying pigs fits the allometric equation with a k value of 0.033 (Petherick, 1983). This, however, does not take into account the extra space needed for activity. On average, over the day, 10-20% of pigs are active at any one time (Ekkel *et al.*, 2003). Although how much space an active pig needs has not been assessed, it is reasonable to assume that for any activity such as exploration, walking to the dunging area or feeder or for social interaction, pigs require an amount of space equivalent to a k value of 0.038 (EFSA, 2005). Taking all these considerations together, it can be suggested that overall k value would be $0.8\times0.033 + 0.2\times0.038 = 0.034$. In order to take into account the additional space required for a separate dunging area, the European Food Safety Authority has recommended a final k value of 0.036 (EFSA, 2005). Interestingly, minimum space allowances in the European Union are significantly lower and approximate k values of 0.028 for grower-finisher pigs (European Community, 2001).

After reviewing more than 10 published peer-reviewed studies on the effect of space allowance on average daily gain, Gonyou *et al.* (2006) concluded that the critical k value, below which a decrease in average daily gain occurred, varied from 0.0317 to 0.0348. Thus, empirical evidence seems to support theoretical considerations on space requirements.

Several factors (including group size, type of floor, health status and temperature) may modify space requirements. It has been suggested that pigs in large groups may need less space per animal than in small groups due to the sharing of free space (McGlone and Newby, 1994). Although some studies provide support for this hypothesis (e.g. Wolter *et al.*, 2000), others have not found any difference in response to crowding between large and small groups (Street and Gonyou, 2005).

Some codes of practice recommend an increase in space allowance if pigs are on partially rather than on fully slatted floors (e.g. AAFC, 1993). Gonyou *et al.* (2006) did not find any difference between the two types of floor. However, this study did find a greater slope for the growth and intake responses in the crowded range of the data for pigs on partially slatted floors compared with pigs on fully slatted floors. This would suggest that, although space requirements are similar on both types of floor, the effects of crowding are more severe on partially slatted floors.

When pigs are likely to be challenged by disease, space allowance needs to be increased if feed intake and performance are to be maintained. The benefits of increasing space allowance and the penalties of decreasing it are greater if certain diseases are present (Whittemore, 1993). The effect of ambient temperature on lying posture and thus space requirements has already been explained.

Mechanisms

Social interactions may have a negative effect of feed intake and growth rate through several mechanisms, including (1) a direct effect on intake by inhibiting or disrupting access to the feeder, (2) an indirect effect on feed intake through a physiological stress response caused by agonistic interactions, and (3) by exacerbating the effects of heat stress when pigs are crowded.

The negative effect of stress on feed intake has long been recognised (Forbes, 1995), although the precise pathways involved are still debated. It is likely, however, that the inhibitory effect of stress on appetite results from a complex interplay among leptin, glucocorticoids and corticotropin releasing hormone (CRH) (Matteri *et al.*, 2000).

Adequate space allowance is crucial for thermoregulation, as pigs modify the amount of heat loss by changing their lying posture. If space allowance is too low, pigs may be forced to lie in close proximity and in sternal recumbency as opposed to lateral recumbency. Prevention of heat loss, if not compensated by other cooling measures such as improved ventilation or showers, may result in heat stress. For example, Götz and Rist (1984) found increased rectal temperatures with decreasing space allowance in fatteners kept at 32 °C. Heat stress is most likely at the end of the fattening period, as heavy pigs have lower surface to volume ratio than lighter pigs. Heat stress causes a decrease in food intake in order to reduce the heat production associated with feeding, digestion, absorption and metabolism and to prevent an excessive increase in body temperature (NRC, 1987).

Areas deserving further research

Most studies on the effects of social behaviour on production and welfare have focused on the negative, stress-inducing consequences of social interaction and this chapter reflects this emphasis. However, social interactions also have positive effects on welfare and, presumably, on production. The term 'social buffering' describes the fact that when animals of a social species are with conspecifics, they show a better recovery from aversive experiences (Kikusui *et al.*, 2006). Social buffering has been found in a variety of species, including rats (Davitz and Mason, 1955), guinea pigs (Hennessy *et al.*, 2000), non-human primates (Levine *et al.*, 1978) and humans (Thorsteinsson *et al.*, 1998). To date, social buffering has been little studied in the domestic pig, although one example of this is the study of Ruis *et al.* (2001). Given their social nature, the possibility exists of pigs showing more widely the phenomenon similar to that described in other species. If this is the case, then social interactions could reduce the negative effects of stress on feed intake. As explained above, the possible effects of group size on behaviour and welfare as well as the possible interaction between group size and space allowance remain controversial and deserve further research.

References

AAFC, 1993. Recommended code of practice for the care and handling of farm Animals: pigs. Agriculture and Agri-Food Canada Publication 1898/E. Ottawa, Canada, 55 pp.

Barnett, J.L., Cronin, G.M., McCallum, T.H. and Newman, E.A., 1994. Effects of food and time of day on aggression when grouping unfamiliar adult pigs. Applied Animal Behaviour Science 39: 339-347.

Baxter, M.R., 1983. Feeding and aggression in pigs. Applied Animal Ethology 11: 74-75.

Baxter, M.R., 1989. Design of a new feeder for pigs. Farm Building Progress 96: 19-22.

Baxter, S., 1984. Intensive Pig Production: Environmental Management and Design. Granada Publishing Ltd., London, UK

Blackshaw, J.K. and Blackshaw, A.W., 1990. Evaluation of two feeder designs for growing pigs. Australian Journal of Experimental Agriculture 30: 491-493.

Bruininx, E.M.A.M., Van der Peet-Schwering, C.M.C., Schrama, J.W., Vereijken, P.F.G., Vesseur, P.C., Everts, H., Den Hartog, L.A. and Benyon, A.C., 2001. Individually measured feed intake characteristics and growthperformance of group-housed weanling pigs: effects of sex, initial body weight, and body weight distribution between groups. Journal of Animal Science 79: 301-308.

Coutellier, L., Arnould, C., Boissy, A., Orgeur, P., Prunier, A., Veissier, I. and Meunier-Salaun, M.C., 2007. Pig's response to repeated social regrouping and relocation during the growing-finishing period. Applied Animal Behaviour Science 105: 102-114.

Davitz, J. and Mason, D.J., 1955. Socially facilitated reduction of fear response in rats. Journal of Comparative and Physiological Psychology 47: 941-947.

D'Eath, R., 2005. Socialising piglets before weaning improves social hierarchy formation when pigs are mixed post-weaning. Applied Animal Behaviour Science 93: 199-211.

De Haer, L.C.M. and Merks, J.W.M., 1992. Patterns of daily food intake in growing pigs. Animal Production 54: 95-104.

De Haer, L.C.M. and De Vries, A.G., 1993. Feed intake patterns and feed digestibility in growing pigs housed individually or in groups. Livestock Production Science 33: 277-292.

Ducreaux, E., Aloui, B., Robin, P., Courboulay, V. and Meunier-Salaun, M. C., 2002. Ambient temperature influences the choice made by pigs for certain types of floor. Journes de la Recherche Porcine 34: 211-216.

Edwards, S.A., Armsby, A.W. and Spechter H.H., 1988. Effects of floor area allowance on performance of growing pigs kept on fully slatted floors Animal Production 46: 453-459.

EFSA, 2005 The welfare of weaners and rearing pigs: effects of different space allowances and floor types. Parma, Italy.

Ekkel, E.D., Spoolder, H.A.M., Hulsegge, I. and Hopster, H., 2003. Lying characteristics as determinants for space requirements in pigs. Applied Animal Behaviour Science 80: 19-30.

European Community, 2001. European Council Directive 2001/88/EC of 23rd October 2001 amending directive 91/630/EC laying down minimum standards for the protection of pigs. Official Journal of the European Union L 316 (2001/12/01).

Forbes, J.M., 1995. Environmental factors affecting intake. In: Forbes, J.M. (ed.) Voluntary Food Intake and Diet Selection in Farm Animals, CAB International, Wallingford, England, pp. 332-353.

Gonyou, H.W. and Lou, Z., 2000. Effects of eating space and availability of water in feeders on productivity and eating behaviour of grower/finisher pigs. Journal of Animal Science 78: 865-870.

Gonyou, H.W., Rohde Parfet, K.A., Anderson, D.B. and Olsson, R.D., 1988. Effects of Amperozide and Azaperone on aggression and productivity of growing-finishing pigs. Journal of Animal Science 66: 2856-2864.

Gonyou, H. W., Brumm, M.C., Bush, E., Deen, J., Edwards, S.A., Fangman, T., McGlone, J.J., Meunier-Salaun, M., Morrison, R.B., Spoolder, H., Sundberg, P.L. and Johnson, A.K., 2006. Application of broken-line analysis to assess floor space requirements of nursery and grower-finisher pigs expressed on an allometric basis. Journal of Animal Science 84: 229-235.

Götz, M. and Rist, M., 1984 Possibilities to avoid heat-stress in pigs. In: Unshelm, J., Van Putten, G. and Zeeb, K. (eds.) Proceedings of the International Congress on Applied Ethology in Farm Animals, Kiel, pp. 209-213.

Graves, H.B., 1984. Behaviour and ecology of wild and feral swine (sus scrofa). Journal of Animal Science 58: 482-492.

Guy, J.H., Burns, S.E., Barker, J.M. and Edwards, S.A., 2009. Synthetic maternal pheromone affects post-mixing aggression and lesions in weaned pigs. Animal Welfare, in press.

Hennessy, M.B., Maken, D.S. and Graves, F.C., 2000. Consequences of the presence of the mother or unfamiliar adult female on cortisol, ACTH, testosterone and behavioral responses of periadolescent guinea pigs during exposure to novelty. Psychoneuroendocrinology 25: 619-632.

Hsia, L.C. and Wood-Gush, D.G.M., 1983. A note on social facilitation and competition in the feeding behaviour of pigs. Animal Production 37: 149-152.

Hsia, L.C. and Wood-Gush, D.G.M., 1984. The temporal patterns of food intake and allelomimetic feeding by pigs of different ages. Applied Animal Ethology 11: 271-282.

Hurnik, J.F. and N.J. Lewis, 1991. Use of body surface area to set minimum space allowance for confined pigs and cattle. Canadian Journal of Animal Science 71: 577-580.

Hyun, Y., Ellis, M., McKeith, F.K. and Wilson, E.R., 1997. Feed intake pattern of group-housed growing-finishing pigs monitored using a computerised feed intake recording system. Journal of Animal Science 75: 1443-1451.

Hyun, Y., Ellis, M. and Johnson, R.W., 1998. Effect of feeder type, space allowance, and mixing on the growth performance and feed intake pattern of growing pigs. Journal of Animal Science 76: 2771-2778.

Kikusui T., Winslow, J.T. and Mori, Y., 2006. Social buffering: relief from stress and anxiety Philosophical Transactions of the Royal Society B 361: 2215-2228.

Kornegay, E.T. and Notter, D.R., 1984. Effects of floor space and number of pigs per pen on performance. Pigs New and Information 5: 23-33.

Kornegay, E.T., Lindemann, M.D. and Ravindran, V., 1993. Effects of dietary lysine levels on performance and immune response of weanling pigs housed at two space allowances. Journal of Animal Science 71: 552-556

Laitat, M., Vandenheede, M., Desiron, A., Canart, B. and Nicks, B., 1999. Comparison of feeding behaviour and performance of weaned pigs given food in two types of dry feeders with integrated drinkers. Animal Science 68: 35-42.

Laitat, M., Vandenheede, M. and Nicks, B., 2005. Facteurs influencant le comportment alimentaire et les performances du porc sevre: l'equipment d'alimentation. Annees de Medicine Veterinaire 149: 61-74.

Levine, S., Coe, C.L. and Smotherman, W.P., 1978. Prolonged cortisol elevation in the infant squirrel monkey after reunion with mother. Physiology and Behavior 20: 7-10.

Lindemann, M.D., Kornegay, E.T., Meldrum, J.B., Schurig, G. and Gwazdauskas, F.C., 1987. The effect of feeder space allowance on weaned pig performance. Journal of Animal Science 64: 8-14.

Magowan, E., McCann, M.E.E. and O'Connell, N.E., 2008. The effect of feeder type and change of feeder type on growing and finishing pig performance and behaviour. Animal Feed Science and Technology 142: 133-143.

Matteri, R. L., Carroll, J. A. and Dyer, C. J., 2000. Neuroendocrine responses to stress. In: Moberg, G. P. and Mench, J.A. (eds.). The Biology of Animal Stress. Basic Principles and Implications for Animal Welfare. CABI Publishing, Cambridge, US, pp. 43-76.

McGlone, J.J., Curtis, S.E. and Banks, E.M., 1987. Evidence for aggression-modulating pheromones in prepuberal pigs. Behavioural Neural Biology 47: 27-39.

McGlone, J.J., Heald, T.E. and Hayden, S.L., 1983. Physical and behavioural measures of feeding space for nursery age swine. Journal of Animal Science 57 (suppl 1): 140.

McGlone, J.J. and Newby, B.E., 1994. Space requirements for finishing pigs in confinement: behavior and performance while group size and space vary. Applied Animal Behaviour Science 39: 331-338.

Meese, G.B. and Ewank, R., 1972. A note on instability of the dominance hierarchy and variations in level of aggression within groups of fattening pigs. Animal Production 14: 359-362.

Nielsen, B.L., Lawrence, A.B. and Whittemore, C.T., 1995. Effect of group size on feeding behaviour, social behaviour and performance of growing pigs using single space feeders. Applied Animal Behaviour Science 44: 73-85.

Nielsen, B.L., Lawrence, A.B. and Whittemore, C.T., 1996. Feeding behaviour of growing pigs using single or multi-space feeders. Applied Animal Behaviour Science 47: 235-246.

NRC, 1987 Predicting Feed Intake of Food-Producing Animals. National Academy Press, Washington, US, 85 pp.

O'Connell, N.E., Beattie, V.E. and Weatherup, R.N., 2002. Influence of feeder type on the performance and behaviour of weaned pigs. Livestock Production Science 74: 13-17.

O'Connell, N.E., Beattie, V.E. and Weatherup, R.N., 2004 Influence of group size during the post-weaning period on the performance and behaviour of pigs. Livestock Production Science 86: 225-232.

Petherick, J.C., 1983. A biological basis for the design of space in livestock housing In: Baxter, S.H., Baxter, M.R. and MacCormack, J.A.S.C. (eds.) Farm Animal Housing and Welfare. Martinus Nijhoff Publisher, Boston, US, pp. 103-120.

Petherick, J.C. and Baxter S.H., 1981. Modelling the static spatial requirements of livestock In: MacCormak, J.A.D. (ed.) Modelling, design and evaluation of agricultural buildings. CIGR Section II Seminar, Scottish Farm Buildings Investigation Unit, Aberdeen, UK, pp. 75-82.

Pluske, J.R., Williams, I.H. and Aherne, F.X., 1995. Nutrition of the neonatal pig. In: Varley, M. A. (ed.). The Neonatal Pig. Development and Survival. CAB International, Wallingford, England, pp. 187-235.

Pluske, J.R. and Williams, I.H., 1996. The influence of feeder type and method of group allocation at weaning on voluntary food intake and growth in piglets. Animal Science 62: 115-120.

Pollmann, D.S., 1993. Effects of nursery feeding programs on subsequent grower-finisher pig performance. In: Martin, J. (ed.) Proceedings of the Fourteenth Western Nutrition Conference. Faculty of Extension, University of Alberta, Edmonton, Canada, pp. 243-254.

Randolph, J.H., Cromwell, G.L., Stahly, T.S. and Kratzer, D.D., 1981. Effects of group size and space allowance on performance and behaviour of swine. Journal of Animal Science 53: 922-927.

Rasmussen, D.K., Weber, R. and Wechsler, B., 2006. Effect of animal/feeding- place ratio on the behaviour and performance of fattening pigs fed via sensor-controlled liquid feeding. Applied Animal Behaviour Science 98: 45-53.

Ruis, M.A.W., De Groot, J., Brake, J.H.A.T., Ekkel, E.D., Van de Burgwal, J.A., Erkens, J.H.F., Emgel, B., Buist, W.G., Blokhuis, H.J. and Koolhaas, J.M., 2001. Behavioural and physiological consequences of acute social defeat in growing gilts: effects of the social environment. Applied Animal Behaviour Science 70: 201-225.

Schmolke, S.A., Li, Y.Z. and Gonyou, H.W., 2003. Effect of group size on performance of growing-finishing pigs. Journal of Animal Science 81: 874-878.

Simonsen, H.B., 1990. Behaviour and distribution of fattening pigs in the multi-activity pen. Applied Animal Behaviour Science 27: 311-324.

Spoolder, H.A.M., Edwards, S.A. and Corning, S., 1999. Effects of group size and feeder space allowance on welfare in finishing pigs. Animal Science 69: 481-489.

Stolba, A. and Wood-Gush, D.G.M., 1989. The behaviour of pigs in a semi-natural environment. Animal Production 48: 419-425.

Stookey, J.M. and Gonyou, H.W., 1994. The effects of regrouping on behaviour and production parameters in finishing pigs. Journal of Animal Science 72: 2804-2811.

Street, B.R. and Gonyou, H.W., 2005. The effects of housing grow-finish pigs at two different group sizes and space allocations. Journal of Animal Science 83 (Suppl. 2): 38.

Taylor, I. and Curtis, S., 1989. Grower feeders: 11 designs reviewed. National Hog Farmer 34: 29-39.

Thorsteinsson, E.B., James, J.E. and Gregg, M.E., 1998. Effects of video-relayed social support on hemodynamic reactivity and salivary cortisol during laboratory-based behavioral challenge. Health Psychology 17. 436-444.

Turner, S.P., Dahlgren, M., Arey, D.S. and Edwards, S.A., 2002. Effect of social group size and liveweight on feeder space requirement of growing pigs fed *ad libitum*. Animal Science, 75: 75-83.

Turner, S.P. and Edwards, S.A., 2004. Housing immature domestic pigs in large social groups: implications for social organization in a hierarchical society. Applied Animal Behaviour Science 87: 239-253.

Turner, S.P., Allcroft, D.J. and Edwards, S.A., 2003. Housing pigs in large social groups: a review of implications for performance and other economic traits. Livestock Production Science 82: 39-51.

Turner, S.P., Farnworth, M.J., White, I.M.S., Brotherstone, S., Mendl, M., Knap, P., Penny, P. and Lawrence, A.B., 2006. The accumulation of skin lesions and their use as a predictor of individual aggressiveness in pigs. Applied Animal Behaviour Science 96: 245-259.

Turner, S.P., Horgan, G.W. and Edwards, S.A., 2001. Effect of social group size on aggressive behaviour between unacquainted domestic pigs. Applied Animal Behaviour Science 74: 203-215.

Turner, S.P., Sinclair, A.G. and Edwards, S.A., 2000. The interaction of liveweight and the degree of competition on drinking behaviour in growing pigs at different group sizes. Applied Animal Behaviour Science 67: 321-334.

Vargus, V.J., Craig, J.V. and Hines, R.H., 1987. Effects of feeding systems on social and feeding behaviour and performance of finishing pigs. Journal of Animal Science 65: 463-474.

Walker, A.J. and Overton, D.C., 1989. Comparison of the performance of finishing pigs fed ad libitum from either conventional or single-place feeders. In: Forbes, J. M., Varley, M. A. and Lawrence, T.L. J. (eds). The Voluntary Food Intake of Pigs. British Society of Animal Production, Edinburgh, pp 114-116.

Walker, N., 1990. A comparison of single- and multi-space feeders for growing pigs fed non-pelletted diets ad libitum. Animal Feed Science and Technology 30: 169-173.

Walker, N., 1991. The effects on performance and behaviour of numbers of growing pigs per mono-place feeder. Animal Feed Science and Technology 30: 169-173.

Wastell, M.E. and Pettigrew, J.E., 2000. Dry vs wet feeders, is there a difference? Proceedings 1st International Conference on Swine Housing. American Society of Agricultural Engineers, St Joseph, pp 103-109.

Wellock, I.J., Emmans, G.C. and Kyriazakis, I., 2003. Predicting the consequences of social stressors on pig food intake and performance. Journal of Animal science 81: 2995-3007.

Whittemore, C., 1993. The science and practice of pig production. Longman Scientific and Technical, Harlow, England. 661 pp.

Wolter, B.F., Ellis, M., Curtis, S.E., Augspurger, N.R., Hamilton, D.N., Parr, E.N. and Webel, D.M., 2001. Effect of group size on pig performance in a wean-to-finish production system. Journal of Animal Science 79: 1067-1073.

Wolter, B.F., Ellis, M., Curtis, S.E., Parr, E.N. and Webel, D.M., 2000. Group size and floor-space allowance can affect weanling pig performance. Journal of Animal Science 78: 2062-2067.

Chapter 13. Anorexia during infection in mammals: variation and its sources

I. Kyriazakis[1,2] *and A. Doeschl-Wilson*[3]
[1]*School of Agriculture, Food and Rural Development, Newcastle University, Newcastle-upon-Tyne, NE1 7RU, United Kingdom; ilias.kyriazakis@newcastle.ac.uk, i.kyriazakis@btinternet.com*
[2]*Veterinary Faculty, University of Thessaly, P.O. Box 199, 43100 Karditsa, Greece*
[3]*Scottish Agricultural College, Bush Estate, Penicuik, Midlothian, EH26 0PH, United Kingdom*

Abstract

A reduction in the voluntary food intake of animals, henceforth called anorexia, accompanies their exposure to pathogens and antigenic challenges. Whilst the phenomenon is well established, its timing, magnitude, duration and recovery from exposure to pathogens are poorly understood. In this paper we consider the sources that contribute to variation in these characteristics of anorexia, and their significance. (1) Pathogen: It is now well established that the characteristics of anorexia are very much dependent on pathogen type and dose. For example, anorexia appears within a few hours from infection with a micro parasite and has a relatively short duration, whereas it appears a few weeks post infection and lasts for several weeks in macro parasitic infections. We suggest that this is due to pathogen life cycle and its consequences on the host. (2) Host nutrition: Since anorexia is linked to the immune response it is reasonable to expect that host nutrition may affect the characteristics of anorexia through its effect on the immune response. Some of these characteristics appear to be more sensitive to host nutrition than others, such as anorexia duration and rate of recovery; an explanation to account for this is offered. It is thus possible to exploit host nutrition to manipulate the impact of pathogen exposure on the infected host. (3) Host genotype: There is significant less information regarding host genotype effects on anorexia. Here we suggest that the characteristics of anorexia depend on the ability of the host to cope with the pathogen and hence its immune response. In the absence of literature information we hypothesise how these host characteristics might affect the phenomenon of anorexia and suggest appropriately designed experiments to address this issue. Finally the timescales of anorexia are not always defined and it is common that the period of reduction in food intake and its recovery are lumped together. We explore how this practice might affect the conclusions one might draw about anorexia and how one should deal with it. Although pathogen-induced anorexia is an unavoidable consequence of infection, some of its features might be subject to manipulation and, therefore, its impact upon the infected animals may also be minimised.

Keywords: anorexia, food composition, food intake, host genotype, pathogen

Introduction

Anorexia, defined as a reduction in the voluntary food intake of animals, is a common feature of infection. It is observed across pathogens and across mammalian and avian hosts (Kyriazakis *et al.*, 1998). Its extent can vary from a small reduction in food intake (10-20%) to complete cessation of eating, and it is considered largely responsible for the consequences of infection on host growth and reproduction. For this reason the phenomenon merits special attention: the development of a framework that will enable its prediction would allow us to deal with it in a more effective manner, and perhaps overcome some of its negative consequences.

Although anorexia has received some experimental and theoretical attention, currently there is some confusion about its occurrence, magnitude and duration, and the phenomenon appears to be characterised by substantial variability between hosts and pathogens. This has undesired consequences on its predictability. We suggest that this is partly because anorexia has not been approached in a systematic manner. A systematic approach would consist of identifying the sources of such variation and by examining their effects on the patterns and characteristics of anorexia. This is the aim of this paper and for this reason the patterns of anorexia development are defined first. The sources of variation in anorexia are expected to arise from the animal itself (i.e. host genetics), the pathogens it is exposed to and the environment it is kept in (Kyriazakis and Houdijk, 2007). Here we focus on the nutritional environment, as this is the aspect of the environment most frequently addressed in relation to anorexia. By approaching anorexia in this manner we should be able to overcome current confusion, identify gaps in knowledge and thus make the phenomenon and its consequences more predictable.

Patterns of anorexia development

The patterns of development of anorexia are shown in Figure 1 during (1A) a bacterial infection of pigs and (1B) a gastrointestinal parasite infection of sheep. The general pattern, very similar across pathogens but very different in its time scales, is characterised by the following features:

(1) There is usually a *lag time* between the point of infection and the first signs of anorexia; in Figure 1 these happen within the day of bacterial infection, whereas they take several weeks to develop for the gastrointestinal parasite infection. (2) The *rate of change* towards a certain reduced level of food intake can also vary between a few hours to a number of weeks; below reasons for these variations will be given. (3) Following this rate of decline, food intake reaches a certain lower level. This has been called the *extent of anorexia*. It has been suggested that for all intents and purposes this 'lowest intake' is approximately 80-85% of the intake of the uninfected controls (Kyriazakis *et al.*, 1998), during subclinical disease. However, when animals develop clinical disease food intake is further reduced towards complete cessation of feeding. (4) The above lowest intake may be maintained for a period of time (*duration of anorexia*) that can vary between a few days to several weeks; extent and duration of anorexia are mainly responsible for the consequences of infection on host performance. (5) At some stage the host is able to overcome the consequences of the pathogen and starts to increase its intake towards that of uninfected control subjects of the same size or weight. The issues here are how quick this *rate of recovery* will be and whether the food intake of the previously infected animal will return to that of the uninfected control animal, or even exceed it.

The influences of pathogen, nutritional environment and host genotype upon the individual characteristics of anorexia are reviewed below. Although the view taken is that by doing so we should be able to reach conclusions about the sources of variation of pathogen induced anorexia, the data presented in the literature does not always allow this. Food intake is often averaged and

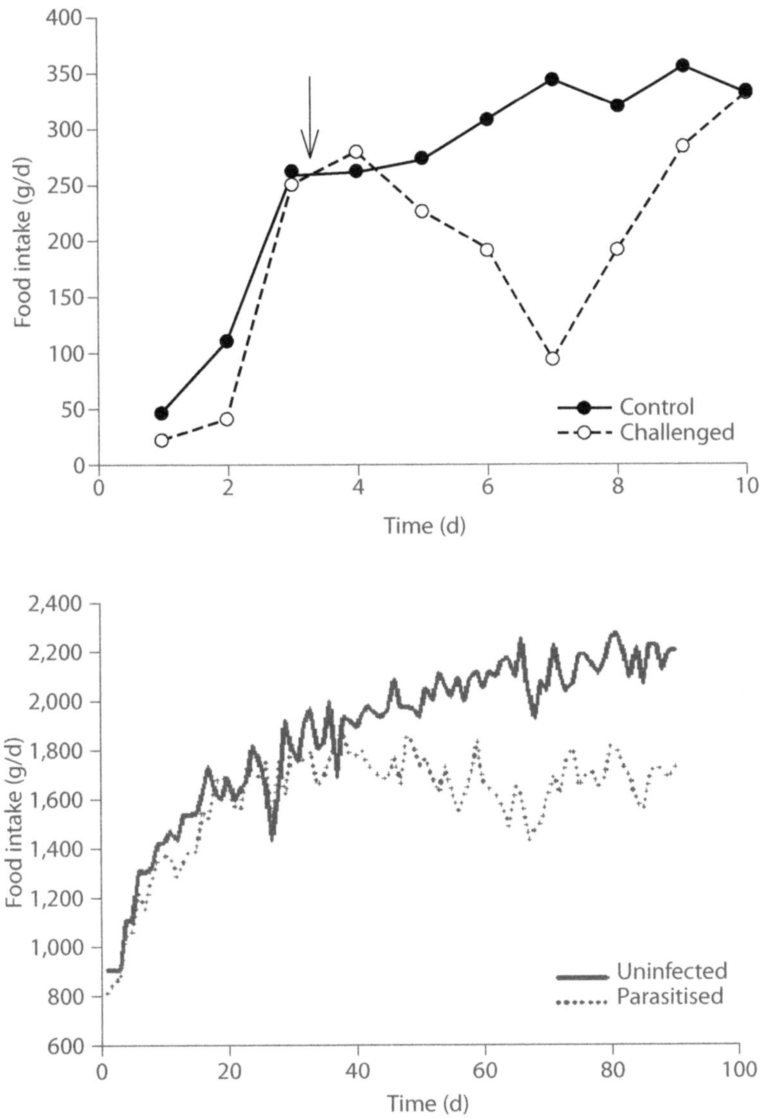

Figure 1. The food intake of weaned pigs infected with enterotoxigenic Escherichia coli *(at the point of arrow) (Figure 1A) and of growing sheep infected with the nematode* Trichostrongylus colubriformis *on day 1 (Figure 1B). Note the difference in the time scale of the development of anorexia: a few days in the former and almost 40 days in the latter case.*

analysed over a time scale that includes both the period of lower critical level of food intake and the period of recovery. For example, in experiments where anorexia was the outcome of LPS challenge (e.g. Spurlock *et al.*, 1997) the reduction in food intake lasts only for a few days and yet food intake is reported and analysed over weekly periods. It is possible that in such experiments any influences on the extent of anorexia are cancelled out by the effects on rate of recovery (see below).

Secondly, whether food intake is analysed over time or at a weight will affect the conclusions drawn. This is because of the different growth trajectories of infected animals, for example on foods of different qualities or genotypes selected for different performance characteristics. During its early stages at least, infection imposes the same metabolic penalties on animals, irrespective of their nutrient intake. Thus, animals on nutritionally poor diets will suffer proportionally more the metabolic consequences of the infection rather than animals on good quality ones (Coop and Kyriazakis, 1999). This will have a disproportionate effect on their growth. In order to overcome the above discrepancies in comparisons of food intake between infected animals subjected to different treatments and between infected animals and their controls, we have attempted to relate food intake during infection either to food intake before infection, or to live weight, i.e. expressing food intake either per unit of live weight or per unit of metabolic live weight.

Pathogen specific effects on anorexia

Anorexia is related to the host response to pathogens, through the stimulation of its immune system. The acute phase response accompanies host exposure to pathogens and this includes the secretion of cytokines, which are highly anorexogenic (Balaji *et al.*, 2002; Petry *et al.*, 2005). In order for anorexia to develop pathogens need to overcome defence barriers and be recognised by the (innate) immune system – hence the lag time between point of infection and the appearance of the phenomenon. Anorexia appears to cease when hosts express acquired immunity in full. Hosts already immune to pathogens do not show anorexia and only when immunity breaks down then anorexia reappears (Zaralis *et al.*, 2008a,b). As a consequence of these, pathogens that affect the immune response in different ways will have different consequences on the features of anorexia. Microparasites, such as bacteria and viruses that replicate within the host at relatively faster rates are likely to be recognised by the host's immune response at an earlier stage (Houdijk *et al.*, 2007). On the other hand macroparasites, such as gastrointestinal nematodes, require time to establish and develop into mature forms within the host and their accumulation depends on the larval intake from the environment (Kyriazakis *et al.*, 1996). In addition macroparasites appear to be more successful in evading host immune responses than many micro-parasites (Coop and Kyriazakis, 1999) and removal of gastrointestinal macroparasites appears to be slower than the removal of bacteria and viruses (Wakelin, 2000).

The above account for the differences in the time scales of the features of anorexia are presented on Figure 1. Anorexia due to microparasites will have a shorter lag time, a faster reduction rate, a shorter duration and perhaps a faster rate of recovery than anorexia due to macroparasites. The first two characteristics may last for a few hours for the former pathogen (Greiner *et al.*, 2000, Balaji *et al.*, 2002), whereas may take several weeks for the latter (Kyriazakis *et al.*, 1996). There has been no information on the rate of recovery from either pathogens, whilst the other conditions (e.g. hosts, nutritional environment) are kept the same, but the assumption that hosts will recover faster from anorexia due to microparasites rather than macroparasites seems a reasonable one. The only anorexia characteristic that appears to be the same between the two classes of pathogens is the *extent* of anorexia. During subclinical infections the lowest level of intake achieved is 80-85% of the uninfected control animals in both micro and macroparasites (Sandberg *et al.*, 2006).

Given the relationship between pathogens and the stimulation of the host immune response, it is to be expected that some features of anorexia will not only depend on pathogen type but also on the pathogen load within the host. Within limits, higher pathogen loads with faster replication rates will lead to a shorter lag time, faster reduction rates and a greater extent of anorexia for both micro and macroparasites (Coop *et al.*, 1982; Houdijk *et al.*, 2007). However, the duration of anorexia seems to be independent of within host pathogen load (Steel *et al.*, 1980; Houdijk *et al.*, 2007). This is probably because the full expression of acquired immunity that leads to the recovery from anorexia is independent of pathogen load (Wakelin, 2000), but it is a function of the host (Coop and Kyriazakis, 1999).

Nutritional environment effects on anorexia

Effects of food composition on lag time and rate of reduction

Given the rapid response of subjects to the challenge by bacteria and viruses as indicated above, it is not surprising that there is no good information on the effects of food composition on the lag time and the rate of reduction in food intake of infected animals. The exception is an experiment of Gidene and Licois (2005) where rabbits were infected with *Escherichia coli*, whilst they were given access to foods that differed in their fibre content. The start of the reduction in the development of anorexia was almost instantaneous and in both cases food intake had reached its minimum within 2 days.

There is significantly more evidence that food composition affects the above two characteristics during challenges with macro parasites and gastrointestinal nematodes in particular; this is mainly a reflection of the time it takes for anorexia to be established (see above). Datta *et al.* (1998) gave sheep access to isoenergetic foods that differed in their CP contents (ranging from 100-220 g/kg DM) whilst they were infected with the abomasal parasite *Haemonchus contortus.* In all dietary treatments food intake started to reduce from week 4 post infection (pi) and reached the maximum reduction by week 6 pi. These findings are in agreement with the observations of Kyriazakis *et al.* (1996) who also gave sheep access to foods of different protein contents whilst infected with *T. colubriformis* and observed an identical pattern in the development of anorexia between foods. A similar picture is also seen in experiment where ruminants have been given access to different forages (e.g. Bawden, 1969; Van Dam *et al.*, 1998). In these cases forage quality had no effect on the time anorexia was observed and the time it took for the maximum reduction in food intake to be established (within 3 weeks pi). On the basis of the above information it is concluded that food composition appears to have no effect on the time anorexia is observed and the time it takes for the maximum reduction in food intake to be achieved in subjects challenged by pathogens.

Effects on extent of anorexia

There is considerable confusion within the literature on whether food composition affects the extent of anorexia. This is despite the fact that this characteristic of anorexia will have an overwhelming influence on food intake consumed over the course of the infection. The sources for this confusion have been discussed above (analysis of the time course and expression of anorexia). Experiments that have addressed the effect of food nutrient content on the extent of anorexia have been performed on both non-ruminant (including laboratory rodents) and ruminant subjects. In the vast majority of cases, the nutrient under investigation has been protein or amino acids. The experiment of Chirase *et al.* (1991) on the effects of food Zn content on the food intake of cattle infected with the infectious bovine rhinotracheitis (IBR) virus is amongst the notable exceptions.

The experiments on non-ruminant species suggest contradictorily that either there is a significant effect or there is no effect of food protein content on the extent of anorexia. In the first category fall the experiments of Williams *et al.* (1997a), Tu *et al.* (2007) and J.C. Coltherd (personal communication, 2008). The data from the experiment of Williams *et al.* (1997a) is reproduced in Table 1. Pigs were offered access to foods of different lysine contents in environments intended to have either high or low levels of 'stimulation of the immune system'; this was achieved by exposure to a 'dirty environment' in the absence of in-feed antimicrobials and through repeated vaccination. Tu *et al.* (2007) and J.C. Coltherd (personal communication, 2009) challenged mice with the intestinal parasite *Heligmosomoides (polygyrus) bakeri* whilst they were given access to foods that differed in their CP contents. In the first three experiments there was a significant interaction between food protein (amino acid) content and the extent of anorexia: as the protein content of the food was reduced the difference in the food intake between the uninfected and challenged animals was also reduced. The food intake of these two groups was identical at the lowest level of protein (Table 1). Tu *et al.* (2007) did not use uninfected mice in their study; however, whilst the food intake of the mice on the protein-sufficient food was reduced

Table 1. The food intake and live weight gain of pigs that were kept either in a clean environment (C) or a high immune system activation environment (IS) and offered foods of different dietary lysine content from 6 to 27 kg live weight (Williams et al., *1997a).*

	Immune activation	Dietary lysine (g/kg food)			
		6	9	12	15
Food intake (g/d)	C	896	1,025	1,052	1,002
	IS	889	954	889	911
	Ratio IS: C	0.99	0.93	0.85	0.91
Gain (g/d)	C	400	556	644	663
	IS	357	495	510	504
	Ratio IS: C	0.89	0.89	0.79	0.76

immediately pi, the food intake of the mice on the protein-deficient food remained unaffected (as analysed in their Figure 2c).

These effects can be accounted for by the fact that the food intake of the uninfected animals was reduced at the lowest level of protein, compared to the intake of the infected animals on the higher levels of intake. Wellock *et al.* (2003) have suggested that the inability of non-ruminant animals to increase their intake on low protein foods is usually due to their inability to cope with the excess energy intake that would accompany this situation. Food intake of animals on the lowest protein content of the uninfected animals could thus be seen as being constrained. The above results suggest that pathogen challenge or antigen stimulation would not reduce any further the already constrained food intake; this suggestion seems also to make teleological sense. On the other hand the experiments by Brailsford and Mapes (1987) on mice challenged with *H. bakeri*, and Van Heugten *et al.* (1994a) on pigs challenged with LPS and Williams *et al.* (1997b) on pigs maintained in a dirty environment, report an absence of effects of food protein content on the extent of anorexia in challenged animals. However, all these experiments suffer from the limitations highlighted above. Food intake is reported over periods that include both the lowest food intake and the periods of recovery. It is likely, therefore, that the contradictory conclusions reached on this issue even by the same authors reflect how data has been analysed and reported.

Experiments that have addressed this issue in ruminant animals have been mainly performed on sheep that have been exposed to gastrointestinal parasites, whilst they were given access to foods that differed in their crude protein (CP) or metabolisable protein (MP) contents. Where data analysis overcomes the limitations reported above, the conclusion drawn is that food protein content does not affect the extent of anorexia in parasitised animals. The data of Greer *et al.* (2009) is a good case in point: when anorexia during *T colubriformis* infection was compared between sheep given access to foods of different MP content, the conclusion was that animals on the low MP food suffered a greater degree of anorexia. However, when the comparisons on food intake were made per unit of body weight or metabolic body weight the opposite conclusion was reached. The same absence of an effect on the extent of anorexia was seen when cattle infected with the IBR virus were given access to foods that differed in Zn content (Chirase *et al.*, 1991). The striking difference between the picture seen in non ruminant and ruminant animals could be a reflection of the role of N in the control of food intake of ruminants. It is likely that sheep are not trying to regulate their N content in the same manner as non-ruminants do, when they are offered access to foods of low nutrient contents (Kyiazakis, 2003).

Surprisingly, there are only few experiments that have investigated the effects of food energy content and source *per se* on the extent of anorexia. The exceptions are those by Van Heugten *et al.* (1994b) and Spurlock *et al.* (1997) on pigs. In both experiments the challenge to the animals was achieved through the administration of LPS and/or other immunostimulants. In the experiments of Van Heugten *et al.* (1994b) a change in the feed energy density, achieved through the substitution of cellulose with fat or starch, had no effect on the extent of anorexia in challenged pigs and chickens. Interestingly, however, the diet dilution was not accompanied by a change in the food intake of the non-challenged animals either. In the experiment of Spurlock *et al.* (1997) the substitution of conventional corn with high-oil corn had similarly no effects. It should be noted, however, that the experiments of Van Heugten *et al.* (1994b) and Spurlock *et al.*

(1997) suffer from the criticism that food intake was reported as an average over relatively long periods that may have included the periods of food intake recovery. On the basis of the above we can not draw strong conclusion on the effects of food energy density and energy source on the extent of anorexia. The seminal experiments to address this question have yet to be performed.

Effects of on duration of anorexia

There is very convincing evidence that food composition affects the duration of anorexia in infected subjects. Such evidence comes from experiments that have lasted sufficiently long for such comparisons to be made and include mainly experiments where animals have been infected with macro-parasites (Kyriazakis *et al.*, 1996; Datta *et al.*, 1998; Knox and Steele, 1999). The food dimension that has been investigated is either food protein or mineral content, and 'forage quality' (mainly supplementation of forages). The consistent conclusion from all such experiments is that duration of anorexia is reduced on high quality foods. In the experiment of Datta *et al.* (1998), for example, where sheep were infected with the nematode *H. contortus,* anorexia lasted only for a couple of weeks in animals given access to high protein foods (19 and 22% CP), whereas it was still evident after 5 weeks in animals given access to low protein foods (10 and 13% CP). The same picture was seen in cattle infected with the IBR virus: anorexia lasted for 3 vs. 8 days in animals whose diet had been supplemented or not respectively with Zn (Chirase *et al.*, 1991).

It is assumed that the above effects are due to effects of nutrients on the degree of expression of immunity (Coop and Kyriazakis, 1999). In some of the above experiments, evidence is also provided that a shorter duration of anorexia is associated with lower pathogen loads (Sandberg *et al.*, 2006), although it is not always possible to associate the exact point of recovery from anorexia with precise pathogen loads. This evidence supports the view that there is a link between the immune response of the host and the development of anorexia.

Effects on rate of recovery

This feature of anorexia would be expected to be underlined by two factors: (a) how quickly the host's immune response reduces and eliminates the pathogen load, and (b) what is the animal trying to achieve once it recovers from anorexia. It is expected that food composition affects the degree of expression of immunity and thus the rate by which the animal affects its resident pathogen population (Coop and Kyriazakis, 1999); this will have consequences on the rate of change in food intake as discussed above. As it is likely that infection has altered the animal's body composition as well as slowing its growth rate, as in the case of pigs infected by the porcine reproductive and respiratory syndrome virus (PRRSV) (Bassaganya-Riera *et al.*, 2001; Escobar *et al.*, 2004), it is possible that the animal might wish to restore the former and regain some of the time lost in its growth (Kyriazakis and Emmans, 1992). As both these factors are expected to be affected by food composition, the further expectation is that food composition will affect the rate of recovery from anorexia.

There are not many papers in the literature that present data of sufficient duration and thus allow conclusions to be drawn on this issue, but the limited evidence is consistent with this. A reanalysis of the data from Kyriazakis *et al.* (1996), by expressing food intake per unit of live weight, suggested that sheep challenged with *T. colubriformis* and given a high protein food recovered at a faster rate (1.5 times) than sheep given access to a lower protein food. In the experiment of Tu *et al.* (2007) mice previously infected with *H. polygyrus* were able to compensate in their live weight when they were given access to a high CP food, but not when to a low protein one, once the parasite was removed by an anthelmintic. Unfortunately, there has not been possible to identify any further suitably designed experiments that have measured food intake during recovery in animals previously exposed to pathogens in order to further support this assertion.

Host genotype effects on anorexia

Evidence for between and within breed variation in anorexia

Although obvious from a heuristic point of view (e.g. Sandberg *et al.*, 2006; Kyriazakis and Houdijk, 2007), experimental evidence for the influence of host genotype on the patterns of pathogen induced anorexia is relatively scarce, precisely for the reasons discussed previously (reporting food intake over the course of infection and recovery, and making the appropriate comparisons between animals). Nevertheless, two recent challenge studies in sheep (Zaralis *et al.*, 2008a) and mice (J.C. Coltherd, personal communication, 2009; Theeruth, 2008) infected with gastrointestinal parasites point towards significant between and within breed variation in the patterns of pathogen induced anorexia. In both studies, food intake and infection characteristics were compared between two breeds deriving from different selection lines. The experiments also included variation in dietary protein intake. Both experiments report that animals from lines intensively selected for growth generally experienced more severe infection characteristics (i.e. faecal egg counts) and anorexia than animals from less intensively selected lines; these breed differences also depended on dietary treatment (Zaralis *et al.*, 2008; J.C. Coltherd, personal communication, 2009). In particular, whereas animals from the fast growing lines showed significant reductions in food intake over prolonged time compared to uninfected control animals of the same lines, most animals from the alternative lines showed little to no reduction in food intake. A non-linear mixed model approach used to describe anorexia time trends of individual mice of the above study, further revealed significant differences in individual anorexia patterns between animals of the same fast growth selection line (Theeruth, 2008). As shown in Figure 2, individual mice differed both in the degree and the duration of anorexia, as well as in their rates of recovery of food intake.

It is noteworthy that breed differences in anorexia patterns were not apparent in a similar trial performed on pregnant and lactating ewes of different genotypes (Zaralis *et al.*, 2008b). Although nematode infection caused a significant reduction in average daily food intake in both breeds during the period of breakdown of immunity, no statistically significant difference in the extent and duration of anorexia was found between the ewes derived from intense selection for high productivity (Greyface cross) and less intensively selected Scottish Blackface ewes, despite significantly higher faecal egg counts in the Greyface crosses.

Although the above studies point towards significant variation both between and within breeds in the characteristics of infection induced anorexia, sample sizes in both trials were not sufficiently large to determine whether the observed variation were caused by host genetic differences.

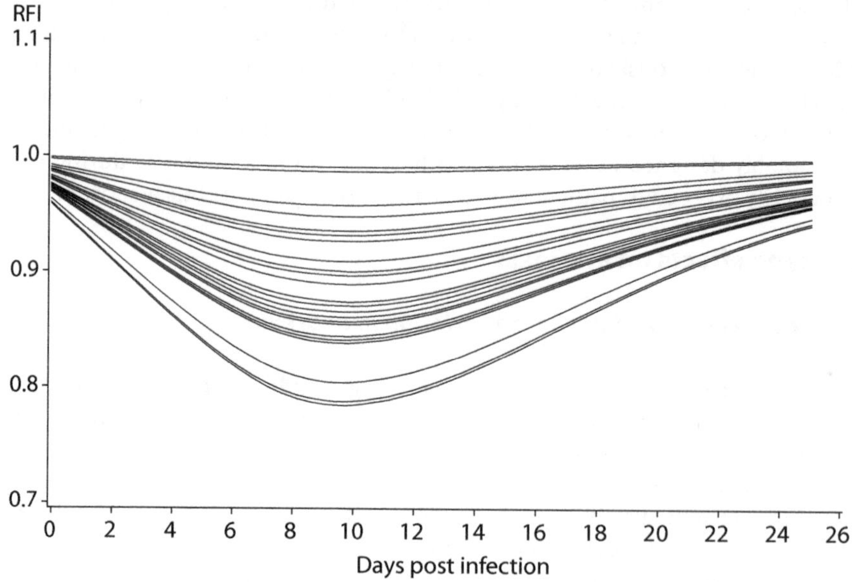

Figure 2. Fitted time trends of relative food intake (RFI) for individual mice intensively selected for growth (Roslin High Line, Bünger et al.*, 2001) after infection with gastro-intestinal parasites (Theeruth, 2008). The relative food intake for each individual infected animal i at time t_j was calculated relative to the average food intake of non-infected control animals of the same line and same dietary treatment at the same time (n=6) (Theeruth, 2008). A mixed model was fitted, which is given by the (Gompertz derivative) function: $RFI_i(t_j) = [1 - (a + \delta_i) \times b \times exp(-exp(-b \times (t_j - t^*)) - b \times (t_j - t^*))] + \varepsilon_{i.}$*

The lack of variation in anorexia in the ewe studies also suggest that caution should be applied when making assumptions about the reduction in food intake as response to infection, based on infection characteristics (e.g. faecal egg counts) alone.

Indirect evidence for host genetic influence on anorexia

1. Genetic variation in the host infection level and immune response

The close relationship between anorexia, host pathogen load (Kyriazakis *et al.*, 1996) and immune response (Escobar *et al.*, 2004; Greer *et al.*, 2005) implies that genetic variation in anorexia should be expected from genetic variation in host immune response. Sandberg *et al.* (2006) proposed that hosts with greater genetic ability to recognise pathogens and launch an effective adaptive immune response, leading to faster recovery from the pathogen, would be expected to reduce their food intake at faster rates and express anorexia for a shorter duration of time. In contrast to the scarce evidence for host genetic influence on pathogen induced anorexia, evidence for genetic differences in the host immune response is rapidly accumulating (e.g. Halbur *et al.*, 1998; Henryon *et al.*, 2001; Vincent *et al.*, 2005, 2006). In particular, Petry *et al.* (2007) compared expression levels of various immune markers, including the known appetite suppressors TNF-α,

IFN-γ, IL-8 and IL-1β (Exton, 1997), of pigs from two genetically diverse breeds after infection with the PRRSV. The study revealed significant differences between both breeds and between pigs classified as high or low responders, according to phenotypic responses in the expression levels of many genetic immune markers and in particular of all of the above mentioned genes associated with eliciting anorexia. Although food intake had not been reported explicitly in these studies, recorded body weight changes during the time course of infection could be considered as crude indicators for food intake. Previous analysis performed on phenotypic measures obtained from the same trial revealed that pigs of the fast lean growth line which were later classified through higher expression levels of TNF-α, IFN-γ, IL-8 and IL-1β also suffered substantially higher reductions in growth rate over a more prolonged time period than pigs of the commercial line, which showed on average less severe infection symptoms and lower gene expression levels (Petry *et al.*, 2005). In a similar study, Vincent *et al.* (2005, 2006) found significant breed differences in the susceptibility to PRRSV based on a variety of *in-vitro* and *in-vivo* measures, which were later complemented by findings of also significant differences in the growth response of the same infected pigs (Doeschl-Wilson *et al.*, 2009). Pigs from lines selected for higher lean growth rate not only had significantly higher virus levels and more severe disease symptoms over prolonged time but also suffered a stronger reduction in body weight growth as a consequence of the infection, pointing thus indirectly towards a direct relationship between genetic differences in infection characteristics and anorexia.

2. Anorexia and leptin concentration

The adipocyte hormone leptin is a potent regulator of both food intake (e.g. Henry *et al.*, 1999; Cameron *et al.*, 2000) and the immune response (Colditz, 2002; La Cava *et al.*, 2004). Also, evidence for genetic variation in serum leptin concentrations and leptin mRNA abundance in adipose tissue, has been provided in various studies (Cameron *et al.*, 2000; Nkrumah *et al.*, 2007), including immunological challenge (LPS) studies (Leininger, 2002). One may therefore hypothesise that genetic variation in anorexia may be partly caused by physiological differences in leptin or in leptin response to immune challenge between hosts. This thesis is however not supported by experimental studies. Faggioni *et al.* (1997), after administering LPS (which induces anorexia and increases leptin levels), to leptin receptor deficient (db/db) and leptin deficient (ob/ob) mice, found that db/db mice were somewhat resistant to LPS induced anorexia, whereas the ob/ob were more sensitive to LPS induced anorexia. Their results would thus suggest that differences in anorexia were not caused by leptin per se. Similar conclusions were reached by Zaralis *et al.* (2008a), after finding no significant breed effects in plasma leptin concentrations of lambs after infection with gastro-intestinal parasites despite of significant breed differences in the reduction of food intake.

3. Host genetic difference in feeding behaviour

Considering anorexia as behavioural trait (Hart, 1988; Exton, 1997), one could speculate that an animal's reduction in voluntary food intake during infection may partly result from changes in its short term feeding behaviour. An animal that rests more as a result of the infection may eat less frequently or have meals of shorter duration, and consequently consume less food. Breed difference in feeding behaviour during infection were indeed observed in the study of Hutchings *et al.* (2007), which showed that sheep from a more resistant breed (classified by on average lower faecal egg counts) had a higher tendency to avoid contaminated patches of grass when exposed to pastures infected with gastro-intestinal parasites than sheep from a more susceptible breed. As food intake had not been recorded in this study, it is however unclear whether the observed differences in the grazing behaviour were also mirrored by differences in actual food intake.

Anorexia as new breeding goal?

Host genetic variation in a trait of interest is key to animal breeding, whose value as disease control strategy is now widely accepted (Wilkie and Mallard, 2000). From a breeding point of view the central question is thus whether host genetic variation in anorexia is solely a by-product of genetic variation in the host physiology, immune response or food consumption behaviour, or whether hosts vary genetically in their ability to express anorexia per se. In the first case, anorexia would be considered as unfavourable by-product and may be influenced indirectly by selection for the causal traits. This assumption however contrasts with evolutionary principles, which regard anorexia to be beneficial to the host as it activates the host immune response, counteract pathogen growth or promote beneficial resting behaviour (Murray and Murray, 1979; Hart, 1988; Exton, 1997; Kyriazakis *et al.*, 1998). Based on this viewpoint of beneficial anorexia, genetic variation in anorexia characteristics of infected animals would be advantageous and could be exploited in animal breeding to improve the animals' genetic ability for coping with infection. Host genetic variation in anorexia that is not induced by components of the immune response but rather enhance an animal's ability to cope with infection would provide substantial practical benefits, as reduction in food intake has been shown to occur already at sub-clinical infection levels (Sandberg *et al.*, 2006). Hence selection for increased genetic ability to cope with infection based on anorexia characteristics could be performed at early stages of infection or for low levels of infection before clinical symptoms appear.

However, to date we are barely at the brink of understanding the role of host genetics on infection induced anorexia. Future experiments would require repeated simultaneous measurements of food intake, immune response and infection levels over the time course of infection to determine causes and consequences of anorexia patterns and how these are influenced by host genotype. Once host genetic influence on anorexia has been established, apparent interactions between host genotype and other factors influencing anorexia such as food composition or pathogen type, strain and dose would need to be addressed (Kyriazakis, unpublished results).

References

Balaji, R.., Wright, K.J.., Hill, C.M., Dritz, S.S., Knoppel, E.L. and Minton, J.E., 2000. Acute phase response of pigs challenged orally with *Salmonella typhimurium*. Journal of Animal Science 78: 1885-1981.

Bassaganya-Riera, J., Hontecillas-Magarzo, R., Bregendahl, K.,Wannesmuchier, M.J. and Zimmerman, D.R., 2001. Effects of dietary conjugated conjugated linoleic acid in nursery pigs of dirty and clean environments on growth, empty body composition and immunocompetence. Journal of Animal Science 79: 714-721.

Bawden, RJ., 1996. Relationships between *Oesophagostomum columbianum* infection and the nutritional status of sheep. Australian Journal of Agricultural Research. 20: 589-599.

Brailsford, T.J. and. Mapes, C.J., 1987. Comparisons of Heligmosomoides polygyrus primary infection in protein-deficient and well-nourished mice. Parasitology 95: 311-321.

Bünger, L., Laidlaw, A., Bulfield, G., Eisen, E.J., Medrano, J.F., Bradford, G.E., Pirchner, F., Renne, U., Schlote, W. and Hill, W.G., 2001. Inbred lines of mice derived from long-term growth selected lines: unique resources for mapping growth genes. Mammalian Genome 12: 678-686.

Cameron, N.D., Penman, J.C., McCullough, E., 2000. Serum leptin concentration in pigs selected for high or low daily food intake. Genetics Research 75: 209-213.

Chirase, N.K., Hutcheson, D.P. and Thompson, G.B., 1991. Feed intake, rectal temperature, and serum mineral concentrations of feedlot cattle fed zinc oxide or zinc methionine and challenged with infectious bovine rhinotracheitis virus. Journal of Animal Science 69: 4137-4145.

Colditz, I., 2002. Effects of the immune system on metabolism: implications for production and disease resistance in livestock. Livestock. Production. Science 75: 257-268.

Coop, R.L. and Kyriazakis, I., 1999. Nutrition-parasite interactions. Veterinary Parasitology. 84: 187-204.

Coop, R.L., Sykes, A.R. and Angus, K.W., 1982. The effect of three levels of intake of *Ostertagia circumncincta* larvae on growth rate, feed intake and body composition of growing lambs. Journal of Agricultural Science 98: 247-255.

Datta, F.U., Nolan, J.V., Rowe, J.B. and Gray, G.D., 1998. Protein supplementation improves the performance of parasitized sheep fed a straw-based diet. International Journal for Parasitology 28: 1269-1278.

Doeschl-Wilson, A.B., Kyriazakis, I., Vincent, A., Rothschild M.F., Thacker E. and Galina-Pantoja L., 2009. Clinical and pathological responses of pigs from two genetically diverse commercial lines to porcine reproductive and respiratory syndrome virus infection. Journal of Animal. Science (in press).

Exton, M.S., 1997. Infection-induced anorexia: active host defence strategy. Appetite 29: 369-383.

Escobar, J., Van Alstine, W.G., Baker, D.., Johnson, R.W., 2004. Decreased protein accretion in pigs with viral and bacterial pneumonia is associated with increased myostatin expression in muscle. Journal of Nutrition 134: 3047-3053.

Faggioni, R., Fuller, J., Moser, A.., Feingold, K.R. and Grunfeld, C., 1997. LPS-induced anorexia in leptin –deficient (ob/ob) and leptin receptor-deficient (db/db) mice. American Journal of Physiology - Regulatory, Integrative and Comparative Physiology 42: R181-186.

Gidenne, T. and Licois, D., 2005. Effect of a high fibre intake on the resistance of the growing rabbit to an experimental inoculation with enteropathogenic strain of *Escherichia coli*. Animal Science 80: 281-288.

Greer, A.W., Stankiewicz, M., Jay, N.P., McAnulty, R.W. and Sykes, A.R., 2005. The effect of concurrent corticosteroid induced immuno-suppression and infection with the intestinal parasite *Trichostrongylus colubriformis* on food intake and utilization in both immunologically naïve and competent sheep. Animal Science 80: 89-99.

Greer, A.W., Sedcole, R.J., Jay, N.P., McAnulty, R.W., Green, R.S., Stankiewicz, M. and Sykes, A.R., 2009. Protein supply influences the nutritional cost of immunity in lambs infected with *Trichostrongylus colubriformis*. Animal (in press).

Greiner, L., Stahly, T.S. and Stabel T.J., 2000. Quantitative relationship of systemic virus concentration on growth and immune response in pigs. Journal of Animal Science 78: 2690-2695.

Halbur, P.G., Rothschild, M.F., Thacker, B.J., Meng, X.J., Paul, P.S. and Bruna J.D., 1998. Differences in the susceptibility of Duroc, Hampshire, and Meishan pigs to infection with a high virulence strain (VR2385) of porcine reproductive and respiratory syndrome virus (PRRSV). Journal of Animal Breeding and Genetics. 115: 181-189.

Hart, B.L., 1988. Biological basis of the behaviour of sick animals. Neuroscience and Biobehavioral Reviews 12: 123-137.

Henry, B.A., Goding, J.W., Alexander, W.S., Tilbrook, A.J., Canny, B.J., Dunshea, F., Rao, A., Mansell, A. and Clarke, I.J.., 1999. Central administration of leptin to ovariectomized ewes inhibits food intake without affecting the secretion of hormones from the pituitary gland: evidence for a dissociation of effects on appetite and neuroendocrine function. Endocrinology 140: 1175-82.

Henryon, M., Berg, P., Jensen, J. and Andersen, S., 2001. Genetic variation for resistance to clinical and subclinical diseases exsits in growing pigs. Animal Science 73: 375-387.

Houdijk, J.G.M., Campbell, F.M., Fortomaris, P.D., Eckersall, P.D. and Kyriazakis, I., 2007. Effects of sub-clinical post-weaning collibacillosis and dietary protein on acute phase proteins in weaner pigs. Livestock Science 108: 182-185

Hutchings, M.R., Knowler, K.J., McAnulty, R. and McEwan, J.C., 2007. Genetically resistant sheep avoid parasites to a greater extent than do susceptible sheep. Proceedings of the Royal. Society Series B 274(1620): 1839-1844.

Knox, M.R. and Steel, J.W., 1999. The effects of urea supplementation on production and parasitological responses of sheep infected with *Haemonchus contortus* and *Trichostrongylus colubriformis*. Veterinary Parasitology 83: 123-135.

Kyriazakis, I., 2003. What are ruminant herbivores trying to achieve through their feeding behaviour and food intake? In: 't Mannetie, L., Ramirez-Aviles, L., Sandoval-Castro, C.A. and Ku-Vera, J.C. (eds.). Matching herbivore nutrition to ecosystems biodiversity. Mérida, Universidad Autonóma de Yucatán, pp. 154-173.

Kyriazakis, I. and Emmans, G.C., 1992. The growth of mammals following a period of nutritional limitation. Journal of Theoretical Biology 156: 485-498.

Kyriazakis, I. and Houdijk, J.G.M., 2007. Food intake and performance of pigs during health disease and recovery. In: J Wiseman, J., Varley, M.A., McOrist, S. and. Kemp, B. (eds.). Paradigms in pig science. Nottingham University Press, Nottingham, pp. 493-513.

Kyriazakis, I., Tolkamp, B.J. and Hutchings, M.R., 1998. Towards a functional explanation for the occurrence of anorexia during parasitic infection. Animal Behaviour 56: 265-274.

Kyriazakis, I., Anderson, D.H., Oldham, J.D., Coop, R.L. and Jackson, F. 1996. Long-term subclinical infection with *Trichostrongylus colubriformis*: effects on food intake, diet selection and performance of growing lambs. Veterinary Parasitology 61: 297-313.

La Cava, A., Alviggi, C. and Matarese, G., 2004. Unraveling the multiple roles of leptin in inflammation and autoimmunity. Journal of Molecular Medicine 82: 4-11

Leininger, M., 2002. Physiological response to acute endotoxemia in swine: effect of genotype on energy metabolites and leptin. Domestic Animal Endocrinology 18: 71-82.

Nkrumah, J.D., Keisler, D.H., Crews, D.H. Jr., Basarab, J.A., Wang, Z., Li, C., Price, M.A., Okine, E.K. and Moore SS., 2007. Genetic and phenotypic relationships of serum leptin concentration with performance, efficiency of gain, and carcass merit of feedlot cattle. Journal of Animal Science 85: 2147-2155.

Petry, D.B., Holl, J.W., Weber, J.S., Doster, A.R.and Osorio, F.A., 2005. Biological responses to porcine reproductive and respiratory syndrome virus in pigs of two genetic populations. Journal of Animal Science 83: 1494-1502.

Petry, D.B., Lunney, J., Boyd, P., Kuhar, D., Blankenship, E. and Johnson, R.K., 2007. Differential immunity in pigs with high and low responses to porcine reproductive and respiratory syndrome virus infection. Journal of Animal Science 85: 2075-2092.

Sandberg, F.B., Emmans, G.C. and Kyriazakis, I., 2006. A model for predicting feed intake of growing animals during exposure to pathogens. Journal of Animal Science 84: 1552-1566.

Spurlock, M.E., Frank, G.R., Willis, G.M., Kuske, J.L. and Cornelius, S.G.., 1997. Effect of dietary energy source and immunological challenge as growth performance and immunological variables in growing pigs. Journal of Animal Science 75: 720-726.

Theeruth, B.K., 2008. Selecting suitable mathematical functions to describe the time course of anorexia during pathogen challenge. Master of Science Thesis, University of Edinburgh, Scotland, UK.

Tu, T., Koski, K.G., Wykes, L.J. and Scott, M.E., 2007. Re-feeding rapidly restores protection against *Heligmosomoides bakeri* (Nematoda) in protein-deficient mice. Parasitology 134: 899-909.

Van Dam, J.T.P., Hofs, P., Tolkamp, B.J. and Zwart, D., 1998. The effect of the quality of roughage on the course of *Trypanosoma vivax* infection in West African dwarf goats - I. Organic matter intake, body weight change and efficiency of nitrogen metabolism. Livestock Production Science 53: 69-80.

Van Heugten, E., Spears, J.W. and Coffey, M.T., 1994a. The effect of dietary protein on performance and immune response in weanling pigs subjected to an inflammatory challenge. Journal of Animal Science 72: 2661-2669.

Van Heugten, E., Coffey, M.T. and Spears, J.W., 1994b. Effects of immune challenge, dieray energy density, and source of energy on performance and immunity in weanling pigs. Journal of Animal Science 74: 2431-2440.

Vincent, A.L., Thacker, B.J., Halbur, P.G., Rothschild, M.F.and Thacker, E.L., 2005. The in vitro susceptibility of macrophages to porcine reproductive and respiratory syndrome virus varies between two genetically diverse lines of pigs. Viral Immunology 18: 506-512.

Vincent, A.L., Thacker, B.J., Halbur, P.G., Rothschild, M.F.and Thacker E.L., 2006. An investigation of susceptibility to porcine reproductive and respiratory syndrome virus between two genetically diverse commercial lines of pigs. Journal of Animal Science 84: 49-57.

Wakelin, D., 2000. Rodent models of genetic resistance to parasitic infections. In: Axford, R.F.E., Bishop, S.C. Nicholas, F.W. and Owen, J.B. (eds). Breeding for disease resistance in farm animals. CABI Publishing, Wallingford, UK, pp. 107-126.

Wellock, I.J., Emmans, G.C. and Kyriazakis, I., 2003. Predicting the consequences of social stressors on pig food intake and performance. Journal of Animal Science 81: 2995-3007.

Wilkie, B.N. and Mallard, B.A., 2000. Genetic aspects of health and disease resistance in pigs. In: Axford, R.F.E., Bishop, S.C. Nicholas, F.W. and Owen, J.B. (eds). Breeding for disease resistance in farm animals. CABI Publishing, Wallingford, UK, pp. 379-396.

Williams, N.H., Stahly, T.S. and Zimmerman, D.R.,. 1997a. Effect of chronic immune system activation on the rate, efficiency and composition of growth and lysine needs of pigs fed from 6 to 27 kg. Journal of Animal Science 75: 2463-2471.

Williams, N.H., Stahly, T.S. and Zimmerman, D.R., 1997b. Effect of level of chronic immune system activation on the growth and dietary lysine needs of pigs fed from 6 to 112 kg. Journal of Animal Science 75: 2481-2496

Zaralis, K., Tolkamp, B.J., Houdijk, J.G.M., Wylie, A.R.G. and Kyriazakis, I., 2008a. Changes in food intake and circulating leptin due to gastrointestinal parasitism in lambs of two breeds. Journal of Animal Science 86: 1891-1903.

Zaralis, K., Tolkamp, B.J., Houdijk, J.G.M., Wylie, A.R.G. and Kyriazakis, I., 2008b. Consequences of protein supplementation for anorexia, expression of immunity and plasma leptin concentrations in parasitized ewes of two breeds. British Journal of Nutrition 101: 499-509.

Chapter 14. Models to predict feed intake

J.L. Black
John L. Black Consulting, Warrimoo, NSW 2774, Australia; jblack@pnc.com.au

Abstract

The value of being able to predict intake of an animal under a wide range of conditions has been recognised for many years because it is one of the primary determinants of growth rate, efficiency of feed use and body composition in animals. The complexity of the interactions between factors known to affect intake means that computer simulation techniques provide a robust approach for predicting intake of pigs under a wide range of animal, dietary and environmental conditions. The philosophy and techniques used for model development and evaluation are discussed briefly. Several conceptual approaches for modelling feed intake in pigs are evaluated including: algorithms fitted statistically to experimental results; potential intake limited by various constraints; maximum efficiency of oxygen utilisation; minimal total discomfort; and short-term meal eating based on nutrient fluxes and gastrointestinal tract distension. Several of the approaches have either been discredited or require considerable more development before they could be used to predict accurately feed intake of pigs under a wide range of environments encountered in commercial pig units. The 'potential-constraint' concept is the most highly developed, and despite several deficiencies, has been demonstrated to predict intake of pigs under many different circumstances. The major components of the 'potential-constraint' models are described, including: methods for defining potential energy intake based on genotype, live weight, maintenance energy requirement and activity; the impact of essential nutrient deficiencies; limitations caused by gastrointestinal tract capacity; cold and hot climates; number of pigs in a pen; stocking density; feeder-space; and disease. Several factors known to affect intake of pigs have not yet been modelled adequately, including: impact of poor air quality; effects of mycotoxins, antinutritional factors and taste aversion; rate of passage of digesta and gastrointestinal tract volume; and the impact of diseases found commonly in commercial pig units.

Keywords: computer modelling, genetic potential, intake constraints

Introduction

Many factors interact to control voluntary feed intake of a pig. Genetic propensity, expressed as the potential daily rate of lean and fat deposition, and the amount of energy used for basal metabolic processes required to sustain life, is the primary determinant of the amount of energy that an animal will strive to consume. Genetic selection of pigs over the last 20 years has resulted in a marked depression in the rate of fat deposition and an increase in lean tissue deposition, which has resulted in faster, more efficient growth, but less total energy being deposited daily. Feed intake has fallen by around 30 g/d annually in association with reduced energy deposition resulting from these genetic changes (Knap, 2009). Similarly, administration of porcine growth hormone, which stimulates protein synthesis, reduces lipogenesis and reduces total energy deposition in the body, has been found to decrease feed intake by approximately 10% (Campbell,

1988). However, it is rare for a pig to consume sufficient feed to satisfy its genetic propensity to use energy, particularly when raised under commercial environments. Several studies summarised by Black and Carr (1993) show that pigs housed individually in research facilities or in boar-test environments eat more and grow 15-25% faster than pigs with similar genetic background raised commercially. Factors known to reduce feed intake of pigs include: number of pigs per pen; floor area and feeder space per pig; air volume per pig; the quality of the air and particularly the concentration of viable bacteria; prevalence of diseases; climatic conditions relative to the thermoneutral temperatures for the pig; temperament of the pig and characteristics of the diet (Black *et al.*, 2001). The impact on feed intake of social (Manteca and Edwards, 2009), climatic, pathogenic (Kyriazakis and Doeschl-Wilson, 2009) and dietary (Zijlstra *et al.*, 2009), including sensory (Klasing and Humphrey, 2009; Roura and Tedó, 2009) and experiential (Bolhuis *et al.*, 2009), factors are examined in detail elsewhere in this publication.

Much is now understood about the physical and biochemical mechanisms that control hunger, satiety and long-term intake of animals (Black *et al.*, 2009; Carroll and Allee, 2009; Forbes, 2009; Roche *et al.*, 2008; Roura and Tedó, 2009). Critical mechanisms influencing the amount of feed eaten by an animal relative to its genetic propensity include: sensory and experiential factors determining preference for a feed; physical characteristics of a feed affecting distension stimulated or inhibited contractions of the stomach and intestines; activation of the ileal/colonic brake caused by undigested nutrients or products of microbial fermentation in the distal intestines; endocrine peptides released in relation to specific nutrients or metabolites from the gastrointestinal tract and pancreas that act on the vagus nerve, midbrain and hypothalamus; two serine-threonine kinases, adenosine monophosphate-activated protein kinase and mammalian target of rapamycin, that monitor the current energy status of an animal; leptin and insulin that monitor adiposity and ghrelin that monitors lean body mass; the melanocortin system within the hypothalamus that interacts with many metabolites, particularly malonyl-CoA, and endocrines to stimulate or inhibit the release of orexigenic and/or anorexigenic peptides, which act either directly or via higher centres of the brain to alter meal consumption and long-term intake. The mechanisms controlling feed intake in an animal are complex, and although much is known, a full understanding is still incomplete.

The value of being able to predict intake of an animal under a wide range of conditions has been recognised for many years because it is one of the primary determinants of growth rate, efficiency of feed use and body composition in animals. However, the difficulty of this task has been recognised by Whittemore *et al.* (2001c) who stated '... it is unrealistic to expect from the scientific literature ... any reasonable prediction of the particular food intake of groups of pigs'. Consequently, some models that predict the performance of pigs require intake of the animals being simulated to be an input to the model, rather than being calculated within the model (Van Milgen *et al.*, 2007). In other animal models, different levels of mathematical representation have been used to predict feed intake. Simple algorithms have been developed relating feed intake to either body weight or age using measurements from one or more experiments. There are no reports of models predicting feed intake of pigs based on the underlying neural-biochemical-humoral mechanisms. This failure to develop a truly mechanistic model is understandable because of the great complexity of the processes involved (Black *et al.*, 2009, Carroll and Allee, 2009). However, there have been several 'conceptual', semi-mechanistic models of feed intake in pigs developed that generally use multiple algorithms to represent many of the factors that are known to influence intake of pigs raised under commercial environments. This chapter provides a brief overview of the value of computer simulation models, the most appropriate level to develop a model and an outline of several models that have been developed to predict feed intake in pigs. Possible future ways of improving intake models for pigs are discussed.

Model development

Modelling philosophy

Development of computer models that simulate biological systems is simply a formalisation of the processes used by most scientists (Black, 1995a; France and Dijkstra, 2000; France and Thornley, 1984). Ideas and concepts about the control of a system under study are first formulated using results from experiments. This process alone often leads to identification of critical experiments and further acquisition of knowledge. The difference between the traditional and simulation approaches to understanding a system arises only because the concepts and data are transformed into mathematical algorithms that can be solved rapidly using computers to provide a quantitative and dynamic appraisal of the concepts and system. A comprehensive, thorough and quantitative appraisal through time of conceived mechanisms regulating complex biological systems is virtually impossible using only intellectual effort because of the magnitude of the interactions. Incorporation of concepts and data into a simulation model allows rapid and rigorous evaluation of the 'suitability' of these concepts for explaining the system when predictions are compared with experimental observations.

Techniques used to evaluate the suitability of a model have been described previously (Black, 1995c; France and Dijkstra, 2000), and include examining general behaviour of the model to known circumstances, identifying variables and parameters within the model to which outputs are sensitive, and making direct comparisons with experimental results. Frequently, models do not predict observations with sufficient accuracy, which leads to revision of the concepts and/or acquisition of new data to define more accurately parameters in algorithms. Models cannot be 'validated', only invalidated. They are simple representations of complex systems and must be 'appropriate' for the purpose of the model. Confidence, that a model is 'suitable' and adequately represents a system, develops over time through its continual use and evaluation. The purpose for which a model is built, determines the level of complexity at which it is formulated.

Complexity of models

Models can be constructed at a range of levels which represent different degrees of detail and complexity. Models predicting outcomes for animal systems can be constructed at the level of the enterprise or flock, individual animal, physiological functions represented by organs, metabolic functions represented by cells and organelles, or biochemical reactions and fluxes of metabolites controlled by enzymes, humoral and gene product/receptor interactions of macromolecules (France and Thornley, 1984). The level of complexity used to develop a model must be appropriate for its perceived function. For example, simple algorithms predicting intake of cattle in relation to weight, reproductive status and pasture quality would be sufficient for a model designed to determine stocking rate, pasture species compositions and fodder conservation strategies to maximum profitability of a grazing beef enterprise. However, if the model was designed to evaluate current concepts on the nutrient-metabolite-neural-humoral control of intake in a pig, the model would need to be made at the macromolecule level. If models are to predict accurately at one level, experience has shown that the model must be constructed at several levels below that required (Black, 1995a). For example, a pig growth model that predicts animal performance and body composition accurately over a wide range of environmental, social and dietary conditions should be constructed at least at the level of metabolic function. This process often requires aggregation of information from the lower levels, such as aggregation of biochemical reactions representing oxidation of amino acids or synthesis of fatty acids.

The complexity of a model often influences the type and number of algorithms used to predict an outcome such as feed intake. High level models would frequently use only a small number of algorithms to predict intake, whereas a model constructed at the macromolecule level may contain many hundreds of algorithms. The algorithms used in high level models are frequently fitted to results from a small number of experiments and are a statistical 'best-fit' to the data. Application of these algorithms outside the range of the data from which they were established often leads to inaccurate predictions, because they do not represent the underlying mechanisms. These models have poor predictive power for new situations. Conversely, models that are based on the fundamental laws of physics and chemistry, and which truly represent the underlying mechanisms, can predict accurately the outcome of new situations not examined before (Black, 1995b). There have been relatively few models of animal systems attempted at this basic mechanistic level (Baldwin, 1995; Nagorcka and Mooney, 1989), because of the complexities involved. Many more models simulating domestic animal performance have been developed at a 'conceptual' level, where the precise physical and chemical mechanisms are too complex or too unknown to put into a model, but the concepts thought to control a process can be represented mathematically. Algorithms are developed to fit the concepts and data from experiments are then used to quantify the parameters in equations (Black, 1995b). Algorithms can be developed to fit almost any concept. The bounds of the predictions are controlled by each algorithm and, unlike statistically fitted equations, the scope of predictions can be controlled.

Concepts used to predict intake of pigs

Different concepts have been used as the basis for predicting voluntary feed intake in animals. The concepts range from: simple statistical fit of algorithms to experimental results (ARC, 1981; NRC, 1998); predicting intake from the sum of the energy needs of an animal for protein and fat deposition, and for maintenance energy requirements in relation to the energy density of the diet (Whittemore *et al.*, 1995), but constrained by various dietary, environmental and social factors (Black *et al.*, 1986; Kyriazakis and Emmans, 1998; Nyachoti *et al.*, 2004); maximisation of the efficiency of oxygen utilisation (Tolkamp and Ketelaars, 1992); the minimal total discomfort theory (Forbes, 2007, 2009); to continual prediction of energy and protein status of cattle and rumen capacity determining meal consumption and long-term intake (Nagorka *et al.*, 2004).

Algorithms fitted to experimental data

Published algorithms predicting feed intake of growing pigs directly from experimental results have been reviewed by Whittemore *et al.* (2001c). The form of equations ranges widely from linear to cubic regressions, exponential and logistic relationships, to more complex representations (Table 1). Feed intake is predicted as kg/day for many equations from either live weight or age. Other equations first predict energy intake, because this is believed by some authors to be the primary determinant of intake, and then adjust the amount of feed eaten in relation to the available energy content of the diet (MJ/kg). Whittemore *et al.* (2001c) provide a strong argument about the futility of predicting feed intake of pigs on the basis of age, unless for a highly defined and constant environment. Intake is more closely related to the weight of the animal than its age, particularly in a highly variable environment. Several authors have provided different equation parameters for pigs of different sex, but few have considered the effect of strain of pig. Whittemore *et al.* (1995) found that intake ranged from 1.87 to 3.15 for 10 genetically diverse pigs reared under similar conditions. Such variation clearly would not be covered by the simple algorithms without changes in the equation parameters.

The predicted feed intake for a pig between 5 and 125 kg live weight is shown for several of the algorithms in Figure 1. A digestible energy content of 13.5 MJ/kg was assumed for the diet where the equations predicted digestible energy intake. The forms of the relationships vary widely between algorithms and the predicted values, particularly for pigs over 80 kg, differ markedly. Predicted values for pigs at 125 kg live weight ranged from 2.96 to 5.78 kg/d. The NRC (1998) equation, for pigs weighing 5-15 kg, predicted negative intake for weights above 40 kg. Figure 1 illustrates the inaccuracies that can occur when simple algorithms are used to predict intake and are extrapolated outside the range of the results for which they were established. The shape and position of the feed intake relationship with live weight can vary widely depending on the

Table 1. Forms of equations fitted to experimental data that have been used to predict either voluntary feed intake or voluntary energy intake of pigs.

Equation form[1]	Reference
$a + bW$	Tullis (1982)
$a + bW + cW^2$	NRC (1998) 5-15 kg, Quiniou *et al.* (2000)
$a + bW + cW^2 + dW^3$	NRC (1998)
$a(1 - e^{-bW})$	ARC (1981), NRC (1998)
aW^b	ARC (1981), Cole *et al.* (1967), Whittemore (1983)
$ae^{-bW-c/W}$	Ramaekers (1996)
$a + b\ln(X + c)$	Eissen *et al.* (1999)
$y_m.e^{(-\ln2.Wn/W).((W-Wm)/(Wn-Wm))**2}$	TMV (1994)
$a/(1 + e^{(b - c.X)})$	Lorenzo-Bermejo *et al.* (2000, 2003)

[1] Where a, b, c and d are constants, W is live weight, Wm is live weight at maximum feed intake, Wn is live weight at half maximum food intake, y_m is expected maximum feed intake and X is days on experiment.

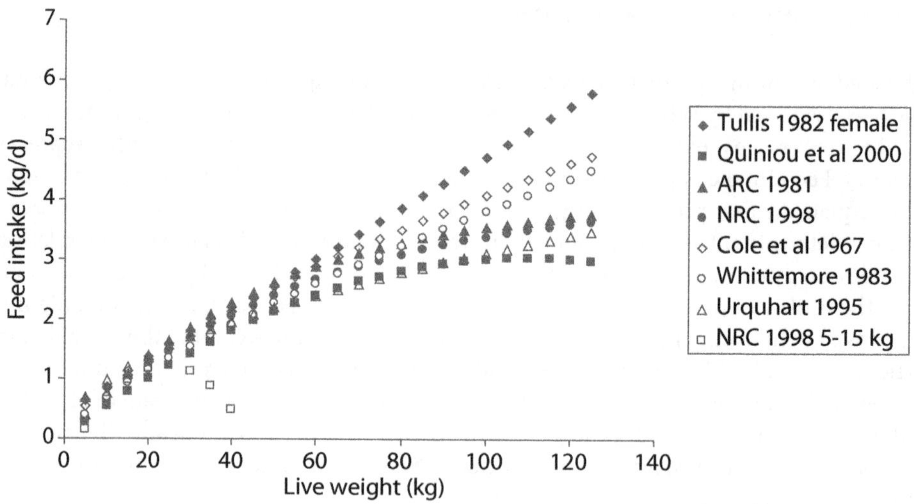

Figure 1. Feed intake of pigs predicted over the live weight range from 5 kg to 125 kg using several simple published algorithms fitted to experimental results.

propensity for a pig to deposit fat and protein as shown by Green *et al.* (2003) when comparing 'lean' type Landrace, with 'meaty' type Pietrain, with 'fatty' type, Meishan pigs.

Several simulation models predicting the growth of pigs use the NRC (1998) exponential equation to either directly predict intake (Pomar *et al.*, 1991) or as the basis for modifying intake measurements made on groups of pigs during use of the PorkMaster model (De Lange *et al.*, 2001). Predictions based on simple algorithms can be considered only a guide to likely intake because they do not attempt to represent the underlying mechanisms controlling intake. These simple algorithms do not allow for differences in genotype of pig or in the dietary, climatic, disease and social environment which are known to alter intake.

Intake controlled by genetic potential to use energy and constraints

The most commonly used conceptual framework for predicting feed intake of animals assumes that an animal will eat sufficient feed to satisfy its genetically determined requirements for nutrients, specifically energy, but environmental factors relating to the diet, climate, disease or housing, may either increase or decrease feed intake from this potential (Black *et al.*, 1986; Emmans, 1981; Kyriazakis and Emmans, 1998; Nyachoti *et al.*, 2004; Poppi *et al.*, 1994; Whittemore *et al.*, 1995). The potential for an animal to consume energy is assumed to be determined by its genetic propensity to deposit protein, fat and other tissues including foetuses and conceptus, to secrete milk and maintain essential body functions. Chemical modifiers of metabolism that change the partition of deposited tissue between protein and fat, such as porcine growth hormone or ractopamine, will also change the 'potential' intake of an animal. The concept assumes that an animal will consume more energy than the potential calculated as above when additional energy is expended during exercise or when it is exposed to cold conditions. However, this desire to consume energy may be overridden by a range of constraints relating to the physical characteristics of the diet, nutrient deficiency or imbalance, antinutritional factors in the diet, climatic conditions above the zone of thermoneutrality, disease, microbial load without acute disease symptoms, number of animals in a group, area per pig and space allocated for eating (Black *et al.*, 2001). The actual amount of feed an animal consumes is assumed to be the minimum of that needed to satisfy the genetic potential plus additional energy expenditure during exercise or cold exposure, and of all other constraints considered. Details of concepts incorporated into models about each constraint and several specific algorithms that have been used are described later in this chapter. However, several authors disagree with this conceptual approach for predicting feed intake and have provided alternatives.

Maximum efficiency of oxygen utilisation

Tolkamp and Ketelaars (1992) believed the 'potential-constraint' hypothesis for predicting intake in ruminants did not adequately describe the observations when animals were consuming low quality roughages. These authors proposed that there are both costs and benefits from consuming feed. The benefits from consuming feed were considered to be the net energy used for maintenance and productive functions, while the costs were associated with oxygen consumption by tissues that cause accumulated damage to cell structures, loss of vitality, aging and a limited life span. The authors speculate that an animal ceases eating when it reaches an optimum efficiency of energy use. The high energy cost of consuming low quality roughages was assumed to override the benefits from the energy derived by the animal and intake of such material would be limited. Tolkamp and Ketelaars (1992) proposed that an animal will cease eating when the intake of net energy per litre of oxygen consumed is maximised. The concept was tested in several experiments by varying characteristics of feed energy density, nitrogen content and physical form, and the model predicted well observed changes in intake. Tolkamp and Ketelaars (1992) suggested that the concept relating feed intake to optimum efficiency of oxygen use, would apply across animal types including monogastric species.

The validity of the Tolkamp and Ketelaars (1992) concept was compared with 'potential-constraint' concept in two experiments using pigs. In one experiment (Whittemore *et al.*, 2001a), the intake was measured of pigs offered either a high-quality control diet or a high-fibre, high-bulk diet, when the animals were moved from a thermoneutral to a cold environment. Both the 'potential-constraint' model and the 'oxygen-efficiency' model predicted feed intake would increase when pigs offered the control diet were exposed to a cold environment. However, the 'oxygen-efficiency' model, but not the 'potential-constraint' model, predicted intake would increase also when pigs offered the high-bulk diet were exposed to cold. The results from the experiment confirmed that feed intake increased when pigs offered the high-quality diet were exposed to cold, but intake did not increase with the high-bulk diet. These results supported the 'potential-constraint' model over the 'oxygen-efficiency' model. Similarly, the second experiment in which a high-quality, medium-bulk or high-bulk diet was fed in two periods supported the 'potential-constraint' model more than the 'oxygen-efficiency' model (Whittemore *et al.*, 2001b).

Minimum total discomfort theory

Forbes (2007, 2009) has several criticisms of the 'potential-constraint' concept for predicting feed intake of animals. First, Forbes believes a concept that has only one feedback mechanism limiting intake and any point in time and then switches to another single mechanism over time does not describe the physiology of an animal. Second, the concept of a fixed limit to an animal function is not supported by experimental evidence. For example, a bird will reduce the frequency of meals and increase the size of a meal 5-fold when the number of pecks on a button required to access food was increased, but total intake remained unchanged. Third, the concept does not allow for the wide daily fluctuation in meal size as is frequency observed for an individual animal, although, when averaged over several days, intake is reasonably constant.

Forbes proposes that an animal learns from experience the causes of discomfort, whether physical or metabolic, and integrates these signals of discomfort from a range of sources to minimise the discomfort experienced. The wide fluctuation in frequency and size of meals is regarded as being part of the learning experience. Rather than selecting a cut-off value for a single resource as proposed by the 'potential-constraint' model, Forbes suggests that the mismatch between the current status of a resource and its optimum value is calculated for every resource being considered. The square of this difference is determined for every resource, with the values for all resources being summed. Finally, the square root of the summed value is obtained to calculate total minimal discomfort. The feed intake corresponding to the minimum total discomfort will then be selected. Forbes (2009) demonstrates that the total discomfort model, similar to the 'potential-constraint' model, could predict the results observed by Whittemore *et al.* (2001a) when animals consuming a low-bulk or high-bulk diet were exposed to cold. Nevertheless, Forbes (2009) recognises that many components of his minimal discomfort model need to be refined and extended before it could become a practical and accurate method for predicting feed intake of animals.

Short-term nutrient deficiency and meal eating

A novel approach has been used by Nagorka (Nagorka *et al.*, 2004, B.N. Nagorcka, personal communications) to predict feed intake by cattle in a mechanistic biochemical model (Nagorka *et al.*, 2000) that is a variation of the dairy cow model developed by Baldwin and colleagues (Baldwin, 1995). The model predicts the fluxes of each major metabolite using Michaelis-Menten kinetics with a short iteration interval of 0.001 day. Feed intake is predicted from two metabolic controls, C_m and C_p which monitor respectively the energy and amino acid: energy status of the animal, and a distension or rumen fill control, C_R. Unlike the 'potential-constraint' model which has a single constraint limiting intake, Nagorcka represents C_m, C_p and C_R as binary variables with values of either 0 or 1. Each control switches from 0 to 1 when the ratio of the current value to a minimal value becomes <1. Alternatively, the controls switch from 1 to 0 when the ratio of the current value to a maximal value becomes >1. An animal is predicted to eat only when the value for C_m, C_p and C_R all equal 1. The model ensures that there is a period of time when an animal will not be eating because this occurs whenever the current value for any control is between the maximum and minimum limits. The model integrates the responses of the animal for two major factors known to affect feed intake, current nutrient status and gut distension. Similar to the concepts proposed by Forbes (2009), the Nagorcka approach could be assumed to monitor the level of 'discomfort' in an animal due to distension of the gut, and over- or under-supply of essential nutrients.

The Nagorka model assesses the current energy status of the animal from the concentration of total metabolisable energy stored as nutrients in blood. The concentration of nutrients in blood is determined by the actual rate of nutrient absorption in relation to the rate of use of individual nutrients to meet current metabolic requirements. The current metabolic requirements for nutrients are set in relation to the genotype, weight and rate of growth of the animal. Nagorka also varied the rate of intake of a feed depending on the hardness of the grain component and the force required to shear the forage component. Hence, both chemical and physical characteristics of the diet alter the initiation and cessation of a meal through changes in rate of eating, rate of digestion and rate of utilisation of individual components of the diet. The Norgaka approach can be applied only to models that simulate at the biochemical level, such as that developed for pigs by Danfaer (2000).

Potential intake

Irrespective of the concepts used to develop a model predicting feed intake in pigs, an understanding is required of factors that determine the potential intake and how this may be modified by genetic, dietary and environmental factors. Pigs with different genetic backgrounds consume different amounts of feed (Green *et al.*, 2003; Knap, 2009; Whittemore *et al.*, 1995). Although the mechanisms controlling feed intake are complex (Black *et al.*, 2009), it is useful, when developing models for practical use by the pig industry, to assume that an animal has a potential voluntary intake that is determined by its genetic capacity to metabolise nutrients, particularly energy. Capacity to metabolise energy depends on the rate of net tissue deposition, the energy content of the tissue deposited and the energy required for maintenance of the animal.

Potential rate of energy deposition

The potential rate of energy deposition can be assessed when pigs are grown from birth to maturity and given free access to a high quality diet where feed intake is not limited by nutrient deficiency or imbalance, or by gut capacity, disease, climatic, social or other factors. Ferguson and Gous (1993a) suggest that the genetic potential for a pig type to deposit energy can be determined when a group of pigs is housed individually, given a choice of both a high- and low-protein diet and has a graded temperature across the pen space. However, the experiment of Hodge (1974), who offered cow's milk from birth to pigs in an ideal environment, suggests that the energy intake of pigs weighing less than about 40 kg can be severely limited when they are offered solid diets. Hodge (1974) found that the energy intake of young pigs offered free access to a liquid diet was substantially higher than what can be obtained from the sow. Thus, high quality liquid diets are needed to define the genetic potential of pigs for several weeks after birth.

Several estimates have been made of the potential rate of energy deposition for different pig genotypes (Black *et al.*, 1986; Ferguson and Gous, 1993b; Siebrits *et al.*, 1986; Whittemore *et al.*, 1995). Frequently, the rate of energy deposition derived from such experiments, and used in models, has been described by a Gompertz equation (Emmans, 1995; Ferguson and Gous, 1993b).

$$dE/dt = DElog_e(E_f/E) \qquad (MJ/d) \tag{1}$$

Where E is current body energy content, E_f is body energy content at maturity and D is a constant representing the decay in relative energy deposition and is a function of genotype. Although it is relatively easy to establish parameters for the Gompertz equation from experimental results, the equation has only one form with the maximum rate of energy deposition occurring at a proportion 1/e of final energy content. Black *et al.* (1995) argue that the Gompertz equation is too inflexible to adequately represent the range in form of energy or protein deposition found in pigs with different genetic backgrounds (Green *et al.*, 2003). The Gompertz equation belongs to a family of curves derived from the Richards equation (France and Thornley, 1984).

$$dE/dt = kE(E_f^n - E^n)/n\,E_f^n \qquad (MJ/d) \tag{2}$$

Where k and n are constants. In the Gompertz representation of the equation, n=0. In another form of the Richards equation, the Logistic equation, also used to describe energy and protein deposition in animals, n=1 and the maximum rate of energy deposition occurs at half final body energy content. Alteration of the values for k and n in the Richards equation allows considerable flexibility in the proportion of final energy content where the maximum occurs. Nevertheless, there are restrictions to the Richards equation because an extended plateau in rate of energy deposition cannot be obtained.

Black *et al.* (1995) describe a more flexible equation than the Richards equation for predicting either the rate of energy or protein deposition in pigs.

$$dE/dt = k[(E_f - E)/E_f](W + W_s)^a \qquad (MJ/d) \tag{3}$$

Where k, a and W_s are constants and W is current body weight. By changing the value for a and adjusting k, it is possible to vary the proportion of final body weight when the maximum rate of energy deposition will occur. In addition, by varying W_s, the equation can predict differences in the 'steepness' of the relationship between rate of energy deposition and body weight. Varying all three parameters plus final energy content can provide a large range in the form of the energy deposition curve to fit the known large variation between pig genotypes (Black *et al.*, 1995). The major drawback with the equation used by Black *et al.* (1995) relates to fitting parameters for the curve from experimental results because of the difficulty in achieving convergence, particularly when selecting initial parameter values, and unstable parameter estimates (Schinckel, 1998).

Maintenance energy requirements

The maintenance energy requirements of pigs has been estimated from many studies and summarised by ARC (1981) and NRC (1998). Traditionally, maintenance energy requirements have been expressed as a function of body weight raised to a power of 0.75 ($W^{0.75}$). The following equations were recommended by ARC (1981) and NRC (1998) for predicting the maintenance requirements for growing pigs:

$$E_m = 458 \text{ kJ/ } W^{0.75} \qquad \text{(kJ/d)} \qquad \text{ARC (1981)} \tag{4}$$

$$E_m = 444 \text{ kJ/ } W^{0.75} \qquad \text{kJ/d)} \qquad \text{NRC (1998)} \tag{5}$$

These representations of the maintenance energy requirements for growing pigs are clearly an oversimplification because the common measure of minimal energy required for maintenance, fasting heat production, is affected markedly by the body composition of a pig, its rate of growth, and the chemical and physical nature of the feed (Black, 1995d; Noblet *et al.*, 1999). For example, Koong *et al.* (1983) offered pigs from either lean or obese genotypes amounts of feed to grow at different growth trajectories over two consecutive 35 day periods. Fasting heat production ranged from 329 to 491 kJ/$W^{0.75}$ and was strongly correlated with the total weight of the gastrointestinal tract and liver. An increase in the intake of a feed or of the fibre content of a diet markedly increases the weight of the gastrointestinal tract and liver (Koong *et al.*, 1983; Stanogias and Pearce, 1985). Visceral tissue has a higher metabolic rate than muscle or fat. Ferrell (1988) demonstrated that although visceral tissue represented only 11% of body weight in sheep, it accounted for 58% of total oxygen consumption. In addition, Ferrell (1988) showed that muscle has a substantially higher maintenance energy requirement than fat. Noblet *et al.* (1999) estimated maintenance requirement of pigs varying widely in genetic propensity to deposit lean and fat tissue and found that it was most closely related to visceral and muscle tissue content, with some variation across genotypes. Visceral tissue contributed about three times more to maintenance energy requirement than muscle tissue. The general equation developed by Noblet *et al.* (1999) to predict the metabolisable energy requirement for maintenance (E_m) was:

$$E_m = 569M^{0.7} + 1788V^{0.7} + G_i \qquad \text{(kJ/d)} \tag{6}$$

Where M is muscle mass (kg), V is visceral mass (kg) and G_i is a constant representing genotype and ranged from -457 to 723 kJ/d. When pigs from the Meishan genotype were removed from the data, Van Milgen *et al.* (1998) produced the following equation to predict fasting heat production:

$$E_m = 508M^{0.66} + 2011V^{0.66} \qquad \text{(kJ/d)} \tag{7}$$

Protein turnover is known to contribute from 15% to 25% of fasting heat production and is strongly correlated with growth rate (Gill and Oldham, 1993). Hence, a representation of the basal energy requirement for maintenance in a pig growth model should involve separate calculations for the effects of mass of viscera, mass of other protein tissues and fat, as well as the rate of net protein deposition.

Energy required for activity

Activity of pigs associated with feeding and group dynamics can account for 20% or more of the total energy required for maintenance (Verstegen *et al.*, 1987). Noblet *et al.* (1993) found that activity accounted for 15% of total heat production in non-breeding sows held in metabolism cages. The values varied widely between individual animals and ranged from 8.1% to 20.3% of fasting heat production. Alternatively, Van Milgen *et al.* (1998) found that physical activity represented less than 10% of fasting heat production in growing pigs housed individually in metabolism cages and kept in a quite environment. Other research has shown that activity of young pigs is greater than older pigs (Halter *et al.*, 1980). Activity also appears to be greater in group housed (Verstegen *et al.*, 1987) than in individually housed pigs (Van Milgen *et al.*, 1998).

Physical activity is reduced when nutrients are fermented in the hind-gut of pigs. Schrama *et al.* (1996) showed that pigs offered a diet containing sugar beet pulp silage compared with tapioca spent 24% less energy on activity. Subsequently, it was shown that energy expenditure on activity decreased by 3.9 kJ/g of fibre from sugar beet pulp silage fermented (Schrama *et al.*, 1998). The reduction in activity was not only related to sugar beet pulp silage, but appears to occur also when other materials such as raw potato starch, which is partially resistant to digestion in the small intestine is incorporated into the diet (Schrama and Bakker, 1999). However, activity was unaffected when highly indigestible wheat straw was included in diets (Schrama and Bakker, 1999).

The energy cost of standing has been estimated to be 30 kJ/$W^{0.75}$ per 100 minutes for pigs ranging in weight from 30 to 200 kg (Noblet *et al.* 1993), and 27 kJ/$W^{0.75}$ per 100 minutes for sows (Noblet *et al.* 1997). These figures are about five times higher than those measured in sheep. Van Milgen *et al.* (1998) found that the actual cost of activity (E_a) for growing pigs in metabolism cages to be related to body protein content expressed as muscle mass (M kg).

$$E_a = 21M^{0.91} \qquad \text{(kJ/hour of activity)} \qquad (8)$$

The energy cost of activity measured by van Milgen *et al.* (1998) appears to be approximately 30% less than that expected for pigs housed in groups. The large white castrate and male pigs had approximately 43% muscle and were active for around 2.65 h/d. Hence, total energy expenditure on activity was approximately 700 kJ/d for pigs weighing 40 kg, compared with 1,033 kJ/d derived from the group-housing results (activity kJ/d = 65 $W^{0.75}$) of Rijnen *et al.* (2003) for pigs of the same weight.

The energy cost of locomotion was measured by Lachica and Aguilera (2000) and found to be approximately 3 J/kg W per metre travelled on level ground irrespective of the weight of pig at 41 kg or 84 kg. A similar energy cost was identified for pigs descending slopes of either 5.25% or 10.5%. However, the energy cost of raising one kg up one metre was found to be 27.1 J/kg W per m for pigs weighing 41 kg and 40.0 J/kg W per m for pigs weighing 84 kg.

The energy cost of eating a diet containing 15.6 MJ/kg of digestible energy was measured at 100 kJ/kg of diet consumed (Noblet *et al.*, 1993). The energy cost of eating is closely related to the physical nature of the diet consumed (Halter *et al.*, 1980), but De Lange *et al.* (2006) did not observe significant differences in the energy cost of feeding a basal diet compared with addition to the basal diet of 20% wheat straw chopped to 10 mm particle size.

A portion of energy expenditure ascribed to activity is associated with eating and includes locomotion to the feeder, standing at the feeder and mastication. Another significant proportion of energy spent on activity is not associated with eating. Giles (1992) estimated the non-eating component of activity to represent approximately 50% of total energy spent on activity for pigs housed in individual crates. This non-eating associated or 'play' activity varied widely between individual pigs. Giles (1992) found that non-feeding associated activity ceased whenever the body temperature of an individual pig increased by only 0.2-0.3 °C above normal temperature of 39 °C. Ideally, predictions of activity associated with feeding behaviour, the energy cost of eating and other non-feeding locomotion activities should be included in pig growth models when predicting potential energy requirements for calculating potential feed intake.

Potential feed intake

Potential energy intake for pigs in a thermoneutral climate is calculated from the sum of the daily energy needs for growth of body fat and protein, basal maintenance requirements predicted from the weight of visceral and muscle tissue, activity for feeding and feeding associated energy expenditure including standing, walking and chewing, and other 'play' activity. The sum of energy required for these metabolic processes is then converted first to metablisable energy and then to digestible energy on the basis of efficiency of utilisation of digested nutrients for different body processes (Black, 1995d). Potential feed intake is then predicted by dividing the requirement for digestible energy by the digestible energy content of the diet.

Changes from potential intake

Rarely is potential intake achieved by a pig raised in commercial environments. Intake can be constrained by a range of factors or stimulated with mild exposure to exercise or cold conditions.

Deficiencies in essential nutrients

Several studies have shown that voluntary intake decreases when pigs are offered diets deficient in protein or essential amino acids, specifically tryptophan (Anderson, 1979; Henry *et al.*, 1992; Likuski *et al.*, 1961; Rogerson and Campbell, 1982). However, other studies suggest that mild deficiencies of protein, lysine or threonine, may lead to small increases in intake (Ferguson and Gous, 1997; Henry, 1995). Observations by R.P. Chapple (personal communications) showed that feed intake of growing pigs was maintained near potential as the protein content of diets with constant energy content was depressed from above requirement to approximately 80% of requirement. However, further depression in protein content of the diets resulted in a linear decline in feed intake, such that extrapolation to a protein-free diet suggested feed intake would be as low as 20% of potential.

Feed intake is depressed in pigs offered diets severely deficient in phosphorus (Mahan, 1982), sodium (Alcantara *et al.*, 1980; Honeyfield *et al.*, 1985) and potassium (Jensen *et al.*, 1961). Increasing deficiencies with each of these minerals appears to reduce growth rate before there is a depression in feed intake. Results from Mahan (1982) suggested that feed intake in pigs is reduced only when phosphorus intake falls below 60% requirement. Similar interpretation can be made for sodium from the results of Alcantara *et al.* (1980).

Effects of deficiencies in protein, essential amino acids, phosphorus, sodium and potassium on feed intake have been incorporated into pig simulation models (Black *et al.*, 1986, J.L. Black, unpublished results). Feed intake is assumed to decline linearly when protein deposition is below 80% of maximum, when sodium or potassium intake is less than 60% of requirement or when phosphorus retained in the body is less than 60% of potential retention. An alternative approach has been taken by Ferguson (2006) who predicts intake will increase to satisfy the protein requirements of a pig.

Physical gastrointestinal tract capacity

Total energy intake is reduced, particularly in young pigs, when the energy density of a diet is reduced by the addition of indigestible, bulky material (Campbell *et al.*, 1975; Cole *et al.*, 1968; Kyriazakis and Emmans, 1995; Owen and Ridgeway, 1967; Taverner *et al.*, 1984). Although pigs weighing 60 kg can increase the capacity of the gastrointestinal tract to compensate within 12 days for a water-filled balloon placed in the stomach (Cunningham, 1968) or for increasing indigestible fibre in the diet (Stanogias and Pearce, 1985), this compensation is often incomplete and excessive bulk in the diet can reduce the intake of energy. Black *et al.* (1986) suggested from a literature review that pigs weighing less than 20 kg could not increase feed intake (kg/day) to compensate for a depression in digestible energy density of the diet below 16 MJ/kg (air dry basis). However, pigs weighing up to 50 kg could increase feed intake to compensate for a depression in energy density for diets containing more than 14 MJ/kg. Heavier pigs weighing 70 kg or more appeared to be able to increase feed intake to maintain energy intake for diets containing as low as 10 MJ/kg of digestible energy. Black *et al.* (1986) used maximum intake values recorded from a range of studies, where pigs ranging in weight from 10 kg to over 200 kg were offered high-bulk diets, to develop a relationship for maximum feed intake limited by physical capacity of the gastrointestinal tract.

$$dF_p/dt = 0.111W^{0.803} \quad \text{(kg/day)} \tag{9}$$

Where F_p is the maximum feed intake set by a physical limit to the gastrointestinal tract and W is live weight (kg). A similar approach has been used by Whittemore (1993), where the maximum intake was related to weight of the pig. However, Kyriazakis and Emmans (1995) and Tsaras *et al.* (1998) showed that intake of young pigs limited by the bulkiness of a feed was described more closely by the water-holding capacity of the diet than by its fibre content. Tsaras *et al.* (1998) predicted maximum feed intake in relation to the water-holding capacity of the feed (C_w, g/g feed) and live weight (W, kg) as follows:

$$dF_p/dt = 0.2346(1/C_w)W \quad \text{(kg/day)} \tag{10}$$

When algorithms 9 and 10 are compared for pigs weighing 15 kg and offered a feed with a water-holding capacity of 8.48 g/g, as measured by Tsaras *et al.* (1998) for sugar-beet pulp, predicted maximum intake is 0.977 kg/day compared with 0.418 kg/day, respectively. The predicted intake of the basal diet used by Tsaras *et al.* (1998), which contained more common ingredients and had a water-holding capacity of 3.86 g/g, was 0.912 kg/day for a pig weighing 15 kg. The predicted intake for the basal diet using the Tsaras *et al.* (1998) algorithm is closer to that predicted from algorithm 9. This comparison suggests that the constituents in feed have a marked effect on intake when restricted by bulk. Water-holding capacity of feed is strongly related to the content of soluble polysaccharides, which swell to form gels that increase volume and viscosity of the digesta. Jenkins *et al.* (1978) fed diets containing a range of fibre sources to humans and observed that the rate of passage of digesta was highest for diets containing wheat bran, and was negatively related to viscosity of the diet. Thus, algorithms based on body weight alone are unlikely to give accurate predictions of limits set by gastrointestinal tract capacity for diets that do not contain ingredients commonly used for pigs. Nevertheless, the addition of enzymes that alter water-holding capacity and reduce viscosity of diets, such as xylanases, glucanases and cellulases, have been shown to markedly increase the intake by young pigs of diets containing some cultivars of wheat (Choct *et al.*, 1999).

Algorithms that predict accurately the maximum intake capacity of a pig must account for the major factors that influence rate of passage of digesta through the gastrointestinal tract and its capacity to increase in volume over time in response to continual distension. The complexity of the mechanisms involved in determining rate of passage of digesta in monogastric animals has been discussed by Black *et al.* (2009). Several models that represent in some detail the concepts influencing the flow and digestion of nutrients along the gastrointestinal tract of pigs have been developed (Bastianelli *et al.*, 1996; Bastianelli and Sauvant, 1998; Rivest *et al.*, 2000; Usry *et al.*, 1991). Although, these models are positive beginnings, a model is needed that represents more of the factors known to control rate of passage of digesta and volume of the gastrointestinal tract in relation to dietary ingredients and weight of pig.

Climatic conditions

Climatic conditions can have a profound effect on the intake of pigs (Black *et al.*, 1999). Energy expenditure of a pig increases to maintain body temperature when ambient temperature falls below the lower critical temperature. Voluntary intake tends to increase, often after a delay of one or two days, to compensate for the additional energy consumed. Conversely, feed intake declines when ambient temperature exceeds the evaporative critical temperature, which is the temperature when energy expenditure is first increased through panting as ambient temperature rises. The magnitude of the rise in feed intake at low temperatures and the fall in intake at high temperatures is influenced by characteristics of the diet, stocking arrangements and weight of the pig.

Several models developed to simulate growth in pigs predict heat exchange between the pig and its environment in relation to the underlying mechanisms (Black *et al.*, 1986; Bruce and Clark, 1979; Knap, 1999). The critical temperatures that trigger changes in feed intake, are calculated for every iteration of the models, in relation to the amount of feed consumed, body weight and composition, rate of protein and fat deposition, energy used for maintaining essential body functions, activity energy expenditure, efficiency of use of absorbed nutrients, ambient temperature, air speed, relative humidity, proportion of skin that is wet, floor type, reflectance characteristics of surrounding surfaces and number of pigs in a pen.

These models predict the amount of energy that is lost from an animal to the environment, and replaced by shivering and other thermogenic reactions, when ambient temperature falls below the lower critical temperature. The amount of feed that must be eaten to replace the additional energy lost to the environment is readily calculated from model estimates of the efficiency of use of energy from absorbed nutrients for all body functions, and from the digestible energy content of the diet. All heat released during metabolic reactions is assumed to be available to warm the animal when it is below the lower critical temperature. However, pigs exposed to cold will not always increase feed intake to maintain the rate of energy deposition that would occur under thermoneutral conditions. Close (1989) showed that feed intake in pigs weighing 90 kg continued to increase as ambient temperature fell below the lower critical temperature, but intake increased little for pigs weighing 18 kg. The inability of lighter weight pigs to increase feed intake when exposed to cold is due primarily to limitations imposed by gastrointestinal tract capacity, and the small increase observed can be attributed to an increase in rate of passage of digesta through the gastrointestinal tract in pigs exposed to cold conditions (Phillips *et al.*, 1982). Conversely, gastrointestinal tract capacity is unlikely to limit the increase in feed intake of pigs weighing 90 kg needed to compensate for the increasing energy lost to the environment when ambient temperature declines below the lower critical temperature.

Group housed pigs spend more time huddling and less time eating when exposed to cold conditions than pigs in a thermoneutral environment (Parker *et al.*, 1980). This observation suggests that feed intake of pigs exposed to cold conditions may be influenced by the number of pigs in a pen. Nienaber *et al.* (1990, 1991) subjected pigs either penned individually or in groups of four to ambient temperatures 12 °C below the estimated lower critical temperature. Feed intake increased over that obtained under thermoneutral conditions by 30% for pigs penned individually compared with only 11% for pigs penned in groups of four. The pigs penned individually spent approximately the same time eating as pigs in thermoneutral conditions, whereas the cold, group-penned pigs spent only half the time eating. These observations suggest that pigs prefer to remain warm and huddled in a group when exposed to cold conditions and eat only when the hunger drive is high, and then for a short period of time compared with pigs penned individually. These concepts have been incorporated into the AUSPIG model (Giles *et al.*, 1998).

The simulation models representing heat exchange between the pig and its environment predict a decline in feed intake under hot conditions that is sufficient to allow total heat production from the pig and heat input from the environment to coincide with that lost to the environment through convection, conduction and evaporation (Black *et al.*, 1986; Knap 1999). The pig ceases non-feeding associated activity once the ambient temperature exceeds the evaporative critical temperature to reduce body heat production (Giles, 1992). This concept of reduced activity has been in incorporated subsequently into the model of Black *et al.* (1986) and is in the model of Knap (1999). In the original model of Black *et al.* (1986), feed intake was assumed to be depressed only when ambient temperature reached the upper critical temperature that coincided with maximum respiration rate and evaporative heat loss. However, observations by Giles (1992) showed that once ambient temperature rose above the evaporative critical temperature, where respiration rate commenced to increase due to heat load, the pig begins to store some of the extra heat input from the environment as an increase in body temperature. Collation of results from many experiments (Giles *et al.*, 1998) showed that feed intake of pigs under various high temperature regimes relative to intake under thermoneutral conditions fell linearly with increasing body temperature from 39.2 °C with intake ceasing at 41.3 °C. The latter temperature was found to coincide closely with the upper critical temperature. Thus, the model of Black *et al.* (1986) has been modified to predict deep body temperature in relation to heat generated by the pig and net gain from the environment. Feed intake under hot conditions is then predicted as a function of body temperature (Black *et al.*, 1999).

Other more simple models that do not predict heat exchange between the pig and its environment make assessments from experimental results of critical temperatures based on weight of the pig and feeding level. Nyachoti *et al.* (2004) summarise results from many experiments and conclude that feed intake increases by 25-39 g/day for every 1 °C ambient temperature is below the lower critical temperature and decreases by 40-80 g/day for every 1 °C above the upper critical temperature.

Number of pigs per pen and stocking density

There is strong evidence that the group penning of growing pigs has a detrimental effect on feed intake and performance. The negative effect on feed intake of placing pigs in groups is exacerbated by reductions in floor space per pig, and when the air volume per pig becomes too small. Chapple (1993) showed that simply by increasing the number of pigs in a pen from one to three or from one to five caused a significant depression in feed intake of 5% and 9% respectively. Similarly, Gonyou *et al.* (1992) found that increasing the number of pigs in a pen from one to five depressed feed intake by 8% despite there being adequate space for each pig. There appears to be little additional effect on performance from increasing the number of pigs in a pen beyond five, but the published results are equivocal. Chapple (1993) found that the intake of pigs was similar irrespective of group size ranging from five to 15 pigs per pen. Similarly, Nielsen and Lawrence (1993) observed no difference in feed intake for pigs housed in groups of five, 10, 15 or 20. Gonyou and Stricklin (1998) examined the effect of group size ranging from three to 15 pigs per pen and found the greatest depression in feed intake occurred between group sizes of three and six pigs, but a small depression in intake continued to 15 pigs per pen. Petherick *et al.* (1989) reported that the intake of pigs housed in groups of 36 was significantly less than that for pigs in groups of either six or 18 per pen. Although McGlone and Newby (1994) observed no difference in the performance of pigs housed in groups of 10, 20 or 40 per pen, animals in the pen with 40 pigs showed the greatest rates of injury and morbidity. Turner *et al.* (2000) found that the growth rate was greater for pigs housed in a group of 20 than for groups of 80 pigs. However, neither Schmolke and Gonyou (2000) nor Wolter *et al.* (2001) observed any differences in the performance of pigs housed in groups of 10, 20, 40 or 80 and 25, 50 and 100, respectively.

Reduction in the intake and performance of pigs when placed into groups occurs even when the animals have been acquainted previously with one another (Black *et al.*, 2001; Gomez *et al.*, 2000). There appears to be little difference in the negative effect of grouping pigs that have been acquainted or are unfamiliar with one another (Kerr *et al.*, 2005). However, the effects on feed intake and performance of grouping pigs are reversible when pigs are sequentially placed individually or in groups within pens (Black *et al.*, 2001; Bornett *et al.*, 2000; Kerr *et al.*, 2005).

In summary, feed intake appears to fall exponentially by approximately 10% as the number of pigs in a pen increased from one to five. Thereafter intake remains relatively unchanged as the number of pigs in a group is increased further. The effect occurs with familiar and unfamiliar pigs and is immediately reversible. These concepts can be represented by the following exponential algorithm as a factor of intake relative to potential intake.

$$F_{pp} = 0.142\ e^{(Pp\ -0.35)} + 0.9 \quad (11)$$

Where F_{pp} is a multiplication factor for potential feed intake and P_p is the number of pigs in a group. Wellock *et al.* (2003) use a logarithmic algorithm to predict the effect of number of pigs in a group on daily growth rate and allow for differences between pigs of different genotypes.

The overcrowding of pigs in a pen reduces feed intake and growth rate. Stocking rate is commonly expressed in relation to floor area allocation (m^2) per live weight (kg), raised to a power ($W^{0.67}$) because it accounts for pigs varying widely in weight. An area of 0.048 $m^2/W^{0.67}$ is required for all pigs in a pen to lie recumbent on their sides, whereas an area of 0.034 $m^2/W^{0.67}$ is required for all pigs to lie on their sternums. A review of the published experiments examining the effect of floor area per pig with groups of five pigs or more indicates that a depression in feed intake commences when the space allocation falls below 0.035 $m^2/W^{0.67}$ and it is depressed linearly by about 20% when that space allowance is 0.020 $m^2/W^{0.67}$ (Black, 1995d). Petherick (1982) suggested that 0.030 $m^2/W^{0.67}$ would be adequate space for pigs housed on fully perforated floors. Whittemore (1998) suggests that floor space should be increased by approximately 25% for pigs housed on solid floors. Wellock *et al.* (2003) use, in a model describing the effects of the social environment on pigs, a value of 0.039 $m^2/W^{0.67}$ as the critical space allowance for when pig performance is depressed, and increase it by 25% for solid floors. Gonyou and Stricklin (1998) examined the effect of group size from three to 15 pigs per pen at three stocking rates of 0.030, 0.039 and 0.048 $m^2/W^{0.67}$ and found that intake and growth rate were depressed only at the lowest area allocated per pig. Hyun *et al.* (1998) found that a space allocation of 0.25 m^2/pig (approximately 0.019 $m^2/W^{0.67}$) reduced growth rate by 16% compared with an allocation of 0.56 m^2/pig (approximately 0.043 $m^2/W^{0.67}$). Space allocations sufficient to reduce pig performance have been shown to increase abnormal behaviours and aggression in pigs (Randolph *et al.*, 1981; Turner *et al.*, 2000).

Models simulating pig performance use a critical stocking rate of either 0.035 $m^2/W^{0.67}$ (Black, 1995d) or 0.039 $m^2/W^{0.67}$ (Wellock *et al.*, 2003), below which feed intake is reduced. Wellock *et al.* (2003) use a log regression algorithm to allow for the effect of floor space on rate of gain, and allow for differences between solid and slatted floors as well as genotype of pig. Black (1995d) assumes intake is depressed linearly by 20% from the critical value when a space allocation of 0.020 $m^2/W^{0.67}$ is reached.

The effect of feeder space on feed intake and performance of pigs has been reviewed by (Nyachoti *et al.*, 2004). Although there are only a limited number of experiments examining specifically the effect of feeder space allocation, the authors conclude that feeder space has little affect on feed intake and performance, because pigs alter feeding behaviour to maintain feed intake when feeder space is limited. Nevertheless, Wellock *et al.* (2003) include a depression in feed intake when feeder space is limiting in their model predicting pig performance. The critical number of feeder spaces per pig is calculated in relation to the number of pigs in a group, the potential feed intake for each pig, maximum feeding rate and the number of minutes in a day. Maximum feeding rate (R_m, kg/min) is a function of mouth capacity ($gW^{1.0}$) and water-holding capacity of the diet. Feeder-space constrained intake (F_{FS}) is predicted as:

$$dF_{FS}/dt = (1440\ S\ R_m)/\ P_p \qquad \text{(kg/day)} \qquad (12)$$

Where S is the number of feeder spaces and P_p is the number of pigs in a group. Wellock *et al.* (2003) calculate the number of feeder spaces available when pigs are offered feed in troughs as a function of the estimated width between the shoulders of pigs and length of trough (T_L).

$$S = T_L/(0.064W^{0.33}) \qquad \text{(number)} \qquad (13)$$

Feed intake and performance of pigs housed under commercial conditions are influenced by the cleanliness of the environment and the quality of the air (Black, 2003; Donham, 1991; Holyoake, 2005). Lee *et al.* (2005) housed weaned pigs in either individual pens or in groups of 10 pigs per pen within a clean or dirty environment in a commercial piggery. The clean environment was created by the daily hosing of pens, flushing the effluent with clean water and spraying the air space with a disinfectant, whereas the room for the dirty environment was not cleaned prior to or during the experiment and the sub-floor area was flushed with recycled effluent. There were large differences in the quality of the air between the two rooms with the clean environment having significantly less ammonia, carbon dioxide and total dust than the dirty environment (Currie *et al.*, 1997). Feed intake for pigs penned individually in the dirty environment was depressed by 13% compared with those reared in the clean environment. Feed intake was depressed by 9.5% when pigs were grouped in pens in the clean environment compared with individually-penned pigs, but intake was not significantly depressed further for group-penned pigs in the dirty environment.

A strong positive relationship between the air volume per pig and the mean growth rate of pigs housed from 10 to 22 weeks of age in single-phase, commercial units has been observed by Murphy *et al.* (2000). There was also a strong negative correlation across pig production units between air volume per pig and the number of viable bacteria in the air, such that the growth rate of the pigs declined significantly as the concentration of bacteria in the air increased (Black *et al.*, 2001). Previous research by Cargill and Banhazi (1998) has demonstrated the importance for pig performance of cleaning sheds and reducing the concentration of viable bacteria in the air. The thorough cleaning of pens and walls between batches of 300 pigs in an all in/all out system reduced the number of air-borne viable bacteria by 53% compared to a continuous-flow management system with the same number of pigs and the same volume of air per pig.

Despite the observed effects of air quality and particularly viable bacteria on feed intake, the concepts have not yet been incorporated into pig simulation models. However, Turner *et al.* (1997) have used fuzzy logic to develop a mathematical model for describing the effects of air quality and immunity status on the incidence, prevalence and severity of atrophic rhinitis in pigs.

Health status

Diseases affect several physiological functions of animals including feed intake, oxygen exchange rate, efficiency of energy utilisation, body protein synthesis and catabolism, body temperature control and tolerance to heat stress (Symons, 1989). The effects of disease have not been incorporated into many of the models developed for predicting pig growth (Danfaer, 2000, Pomar *et al.*, 1991; Van Milgen *et al.*, 2007), whereas they have been added in a simple form to other models. For example, Moughan *et al.* (1995) incorporated the overall effect of the environment, including disease, by reducing potential protein deposition to an 'operational maximum' rather than using the 'genetic maximum'. Similarly, for the model of Black *et al.* (1986), it is recommended that the potential rate of protein deposition and feed intake are depressed, while maintenance requirement is increased when simulating the effects of disease (Mullan *et al.*, 1993) However, recognition of the importance of diseases in commercial pig production has resulted in several concerted efforts to develop approaches for simulating the effects of diseases on feed intake and performance within pig models (Black *et al.*, 1999; Black, 2002; Sandberg *et al.*, 2006; Kyriazakis and Doeschl-Wilson, 2009).

Each of these more detailed approaches for modelling disease assume: that there is usually a lag between the time of infection and discernable physiological responses; the magnitude of the responses depends on the severity of the disease as determined by the pathogen load and virulence; there may or may not be a recovery phase depending on the disease and its severity; recovery in the physiological function may or may not reach pre-infection levels; time to recovery depends on nutritional status of the animal; the magnitude of the response is influenced by pig genotype; and animals previously infected with a specific disease may be immune to further infection or have a diminished response. Many of these concepts are well developed by Sandberg *et al.* (2006) in relation to predicting the effects of disease on feed intake. The principles incorporated into the model have been shown to realistically resemble the observed pattern of disease on feed intake in pigs. However, more effort is required to develop parameters for the algorithms for each of the major diseases affecting pigs before they can be used effectively in practical pig models.

The effects of the major pig diseases in Australia (mycoplasma, pleuropneumonia, Glassers disease, pasteurella, *E. coli*, swine dysentery, Lawsonia, coccidiosis, mange and erysipelas) have been incorporated into the research version of the AUSPIG model (Black, 2002, J.L. Black, unpublished data). The severity of each disease is assessed from clinical observations and/or from predictions of parasite load and virulence. A severity score from 0 to 10, where 0 represents no disease and 10 represents all pigs in the group dying, is predicted for each disease in the herd. The pattern of changes from the point of infection in the severity score for each disease is predicted over time in relation to pathogen load and virulence, lag in response, time to recovery and magnitude of recovery in relation to initial values. The effects on the severity score of immunity, vaccination or other treatments and the time course to recovery for each disease are also predicted. The impact of each disease is predicted on major physiological functions including feed intake, non-feeding activity, fasting heat production, net protein deposition, nutrient digestibility, respiration rate, tissue insulation and heat loss from the lungs. Finally, a severity score is determined for each physiological function, either additive or multiplicative depending on the specific diseases, for all the diseases within a group of pigs. The severity score for each physiological function is then used to predict the changes in that function resulting from all the diseases carried by a group of pigs. Although these concepts have been demonstrated to predict realistic changes in the performance of diseased pigs, sufficient information to parameterise the model equations is available only for pleuropneumonia.

Although concepts have now been developed for representing disease in a mechanistic manner in pig growth models, more effort is required to develop parameters for algorithms describing specific diseases. Accurate prediction of the effects of disease on feed intake will result only when the algorithms are fully parameterised and evaluated for each of the major diseases.

Conclusions

The ability to predict accurately feed intake in simulation models for pigs is important because intake is a major determinant of growth, efficiency of feed utilisation and body composition. Several approaches have been used to predict intake in pigs, with the 'potential-constraint' concept being most highly developed (Black, 1995d; Kyriazakis and Emmans, 1998; Poppi *et al.*, 1994; Whittemore *et al.*, 1995). Although this approach has been criticised for not strictly representing animal physiology, for being unable to predict the commencement and cessation of meals and for being unable to predict daily variation in intake of individual animals, other approaches such as the maximum efficiency of oxygen use (Tolkamp and Ketelaars, 1992) and the minimal total discomfort theory (Forbes, 2007, 2009) have either been discredited (Whittemore *et al.*, 2001a,b) or require a great deal of further development to be of practical value. The binary approach used by Nagorcka *et al.* (2004), which simulates short-term energy and amino acid availability and intestinal tract distension is novel and predicts the commencement and cessation of meals as well as day-to-day variation in feed intake. However, the approach has not been attempted for predicting feed intake by pigs and is restricted to models that are based on the biochemical fluxes such as that of Danfaer (2000). Prediction of intake by pigs based on algorithms developed as statistical 'best fits' to experimental results have generally been discredited because they fail to be accurate once extrapolated outside the range of the data used in their development (Black, 1995b; France and Dijkstra, 2000; Whittemore *et al.*, 2001).

Several factors known to affect feed intake by pigs either have not been simulated or require considerable refinement to improve the accuracy of prediction. Several pig simulation models contain multiplication factors that can be varied by users of the models to adjust feed intake to overcome these deficiencies. The negative impact on feed intake of poor air quality (Cargill and Banhazi, 1998) has not has not been modelled, nor has the negative effect of mycotoxins, other antinutritional factors (Clasadonte and Van der Poel, 2009) or taste aversion (Roura and Tedó, 2009). Several attempts have been made to model the constraints to intake resulting from gastrointestinal tract capacity (Bastianelli and Sauvant, 1998; Black *et al.*, 1986; Tsaras *et al.*, 1998), but these are still rudimentary and require a more mechanistic approach based on the mechanisms known to control rate of passage of digesta and gastrointestinal tract volume in pigs. Similarly, concepts and approaches have been developed for predicting the effects of disease on feed intake and pig performance (Black, 2002; Sandberg *et al.*, 2006). However, these approaches need a great deal of additional work to parameterise algorithms for each of the major diseases affecting pigs raised under commercial environments. Despite these deficiencies, several of the pig simulation models that have been developed represent well many of the factors that are known to determine feed intake of pigs.

Acknowledgements

The review was supported in part by funds from the Australian Co-operative Research Centre for an Internationally Competitive Pork Industry and Australian Pork Limited.

References

Agricultural Research Council, 1981. The nutrient requirements of pigs. Commonwealth Agricultural Bureaux, Farnham Royal, UK.

Alcantara, P.F., Hanson, L.E. and Smith, J.D., 1980. Sodium requirements, balance and tissue composition of growing pigs. Journal of Animal Science 50: 1092-1101.

Anderson, G.H., 1979. Control of protein and energy intake: role of plasma amino acids and brain neurotransmitters. Canadian Journal of Physiology and Pharmacology 57: 1043-1057.

Baldwin, R.L., 1995. Modelling ruminant digestion and metabolism. Chapman and Hall, London, UK, 578 pp.

Bastianelli, D. and Sauvant, D., 1998. Digestion absorption and excretion. In: Kyriazakis, I. (ed.) A quantitative biology of the pig. CAB International, Wallingford UK, pp. 249-273.

Bastianelli, D., Sauvant, D. and Rerat, A., 1996. Mathematical modelling of digestion and nutrient absorption in pigs. Journal of Animal Science 74: 1873-1887.

Black, J.L., 1995a. The evolution of animal growth models. In: Moughan, P.J., Verstegen, M.W.A. and Visser-Reyneveld (eds.) Modelling growth in the pig. Wageningen Pers, Wageningen, the Netherlands, pp. 3-9.

Black, J.L., 1995b. Approaches to modelling. In: Moughan, P.J., Verstegen, M.W.A. and Visser-Reyneveld (eds.) Modelling growth in the pig. Wageningen Pers, Wageningen, the Netherlands, pp. 11-22.

Black, J.L., 1995c. The testing and evaluation of models. In: Moughan, P.J., Verstegen, M.W.A. and Visser-Reyneveld (eds.) Modelling growth in the pig. Wageningen Pers, Wageningen, the Netherlands, pp. 23-31.

Black, J.L., 1995d. Modelling energy metabolism in the pig – critical evaluation of a simple reference model. In: Moughan, P.J., Verstegen, M.W.A. and Visser-Reyneveld (eds.) Modelling growth in the pig. Wageningen Pers, Wageningen, the Netherlands, pp. 87-102.

Black, J.L., 2002. Pig growth models: Future. Journal of Animal Science 80 (Suppl. 1): 174.

Black, J.L., 2003. Effect of air quality on pig performance and health of piggery workers. Australian Pork Limited, Canberra, Australia, 40 pp.

Black, J.L., Bray, H.J. and Giles, L.R., 1999. The thermal and infectious environment. In: Kyriazakis, I. (ed.) A quantitative biology of the pig. CAB International, Wallingford UK, pp. 71-97.

Black, J.L., Campbell, R.G., Williams, I.H., James, K.J. and Davies, G.T., 1986. Simulation of energy and amino acid utilisation in the pig. Research and Development in Agriculture 3: 121-145.

Black, J.L. and Carr, J.R., 1993. A symposium – Stocking density and pig performance. In: Batterham, E.S. (ed.) Manipulating pig production IV. Australasian Pig Science Association, Attwood, Victoria, Australia. p. 84.

Black, J.L., Davies, G.T., Bray, H.R. and Chapple, R.P., 1995. Modelling the effect of genotype, environment and health on nutrient utilisation. In: Danfaer, A. and Lescoat, P. (eds.) Proceedings of the IVth international workshop on modelling nutrient utilisation in farm animals. National Institute of Animal Science, Tjele, Denmark, pp. 85-105.

Black, J.L., Giles, L.R., Wynne, P.C., Knowles, A.G., Kerr, C.A., Jones, M.R., Strom, A.D., Gallagher, N/L. and Eamens, G.J., 2001. A review – Factors limiting the performance of growing pigs in commercial environments. In: Cranwell, P.D. (ed.) Manipulating pig production VIII. Australasian Pig Science Association, Werribee, Victoria, Australia, pp. 9-36.

Black, J.L., Williams, B.A. and Gidley, M.J., 2009. Metabolic regulation of feed intake in monogastric mammals. In: Torrallardona, D. and Roura, E. (eds.) Voluntary feed intake in pigs. Wageningen Academic Publishers, the Netherlands, pp. 189-213.

Bolhuis, J.E., Oostindjer, M., Van den Brand, H., Gerrits, W.J.J. and Kemp, B., 2009. Voluntary feed intake in piglets: potential impact of early experience with flavours derived from the maternal diet. In: Torrallardona, D. and Roura, E. (eds.) Voluntary feed intake in pigs. Wageningen Academic Publishers, the Netherlands, pp. 37-60.

Bornett, H.L.I., Morgan, C.A., Lawrence, A.B. and Mann, J., 2000. The effect of group housing on feeding patterns and social behaviour of previously individually housed growing pigs. Applied Animal Behaviour Science 70: 127-141.

Bruce, J.M. and Clark, J.J., 1979. Models of heat production and critical temperature for growing pigs. Animal Production 28: 353-369.

Campbell, R.G., 1988. Nutritional constraints to lean tissue accretion in farm animals. Nutrition Research Reviews 1: 233-253.

Campbell, R.G., Taverner, M.R. and Mullaney, P.D., 1975. The effect of dietary concentrations of digestible energy on the performance and carcass characteristics of early-weaned pigs. Animal Production 21: 285-294.

Cargill. C. and Banhazi, T., 1998. The importance of cleaning in all-in all-out management systems. Proceedings of the 15th International Pig Veterinary Society Congress, Birmingham, England, p. 15.

Carroll, J.A. and Allee, G.L., 2009. Hormonal control of feed intake in swine. In: Torrallardona, D. and Roura, E. (eds.) Voluntary feed intake in pigs. Wageningen Academic Publishers, the Netherlands, pp. 155-187.

Chapple, R.P., 1993. Effect of stocking arrangement on pig performance. In: E.S. Batterham (ed.) Manipulating Pig Production IV, Australasian Pig Science Association, Werribee, Australia, pp. 87-97.

Choct, M., Cadogan, D.J., Campbell, R.G. and Kershaw, S., 1999. Enzymes can eliminate the differences in the nutritive value of wheats for pigs. In: Cranwell, P.D. (ed.) Manipulating Pig Production VII. Australian Pig Sciences Association, Werribee Victoria, Australia. p. 39.

Clasadonte, L. and van der Poel, A.F.B., (2009). Antinutritional factors: impact on pig's voluntary feed intake. In: Torrallardona, D. and Roura, E. (eds.) Voluntary feed intake in pigs. Wageningen Academic Publishers, the Netherlands, pp. 243-275.

Close, W.H., 1989. The influence of the thermal environment on the voluntary food intake of pigs. In: Forbes, J.M., Varley, M.A. and Lawrence T.L.J. (eds.) The voluntary food intake of pigs. Animal Production Occassional Publication No. 13, London, UK, pp. 87-96.

Cole, D.J.A., Duckworth, J.E. and Holmes, W., 1967. Factors affecting voluntary feed intake in pigs. 1. The effect of digestible energy content of the diet on the intake of castrated male pigs housed in holding pens and in metabolism crates. Animal Production 9: 143-150.

Cole, D.J.A., Duckworth, J.E. and Holmes, W. and Cuthbertson, A., 1968. Factors affecting voluntary feed intake in pigs. 3. The effect of period of restriction, nutrient density of the diet and sex on intake, performance and carcass characteristics. Animal Production 10: 345-357.

Cunningham, H.M., 1968. Compensatory response to balloons inflated in the stomach of pigs. Journal of Animal Science 27: 405-407.

Currie, E., Lee, C., Golden, S.E., Harrison, D.T., Giles, L.R. and Connaughton, I.D., 1997. Measurement of air quality and weaner pig performance in two different environments. In: P.D.Cranwell (ed.) Manipulating Pig Production VI, Australasian Pig Science Association, Werribee, Australia, p. 297.

Danfaer, A., 2000. A pig model for feed evaluation. In: McNamara, J.P., France, J. and Beever, D.E. (eds.) Modelling nutrient utilization in farm animals. CAB International, Wallingford, UK, pp. 393-408.

De Lange, C, van Milgen, J., Dubois, S. and Noblet, J., 2006. Energy cost of ingesting and excreting indigestible material in growing pigs is minimal. Animal Research 55: 551-562.

De Lange, C.F.M., Marty, B., Birkett, S., Moral, P. and Szkotnicki, B., 2001. Application of pig growth models in commercial pork production. Canadian Journal of Animal Science 81: 1-8.

Donham, K.J., 1991. Association of environmental air contaminants with disease and productivity in swine. American Journal of Veterinary Research 52: 1723-1730.

Eissen, J.J., 2000. Breeding for feed intake capacity in pigs. PhD thesis, University of Wageningen.

Emmans, G.C., 1981. A model of growth and feed intake of *ad libitum* fed animals, particularly poultry. In: Hillyer, G.M., Whittemore, C.T. and Gunn, R.G. (eds.) Computers in animal production. Occasional Publicaiton No. 5. British Society of Animal Production, pp. 103-110.

Emmans, G.C., 1995. Energy systems and the prediction of energy and feed intakes. In: Moughan, P.J., Verstegen, M.W.A. and Visser-Reyneveld (eds.) Modelling growth in the pig. Wageningen Pers, Wageningen, The Netherlands, pp. 115-122.

Ferguson, N.S., 2006. Basic concepts describing animal growth and feed intake. In: R. Gous, T. Morris and C. Fisher (eds.) Mechanistic modelling in pig and poultry production. CAB International, Wallingford, UK, pp. 22-53.

Ferguson, N.S. and Gous, R.M., 1993a. Evaluation of pig genotypes. 1. Theoretical aspects of measuring genetic parameters. Animal Production 56: 233-243.

Ferguson, N.S. and Gous, R.M., 1993b. Evaluation of pig genotypes. 2. Testing experimental procedure. Animal Production 56: 245-249.

Ferguson, N.S. and Gous, R.M., 1997. The influence of heat production on voluntary food intake in growing pigs given protein-deficient diets. Animal Science 64: 365-378.

Ferrell, C.L., 1988. Contribution of visceral organs to animal enegy expenditures. Journal of Animal Science 66 Suppl. 3: 23-34.

Forbes, J.M., 2007. A personal view of how ruminant animals control their intake and choice of food: Minimal Total Discomfort. Nutrition Research Reviews 20: 132-146.

Forbes, J.M., 2009. Integration of pre- and post-absorptive factors in feed intake regulation and prediction with particular respect to the pig. In: Torrallardona, D. and Roura, E. (eds.) Voluntary feed intake in pigs. Wageningen Academic Publishers, the Netherlands, pp. 61-86.

France, J., and Dijkstra, J., 2000. Scientific progress and mathematical modelling: different approaches to modelling animal systems. In: McNamara, J.P., France, J. and Beever, D.E. (eds.) Modelling nutrient utilization in farm animals. CAB International, Wallingford, UK, pp. 6-21.

France, J., and Thornley, J.H.M., 1984. Mathematical models in agriculture. Butterworths, London, UK, 335 pp.

Giles, L.R., 1992. Energy expenditure of growing pigs at high ambient temperatures. PhD thesis, University of Sydney, Australia.

Giles, L.R., Lorschy, M.L., Bray, H.J. and Black, J.L. (1998). Predicting feed intake in growing pigs. In: Wiseman, J., Varley, M.A. and Chadwick, J.P., 1998 (eds.) Progress in Pig Science. Nottingam University Press, UK, pp. 209-228.

Gill, M. and Oldham, J.D., 1993. Growth. In: Forbes, J.M. and France, J. (eds.) Quantitative aspects of ruminant digestioin and metabolism. CAB International, Wallingford, U.K., pp. 383-403.

Gomez, R.S., Lewis, A.J., Miller, P.S. and Chen, H.-Y., 2000. Growth performance and digestive and metabolic responses of gilts penned individually or in groups of four. Journal of Animal Science 78: 597-603.

Gonyou, W.H., Chapple, R.P. and Frank, G.R., 1992. Productivity, time budgets and social aspects of eating in pigs penned in groups of five or individually. Applied Animal Behaviour Science 34: 291-301.

Gonyou, W.H., and Stricklin, W.R., 1998. Effects of floor area allowance and group size on the productivity of growing/finishing pigs. Journal of Animal Science 76: 1326-1330.

Green, D.M., Brotherstine, S., Schofield, C.P. and Whittemore, C.T., 2003. Food intake and live growth performance of pigs measured automatically and continuously from 25 to 115 kg live weight. Journal of the Science of food and Agriculture 83: 1150-1155.

Halter, H.M., Wenk, C. and Schurch, A., 1980. Effect of feeding level and feed composition on energy utilisation, physical activity and growth performance of piglets. In: Mount (ed.) Energy Metabolism. Butterworths, London, European Association for Animal Production Publication 26: 195-198.

Henry, Y., 1995. Influence d'un déficit ou d'un déséquilibre alimentaire en acides aminés pendant une phase initiale de la croissance sur les performances du porc en finition [Effect of a dietary amino-acid deficiency or imbalance during the initial period of growth in pigs on subsequent performance at slaughter]. Annales de Zootechnie 44: 3-28.

Henry, Y., Seve, B., Colleaux, Y., Ganier, P., Saligaut, C and Jego, P., 1992. Interactive effects of dietary levels of tryptophan and protein on voluntary feed intake and growth performance in pigs. Journal of Animal Science 70: 1873-1887.

Hodge, R.W., 1974. Efficiency of food conversion and body composition in the preruminant lamb and the young pig. British Journal of Nutrition 32: 113-126.

Holyoake, P.K., 2005. The impact of air quality on animal health. In: Paterson, J.E. (ed.) Manipulating pig production X. Australasian Pig Science Association, Werribee, Australia, pp. 120-131.

Honeyfield, D.C., Froseth, J.A. and Barke, R.J., 1985. Dietary sodium and chlorine levels for growing-finishing pigs. Journal of Animal Science 60: 691-698.

Hyun, Y. Ellis, M., Riskowski, G. and Johnson, R.W., 1998. Growth performance of pigs subjected to multiple concurrent environmental stressors. Journal of Animal Science 76: 721-727.

Jenkins, D.J.A., Wolever, T.M.S., Leeds, A.R., Gassull,M.A., Haisman, P., Dilawari, J., Goff, D.V., Metz, G.L. and Alberti, K.G.M.M., 1978. Dietary fibres, fibre analogues and glucose tolerance: importance of viscosity. British Medical Journal 1: 1392-1394.

Jensen, A.H., Terrill, S.W. and Becker, D.E., 1961. Response of the young pig to levels of dietary potassium. Journal of Animal Science 20: 464-467.

Kerr, C.A., Giles, L.R., Jones, M.R. and Reverter, A., 2005. Effects of grouping unfamiliar cohorts, high ambient temperature and stocking density on live performance of growing pigs. Journal of Animal Science 83: 908-915.

Klasing, K.C. and Humphrey, B.D., 2009. Taste receptor biology. In: Torrallardona, D. and Roura, E. (eds.) Voluntary feed intake in pigs. Wageningen Academic Publishers, the Netherlands, pp. 87-104.

Knap, P.W., 1999. Simulation of growth in pigs: evaluation of a model to relate thermoregulation to body protein and lipid content and deposition. Animal Science 68: 655-679.

Knap, P.W., 2009. Voluntary feed intake and pig breeding. In: Torrallardona, D. and Roura, E. (eds.) Voluntary feed intake in pigs. Wageningen Academic Publishers, the Netherlands, pp. 13-35.

Koong, L.J., Nienabar, J.A. and Mersmann, H.J., 1983. Effect of plane of nutrition on organ size and fasting heat production in genetically obese and lean pigs. Journal of Nutrition 113: 1626-1631.

Kyriazakis, I. and Doeschl-Wilson, A., 2009. Anorexia during infection in mammals: variation and its sources. In: Torrallardona, D. and Roura, E. (eds.) Voluntary feed intake in pigs. Wageningen Academic Publishers, the Netherlands, pp. 307-321.

Kyriazakis, I. and Emanns, G.C., 1995. The voluntary feed intake of pigs given feeds based on wheat bran, dried citrus pulp and grass meal, in relation to measurements of feed bulk. British Journal of Nutrition 73: 191-207.

Kyriazakis, I. and Emanns, G.C., 1998. Voluntary feed intake and diet selection. In: Kyriazakis, I. (ed.) A quantitative biology of the pig. CAB International, Wallingford, UK, pp. 229-248.

Lachica, M. and Aguilera, J.F., 2000. Estimation of the energy costs of locomotion in the Iberian pig. British Journal of Nutrition 83: 35-41.

Lee, C., Giles, L., Bryden, W., Downing, J., Owens, P., Kirby, A. and Wynn, P., 2005. Performance and endocrine responses of group housed weaner pigs exposed to the air quality of a commercial environment. Livestock Production Science 93: 255-262.

Likuski, H.J.A., Bowland, J.P. and Berg, R.T, 1961. Energy digestibility and nitrogen retention by pigs and rats fed diets containing non-nutritive diluents and varying protein level. Canadian Journal of Animal Science 41: 89-101.

Lorenzo Bermejo, J., Roehe, R., Rave, G. and Kalm, E., 2000. Optimisation of feed intake curve in growing pigs using linear and nonlinear models. In: van Arendonk, J.A.M. (ed.) Proceedings of the 51st meeting of the European Association of Animal Production, book of abstracts no. 6, Wageningen Pers, Wageningen, The Netherlands, pp. 44.

Lorenzo Bermejo, J., Roehe, R., Rave, G. and Kalm, E., 2003. Comparison of linear and non-linear functions and covariance structures to estimate feed intake pattern in growing pigs. Livestock Production Science 82: 15-26.

Mahan, D.C., 1982. Dietary calcium and phosphorus levels for weanling swine. Journal of Animal Science 54: 559-564.

Manteca, X. and Edwards, S., 2009. Feeding behaviour and social influences on feed intake. In: Torrallardona, D. and Roura, E. (eds.) Voluntary feed intake in pigs. Wageningen Academic Publishers, the Netherlands, pp. 293-306.

McGlone, J.J. and Newby, B.E., 1994. Space requirements for finishing pigs in confinement: behaviour and performance while group size and space vary. Applied Animal Behaviour Science 39: 331-338.

Moughan, P.J., 1995. Modelling protein metabolism in the pig – critical evaluation of a simple reference model. In: Moughan, P.J., Verstegen, M.W.A. and Visser-Reyneveld (eds.) Modelling growth in the pig. Wageningen Pers, Wageningen, the Netherlands, pp.103-112.

Mullan, B.P., Davies, G.T. and Cutler, R.S., 1993. Simulation of the economic impact of transmissible gastroenteritis on commercial pig production in Australia. Australian Veterinary Journal 71: 151-154.

Murphy, T., Cargill, C. and Carr, J. (2000). The effects of stocking density on air quality. In: Proceedings of the 16th International Pig Veterinary Society Congress, Melbourne, Australia, p. 326.

Nagorka, B.N., Charmley, E. and Duynisveld, J.L., 2004. Using a dynamic ruminant model to understand the differences in performance of cattle fed rations based on barley and/or a potato processing by-product. Journal of Animal Science 82 (Supp. 1): 159 (Abstr.)

Nagorka, B.N., Gordon, G.L.R and Dynes, R.A., 2000. Towards a more accurate representation of fermentation in mathematical models of the rumen. In: McNamara, J.P., France, J. and Beever, D.E. (eds.) Modelling nutrient utilisation in farm animals. CABI Publishing, Wallingford, UK, pp.37-48.

Nagorcka, B.N. and Mooney, J.R., 1989. The reaction-diffusion system as a spatial organiser during initialisation and development of hair follicles and formation of the fibre. In: Rogers, G.E., Reis, P.J., Ward, K.A. and Marshall, R.C. (eds.) The biology of wool and hair. Chapman and Hall, London, pp. 365-379.

National Research Council, 1998. Nutrient requirements of swine. Tenth revised edition. National Academy Press, Washington DC, USA, 189 pp.

Nielsen, B.L. and Lawrence, A.B. (1993). Effect of group size on the behaviour and performance of growing pigs. In: E.S. Batterham (ed.) Manipulating Pig Production IV, Australasian Pig Science Association, Werribee, Australia, pp. 85-87.

Nienaber, J.A., McDonald, T.P., Hahn, G.L. and Chen, Y.R., 1990. Eating dynamics of growing-finishing swine. Transactions of the American Society of Agricultural Engineers 33: 2011-2018.

Nienaber, J.A., McDonald, T.P., Hahn, G.L. and Chen, Y.R., 1991. Group feeding behaviour of swine. Transactions of the American Society of Agricultural Engineers 34: 289-294.

Noblet, J., Dourmad, J.Y., Etienne, M. and Le Dividich, J., 1997. Energy metabolism in pregnant sows and newborn pigs. Journal of Animal Science 75: 2705-2743.

Noblet, J., Karege, S., Dubois, S. and van Milgen, J., 1999. Metabolic utilization of energy and maintenance requirements in growing pigs: effects of sex and genotype. Journal of Animal Science 77: 1208-1216.

Noblet, J., Shi, X.S. and Dubois, S., 1993. Energy cost of standing activity in sows. Livestock Production Science 34: 127-136.

Nyachoti, C.M., Zijlstra, R.T., de Lange, C.F.M. and Patience, J.F., 2004. Voluntary feed intake in growing-finishing pigs: A review of the main determining factors and potential approaches for accurate predictions. Canadian Journal of Animal Science 84: 549-566.

Owen, J.B. and Ridgeway, W.J., 1967. The effect of dietary energy content on the voluntary intake of pigs. Animal Production 9: 107-113.

Parker, R.O., Williams, P.E.V., Aherne, F.X. and Young, B.A. 1980. Serum concentration changes in protein, glucose, urea, thyroxine and triiodothyrnine and thermostability of neonatal pigs farrowed at 25 and 10 °C. Canadian Journal of Animal Science 60: 503-511.

Petherick, J.C. (1982). A biological basis for the design of space in pig housing. MSc Thesis. University of Aberdeen.

Petherick, J.C., Neattie, A.W. and Bodero, D.A.V., 1989. The effect of group size on the performance of growing pigs. Animal Production 49: 497-502.

Phillips, P.A., Young, B.A. and McQuitty, J.B., 1982. Liveweight, protein deposition amd digestibility responses in growing pigs exposed to low temperature. Canadian Journal of Animal Science 62: 95-108.

Pomar, C., Harris, D.L. and Minvielle, F., 1991. Computer simulation model of swine production systems: I. Modeling the growth of young pigs. Journal of Animal Science 69: 1468-1488.

Poppi, D.P., Gill, M. and France, J., 1994. Integration of theories of intake regulation in growing ruminants. Journal of Theoretical Bology 167: 129-145.

Quiniou, N., Dubois, S. and Noblet, J., 2000. Voluntary feed intake and feeding behaviour of group-housed pigs are affected by ambient temperature and body weight. Livestock Production Science 63: 245-253.

Ramaekers, P.J.L., 1996. Control of individual daily growth in group-housed pigs using feeding stations. PhD thesis, Wageningen Agricultural University, the Netherlands.

Randolph, J.H., Cromwell, G.L., Stahly, T.S. and Kratzer, D.D., 1981. Effects of group size and space allowance on performance and behavior of swine. Journal of Animal Science 53: 922-927.

Rijnen, M.M.J.A., Verstegen, M.W.A., Heetkamp, M.J.W. and Schrama, J.W., 2003. Effects of two different dietary fermentable carbohydrates on activity and heat production in group-housed growing pigs. Journal of Animal Science 81: 1210-1219.

Rivest, J., Bernier, J.F. and Pomar, C., 2000. A dynamic model of protein digestion in the small intestine of pigs. Journal of Animal Science 78: 328-340.

Roche, J.R., Blanche, D., Kay, J.K., Miller, D.R., Sheahan, A.J. and Miller, D.W., 2008. Neuroendocrine and physiological regulation of intake with particular reference to domesticated ruminant animals. Nutrition Research Reviews 21: 207-234.

Rogerson J.C. and Campbell, R.G., 1982. The response of early-weaned piglets to various levels of lysine in diets of moderate energy content. Animal Production 35: 335-339.

Roura, E. and Tedó, G., 2009, Feed appetence in pigs: an oronasal sensing perspective. In: Torrallardona, D. and Roura, E. (eds.) Voluntary feed intake in pigs. Wageningen Academic Publishers, the Netherlands, pp. 105-140.

Sandberg, F.B., Emanns, G.C. and Kyriazakis, I., 2006. A model for predicting feed intake of growing animals during exposure to pathogens. Journal of Animal Science 84: 1552-1566.

Schinckel, A.P., 1998. Describing the pig. In: Kyriazakis, I. (ed.) A quantitative biology of the pig. CABI Publishing, Wallingford, UK, pp. 7-38.

Schmolke, S.A. and Gonyou, H.W., 2000. Effect of group sizes 10, 20, 40, and 80 pigs on productivity during the grower-finisher period. Advances in Pork Production 11: 44.

Schrama, J.W. and Bakker, G.C.M., 1999. Changes in energy metabolism in relation to physical activity due to fermentable carbohydrates in group-housed growing pigs. Journal of Animal Science 77: 3274-3280.

Schrama, J.W., Bosch, M.W., Verstegen, M.W.A., Vorselaar, A.H.P.M., Haaksma, J. and Heetkamp, M.J.W., 1998. The energetic value of nonstarch polysaccharides in relation to physical activity in group-housed, growing pigs. Journal of Animal Science 76: 3016-3023.

Schrama, J.W., Verstegen, M.A.W., Verboeket, P.H.J., Schutte, J.B. and Haaksma, J., 1996. Energy metabolism in relation to physical activity in growing pigs as affected by type of dietary carbohydrate. Journal of Animal Science 74: 2220-2225.

Siebrits, F.K., Kemm, E.H., Ras, M.N. and Barnes, P.M., 1986. Protein deposition in pigs as influenced by sex, type and live mass. 1. The pattern and composition of protein deposition. South African Journal of Animal Science 16: 23-27.

Stanogias, G. and Pearce, G.R., 1985. The digestion of fibre by pigs. 3. Effects of the amount and type of fibre on physical characteristics of segments of the gastrointestinal tract. British Journal of Nutrition 53: 537-548.

Symons, L.E.A., 1989. Pathophysiology of endoparasitic infection compared with ectoparacitic infection and microbial infection. Academic Press, Sydney, Australia.

Taverner, M.R., Campbell, R.G. and Biden, R.S., 1984. The effects of energy concentration on voluntary intake of grower pigs fed a high fibre diet. Proceedings of the Australian Society of Animal Production 15: 757.

Technisch Model Varkensvoeding Werkgroep, 1994. Informatiemodel TMV. Report P1-117. Research Institute for Pig Husbandry, Rosmalen, the Netherlands.

Tolkamp, B.J. and Ketelaars, J.J.M.H., 1992. Towards a new theory of feed intake regulation in ruminants 2. Costs and benefits of feed consumption: an optimization approach. Livestock Production Science 30: 297-317.

Tsaras, L.N., Kyriazakis, I. and Emmans, G.C., 1998. The prediction of the voluntary food intake of pigs on poor quality foods. Animal Science 66: 713-723.

Tullis, J.B., 1982. Protein growth in pigs. PhD thesis, University of Edinburgh, UK.

Turner, L., Gates, R., Wathes, C. and Audsley, E., 1997. Using fuzzy logic for assessment of dust and ammonia effects on swine respiratory disease. Proceedings of the 5th international symposium of the Americal Society of Agricultural Engineers 2: 638-646.

Turner, S.P., Ewen, M., Rooke, J.A. and Edwards, S.A., 2000. The effect of space allowance on performance, aggression and immune competence of growing pigs housed on straw deep-litter at different group sizes. Livestock Production Science 66: 47-55.

Usry, J.L., Turner, L.W., Stahly, T.S., Bridges, T.C. and Gates, R.S., 1991. GI tract simulation model of the growing pig. Transactions of the American Society of Agricultural Engineers 34: 1879-1890.

Van Milgen, J., Bernier, J.F., Lecozler, Y., Dubois, S. and Noblet, J., 1998. Major determinants of fasting heat production and energetic cost of activity in growing pigs of different body weight and breed/castration combination. British Journal of Nutrition 79: 509-517.

Van Milgen, J., Valancogne, A., Dubois, S., Dourmad, J-Y, Seve, B. and Noblet, J., 2007. InraPorc: A model and decision support tool for the nutrition of growing pigs. Animal Feed Science and Technology 143: 387-405.

Verstegen, M.W.A., Verhagen, J.M.F. and den Hartog, L.A., 1987. Energy requirements of pigs during pregnancy: a review. Livestock Production Science 16: 75-89.

Wellock, I.J., Emmans, G.C. and Kyriazakis, I., 2003. Predicting the consequences of social stressors on pig feed intake and performance. Journal of Animal Science 81: 2995-3007.

Whittemore, C.T., 1983. The science and practice of pig production. Longman Ltd, London.

Whittemore, C.T., 1998. The science and practice of pig production. 2nd Edition, Longman Ltd, London.

Whittemore, C.T., Green, D.M. and Knap, P.W., 2001c. Technical review of the energy and protein requirements of growing pigs: food intake. Animal Science 73: 3-17.

Whittemore, C.T., Kerr, J.C. and Cameron, N.D., 1995. An approach to prediction of feed intake in growing pigs using simple body measurements. Agricultural Systems 47: 235-244.

Whittemore, E.C., Emanns, G.C., Tolkamp, B.J. and Kyriazakis, I., 2001b. Tests of two theories of food intake using growing pigs 2. The effect of a period of reduced growth rate on the subsequent intake of foods of differing bulk content.

Whittemore, E.C., Kyriazakis, I., Emanns, G.C. and Tolkamp, B.J., 2001a. Tests of two theories of food intake using growing pigs 1. The effect of ambient temperature on the intake of foods of differing bulk content. Animal Science 72: 351-360.

Wolter, B.F., Ellis, M., Curtis, S.E., Augspurger, N.R., Hamilton, D.N., Parr, N.E. and Webel, D.M., 2001. Effect of group size on performance in a wean-to-finish production system. Journal of Animal Science 79: 1067-1078.

Zijlstra, R.T., Tibble, S. and van Kempen, T.A.T.G., 2009. Feed manufacturing technology and feed intake in young pigs. In: Torrallardona, D. and Roura, E. (eds.) Voluntary feed intake in pigs. Wageningen Academic Publishers, the Netherlands, pp. 277-292.

Author profiles of main authors

John L. **Black**, Ph.D. Research Management Consultant and Director of John L Black Consulting, a company managing research programs for a wide range of companies with interests across many rural industries including the pork, beef cattle, dairy, broiler, grains, fodder and honeybee industries. Major interests are in comparative physiology, integration of information in computer simulation models and application of science to agriculture. Previous positions were as a Chief Research Scientist, Assistant Chief and Offer-in-Charge of the Ian Clunies Ross laboratory in the Commonwealth Scientific and Industrial Research Organisation of Australia and Adjunct Professor in the faculty of Veterinary Science, University of Sydney, Australia.

J.E. (Liesbeth) **Bolhuis**, Ph.D. Assistant professor of the Adaptation Physiology group of Wageningen University. She has a background in Animal Science. She did her PhD on 'Personalities in pigs – Individual characteristics and coping with environmental challenges' at the Ethology group and Human and Animal Physiology group of Wageningen University. After that she joined the multidisciplinary Adaptation Physiology group, chaired by Prof. Dr. Bas Kemp. The Adaptation Physiology group aims to unravel the determinants of adaptive capacity in animals, with a major focus on the influence of early life conditions. Liesbeth Bolhuis is, as an expert in pig behaviour and welfare, involved in several research projects within the group. One of these projects, carried out in collaboration with the Animal Nutrition group, addresses weaning-related problems in piglets and aims at 'Facilitating the development of solid feed intake in piglets by means of vertical (mother-offspring) information transfer'.

Jeff **Carroll**, Ph.D. Research Leader for the USDA-Agricultural Research Service's Livestock Issues Research Unit. He is a Diplomat of the American College of Animal Physiology and holds adjunct appointments at the University of Missouri, Texas Tech University, Texas A&M University and Mississippi State University. After completing his Ph.D. at Texas A&M University, Dr. Jeff Carroll joined the USDA-Agricultural Research Service's Animal Physiology Research Unit serving as Research Physiologist. His research efforts focus on improving livestock performance and well-being through the investigation of the interactions associated with growth, stress, nutrition, and immune regulation. He has authored/co-authored 297 publications consisting of 74 refereed journal articles, 1 review article, 4 book chapters, 162 abstracts, and 57 technical reports. He has presented 5 invited international presentations on environmental and nutritional modulation of immune function in young pigs and growth and immunity in cattle.

J. Michael **Forbes**, Ph.D. Emeritus Professor, Institute of Integrative and Comparative Biology, University of Leeds, United Kingdom. He has a background in Animal Physiology from Nottingham University and completed his Ph.D. at Leeds University were he was successively appointed Research Assistant, Assistant Lecturer, Lecturer, Senior Lecturer, Reader, Professor in the Departments of Agriculture, Animal Physiology and Nutrition, Biological Sciences; now retired. His main research interest is in the control and prediction of voluntary food intake and diet selection in farm animals: sheep, cattle, goats, pigs, poultry. Author of 120 refereed and invited papers and chapters in internationally-recognised publications. He has been invited at numerous international conferences, including four of the 5-yearly International Conferences on Ruminant Physiology of Digestion and Metabolism. Former President of the British Society of Animal Science; former Editor in Chief of Nutrition Research Reviews, published by the Nutrition Society; former Chair of the UK Conference of Professors in Agriculture and Cognate Subjects; former Chair of the Computer User Group at the University of Leeds.

Kirk C. **Klasing**, Ph.D. Professor of the Department of Animal Science, University of California, Davis. His pedigree includes time spent at Purdue University, Cornell University, University of Illinois, and USDA, Beltsville. His research explores the nutritional costs of immunocompetence. He teaches undergraduate classes in Comparative Animal Nutrition and a graduate class in Nutrition and Immunity. His research encompasses three main areas: First the impact of an immune response against infectious diseases on growth and reproduction; second we are also investigating the impact of an animal's diet on the immune response; and third we are exploring the diverse nutritional and immune strategies of omnivorous, carnivorous, nectarivorous, herbivorous and granivorous animals that includes the characterization of mammalian and avian taste biology.

Pieter **Knap**, Ph.D. Genetic Strategy Manager for PIC. He studied Animal Science in Wageningen (the Netherlands) and has worked in commercial pig breeding since 1981 – in the Netherlands (for the Dutch Pig Herdbook Societies, now Topigs), Norway (for Norsvin International), England, Scotland and Germany (currently for PIC). His scientific work has mainly focused on breeding program design, international breeding goals, breeding value estimation, performance testing strategies, and marker-assisted selection strategies. In 2000 he received a PhD from Wageningen University on research on stochastic growth modelling. He is currently president of the Commission on Pig Production of EAAP.

Ilias **Kyriazakis**, Ph.D. Professor at the Veterinary Faculty, University of Thessaly, in Greece and the School of Agriculture, Newcastle University, in the United Kingdom. He also holds a number of Visiting Professorships and Honours in European and North American institutions. He is a veterinarian by training and a *de facto* Diplomat of the European College of Veterinary and Comparative Nutrition. His research interests are in the consequences of management on animal responses; these include the interactions between host nutrition and responses to pathogens in laboratory and farm animals, and modelling the responses of animals to their management treatments. He has worked with most farm animal species, starting with pigs and more recently making increasing use of laboratory rodents to ask the above questions at a more fundamental level. He has authored a number of publications on the phenomenon of pathogen-induced anorexia, including ones that address the sources of variation for the phenomenon and its evolutionary context.

F. Xavier **Manteca**, Ph.D. He obtained his Veterinary degree in 1987 and his Ph.D. in 1992 from the Animal Physiology Unit of the Veterinary School of the Universitat Autònoma de Barcelona (UAB), Spain. He is currently an Associate Professor at the UAB teaching Veterinary Clinical Ethology as well as courses on animal ethology and welfare. Dr. Manteca has been a visiting researcher in many top universities worldwide such as the University of Edmonton (Canada), the University of Edinburg (United Kingdom), the University of Alaska-Fairbanks and Arizona State University (USA), among others. He is a member of the Commission for Ethics in Human and Animal Research of the UAB, the Spanish Society of Ethology, the British Nutrition Society and the International Society for Applied Ethology. His main research interest is currently focused in the 'Integration of animal welfare in the food quality chain' funded by the EU, with emphasis in pigs. He is also interested in other domestic and wild animal behaviour projects such as the dog, the Eurasian otter and the roe deer. He has authored a large number of publications related to Animal Behaviour and participated in many international scientific events.

Eugeni **Roura**, Ph.D. R&D Manager of the Feed Additives Division of Lucta SA, Barcelona, Spain. His academic background includes a veterinary degree and a Ph.D. from the Universitat Autònoma de Barcelona (UAB) and a M.Sc. in Nutrition from the University of California (Davis). His career has been developed around technical service and feed additives research and development related to the global feed market including research collaborations with most of the top feed/premix corporations worldwide. He is author of many publications related to animal nutrition and the immune system, physiology of feed intake in farm animals with emphasis in the peripheral sensing mechanisms and development and application of feed additives in farm animals with emphasis in non-ruminant species (pigs and poultry). He has participated in many international scientific events.

Andrea **Sbarbati**, M.D., Ph.D. Professor of Normal Human Anatomy in the Medical faculty of the University of Verona, Italy. He graduated in Medicine with full marks on 1982 (University of Ancona, Italy). In 1986 he obtained his Ph.D. in Neurological Sciences. He wrote many scientific papers, many of which were published in international scientific journals. His broad expertise in the morphological sciences, in particular chemoreceptor studies, comes from the personal utilization of diverse morphological techniques: electron microscopy, freeze-fracturing, electron spectroscopy imaging, X-ray microanalysis, ultra-structural histochemistry, morphometry, immunohisto- and cytochemistry. In the last 10 years, he paid special attention to the potential of nuclear magnetic resonance (NMR) in the morphological sciences. He has been the driving force for establishing a NMR laboratory in Verona. This laboratory has been endowed with up-to-date instrumentation for NMR spectroscopy and magnetic resonance imaging of small animals. He published several NMR papers in international journals.

David **Torrallardona**, Ph.D. Pig Nutrition Manager at the Animal Nutrition Department of IRTA, Spain. He received a veterinary degree from the Universitat Autònoma de Barcelona in 1988. He completed his PhD studies on the contribution of the gastrointestinal microflora to the amino acid supply in non-ruminants at the Rowett Research Institute in Aberdeen (Scotland) in 1994. He is currently serving as vice-president of the Commission on Pig Production of the European Association of Animal Science (EAAP). Current research interest focuses on the control of the piglet's intestinal health by nutritional means. He has recently been involved in projects to test the effects of different feed ingredients on feed palatability and gut health and on the efficacy of alternative products to antibiotic growth promoters. Dr. Torrallardona has published many scientific papers and has been invited speaker to many international scientific events worldwide.

A.(Thomas) F.B. **Van der Poel**, Ph.D. Associate Professor, Animal Nutrition Group, Wageningen University, Wageningen (the Netherlands). He obtained his MSc and Ph.D. in Animal Nutrition from the Wageningen University. After obtaining his PhD degree in 1990, his research has focused on the research theme 'effects of processing on nutrient reactivity' as related to several animal species, currently especially pets. In addition to his work in these research themes, he has the responsibility for the research conducted at the Wageningen Feed Processing Centre since 1993. In these facilities he has worked, together with the operators on improvement of these facilities, initiating the use of expander techniques and recently – 2007 – extrusion technology.

Ruurd **Zijlstra**, Ph.D. He is currently the Feed Industry Research Chair at the University of Alberta, Edmonton, Alberta, Canada. He was born and raised in the Netherlands where he completed a B.Sc. degree, and a M.Sc.-degree at Wageningen University, and worked for a year in the feed industry. In 1996, he completed a Ph.D. at the University of Illinois and moved to Canada. Until 2004, he was the Research Scientist - Nutrition at Prairie Swine Centre in Saskatoon, Saskatchewan, Canada. He has produced 42 scientific papers in peer-reviewed journals and 7 book chapters in feed quality evaluation and other pig nutrition topics. His current research program is focused on feed quality evaluation techniques, nutritional quality of fractionated crops and other alternative feedstuffs, and unique aspects of carbohydrate nutrition in pigs.

Index

G

H

I

T

U

V

W